Aquatic Oligochaete Biology VI

Developments in Hydrobiology 115

Series editor

H. J. Dumont

Aquatic Oligochaete Biology VI

Proceedings of the VI International Symposium
on Aquatic Oligochaetes held in Strömstat, Sweden,
September 5–10, 1994

Edited by

Kathryn A. Coates, Trefor B. Reynoldson
& Thomas B. Reynoldson

Reprinted from Hydrobiologia, vol. 334 (1996)

Springer Science+Business Media, B.V.

Library of Congress Cataloging-in-Publication Data

A C.I.P. Catalogue record for this book is available from the Library of Congress.

ISBN 978-0-7923-3999-1   ISBN 978-94-011-5452-9 (eBook)
DOI 10.1007/978-94-011-5452-9

Printed on acid-free paper

Contents

Hydrobiologia **334**: ix, 1996.
K.A. Coates, Trefor B. Reynoldson & Thomas B. Reynoldson (eds), Aquatic Oligochaete Biology VI.

Preface

This volume reflects the proceedings of the Sixth International Symposium on Aquatic Oligochaetes. The symposium was hosted by Dr Christer Erséus (Swedish Museum of Natural History) during September 5–10, 1994 at the Tjärnö Marine Biological Laboratory, near Strömstad, Sweden. The meeting was open to anyone interested in the morphology, systematics, evolution, biogeography, ecology, pollution biology or physiology of the Oligochaeta of surface or marine waters, and it was attended by over 60 scientists from 21 different countries.

Tjärnö is an island in the middle of an extensive archipelago near the Norwegian border, on the Swedish west coast. With its splendid location and facilities, the Tjärnö Laboratory proved to be an excellent venue for scientific exchange and social interaction, although some participants could not resist the temptation of sneaking out from time to time to collect mushrooms, which were abundant in the surroundings. The program also included proper excursions, both on the sea with bird and seal watching, and by coach through the northern part of the province Bohuslän. The bus tour was sponsored by the town of Strömstad, which also generously hosted a dinner with locally flavored entertainment. We are all very much undebted to Dr Lars Afzelius (Director) and Mr Lars Hagström, both at the Tjärnö Laboratory, for all their efforts and assistance during the symposium, and to Mr Torsten Torstensson (chairman of the Strömstad town council) for all his support. Financial aid to the organization of the symposium was provided by the International Science Foundation, the Wenner-Gren Center Foundation, the Swedish Natural Science Research Council and the Swedish Museum of Natural History. In addition, Armeniska Kulturföreningen i Södertälje and numerous individuals among the participants provided contributions towards the travel expenses of several colleagues from East European countries. During the week at Tjärnö, Mr S. Lundberg, Mrs K. Rignéus, Mr L. Sandberg, Mrs E. Sigvaldadottir and Ms K. Sindermark (all at the Swedish Museum of Natural History) assisted in all kinds of practical matters. Their help and friendly attitude were very much appreciated by everybody.

All papers in this volume have been rigorously peer reviewed. All reviewers are thanked for their constructive comments and criticisms.

CHRISTER ERSÉUS
Swedish Museum of Natural History

Hydrobiologia **334**: xi–xii, 1996.
K.A. Coates, Trefor B. Reynoldson & Thomas B. Reynoldson (eds), Aquatic Oligochaete Biology VI.

List of Participants at the VIth International Symposium on Aquatic Oligochaetes

Sergey A. AFANASYEV, Institute of Hydrobiology, Academy of Sciences of Ukraine, Geroyev Stalingrada, 12, Kiev 25210, Ukraine

Carla BONACINA, C.N.R. Istituto Italiano di Idrobiologia, Largo Tonolli, 50, I-28048 Pallanza, Italy

Giuliano BONOMI, Dipartimento di Biologia Evoluzionistica Sperimentale, Universita degli Studi di Bologna, Via San Giacomo, 9, I-40126 Bologna, Italy

Regine BÖNSCH, Institut für Angewandte Ökologie, Borenweg 3, D-18057 Rostock, Germany

Trond BREMNES, Freshwater Ecology and Inland Fisheries Laboratory, LFI, Zoological Museum, University of Oslo, Sars Gate 1, N-0562 Oslo, Norway

Ralph O. BRINKHURST, Aquatic Resources Center, P.O. Box 680818, Franklin, TN 37068, USA

Eugene M. BURRESON, Virginia Institute of Marine Science, College of William and Mary, P.O. Box 1346, Gloucester Pt., VA 23062, USA

Sandra CASELLATO, Dipartimento di Biologia, Universitá degli Studi di Padova, Via Trieste, 75, I-35121 Padova, Italy

Kathryn A. COATES, Invertebrate Zoology, Royal Ontario Museum, 100 Queen's Park Toronto, Toronto, Ontario M5S 2C6, Canada. Present address: Bermuda Biological Station for Research, Inc., 17 Biological Lane, Ferry Reach, GE01, Bermuda

Rut COLLADO, Departamento Bioloxía Animal e Bioloxía Vexetal, Fac. Ciencias, Univ. A Coruña, Campus A Zapateira s/n, E-15071 A Coruña, Spain

Robert J. DIAZ, Virginia Institute of Marine Science, College of William and Mary, P.O. Box 1346, Gloucester Pt., VA 23062, USA

Andreina DICHIARA PAOLETTI, Dipartimento di Biologia, sez. Ecologia, Università degli Studi di Milano, Via Celoria 26, I-20133 Milano, Italy

Elzbieta DUMNICKA, Institute of Freshwater Biology, Polish Academy of Sciences, ul. Slawkoska 17, PL-31-016 Krakow, Poland

Marianne C. ECKROTH, 8512 Hickory Hill Lane, Huntsville, AL 35802, USA

Magda DE EGUILEOR, Dipartimento di Biologia, sez. Zoologia e Citologia, Università degli Studi di Milano, Via Celoria 26, I-20133 Milano, Italy

Christer ERSÉUS, Department of Invertebrate Zoology, Swedish Museum of Natural History, Box 50007, S-104 05 Stockholm, Sweden

Marco FERRAGUTI, Dipartimento di Biologia, sez. Zoologia e Citologia, Università degli Studi di Milano, 26, Via Celoria, I-20133 Milano, Italy

Nonna P. FINOGENOVA, Zoological Institute of Russian Academy of Sciences, St. Petersburg 199034, Russia

Stuart R. GELDER, Department of Biology, University of Maine at Presque Isle, 181 Main Street, Presque Isle, ME 04769-2888, USA

Narcisse GIANI, Laboratoire d'Hydrobiologie, Université Paul Sabatier, 118, route de Narbonne, F-31062 Toulouse Cedex, France

Olav GIERE, Zoological Institute and Zoological Museum, University of Hamburg, Martin-Luther-King-Platz 3, D-20146 Hamburg, Germany

Reinmar GRIMM, Zoological Institute and Zoological Museum, University of Hamburg, Martin-Luther-King-Platz 3, D-20146 Hamburg, Germany

Lena GUSTAVSSON, Department of Zoology, University of Göteborg, Medicinaregatan 18, S-413 90 Göteborg, Sweden

Brenda HEALY, Department of Zoology, University College, Belfield, Dublin 4, Ireland

Karen JENDEREDJIAN, Sevan National Park, Nairian 187, Sevan 378610, Armenia

Michel LAFONT, CEMAGREF, B.E.A. Division, 3 bis Quai Chauveau CP 220, F-69336 Lyon Cedex 09, France

Claude LANG, Conservation de la faune, Marquisat 1, CH-1025 St-Sulpice, Switzerland

Giulio LANZAVECCHIA, Dipartimento di Biologia, Sez. Zoologia e Citologia, Università degli Studi di Milano, Via Celoria 26, I-20133 Milano, Italy

Boris LÖHLEIN, Ecosystem Research Centre, Schauenburgerstrasse 112, D-24118 Kiel, Germany

Stefan LUNDBERG, Service Center for Taxonomic Zoology, Swedish Museum of Natural History, Box 50007, S-104 05 Stockholm, Sweden

Svein B. MANUM, Department of Geology, University of Oslo, P.O. Box 1047, Blindern, N-0316 Oslo 3, Norway

Patrick MARTIN, Freshwater Biology, Royal Belgian Institute of Natural Sciences, 29, Rue Vautier, B-1040 Brussels, Belgium

Enrique MARTINEZ-ANSEMIL, Departamento de Bioloxía Animal e Bioloxía Vexetal, Facultade de Ciencias, Universidade da Coruña, Campus da Zapateira s/n, E-15071 A Coruña, Spain

Maite MARTINEZ-MADRID, Departamento Biologia Animal y Genetica, Facultad de Ciencias, Universidad del Pais Vasco, Apdo 644, E-48080 Bilbao, Spain

Göran MILBRINK, Department of Zoology, Uppsala University, Villavägen 9, S-752 36 Uppsala, Sweden

Michael R. MILLIGAN, Center for Systematics and Taxonomy, P.O. Box 37534, Sarasota, FL 34278, USA

László MOLNÁR, Department of Zoology, Janus Pannonius University, Ifjúság u. 6, H-7601 Pécs, Hungary

Alex I. Muir, Marine Biological Services Division, Zoology Department, The Natural History Museum, Cromwell Road, South Kensington, London, SW7 5BD, United Kingdom

Pietro OMODEO, Dipartimento Biologia Evolutiva, Universitá di Siena, Via Mattioli 4, I-53100 Siena, Italy

Andrea PASTERIS, Dipartamento di Biologia Evoluzionistica Sperimentale, Universita degli Studi di Bologna, Via San Giacomo 9, I-40126 Bologna, Italy

Juan Ignacio PEREZ-IGLESIAS, Departamento Biologia Animal y Genetica, Facultad de Ciencias, Universidad del Pais Vasco, Apdo 644, E-48080 Bilbao, Spain

Montserrat REAL, Department of Zoology, University of Cambridge, Downing Street, Cambridge CB2 3EJ, United Kingdom

Trefor B REYNOLDSON, National Water Research Institute, 867, Lakeshore Road, Burlington, Ontario L7R 4A6, Canada

Kerstin RIGNÉUS, Department of Invertebrate Zoology, Swedish Museum of Natural History, Box 50007, S-104 05 Stockholm, Sweden

Pilar RODRIGUEZ, Departamento Biologia Animal y Genetica, Facultad de Ciencias, Universidad del Pais Vasco, Apdo 644, E-48080 Bilbao, Spain

Agnés ROSSO. CEMAGREF, 3 bis Quai Chauveau, CP 220, F-69336 Lyon Cedex 09, France

Emilia ROTA, Dipartimento Biologia Evolutiva, Universitá di Siena, Via Mattioli 4, I-53100 Siena, Italy

Beatrice SAMBUGAR, Centro Richerche S.A.R. Sistema Ambiente Risorse, Vicolo Chiodo, 10, I-37121 Verona, Italy

Lennart SANDBERG, Department of Invertebrate Zoology, Swedish Museum of Natural History, Box 50007, S-104 05 Stockholm, Sweden

Georg SAUTER, Institut für Seenforschung, Untere Seestrasse 81, D-88085 Langenargen, Germany

Rüdiger M. SCHMELZ, FB 5/Spezielle Zoologie, Fachbereich Biologie/Chemie, Universität Osnabrück, Postfach 4469, D-49069 Osnabrück, Germany

Takashi SHIMIZU, Molecular and Cellular Interactions, Division of Biological Sciences, Graduate School of Science, Hokkaido University, Sapporo 060, Japan

Mark E. SIDDALL, Virginia Institute of Marine Science, College of William and Mary, P.O. Box 1346, Gloucester Pt., VA 23062, USA

Elin SIGVALDADOTTIR, Service Center for Taxonomic Zoology, Swedish Museum of Natural History, Box 50007, S-104 05 Stockholm, Sweden

Karin A. SINDEMARK, Department of Invertebrate Zoology, Swedish Museum of Natural History, Box 50007, S-104 05 Stockholm, Sweden

Tatyana D. SLEPUKHINA, Institute of Lake Research, Sevastyanova 9, St. Petersburg 196199, Russia

Svein-Erik SLOREID, Norsk Institutt for Naturforskning, P.O. Box 1037, Blindern, N-0315 Oslo, Norway

Ferdinand SPORKA, Institute of Zoology and Ecosozology, Slovak Academy of Sciences, Drienová 3, SK-821 02 Bratislava, Slovakia

Donald F. STACEY, Department of Invertebrate Zoology, Royal Ontario Museum, 100 Queens Park, Toronto, Ontario M5S 2C6, Canada

András SZÍTÓ, Fish Culture Research Institute, P.O. Box 47, H-5541 Szarvas, Hungary

Tarmo TIMM, Vörtsjärv Limnological Station, EE-2454 Rannu, Tartumaa, Estonia

Viivi TIMM, Vörtsjärv Limnological Station, EE-2454 Rannu, Tartumaa, Estonia

Piet F.M. VERDONSCHOT, Instituut voor Bos- en Natuuronderzoek, Ministerie van Landbouw, Natuurbeheer en Visserij, P.O. Box 23, NL-6700 AA Wageningen, The Netherlands

Wilfried WESTHEIDE, FB 5/Spezielle Zoologie, Fachbereich Biologie/Chemie, Universität Osnabrück, Postfach 4469, D-49069 Osnabrück, Germany

Mark WETZEL, Center for Biodiversity, Illinois Natural History Survey, 172 Natural Resources Building, 607 East Peabody Drive, Champaign, IL 61820, USA

Hydrobiologia **334**: 1–9, 1996.
K. A. Coates, Trefor B. Reynoldson & Thomas B. Reynoldson (eds), Aquatic Oligochaete Biology VI.
© 1996 *Kluwer Academic Publishers.*

1

Tenagodrilus musculus, a new genus and species of Lumbriculidae (Clitellata) from a temporary pond in Alabama, USA

Marianne C. Eckroth[1] & Ralph O. Brinkhurst[2]
[1]*Department of Biology, University of Alabama in Huntsville, Huntsville, AL 35899, USA*
[2]*Aquatic Resources Center, Box 680818, Franklin, TN 37068-0818, USA*

Key words: Tenagodrilus, Lumbriculidae, Clitellata, hyporheic fauna

Abstract

A new hyporheic lumbriculid clitellate, *Tenagodrilus musculus* g.n., sp.n. from Huntsville, Alabama, USA is described. The plan of the reproductive system is plesiomorphic, as in *Stylodrilus* and other genera, but there is a pair of pendant penes like those of *Lumbriculus*. The massive penial and spermathecal pores open into deep sacs, and the penial musculature is both complex and unique. The posterior segments have commissural vessels with elaborate blind-ending lateral processes. In developing worms the ciliated part of the vas deferens appear to grow toward the atria, but another, unciliated part of the vas deferens grows out from the atria. In partially mature forms, a rudimentary second pair of ovaries without female funnels was observed, which is not present in mature worms. Mature worms are found throughout the period when the temporary pond in which the taxon was found is full, suggesting that it is a hyporheic form.

Introduction

During a study of a temporary pond near the campus of the University of Alabama in Huntsville, aquatic clitellates and other invertebrates were collected from 1980 to 1987 (Modlin & Bush, 1988). The pond is located next to a radio station with the call sign WEUP, and is therefore designated as the WEUP pond. A large, fragile lumbriculid species was found each year, and after serial sections were prepared and examined, it was determined that the taxon could not be accommodated within known species, or even genera. This new taxon is described and some information regarding the ontogeny of the reproductive system will be presented. The available information regarding the life history is related to the periods when the pond was full.

Methods and materials

Animals were collected from the WEUP pond between 1980 and 1987. Collections were taken weekly from the time the pond filled in the fall until it dried in the spring. The worms were fixed in the field with formalin, and were later transferred to 70% alcohol. The lumbriculids were examined whole in alcohol and in CMCP, as dissections, and as transverse and longitudinal sagittal sections. The specimens to be sectioned were embedded in Paraplast; sections were stained with the PAS technique with a counterstain of haematoxylin and malachite green.

Results

Taxonomy

Tenagodrilus g.n.
Genus of family Lumbriculidae with testes paired in IX and X, ovaries paired in XI, atria paired in X, spermathecae one pair in IX, sometimes an additional pair in VIII. Posterior vasa deferentia enter XI before reentering X. Atria with complex penial musculature and ejaculatory ducts between ampullae and pendant penes. Male and spermathecal pores very large, located immediately behind the ventral setae. Blind-ending lateral

extensions of the commissural blood vessels present in posterior segments.

Type species: Tenagodrilus musculus, by monotypy.

Etymology: Tenago – Gr., living in pools, -*drilus* – worm.

Tenagodrilus musculus *sp.n.*

Holotype: USNM 171060, a whole animal in 70% alcohol, from the type locality, coll. R. Modlin, 02/01/1987.

Type locality: WEUP Pond, intersection of Sparkman and Jordan, Huntsville, Alabama, USA, RIW-T3S,-section 21. Madison Co. AL, coll. Richard Modlin, December 1983 to April 1987, coll. M.C.E. December through April, 1991 to 1993.

Paratypes: USNM 171061, 171062, 10 whole animals in fluid with one pair of spermathecae, two whole animals with two pairs of spermathecae, from the same collection as the holotype; M.C.E. collection: 17 specimens, 2 of which mature, from the same collection.

Other material: USNM 171063-5, 3 specimens, longitudinally sectioned, 1 with spermathecae protruded forward, on 13 slides (coll. 02.01.1987), 1 with spermatheca reflected posteriad, half worm sectioned, the rest mounted whole, on 10 slides (coll. 02.15.1984) and 1 with developing genitalia, on 13 slides (coll. 04.24.1984), all coll. R. Modlin from the type locality. M.C.E. collection: over 250 specimens, transverse and longitudinally sectioned, dissected mounts and whole mounts, and worms in 70% alcohol, from the type locality, collected at various dates from 1980 to 1987 by R. Modlin. R.O.B. collection: 10 mature specimens in 70% alcohol, 2 longitudinal serially sectioned specimens. Department of Animal Biology and Genetics, University of Basque Country, Bilbao, Spain, 10 mature specimens in alcohol, LUM001.

Description: Live specimens grayish red-brown, opaque. Body length, mature individuals (preserved) 22–82 mm, width 0.5–1.3 mm at the clitellum, number of segments 90–297. Prostomium bullet-shaped, annulation slight, not doubled (Fig. 1). Setae paired, simple-pointed, sigmoid with nodulus. Some ventral setae with very slight indication of a rudimentary upper tooth, slightly exaggerated in the illustration (Fig. 2). Setae become progressively longer and thinner to mid body, but are as short as in II in posterior segments. Setal formula (X) 50:2:7:105 (Reynolds, 1977). Cuticle colorless and thin. Clitellum inconspicuous,

extending over IX–XII or parts thereof. Spermathecal pores large (32 μm anterior to posterior; 56 μm side to side), posterior to, and in line with, ventral setae, close to posterior margin of IX (Fig. 1). In 10% of specimens, a second pair of spermathecal pores in VIII, the same in position and size. Equally large pair of male pores in X posterior to and in line of ventral setae (Fig. 1). One pair inconspicuous female pores (8 μm anterior to posterior, 40 μm side to side) in intersegmental furrow 11/12.

Eversible pharyngeal pad thick dorsally, muscles slightly developed, cuticulate anteriorly but pharynx heavily ciliated from III to V, pharyngeal glands from II to V. Pharynx floor thin and cuticulate. Chloragogen begins in V. Dorsal blood vessel divides in peristomium to form two ventro-lateral vessels that unite in V. Two vessels leave dorsal vessel to supply supraoesophageal ganglia. In anterior segments to X, commissural vessels long and convoluted (Fig. 3A). In post reproductive segments, one pair of thick commissural vessels per segment to XIV (Fig. 3B). Hearts from XIV to XVI. After XIX, commissural vessels two pairs per segment (Fig. 3C); by XXX both pairs of commissural vessels bear blind-ending lateral vessels which enter the muscular body wall and return to terminate in the coelom as small bulbs (Fig. 3D). Mature specimens with long vessels penetrating sperm and egg sacs.

Testes paired in IX and X, paired sperm sacs penetrate to XVI. Sperm funnels conspicuous in IX and X, first pair anterior on 9/10 behind spermathecal ampullae, second pair in sperm sacs of X, located within XI. Anterior sperm funnels penetrate 9/10 laterally, vasa deferentia follow ventral wall of X, and run posteriad to medial aspects of penis musculature (Fig. 4A). Each anterior vas deferens ascends through penial musculature to penetrate atrial ampulla just above base, opening into atrium subapically after running through atrium wall muscles. Posterior prosoporous vasa deferentia begin on anterior of 10/11, grow through the septa, descending ventrally. From there, each runs along ventral wall to reenter X close to the penial musculature, then enters the atrium following a path similar to that of the anterior pair but on lateral side of atrium (illustrated diagrammatically in Fig. 4A). The posterior sperm funnels often drawn posteriad into the sperm sacs (Fig. 4B). Atrial ampullae spherical with thin muscular walls and thin outer layer of prostate cells. The rest of each male duct, consisting of ejaculatory duct and long pendant penis in large sac, enclosed in a muscular complex that surrounds the atrial ampulla (Fig. 4A-B). Retractor muscles originate in dorsal

3

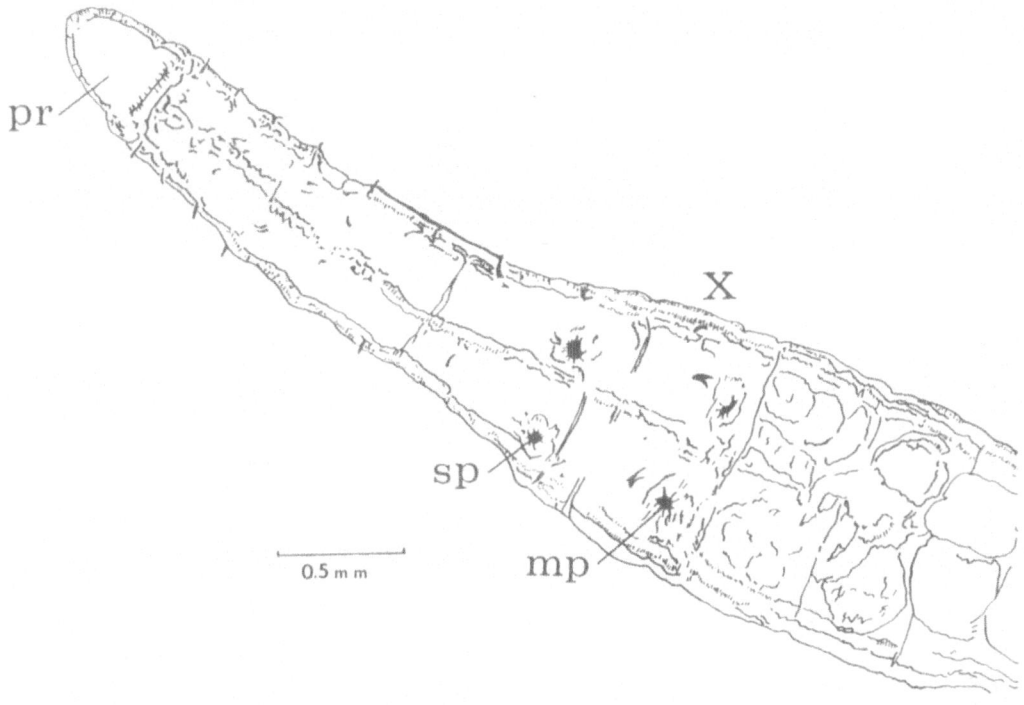

Fig. 1. Tenagodrilus musculus g.n., sp.n. Ventral view of anterior end of whole mounted specimen showing prostomium and gonopores. mp = male pore, pr = prostomium, sp = spermathecal pore.

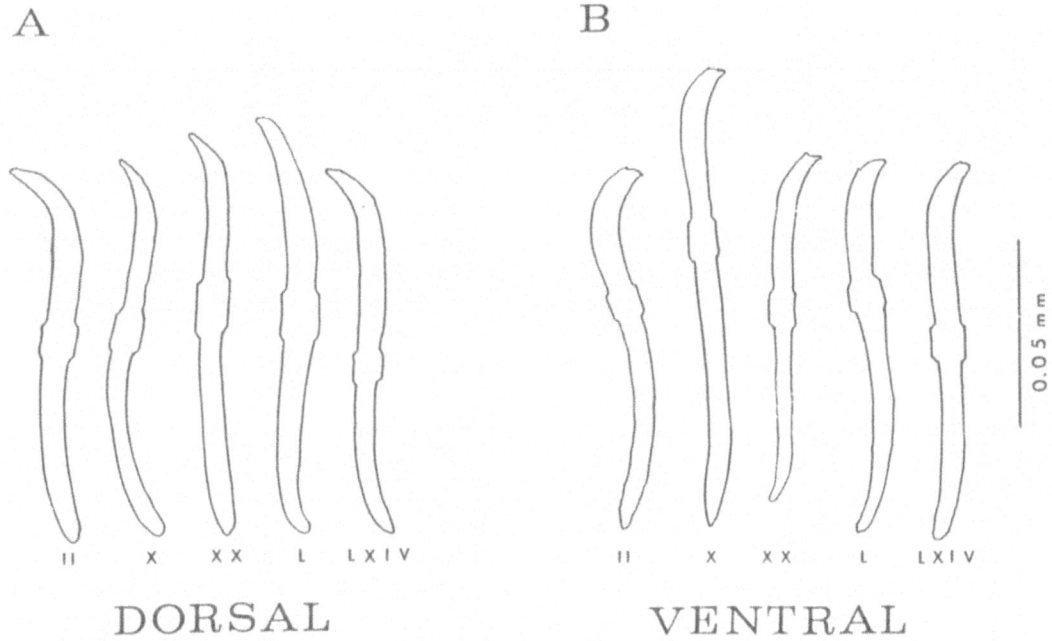

Fig. 2. Tenagodrilus musculus g.n., sp.n. Setae, A: setae of dorsal bundles, B: setae of ventral bundles, segments as indicated. The slight indication of an upper tooth is exaggerated here as it is otherwise too subtle to draw.

4

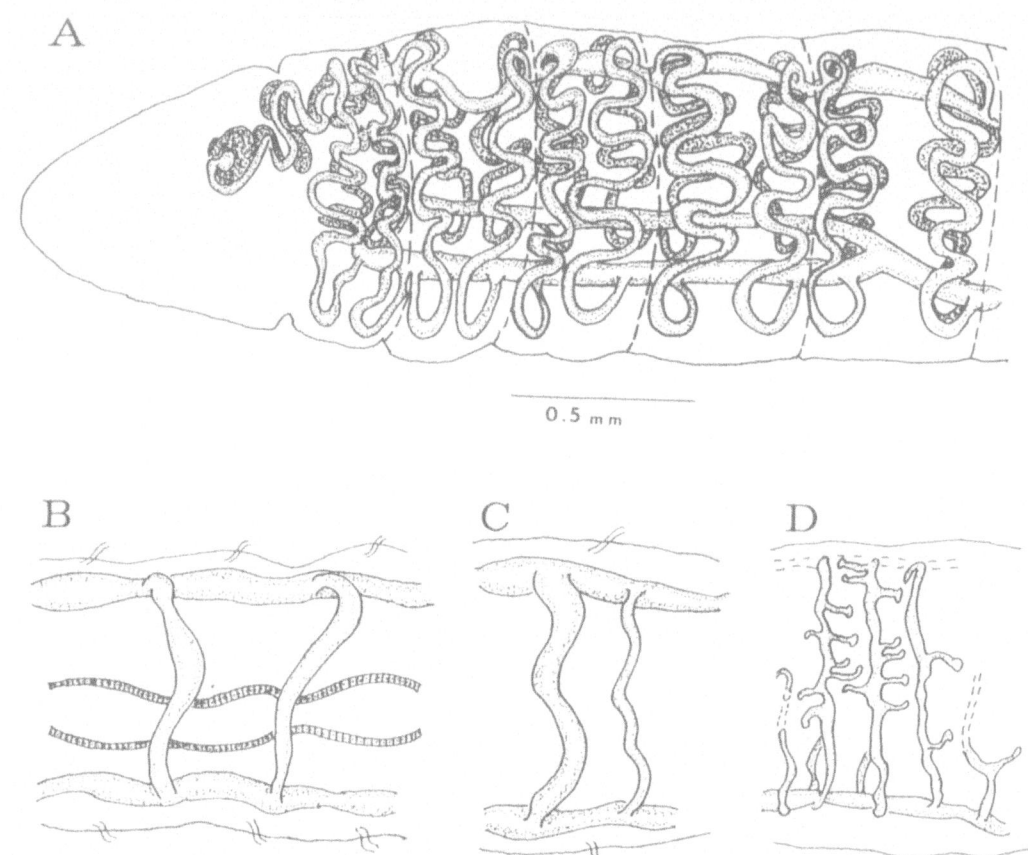

Fig. 3. *Tenagodrilus musculus* g.n., sp.n. Vascular system. A: anterior end, lateral view, the commissurals being even longer and coiled in life, B: lateral view, one side, segments XII and XIII, single commissurals, C: lateral view, one side, segment XXV, showing paired commissurals, D: lateral view, one side, segment XXXV, commissurals with blind-ending laterals. See text for details.

body wall and insert between ampullae and penis sacs. When retractor muscles relaxed, the ampullae lie above penes with retractor muscles on each side (Fig. 4A) when contracted, atrial ampullae are reflected posteriad (Fig. 4B). This also results in coiling of the ejaculatory duct. Protractor muscles originate on the ventral body wall, and insert between the atrial ampulla and the ejaculatory duct. When contracted, the penes protrude and ejaculatory ducts are straightened. Penes about four times longer than broad, without visible cell bodies or nuclei (Fig. 4C).

Ovaries paired in XI, extending to XII. Egg sacs from XI extending to XVII or XIX. Female funnels large and obvious, opening in 11/12. In immature individuals there is a second pair of rudimentary ovaries but no female funnel. These ovaries are not present in the mature specimens.

Spermathecae paired in IX, a second pair in VIII in 10% of worms. Spermathecal ducts leave posterior

subapical walls of large vestibules to enter voluminous ampullae anteriorly (Fig. 4). Ampullae may be reflected through 9/10 into anterior part of X, the atrial complex lying well posterior in X; alternatively ducts reflected anteriad along the floor of IX with ampullae reflected over them dorsally and posteriad.

Etymology: musculus – referring to muscular complex of male ducts.

Development

Sections of several semi-mature specimens were examined. At the stage when there are rudimentary spermathecae and atria, a pair of developing ovaries was observed in XII, the post ovarian segment (Fig. 5). These are absent in mature specimens, and are never associated with female ducts. The spermathecae consist of a pair of club-shaped structures, the future ampullae without a lumen, each located above a col-

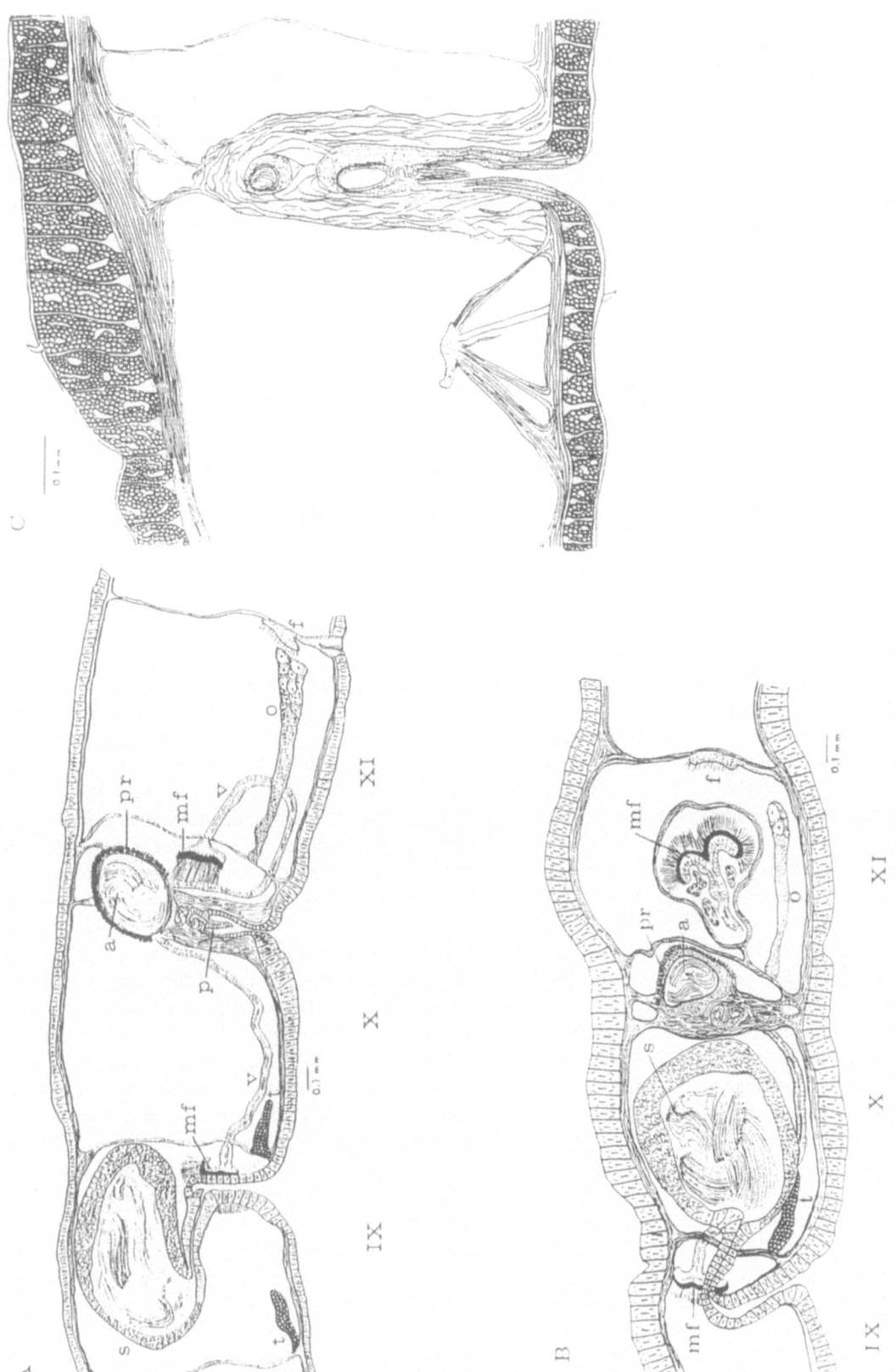

Fig. 4. Tenagodrilus musculus g.n., sp.n. Three reconstructed lateral views of reproductive system in IX–XI. A: with spermatheca in situ in IX, B: with spermathecal ampulla reflected into X and posterior sperm funnel in sperm sac, C: penis in muscular complex. a = atrium, f = female funnel, mf = male funnel, o = ovary, p = penis, pr = prostate gland, s = spermatheca, t = testis, v = vas deferens.

6

umn of tissue that will become the spermathecal duct. The ciliated portions of the vasa deferentia are associated with the sperm funnels and seem to have developed from the ciliated mesodermal cells of the sperm funnels. The ciliated vas deferentia do not appear to make any connection with the unciliated portions, which appear to develop from the ectodermal cells forming the atria.

Habitat

The WEUP pond fills a depression with an Oak Flat Forest located in the northwestern part of the City of Huntsville, Madison County, Alabama, USA. The six tree species (four oak species, a black gum and a red maple) within the depression differ from those in the general surroundings, which include two hickory species, a loblolly pine and greenbrier. The deciduous species within the pond create a leaf layer 10–30 cm deep in the pond. Beneath the leaf layer is a 5 to 10 cm layer of fine root mat, beneath which lies Abernathy silt loam, a recent local alluvium that occupies nearly all gentle depressions and sinks in areas of Cumberland and Decatur soils. The bedrock is limestone. At full capacity the pond covers about 2600 m^2. Depth averages 0.2 m with a maximum of 0.8 m. The pond begins to fill at about the time of leaf fall in November. It usually remains wet until the trees leaf out in spring in April or May. In the 1985/1986 season, the pond dried for a while during February (Modlin & Bush, 1988).

Cladocera and the isopod *Caecidotea obtusus* (Williams, 1970), *Crangonyx* sp., a decapod and (occasionally) two anostracans and a leech were found in the pond (1988) in addition to the oligochaetes. The tubificids *Quistadrilus multisetosus* (Smith, 1900), *Varichaetadrilus fulleri* Brinkhurst and Kathman, 1983, and *V. angustipenis* (Brinkhurst & Cook, 1966) and the naidid *Dero digitata* (Müller, 1773) are present with the lumbriculid. Two other tubificids were present as unidentifiable immature worms.

Life history

Over 500 specimens were examined. These were collected in the 1983/1984, 1985/1986 and 1986/1987 seasons when the pond was full of water. Worms observed in other years, December through April 1991 to 1993 (coll. M.C.E.) were returned to the pond. Newly hatched worms (1.5 mm to 5 mm in length), immature and mature worms were present as soon as the

pond was full, and persisted through most months. By late April or May, when the pond was drying, only single specimens were found, and these were mature. The life history is not timed to coincide with the filling of the pond. A few fragments of worms were observed. These appeared to be regenerating new segments at both ends, and may indicate that asexual reproduction is possible. No drought resistant cysts were found. Thirty large cocoons, most containing four worms with paired, simple-pointed setae were found on the 13th of April, 1984 (Fig. 6).

Discussion

Brinkhurst (1989) presented an evolutionary hypothesis for the Lumbriculidae based on an analysis using a parsimony method. When *Tenagodrilus* is coded according to the character states used in that analysis, the result places it in an unresolved trichotomy with (a) *Bythonomus* and (b) *Stylodrilus* plus *Lumbriculus*. The only character separating (a) and (b) in that analysis is the slight difference in the setae (character 2). When the setae of lumbriculids are bifid, the upper tooth is usually small, even indistinct, and the character state may be ambiguous. Other lumbriculid genera include species with simple-pointed setae and others with bifid setae (Timm & Rodriguez, 1994). The new taxon has simple-pointed setae, although there is the faintest suggestion of a notch on the upper side of some. This setal characteristic is not considered to be a significant generic character. If that character is deleted from the analysis, codings for *Bythonomus* and *Stylodrilus* are identical. In fact, Brinkhurst (1965) had already merged *Bythonomus* in *Stylodrilus* for this reason.

The characters separating *Stylodrilus* from *Lumbriculus* in the cladistic analysis (Brinkhurst, 1989) are the position of the spermathecal pores (character 9) and the penetration of the posterior vasa deferentia through the septum separating testes (GII) and ovaries (GIII) (character 21). In terms of these two characters, *Tenagodrilus* is coded the same as *Stylodrilus*. In *Lumbriculus* the spermathecae are behind the gonadal segments, and the posterior vasa deferentia do not penetrate the septa. In respect to other characters, *Tenagodrilus* is unique among lumbriculids. These characters involve the nature of the posterior blood vascular system and the detailed structure of the male ducts.

In the phylogenetic analysis by Brinkhurst (1989), the posterior lateral blood vessels were coded present or absent. The situation is actually more complex,

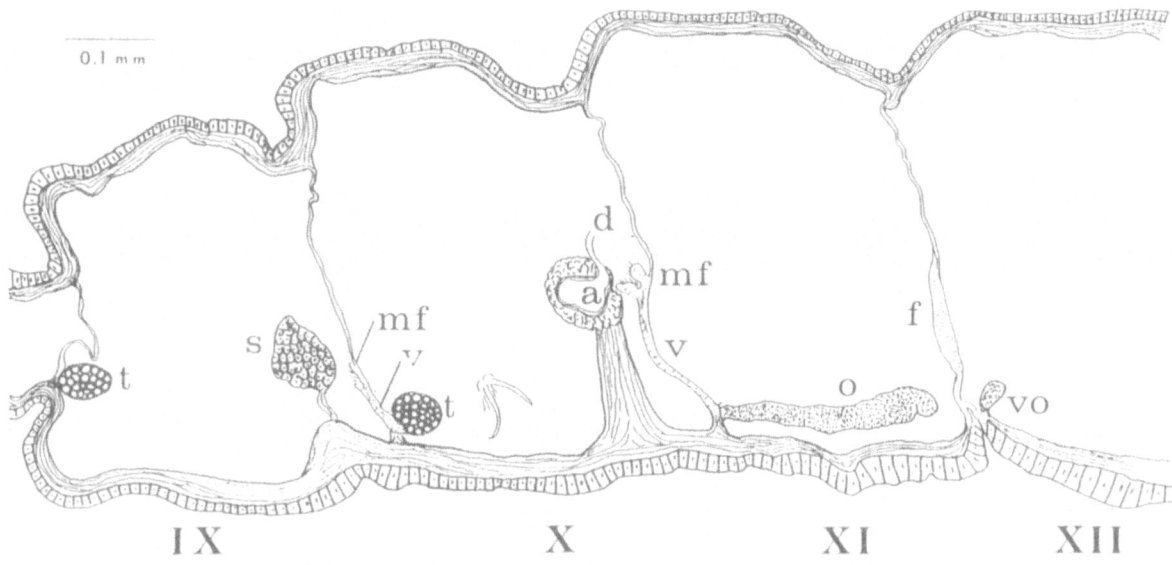

Fig. 5. *Tenagodrilus musculus* g.n., sp.n. Developing reproductive structures in partially mature specimen. d = non-ciliated duct from atrium, vo = vestigial ovary in XII, other legends as in Fig. 4. See text for details.

and the coding requires expansion to include these distinct differences. The so-called lateral blood vessels of several genera are usually blind-ending extensions of the dorsal vessel, which may connect with the intestinal perivisceral sinus, as in *Trichodrilus, Stylodrilus, Lamprodrilus, Lumbriculus* (Cook, 1971). *Stylodrilus aurantiacus* (Pierantoni, 1904) and *Stylodrilus lankesteri* (Vejdovsky, 1877) have commissural vessels, with blind lateral branches, connecting the dorsal and ventral vessels. These taxa are regarded as species dubia (Brinkhurst, 1965), and so their generic classification is uncertain. *Rhynchelmis* species have commissurals with blind laterals in median segments (e.g. XXX–XXXV) but the branched vessels in posterior segments run from the dorsal vessel, and have no contact with the ventral vessel (Hrabe, 1961). In *Eclipidrilus*, there are long commissural vessels with blind-ending lateral projections, but the commissural vessels are said to end without connection to the ventral vessel. An illustration by Cook (1967, Fig. 6c) suggests that these commissurals reach well down toward the ventral vessel. Even if they do connect, they resemble the commissural vessels of *Tenagodrilus*, which clearly connect to the ventral vessel. The blind-ending, truly lateral, vessels of *Eclipidrilus* and *Tenagodrilus* are very similar in that they penetrate the muscles of the body wall and then re-enter the coelom where they end in small expansions (Cook, 1967, Fig. 6d). The nature of the blind-ending lateral blood vessels may

reflect environmental conditions and may vary within species, and there is considerable variation within genera (Rodriguez & Giani, 1994). There does appear to be a difference between *Tenagodrilus* and other well-established taxa, but the nature of the vascular system would not be enough, in itself, to establish generic ranking for this taxon.

The male ducts of *Tenagodrilus* are so distinctive that it seems appropriate to erect a new genus for this species. More weight has been given to characters dealing with the male ducts than any others in older studies of oligochaetes. In *Stylodrilus,* penes are absent in some species, present in others. Where they are present, they are usually short, but they are long in *S. heringianus* Claparède, 1862. In all these older species the penes are exposed on the ventral body wall as there are no penis sacs. In two recently-described species (*S. glandulosus* Giani & Martinez-Ansemil, 1984, and *S. curvithecus* Collado, Martinez-Ansemil & Giani, 1993) there are small penis sacs and also ejaculatory ducts, but these do not approach the dimensions of those in *Tenagodrilus*. Both of these species strain the existing limits of *Stylodrilus*, which needs to be assessed with cladistic methods. *Lumbriculus* is a particularly difficult taxon because of the predominance of asexual reproduction in, at least, some species, that creates great variability in the number and position of reproductive organs. This variation has variously been regarded as intra-specific or, alternatively, as the basis

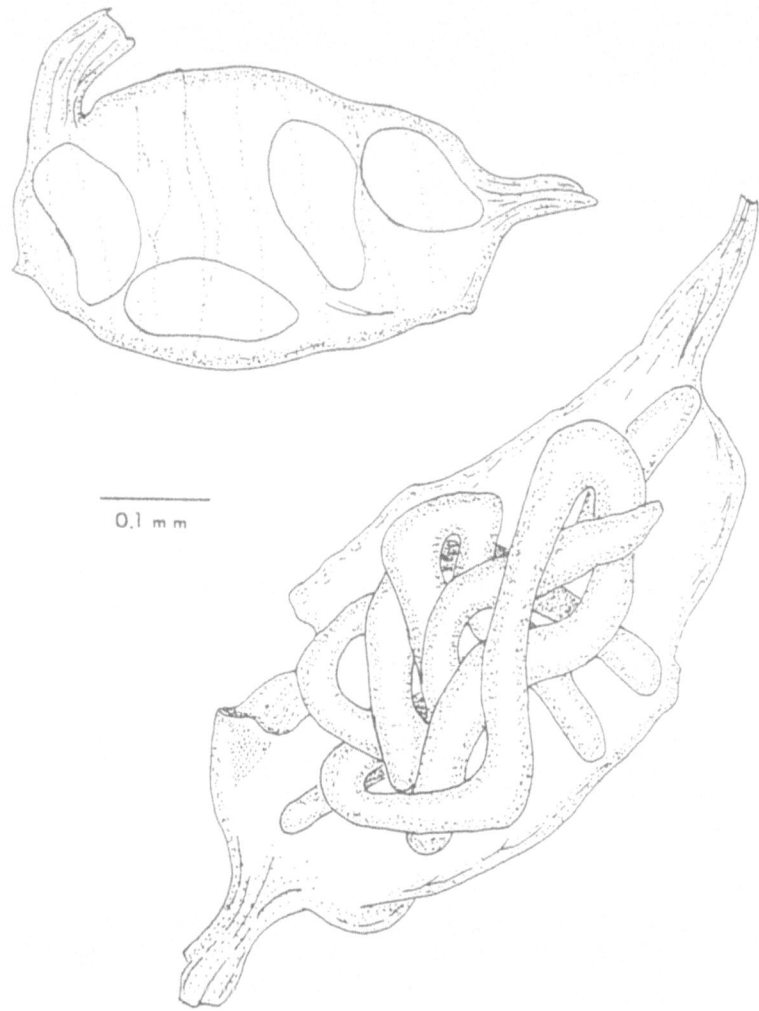

0.1 mm

Fig. 6. Tenagodrilus musculus g.n., sp.n. Cocoons with eggs or developing worms.

of separation of genera (Holmquist, 1976; Timm & Rodriguez, 1994). The latter authors report difficulty in separating *Lumbriculus* and *Trichodrilus*. The nature of the penes and their penis sacs were thought to be characteristic of *Lumbriculus*, but Rodriguez & Giani (1994) describe some *Trichodrilus* species with quite long penes formed from cells in the ectal part of the atrium as opposed to normal pendent penes that develop as folds of the penis sacs. The former might be better termed protrusible penes. The penes in *Tenagodrilus* appear to have no cell boundaries within the body of the organs, but they are true pendent penes in sacs.

None of the existing genera have the complex of protractor and retractor muscles that evert the penes and straighten the ejaculatory ducts in *Tenagodrilus*. We rely on this apomorphic condition of the male duct to delimit the genus. This seems preferable to extending the limits of *Stylodrilus* even further. This position may have to be revised once a careful cladistic study of the whole constellation of apparently related genera is completed at the species level.

Two other aspects of the anatomy of *Tenagodrilus* deserve comment. The rudimentary ovaries in segment XII of partly mature specimens lack funnels and are presumably functionless. They are lost during further development. The hypothesis of the junior author is that the ancestors of the oligochaetes had four pairs of gonads. Traces of gonads that are non-functional and not associated with gonoducts are found in this species and other taxa [*Aulodrilus* (Tubificidae) for example]. They lie in segments of a four-segment sequence during development, which is considered to be consistent with an hypothesis that clitellate evolution proceeds from forms with four pairs of gonads to those with only two pairs, rather than the reverse (Brinkhurst, 1991).

The observations regarding the vasa deferentia require substantiation. It is generally held that the vasa deferentia develop from the mesodermal elements of the septa and sperm funnel, and grow back to reach the developing atria, which are ectodermal, a concept based on very limited study (Hrabe, 1939). We could not trace a connection between the ciliated portion of the vas deferens developing from the sperm funnel and the unciliated duct that appears to be growing out from the atrium. These two sections are distinguished in many tubificid vasa deferentia (Jamieson, 1981), but there has been no suggestion that they have separate origins.

The life history of the lumbriculid does not seem to be related to the filling of the pond, unlike that of the isopod investigated by Modlin & Bush (1988). There is no sign of resistant cyst formation by these worms, although these are known to occur in other species in the family (Brinkhurst & Gelder, 1991). From this we conclude that the species is probably an inhabitant of the hyporheos, which would not be surprising in this limestone district. The period when the pond is dry coincides with the rainy season in this area. The only exception was the dry period in February 1986 during a drought. Drying in normal years is probably due to evapotranspiration, as the dry period corresponds to the time when the deciduous trees are in leaf.

Acknowledgments

The material for this paper was made available by Richard F. Modlin, access to the pond was kindly permitted by Hundley Batts, Sr. of WEUP, Huntsville, AL. William Gartska provided facilities for the histological work, and Robert Lawton and Susan Weber provided botanical information. Robert L. Thibeault provided financial support.

References

Brinkhurst, R. O., 1965. A revision of the genera *Stylodrilus* and *Bythonomus* (Oligochaeta, Lumbriculidae). Proc. Zool. Soc. Lond. 144: 431–444.

Brinkhurst, R. O., 1989. A phylogenetic analysis of the Lumbriculidae (Annelida, Oligochaeta). Can. J. Zool. 67: 2731–2739.

Brinkhurst, R. O., 1991. Ancestors (Oligochaeta). Mitt. hamb. zool. Mus. Inst. 88: 97–110.

Brinkhurst, R. O. & S. R. Gelder, 1991. Annelida: Oligochaeta and Branchiobdellida. In J. H. Thorp & A. P. Covich (eds), Ecology and Classification of North American Freshwater Invertebrates. Academic Press, New York: 401–435.

Cook, D. G., 1967. Studies on the Lumbriculidae (Oligochaeta) in Britain. J. Zool., Lond. 153: 353–368.

Cook, D. G., 1971. Anatomy: Microdriles In R. O. Brinkhurst & B. G. M. Jamieson (eds), Aquatic Oligochaeta of the World. University of Toronto Press, Toronto: 8–41.

Finogenova, N. P. & N. R. Arkhipova, 1994. Morphology of some species of the genus *Aulodrilus* Bretscher. Hydrobiologia 278: 7–15.

Hrabe, S., 1939. O vyvoji samcího vyvodného aparátu u nekterych nítenek a zízalic. Sb. prír. Klubu Trebici 6–7: 56–65.

Hrabe, S., 1961. Dva Nove Druhy Rodu *Rhynchelmis* ze Slovenska. Publ. Fac. Sci. Univ. Brno. 421: 129–145.

Holmquist, C., 1976. Lumbriculids (Oligochaeta) of Northern Alaska and Northwest Canada. Zool. Jb. (Syst.) 103: 377–431.

Jamieson, B. G. M., 1981. The Ultrastructure of the Oligochaeta. Academic Press, London.

Modlin, R. F. & K. C. Bush, 1988. Life cycle during a drought year of *Caecidotea obtusus* from a temporary pond in North Alabama (Crustacea, Isopoda). J. Alabama Acad. Sci. 59: 237–246.

Reynolds, J. W., 1977. The Earthworms (Lumbricidae and Sparganophilidae) of Ontario. Royal Ontario Museum Publications in Life Sciences, Ontario, 139 pp.

Rodriguez, P. & N. Giani, 1994. A preliminary review of the taxonomic characters used for the systematics of the genus *Trichodrilus* Claparede (Oligochaeta, Lumbriculidae). Hydrobiologia 278: 35–51.

Timm, T. & P. Rodriguez, 1994. Description of a new *Lumbriculus* species (Oligochaeta, Lumbriculidae) from the Russian far east. Ann. Limnol. 30: 95–100.

Hydrobiologia **334**: 11–15, 1996.
K. A. Coates, Trefor B. Reynoldson & Thomas B. Reynoldson (eds), Aquatic Oligochaete Biology VI.

Kathrynella, a new oligochaete genus from Guyana

Pietro Omodeo
Dipartimento di Biologia, Il Università di Roma 'Tor Vergata', Via O. Raimondo, I-00173 Rome, Italy
Mail address: Dipartimento di Biologia Evolutiva, Università di Siena, Via P.A. Mattioli 4, I-53100 Siena, Italy

Key words: Kathrynella, new genus, Alluroididae, Guyana, genital apparatus, vascular apparatus

Abstract

Kathrynella guyanae gen. n., sp. n. from a limicolous habitat in Guyana is described. It is distinguished from the known genera of Alluroididae by having testes in XI, male pores in XIV, and one pair of spermathecae opening laterally at 6/7. The circulatory apparatus appears remarkably developed. The diagnosis of the family is emended to include the new species.

Introduction

Three years ago, Dr Kathryn Coates kindly entrusted me with a few specimens of an unknown species of aquatic megadrile collected in Guyana (cf. Coates & Stacey, 1994). These specimens belong to a hitherto undescribed genus of Alluroididae showing some unusual characters.

Materials and methods

Two whole-mounted and three alcohol-preserved specimens were entrusted to the author. One specimen was dissected; it was then embedded in paraffin and its 25 anterior segments were sectioned transversely. The first 20 segments of the holotype, previously mounted in Canada balsam, were sectioned longitudinally. All sections were stained with ferric haematoxylin.

Type material is deposited at Invertebrate Zoology, Royal Ontario Museum, Toronto (ROMIZ).

Upper case roman numerals denote segment numbers, whilst intersegmental furrows or septa are denoted by the arabic numerals of the segments on either side.

Systematic account

Family **Alluroididae** *Michaelsen, 1900, emendavit Jamieson, 1968*

The allocation of the new genus *Kathrynella* to the Alluroididae requires the following changes in Jamieson's (1968) diagnosis of the family: Male pores intraclitellar one pair, either at the anterior border of XIII, in XIII, *or in XIV*; holoandric, proandric, *or metandric* male apparatus.

Genus **Kathrynella** *gen. n.*

Diagnosis: Alluroid megadriles with a cylindrical muscular atrium (euprostate); male pores paired, intraclitellar, opening ventrally and in equatorial position in XIV. Spermathecal pores paired, opening in intersegmental furrow 6/7. Modified setae *a* and *b* in VII; grooved penial setae ventral in XIV. Circulatory apparatus strongly developed, showing in the anterior segments long lateral commissures between dorsal and ventral vessels rolled into a ball.

Observations: The new genus differs from other genera of Alluroididae because of the metandric condition of the male apparatus. Also the presence of sperm sacs and the absence of glandular lining of the atria may be diagnostic.

Derivatio nominis: The genus is affectionately dedicated to Dr Kathryn A. Coates, distinguished student of oligochaetes.

Kathrynella guyanae *sp. n.*

Type material. Holotype, ROMIZ I3234 (longitudinally sectioned specimen; unsectioned caudal portion in alcohol). Paratypes, ROMIZ I3235 (transversely sectioned specimen; unsectioned caudal portion in alcohol); ROMIZ I3236 (two specimens in alcohol and a whole mounted specimen in Canada balsam).

Type locality. Kurupukari, Guyana, adjacent to Essequibo River, in sediment of a small pool, between roots and twigs. Donald F. Stacey coll., October 8, 1990.

External characters. Body thread-like, flattened in clitellar and caudal regions (Fig. 1F), with deep intersegmental furrows conferring a moniliform appearance to the posterior half of worms. Longitudinal, lateromuscular grooves of body wall not observed. Length of entire specimens 85–89 mm; diameter 600 μm at X, 520 μm at mid-body. Segment number 282, 259 (three specimens were amputated). No cutaneous pigmentation, integument transparent. Prostomium zygolobous, conical; peristomium short. Setae paired with lumbricine arrangement (Fig. 1A, E, F). Pygidium small, anus as a vertical cleft; in one specimen a growth blastema was visible immediately anterior to the pygidium, as is common in limicolous worms (Fig. 1B).

Clitellum cylindrical, extending from the posterior part of XI to XVII (Fig. 1G; Figs 3 & 4); the ventral side of XIV bears two large, circular papillae with swollen borders and a very thick, whitish, central part crossed by 4–5 transverse furrows; behind and beside these papillae two crescent-shaped fields are visible, each bearing a single, large, grooved, penial seta (Fig. 1A). In all specimens examined, penial setae were broken but their length may be estimated longer than 300 μm. No other ventral setae were visible in XIV. Dorsal setal couples of midclitellar segments do not appear perpendicular to the long body axis but slanting, in that setae c are located more posteriorly than setae d.

Male pore as a slit on hinder portion of each crescent-shaped field in XIV (Fig. 1A); paired female pores on anterior border of XIV, near midventral line. No dorsal coelomic pores. Nephridiopores inconspicuous. Spermathecal pores in furrow 6/7 on setal line a.

Setal ratio $aa : ab : bc : cd : dd = 5.6 : 1.1 : 5.6 : 1 : 10$ at segment X; $2.5 : 1 : 1.5 : 1 : 12$

in posterior segments (where all setae are displaced ventrad) (Fig. 1E, F). The length of setae increases from II, where it is 37 μm, to VII where setae a and b (Fig. 1C) measure 92 μm and setae c and d are 43 μm. Further posteriorly, the setal length decreases gradually (penial setae of XIV excepted), to become constant (ca. 50 μm) behind the clitellum. Only caudal setae show distal ornamentation (Fig. 1D).

Internal organization. Body wall rather thin; cuticle approximately 0.5 μm thick; layer of circular muscles very thin. Septa 5/6-10/11 thickened, funnel-shaped (Fig. 2). No pharyngeal pad distinguishable. Neither a gizzard, nor any apparent distinction between oesophagus and intestine are present. Ciliated epithelium of intestine ending in clitellar region. Perintestinal blood sinus occurring from VII backwards (Fig. 1I; Fig. 5). The holonephridia, apparently avesiculate, begin in XII or XIII; they are initially poorly developed and acquire normal size only behind clitellum.

The circulatory apparatus shows an exceptionally great development. Dorsal and ventral vessels run the whole length of the worm and have very large diameter, only slightly less than that of the intestine. Seven pairs of long commissural vessels occur in V–XI, wound in balls whose size increases backwards; each ball is covered with chloragocytes (in one specimen the two vascular balls of XI completely fill the coelomic cavity, representing perhaps a pathological condition) (Fig. 2). A pair of large vessels originates in VII from the perintestinal blood sinus, and from XII to XVI run along the sperm sacs (Fig. 1O). Many lesser vessels are to be seen everywhere, but no capillaries.

Testes are located in the anterior part of XI and face the large seminiferous funnels, which are covered with sperm and bulge into segment XII (Fig. 1G). Sperm ducts are coiled in XII in their proximal course, then seem to run laterally within the body wall till XVII; here they emerge again in the coelomic cavity, become thicker, and run frontwards (Figs 1G, M, O; Figs 4 & 5) till XIV, where they form the atria. Atria possess thick muscular walls and appear as shining, bent, spindles which open in the middle of XIV, behind the penial setae; atria (which correspond to the 'euprostates' of Eudrilidae) have a thin external coating of non-glandular cells (Fig. 1N). Other structures of the male apparatus are two long sperm sacs (Fig. 1G; Fig. 3), which begin from septum 11/12 and run till XX, parallel to the egg sacs and to the posterior course of vasa deferentia.

Ovaries are located in the ventral forepart of XIII (Fig. 1G, L; Fig. 3); two long cylindrical egg sacs

13

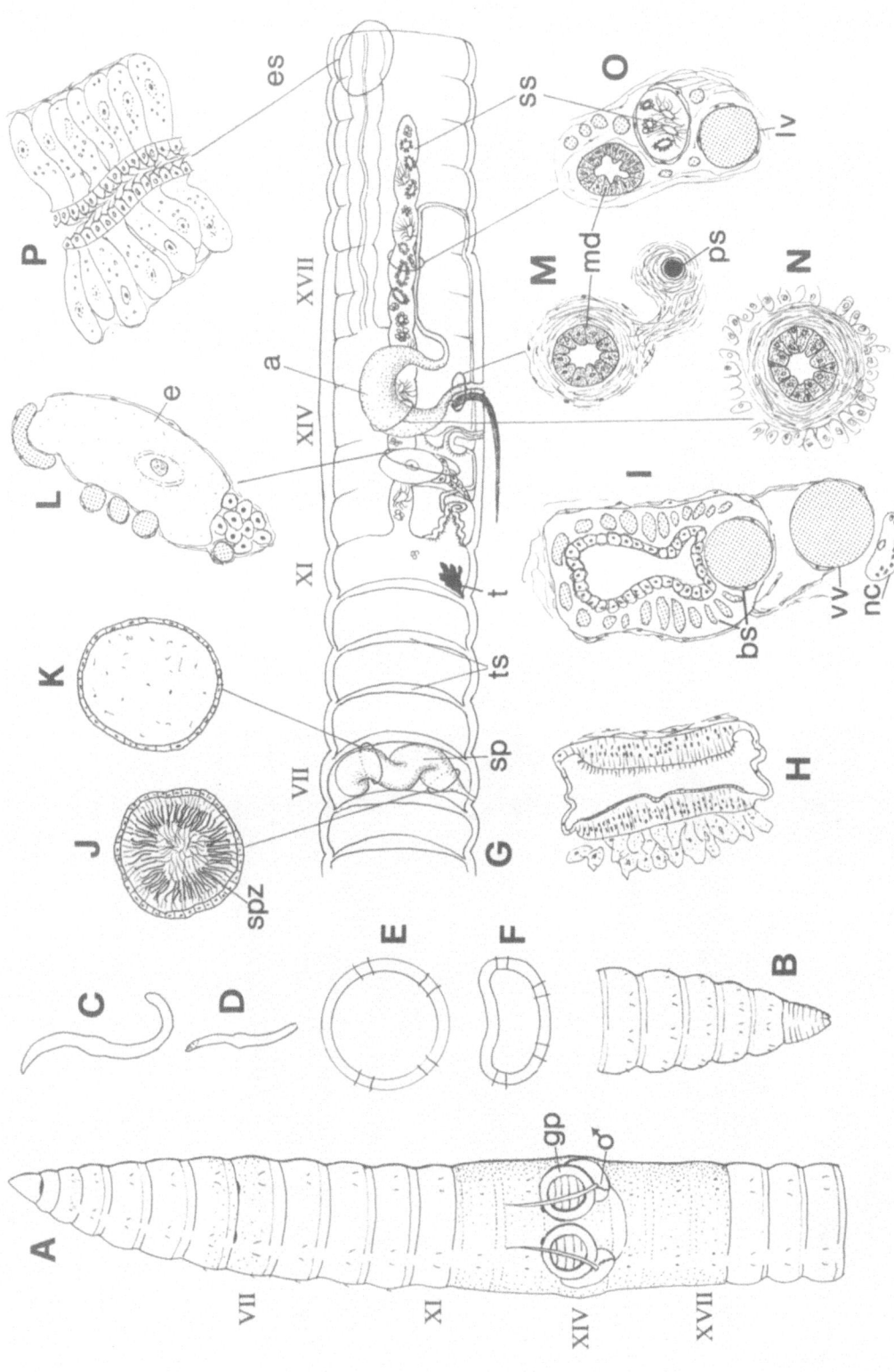

Fig. 1. Morphology and anatomy of *Kathrynella guyanae* sp. n. A. Ventral view of anterior segments; B. Caudal region and pygidium; C. Modified seta from segment VII; D. Ornamented seta from caudal segment; E. Cross section of midbody showing setal arrangement; F. Cross section of flattened caudal region; G. Scheme showing arrangement of genital organs; H. Section through pharynx showing ciliated pavement on the right and brush bordered ceiling on the left; I. Section through intestine at XII; note the perintestinal blood sinus, the ventral vessel, and the nerve cord containing a giant axon; J. Section through sperm-containing portion of spermathecal ampulla; K. Section through distal portion of ovary showing a large oocyte and abundant blood vessels; L. Section through ental part of spermathecal ampulla; M. Section through ectal portion of male duct and penial seta; N. Section through the middle of atrium; O. Section through sperm sac, male duct and lateral blood vessel; P. Section through egg sac. Legend: a = atrium; bs = blood sinus; e = oocyte; es = egg sac; gp = genital papilla; lv = lateral vessel; md = male duct; nc = nerve cord; ps = penial seta; sp = spermatheca; spz = spermatozoa; ss = sperm sac; t = testis; ts = thickened septum; vv = ventral vessel; ♂ = male pore.

14

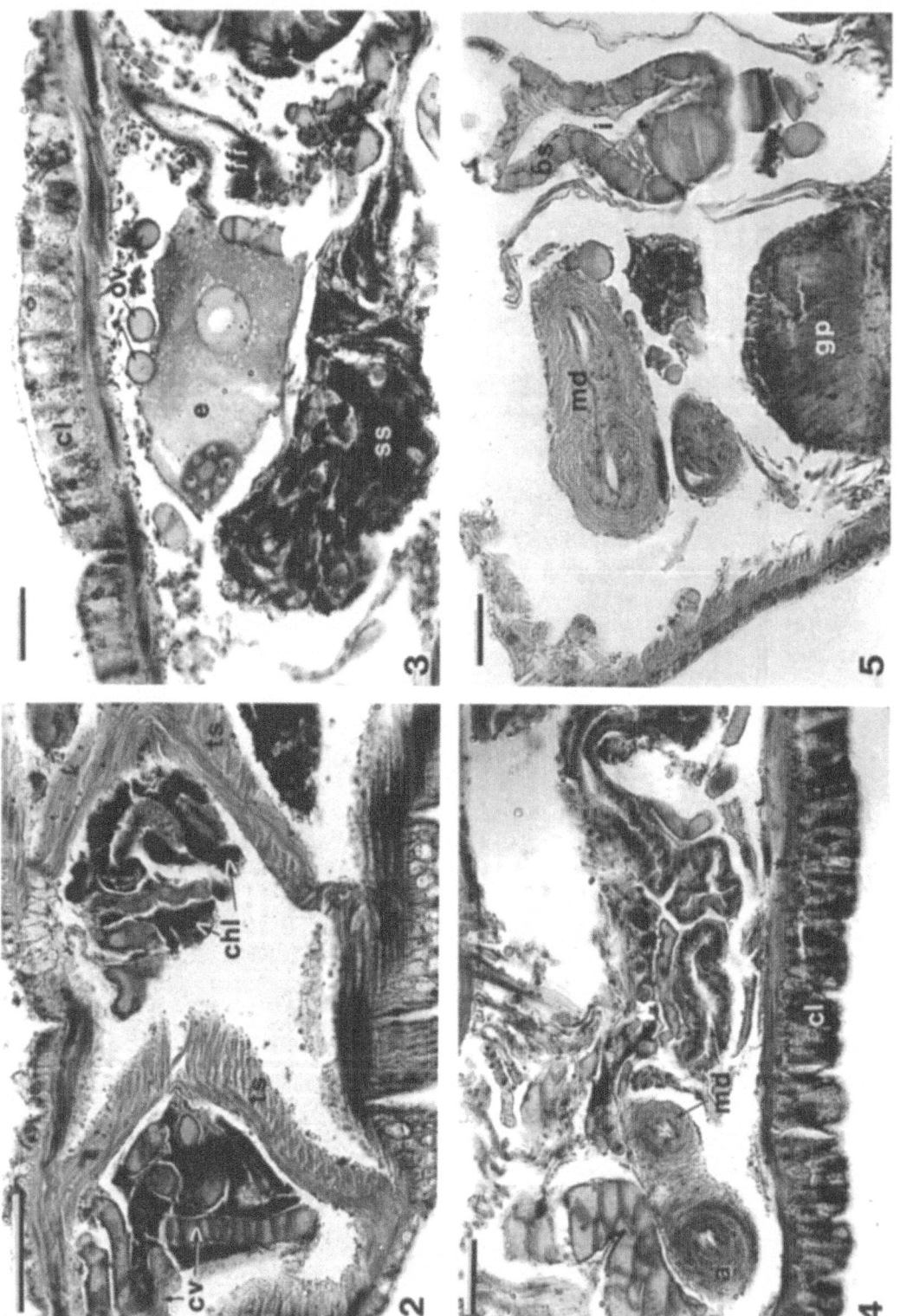

Figs. 2–5. Details of the anatomy of *Kathrynella guyanae* sp. n. (2) Parasagittal section through V–VI, showing the vascular balls formed by commissural blood vessels. Scale = 100 μm. (3) Parasagittal section through XIII, showing an oocyte surrounded by blood vessels (see also Fig. 1L). Scale = 50 μm. (4) Parasagittal section through XIV–XV, showing the male duct in its distal, cephalad course as approaching the atrium. Scale = 50 μm. (5) Transverse section through male organs (XIV) and intestine (bulging from XIII). Scale = 50 μm. a = atrium; bs = blood sinus; chl = chloragogen tissue; cl = clitellum; cv = commissural vessel; e = oocyte; ff = female funnel; gp = genital papilla; i = intestine; md = male duct; ov = blood vessels surrounding the oocyte; ss = sperm sac; ts = thickened septum.

(Fig. 1G, P), apparently beginning in XVII, extend as far back as XXXV; they possess thick glandular walls and contain up to a dozen large ripe eggs (ca. 530 μm wide) between XXV–XXXV.

Two large, convoluted spermathecae occur in VII (Fig. 1G) and consist of a short, partly ciliated duct and elongate ampulla; the ectal section of the latter contains the bulk of spermatozoa (Fig. 1J), whereas the ental section is filled with a hyaline secretion (Fig. 1K). The spermathecal length exceeds the body diameter and possibly the length of penial setae. It can be inferred that during copulation the sucker-shaped genital papillae in XIV help to fasten the two partners together, while at the same time the penial grooved setae are introduced into the spermathecae for sperm transfer.

Systematic position. Kathrynella guyanae sp. n. shows traits which are reminiscent of various microdrile taxa. The body thinness and the large blood vessels are typical of microdriles living in anoxic mud; the long egg sacs and sperm sacs, as well as the absence of capillaries, resemble those of some Haplotaxidae. Penial setae and the complex structure of vasa deferentia are similar to those of the megadriles *Brinkhurstia* (Brinkhurst, 1964; Righi *et al.*, 1978) and *Standeria* (Jamieson, 1968), and so is the anatomy of nephridia. There are, however, two main differences (autapomorphies) from the Alluroididae hitherto described i.e. the location of male pores in XIV and the metandric condition of the male apparatus which contrasts with the proandric condition of the Alluroidinae and the holoandric, supposedly plesiomorphic (cf. Jamieson, 1980), condition of the Syngenodrilinae. Yet, if we consider that parallel modifications of male apparatus, spermathecae and male pores have appeared in several genera (*Benhamia, Dichogaster, Microscolex, Parachilota*), then the differences observed within the Alluroididae, as here considered, seem to be less relevant. As stated by Pickford (1937: 27): 'A variability of the intensity of meristic suppression [of prostates and spermathecae] is probably associated with the loss of the anterior or posterior pair of testes'. That seems to be the present case.

Lastly, there is a noteworthy resemblance of the 'atria' of the new species to the typical 'euprostates' of the Eudrilidae, an African family of unknown kinship.

Distribution and habitat. Only known from the type locality.

Acknowledgements

I thank heartily K. Coates and D. Stacey, who gave me the occasion for studying such interesting material. Thanks are also due to E. Rota for help during the preparation of the manuscript. This study was partly supported by a M.U.R.S.T. grant to the author.

References

Brinkhurst, R. O., 1964. A taxonomic revision of the Alluroididae (Oligochaeta). Proc. zool. Soc. Lond. 142: 527–536.

Coates, K. A. & D. F. Stacey, 1994. Oligochaetes (Naididae, Tubificidae, Enchytraeidae and Alluroididae) of Guyana, Peru and Ecuador. Hydrobiologia 278: 79–84.

Jamieson, B. G. M., 1968. A taxonometric investigation of the Alluroididae (Oligochaeta). J. Zool., Lond. 155: 55–86.

Jamieson, B. G. M., 1980. Preliminary discussion of an Hennigian analysis of the phylogeny and systematics of opisthoporous Oligochaetes. Rev. Ecol. Biol. Sol 17: 261–275.

Michaelsen, W., 1900. Oligochaeta. Das Tierreich, 10. Lief. Friedländer und Sohn, Berlin, 575 pp.

Pickford, G. E., 1937. A Monograph of the Acanthodrilinae Earthworms of South Africa. Heffer & Sons Ltd., Cambridge, 612 pp.

Righi, G., I. Ayres & E. C. R. Bittencourt, 1978. Oligochaeta (Annelida) do Instituto Nacional de Pesquisas da Amazônia. Acta Amazonica 8: 1–49.

Hydrobiologia **334**: 17–29, 1996.
K. A. Coates, Trefor B. Reynoldson & Thomas B. Reynoldson (eds), Aquatic Oligochaete Biology VI.
© 1996 *Kluwer Academic Publishers.*

Oligochaetes (Naididae, Tubificidae, Opistocystidae, Enchytraeidae, Sparganophilidae and Alluroididae) of Guyana

Donald F. Stacey[1] & Kathryn A. Coates[1,2]
[1] *Invertebrate Zoology, Royal Ontario Museum, 100 Queen's Park, Toronto, Ontario, Canada M5S 2C6*
[2] *Bermuda Biological Station for Research, Inc., GE 01, Bermuda*

Key words: Aquatic oligochaetes, Enchytraeidae, Guyana, distribution, taxonomy

Abstract

About 50 species in more than 20 genera of the microdrile oligochaete families Tubificidae, Naididae, Opistocystidae, and Enchytraeidae and the freshwater megadrile families Sparganophilidae and Alluroididae were identified in recent collections made in Guyana. Only seven species in these families were previously recorded from Guyana. The aquatic oligochaete fauna has similar components to those of the southeastern United States, other locations in South America, and across lower latitudes in the northern hemisphere. A high diversity of species is found in the naidids especially in the genera *Pristina* and *Pristinella*, especially considering the small number of locations that have been sampled. The collections include approximately 12 new species of rhyacodrilines (Tubificidae), *Dero*, *Pristina*, *Pristinella* (Naididae), *Brinkhurstia* (Alluroididae), *Hemienchytraeus* and *Aspidodrilus* (Enchytraeidae). This is the first record of *Aspidodrilus* from outside of Africa and the first new record since 1952.

Introduction

This study represents a continuation of taxonomic and biogeographic projects on the aquatic oligochaetes of South America that have been undertaken at the Royal Ontario Museum. A previous study (Coates & Stacey, 1994) based on collections from three countries of South America, Peru, Guyana, and Ecuador, represented only new species records for the last two countries simply because there were no earlier studies or they had been very limited in taxonomic and geographic scope. In that study, about 41 species in 23 genera, compared to approximately 120 species of aquatic oligochaetes of the same families that are reported from South America (Brinkhurst & Marchese, 1989; Varela, 1990; Juget & Lafont, 1994), were recorded but only a few, about seven, of these species were found in collections from Guyana (Coates & Stacey, 1994). Even though a large number of species are reported from South America, the distribution records are very incomplete for the continent, with some countries and larger regions remaining unexplored for these animals (Di Persia, 1980).

Guyana lies on the northeastern shoulder of South America, just north of the equator between approximately 2° and 8° latitude. It is bounded by the Atlantic Ocean to the north, Venezuela to the west, Brazil to the south, and Surinam to the east. It is within the South American biogeographic track, linking sister taxa from South to central and North America described by Rosen (1976; see also Humphries & Parenti, 1986).

Guyana displays a great variety of habitat types. It is 'a mosaic of many different types of forest and savanna' (Gillespie & Funk, 1993: 9); there are coastal swamp and mangrove forests, savanna and dry evergreen forest, lowland rainforest, montane forest and upland savannas. As each of these reflects topographic and elevational differences, among other things, the diversity of aquatic and sediment habitats is just as great.

While parts of Guyana have been seriously affected by human activity, much of the natural vegetation is relatively untouched. The population density is the third lowest in South America, thus overpopulation and the resulting human pressure to cut tropical forests for agricultural use is not a major problem, as it is in much of the Amazon Basin. Large-scale, multinational lumber and mining operations, however, pose major future threats to forest and fresh-water ecosystems in the country (Gillespie & Funk, 1993).

Materials and methods

Field methods

River sediments were primarily collected using an Ekman grab, while terrestrial sediments were collected with a small folding shovel. The sediment, in both cases, was sieved through a 300 μm screen and the material left on the screen was transferred to a plastic sorting tray. Oligochaetes were removed from the tray and preserved in Kahle's solution. Specimens were placed in 75% alcohol at the conclusion of the field trip.

Taxonomic methods

All mature specimens and other representative specimens of naidids and opistocystids were sorted from the collections under a dissecting microscope, stained with borax carmine, dehydrated in an alcohol through xylene graduated series, and mounted permanently in Canada balsam.

Identifications were made on a compound photomicroscope with interference contrast optics. Videomicroscopy was used so that both investigators could simultaneously observe a specimen and to make automated measurements of setae and other structures. Thirty-five mm black and white photography was done with a Wild MPS 46/52 Photoautomat on Kodak Techpan at ASA 100.

Reference was made to Brinkhurst and Jamieson (1971) and Brinkhurst and Marchese (1989) for routine species identifications. For resolution of other taxonomic issues, reference was made to a number of sources (Varela, 1990; Gluzman, 1990a, 1990b; Coates, 1990; Harman *et al.*, 1988; Brinkhurst & Marchese, 1987; Di Persia, 1980; Nielsen & Christensen, 1959). Author, date extensions for all the identified species are given in Table 2.

Collections

The most recent collections from Guyana were made in September and October 1992. The samples were collected by D.F. Stacey and Youth Challenge Interna-

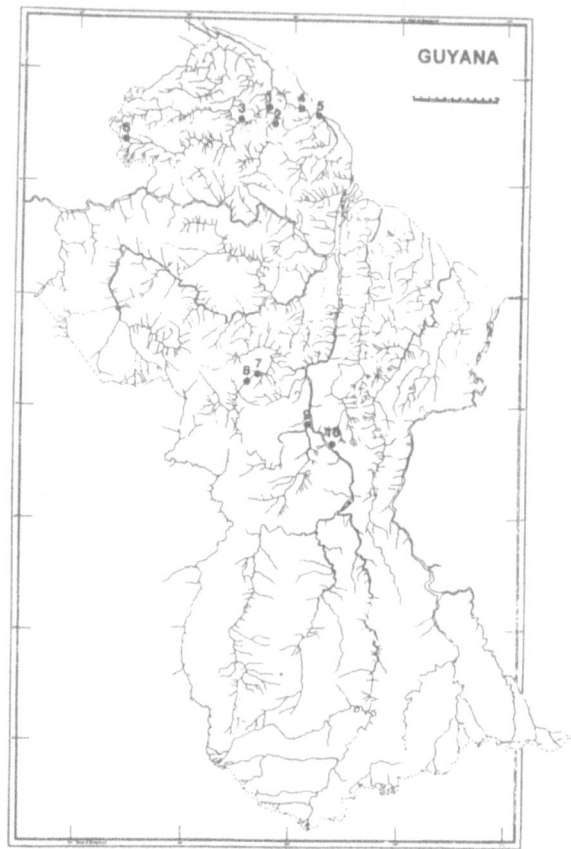

Fig. 1. General collecting sites in Guyana. 1, Santa Cruz; 2, Kwabanna; 3, Chinese Landing; 4, Santa Rosa Mission; 5, beach at Atlantic Ocean; 6, Baramita; 7, Amatuk Falls; 8, Tukeit Falls; 9, Siparuni River; 10, Kurupukari. Sites 1 through 6 were visited in 1992 and 7 through 10 in 1990.

tional participants at: Site 1) Santa Cruz (Waini River), Site 2) Kwabanna (Waini River and a tributary of the Barawa River), Site 3) Chinese Landing (Barama River), Site 4) Santa Rosa Mission (Bara Bara River), and Site 5) Atlantic Ocean shore near mouth of Pomeroon River and just upstream from the mouth of the Maruka River; and collected by R. MacCulloch at Site 6) Baramita (in a creek) (Fig. 1). Samples at each of the sites were taken from both terrestrial and freshwater habitats as well as a coastal, estuarine habitat at Site 5 (Fig. 1, Table 1). The accession numbers of the collections are Royal Ontario Museum, Department of Invertebrate Zoology (ROMIZ) 1992–025 for Sites 1 through 5 and 1992–044 for Site 6.

Table 1. Collecting areas, sample locations and descriptions and species assemblages.

Location	Lat/Long	Date	Site description	Substrate type/depth (m)	Species found
Santa Cruz: Site 1	7° 39.97' N 59° 14.23' W	28 Sep 1992	Stn #7 – Creek southeast of Santa Cruz	mud, organic material, leaf litter depth 0.8 m	*Pristina biserrata*
	7° 39.97' N 59° 14.23' W	29 Sep 1992	Stn #11 – Waini River downstream from Santa Cruz, at mouth of creek	silt depth 6 m	*Limnodrilus hoffmeisteri* *Tubifex siolii*
	7° 40.18' N 59° 14.30' W	29 Sep 1992	Stn #12 – creek downstream from Santa Cruz at stream edge	mud, silt, organics depth 0.2 m	*Achaeta*
	7° 39.97' N 59° 14.23' W	29 Sep 1992	Stn #14 – Waini River downstream from Santa Cruz	mud, silt, little organic material depth 2–3m	*Pristinella notopora* *Tubifex siolii* *?Tubifex* sp.
	7° 39.97' N 59° 14.23' W	29 Sep 1992	Stn #15 – Santa Cruz, near ditch east of basecamp	mud	*Pristinella notopora*
	7° 39.97' N 59° 14.23' W	29 Sep 1992	Stn #16–180 m east of basecamp	mud, roots, dense vegetation	*Hemienchytraeus*
	7° 39.35' N 59 15.00' W	30 Sep 1992	Stn #19 – Barama River, upstream from junction with Waini River	mud, silt max depth 15 m	*?Tubifex* sp.
	7° 39.97' N 59° 14.23' W	30 Sep 1992	Stn #20 – Waini River downstream from Santa Cruz (near Stn #14)	mud, silt, little organic material	*Tubifex siolii*
	7° 39.97' N 59° 14.23' W	30 Sep 1992	Stn #23 – Santa Cruz, ~100 m east of basecamp	mud	*Pristinella longidentata* *Pristinella* sp. 2
	7° 39.97' N 59° 14.23' W	01 Oct 1992	Stn #27 – Waini River at Santa Cruz, on mudflats at low tide	mud, silt, detritus	*Opistocysta funiculus* *Bothrioneurum americanum* *?Tubifex* sp.
	7° 39.97' N 59° 14.23' W	01 Oct 1992	Stn #28 – Creek flowing into Waini River downstream from Santa Cruz, at creek's edge	mud, silt, detritus	*Dero indicus* *Dero lodeni* *Pristinella notopora*
	7° 39.97' N 59° 14.23' W	02 Oct 1992	Stn #30 – Waini River downstream from Santa Cruz, on mudflats at low tide	mud, silt, detritus depth 3 m	*Tubifex siolii*
	7° 39.97' N 59° 14.23' W	02 Oct 1992	Stn #32 – Waini River at Santa Cruz, off dock at basecamp	mud, silt, detritus depth 1 m	*Dero* sp. 2 *Nais elinguis* *Limnodrilus hoffmeisteri*

Earlier collections were made in 1990, ROMIZ 1990–115, on the Potaro and Essequibo Rivers (Fig. 1).

Results

Taxonomy

These collections from Guyana bring the total number of aquatic oligochaetes reported from the country to about 50 (Table 2). The most recent collections from lowland areas in the coastal plain and a northern highland area included 25 species of naidid, with possible new species of *Dero*, *Pristina*, and *Pristinella* (Figs 2 & 3) among them. Identities of the new species are presently under discussion with experts on the group. The presence of *Uncinais uncinata* in South America (? *Uncinais*, in Coates & Stacey, 1994) was confirmed based on specimens from Guyana (R. Grimm, pers. comm.). The species diversity for naidids is comparable to what is known for Italy (Sambugar, 1986) and the

Table 1. Continued.

Location	Lat/Long	Date	Site description	Substrate typedepth (m)	Species found
Chinese Landing: Site 3	7° 31.00′ N 59° 33.00′ W	04 Oct 1992	Stn #35 – Barama River, ~100 m upstream from base-camp at Chinese Landing	mud, silt, clay depth 3 m	*Brinkhurstia* sp. 2 *Opistocysta funiculus* *Pristina americana/peruviana* *Tubifex siolii* *Slavina appendiculata*
	7° 31.00′ N 59° 33.00′ W	04 Oct 1992	Stn #38 – ~50 m from base-camp off trail, in forest	humus	*Hemienchytraeus*
	7° 31.00′ N 59° 33.00′ W	04 Oct 1992	Stn #39 – ~50 m from base-camp off trail, in forest	humus	*Hemienchytraeus*
	7° 31.00′ N 59° 33.00′ W	05 Oct 1992	Stn #40 – near Stns 38/39	humus	*Hemienchytraeus*
	7° 31.00′ N 59° 33.00′ W	05 Oct 1992	Stn #41 – ~200 m from basecamp, off trail, in forest	humus	*Aspidodrilus*
	7° 31.00′ N 59° 33.00′ W	05 Oct 1992	Stn #43 – at basecamp, off trail, in forest	humus, sand, clay beneath leaf litter	*Hemienchytraeus*
	7° 31.00′ N 59° 33.00′ W	05 Oct 1992	Stn #44 – clearing at basecamp	sand, clay, mud	*Hemienchytraeus*
	7° 31.00′ N 59° 33.00′ W	06 Oct 1992	Stn #47 – Barama River, directly across from basecamp at Chinese Landing	mud, sand depth 4 m	*Opistocysta funiculus* *Rhizodrilus* sp. *Tubifex siolii*
	7° 31.00′ N 59° 33.00′ W	06 Oct 1992	Stn #48 – near Stn 44, clearing at basecamp	sand, clay, mud	*Hemienchytraeus* ?*Tupidrilus*
	7° 31.00′ N 59° 33.00′ W	07 Oct 1992	Stn #51 – Barama River, across from basecamp at Chinese Landing	sand depth 5 m	*Tubifex siolii*
	7° 31.00′ N 59° 33.00′ W	07 Oct 1992	Stn #52 – Barama River, ~500 m downstream from basecamp at Chinese Landing	mud, sand depth 3 m	*Pristina americana/peruviana* *Tubifex siolii*
	7° 31.00′ N 59° 33.00′ W	07 Oct 1992	Stn #53 – Barama River, ~600 m downstream from basecamp at Chinese Landing	mud, silt, detritus depth 1–3 m	*Opistocysta funiculus* *Tubifex siolii*
	7° 31.00′ N 59° 33.00′ W	07 Oct 1992	Stn #54 – clearing upstream from basecamp, near Barama River	humus	*Hemienchytraeus*

Iberian Peninsula of Spain (Martinez Ansemil, 1993) but with notably few species of the widespread genus *Nais* and absence of the genus *Chaetogaster*.

The tubificids were not as diverse (Table 2, Fig. 4), represented by only seven species from freshwater habitats, and one from a coastal, estuarine habitat. However, they include two unidentified (new) species of the subfamily Rhyacodrilinae, one a species of *Rhizodrilus* (M. Milligan, pers. comm.) (Fig. 4), as well as numerous mature specimens of the poorly known species *Tubifex siolii* (Fig. 4).

The enchytraeids were more diverse, including representatives from damp and wet soils as well as freshwater habitats. As the taxonomy of South American species is not well-documented, the species numbers remain approximate at about 13 in eight genera. Included are five species of *Hemienchytraeus* and a species of the genus *Aspidodrilus*; the latter genus was previously known only from Africa (Coates, 1990).

Among the megadriles, three aquatic species were found in our recent collections – two species of *Brinkhurstia* (Alluroididae) (Table 2, Fig. 4) and one as yet unidentified species of sparganophilid (Table 2). In

Table 1. Continued.

Location	Lat/Long	Date	Site description	Substrate type/depth (m)	Species found
Kwabanna: Site 2	7° 34.00′ N 59° 09.00′ W	09 Oct 1992	Stn #55 – Waini River, ~1/2 km upstream from Kwabanna	mud, silt, detritus depth 3–4 m	*Pristinella sima* *Tubifex siolii*
	7° 34.00′ N 59° 09.00′ W	09 Oct 1992	Stn #56 – Waini River, ~1 km upstream from Kwabanna	mud, silt depth 5 m	*Tubifex siolii*
		10 Oct 1992	Stn #58 – Waini River, ~3 km downstream from Kwabanna, at mouth of creek	mud, sand depth 5–7 m	*Pristinella jenkinae* *Pristinella notopora* *Tubifex siolii*
		10 Oct 1992	Stn #59 – Waini River, ~2 1/2 km downstream from Kwabanna, off reed bed	mud, detritus, wood chips depth 3–5 m	*Pristina biserrata*
	7° 34.00′ N 59° 09.00′ W	10 Oct 1992	Stn #60 – ~1/2 km downstream from Kwabanna	mud, silt depth 5–6 m	*Tubifex siolii*
	7° 34.00′ N 59° 09.00′ W	11 Oct 1992	Stn #61 – Kwabbana, off trail from logging road into swampy forested area	sandy soil, roots, twigs	*Hemienchytraeus*
	7° 34.00′ N 59° 09.00′ W	11 Oct 1992	Stn #62 – Kwabanna, in swampy forest ~1/2 km up logging road to Maruka	sand, wet leaf litter	*Brinkhurstia americanus* Sparganophilidae
	7° 33.18′ N 59° 06.32′ W	11 Oct 1992	Stn #64 – Barawa River tributary, ~6.5 km from Kwabanna up logging road to Maruka	sand depth 0.7 m	*Pristina americana/peruviana* *Pristina* sp. 1 *Slavina evelinae* *Aulodrilus pigueti*
	7° 34.00′ N 59° 09.00′ W	12 Oct 1992	Stn #65 – Waini River, just upstream from Barawa River mouth	mud, silt depth 7–8 m	*Tubifex siolii*
		12 Oct 1992	Stn #66 – Waini River, ~2 km downstream from Kwabanna	mud, clay, leaf litter depth 2–4 m	*Slavina evelinae* *Tubifex siolii*
	7° 34.00′ N 59° 09.00′ W	12 Oct 1992	Stn #67 – Waini River, ~1/2 km downstream from Kwabanna	fine silt depth 5 m	*Tubifex siolii*
		12 Oct 1992	Stn #68 – Waini River, ~3 km downstream from Kwabanna (near Stn #58)	fine silt depth 5 m	*Tubifex siolii* ?*Tubifex* sp.
	7° 33.18′ N 59° 06.32′ W	13 Oct 1992	Stn #70 – Barawa River tributary, ~7 km from Kwabanna up logging road to Maruka	sand, aquatic vegetation depth 0.2 m	*Opistocysta funiculus* *Dero digitata* *Dero sawayai* *Pristina americana/peruviana* *Pristina biserrata* *Pristina synclites*

the 1990 collections two other members of the family Alluroididae were found – a third species of *Brinkhurstia* and one species in the new genus *Kathrynella* (Omodeo, 1996).

Habitat and species associations

The locations sampled consistently had sediments recognized as mud/silt with some detritus or associated aquatic vegetation (Table 1). In contrast, the species composition varied from one location to anoth-

Table 1. Continued.

Location	Lat/Long	Date	Site description	Substrate type (depth (m)	Species found
	7° 33.18′ N 59° 06.32′ W	13 Oct 1992	Stn #71 – Barawa River tributary, ~6.5 km from Kwabanna up logging road to Maruka (same as Stn #64)	sand, aquatic vegetation depth 0.1–0.3 m	*Opistocysta funiculus* *Pristina americana/peruviana* *Pristinella longidentata* *Slavina evelinae* *Aulodrilus pigueti* *Paranadrilus descolei*
		13 Oct 1992	Stn #72 – Waini River, ~5 km upstream from Kwabanna	mud, heavy silt depth 3.5 m	*Opistocysta funiculus* *Dero palmata* *Pristina americana/peruviana* *Pristina macrochaeta* ?*Tubifex* sp.
	7° 34.00′ N 59° 09.00′ W	13 Oct 1992	Stn #73 – Waini River, Kwabanna landing	mud, wood, aquatic vegetation depth 0.3 m	*Opistocysta funiculus* *Dero palmata* *Pristina americana/peruviana* *Pristina* sp. 1 *Slavina evelinae* *Aulodrilus pigueti*
Santa Rosa Mission: Site 4	7° 39.00′ N 58° 56.00′ W	15 Oct 1992	Stn #74 – Bara Bara River, at Santa Rosa Mission Landing	dark mud, aquatic vegetation depth 0.2 m	*Opistocysta funiculus* *Dero digitata* *Dero lodeni* *Dero* sp. 1 *Pristina americana/peruviana* *Bothrioneurum americanum*
	7° 39.00′ N 58° 56.00′ W	15 Oct 1992	Stn #75 – Bara Bara River, beach near Santa Rosa Mission	sand, mud, vegetation	*Pristina americana/peruviana* *Pristinella jenkinae* *Bothrioneurum americanum*
Atlantic Ocean: Site 5	7° 38.28′ N 58° 46.19′ W	17 Oct 1992	Stn #82 – Atlantic Ocean, beach west of Pomeroon River, low water, lower intertidal	sand	Rhyacodrilinae
		18 Oct 1992	Stn #83 – Maruka River, ~1 km up-river from mouth	fine silt depth 5 m	?*Tubifex* sp. ?*Tupidrilus*
		18 Oct 1992	Stn #84 – Maruka River, ~3/4 km up-river from mouth	fine silt depth 2 m	*Dero* sp. 2
Baramita: Site 6	7° 22.00′ N 60° 29.00′ W	October 1992	Stn #3 – stream near basecamp	mud, silt	*Dero borellii* *Pristina aequiseta* *Pristina americana/peruviana* *Pristinella longidentata* *Pristinella* sp. 1 *Pristinella* sp. 2 *Uncinais uncinata* *Cognettia* cf. *glandulosa* *Guaranidrilus* sp. *Marionina* sp.1 *Marionina* sp.2

Table 2. List of taxa now known from Guyana; *, previously recorded in Guyana (Coates and Stacey, 1994); ! not found in recent collections.

Family Genus/Species	South and Central American Distribution (in addition to Guyana)	Worldwide Distribution
Sparganophilidae		
?immature		nearctic and palearctic
Alluroididae		
*!*Brinkhurstia* sp. 1		S. America
Brinkhurstia sp. 2		S. America
Brinkhurstia americanus (Brinkhurst, 1964)	Argentina, Brazil	S. America
!*Kathrynella guyanae* (Omodeo, 1995)		S. America
Enchytraeidae		
Achaeta sp.	Argentina, Brazil	cosmopolitan
Aspidodrilus sp.		Africa
Cognettia cf. *glandulosa*		widespread, arctic
Guaranidrilus	Argentina, Brazil, Peru	?Africa, Europe, N. & S. America
Hemienchytraeus, 5 spp.	Argentina, Bolivia, Brazil, Ecuador, Paraguay, Peru, Venezuela	circumsubtropical, Europe, N. America
Marionina, 2 spp.	Brazil, Ecuador, Peru	cosmopolitan
*!*Mesenchytraeus*		widespread, arctic
Tupidrilus	Brazil, Peru	S. America
Naididae		
Dero borellii Michaelsen, 1900	Argentina, Brazil, Ecuador,	?N. & S. America Paraguay
Dero digitata (Müller, 1773)	Argentina, Bolivia, Brazil, Costa Rica, Ecuador, Haiti, Paraguay, Peru, Surinam, Uruguay	cosmopolitan
*!*Dero furcatus* (Müller, 1773)	Argentina, Barbados, Bolivia, Brazil, Costa Rica, Dutch West Indies, Haiti, Surinam, Trinidad, Uruguay, Venezuela	cosmopolitan
Dero indicus Michaelsen, 1911	Argentina	S. America, southern India
Dero lodeni Brinkhurst, 1986	Argentina, Brazil, Paraguay, Surinam	S. America, USA
Dero palmata Aiyer, 1929	Argentina, Costa Rica	S. & Central America, southern India
Dero sawayai Marcus, 1943	Anegada, Argentina, Bolivia, Bonaire, Brazil, Ecuador, Haiti, St. Thomas, Surinam, Tortola, ?Venezuela, West Indies	S. & Central America and India
Dero (Aulophorus) sp. 1		
Dero sp. 2		

Table 2. Continued.

Family Genus/Species	South and Central American Distribution (in addition to Guyana)	Worldwide Distribution
Nais elinguis Müller, 1773	Argentina, Bolivia, Guatemala, Peru	cosmopolitan
Pristina aequiseta Bourne, 1891	Argentina, Bolivia, Brazil, Chile, Colombia, Costa Rica, Dutch West Indies, Ecuador, Haiti, Nicaragua, Peru, Surinam, Venezuela	cosmopolitan
Pristina americana Cernosvitov, 1937 (also see *peruviana*)	Argentina, Brazil, Dutch West Indies, Haiti, Peru, Surinam	Africa, S. America
Pristina biserrata Chen, 1940	Argentina, Chile, Ecuador	E. Asia, S. America
Pristina macrochaeta Cernosvitov, 1939	Argentina, Brazil, Paraguay, Peru, Surinam, Uruguay	S. America, Afghanistan
Pristina synclites Stephenson, 1925	Argentina, Ecuador	Africa, S. America, India, USA
Pristina sp. 1		
Pristinella jenkinae (Stephenson, 1931)	Argentina, Brazil, Costa Rica, Guatemala, Peru	cosmopolitan
Pristinella longidentata (Harman, 1965)	Argentina, ?Haiti, Montserrat Is., Peru, Surinam, Venezuela	Europe, S. & Central America, ?New Zealand, USA
Pristinella notopora Cernosvitov, 1937	Argentina, Peru	Africa, Europe, S. America
Pristinella sima (Marcus, 1944)	Argentina, Brazil, Chile	Africa, S. America, USA
Pristinella sp. 1		
Pristinella sp. 2		
Slavina appendiculata d'Udekem, 1855	Argentina, Brazil, Colombia, Costa Rica	Africa, S. & E. Asia, Europe, N. & S. America, New Zealand
Slavina evelinae (Marcus, 1942)	Antigua, Argentina, Brazil, Costa Rica, Surinam, Venezuela	S. & Central America
Uncinais uncinata (Oersted, 1842)		Europe, N. & S. America
Opistocystidae		
*Opistocysta funiculus Cordero, 1948	Argentina, Brazil	?Africa, S. America
Tubificidae		
Aulodrilus pigueti Kowalewski, 1914	Argentina, Brazil, Peru	cosmopolitan
Bothrioneurum americanum Beddard, 1894	Argentina, Paraguay, Peru, Venezuela	S. America, USA
Limnodrilus hoffmeisteri Claparède, 1862	Argentina, Brazil, Peru	cosmopolitan
Paranadrilus descolei Gavrilov, 1955	Argentina, Peru	S. America
Rhizodrilus sp.		
*Tubifex siolii (Marcus, 1947)	Brazil	S. America
Tubifex sp.		
marine tubificid-Rhyacodrilinae		

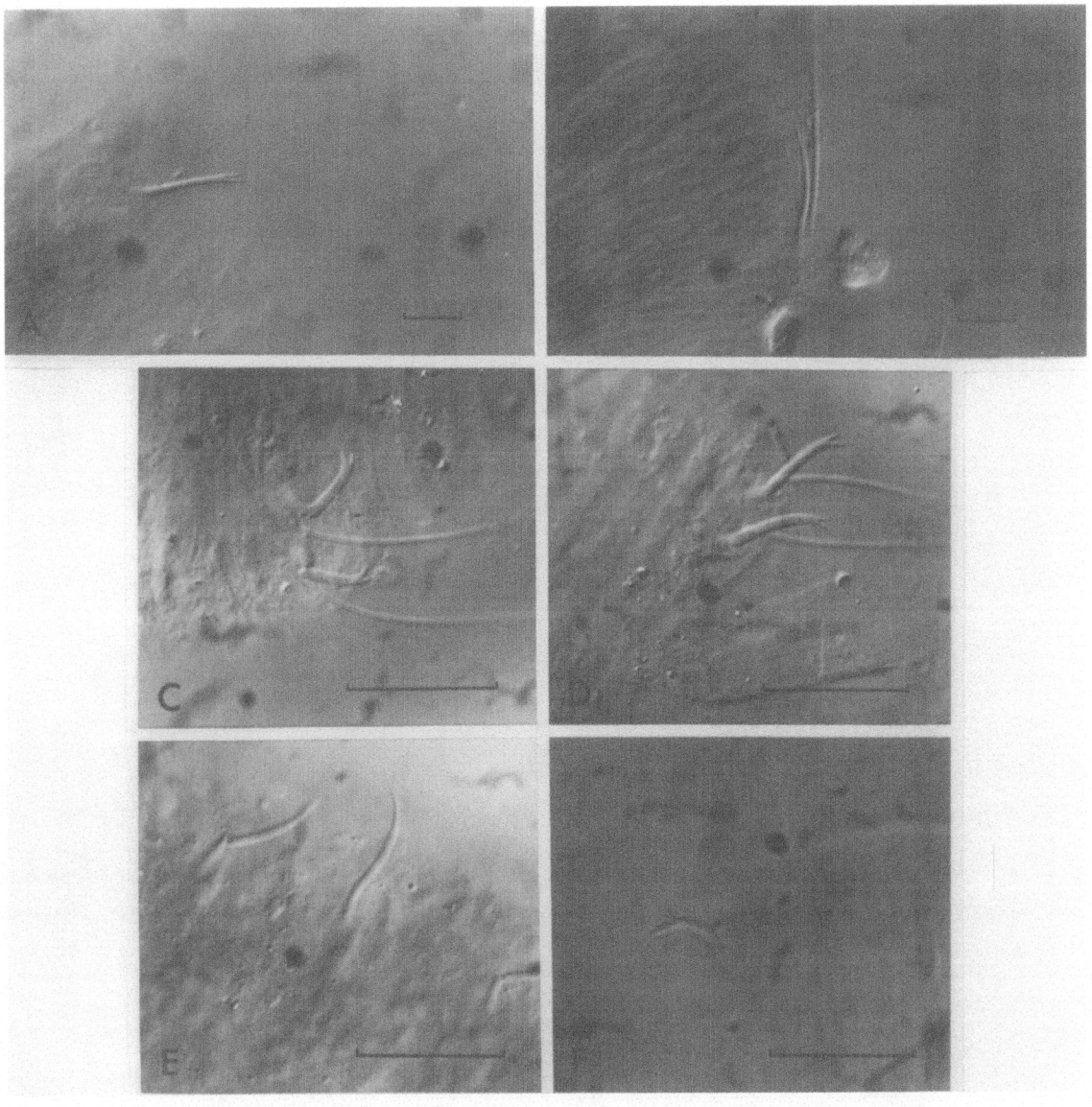

Fig. 2. Naididae: A. *Pristina* sp. 1, needle seta; B. *Pristina americana*, needle seta; C & D. *Pristinella* sp. 1, anterior needle setae; E. *Pristinella* sp. 1, posterior needle seta; F. *Pristinella* sp. 2, needle seta. Scale bars equal to 25 μm.

er. Twenty-one of the taxa identified were recorded from only one site, suggesting that they have very specific habitat requirements or patchy distributions. Generally, species were found in similar habitats from which they are known elsewhere, given that many of the species are not well-known. Records of additional or more specific habitat details could allow discrimination of species-habitat associations within the mud/silt sediment type recorded. Only one sample contained

species typical of organically loaded sites, Stn #32, with *Nais elinguis* and *Limnodrilus hoffmeisteri*.

A few locations had greater proportions of coarser, sand, particles or were fine silts. *Tubifex siolii* was found in samples including the entire range of sediment types, another unidentified species of *Tubifex* was frequently found in siltier sediments but not in sandy locations. The tubificid *Paranidrilus descolei* was found only in one sandy site. *Pristina americana* was also found in a number of sandier samples, as well

26

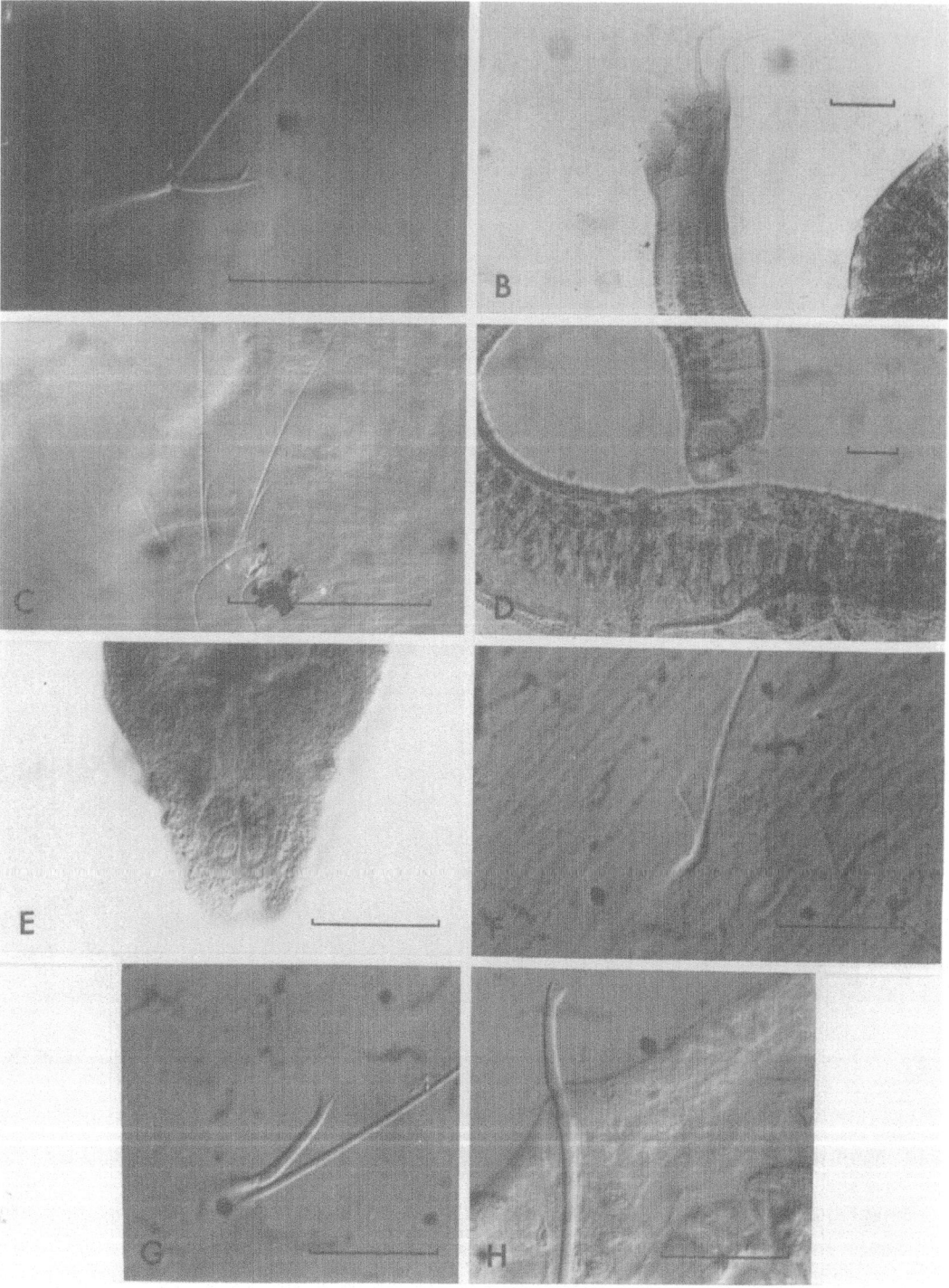

Fig. 3. Naididae: A. *Dero* sp. 1, needle seta; B. *Dero* sp. 1, gills and palps; C. *Dero* sp. 2, needle setae; D. *Dero* sp. 2, gills & palps; E. *Dero sawayai*, gills; F. *Dero palmata*, needle seta; G. *Dero lodeni*, needle seta; H. *Dero indicus*, needle seta. Scale bars for A, C, D and E equal to 100 μm; for B bar equals 500 μm; and for F, G and H bar equals 25 μm.

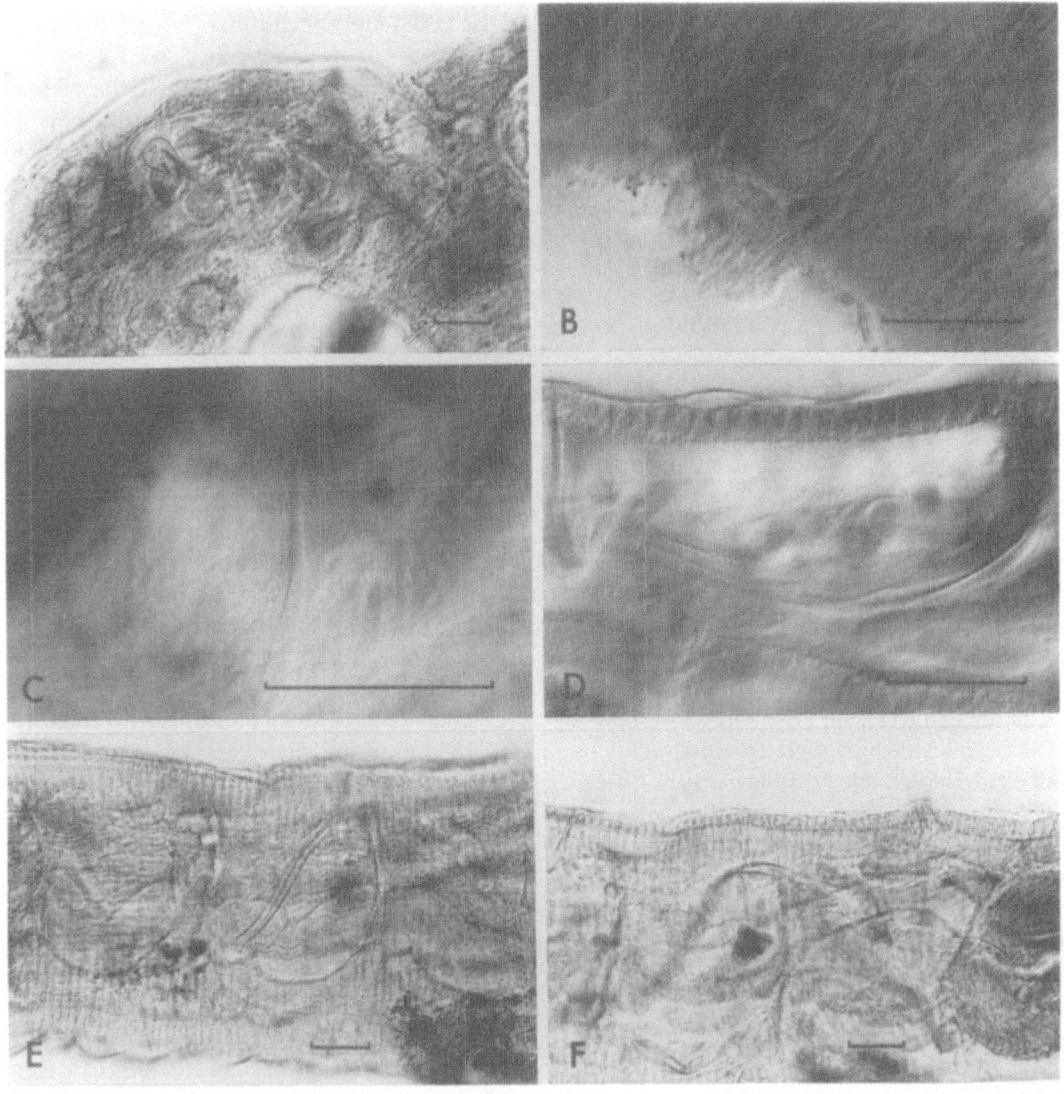

Fig. 4. Alluroididae and Tubificidae: A. *Tubifex siolii*, spermatozeugmata; B. *T. siolii*, male duct and atrium; C. *Rhizodrilus* sp., penial seta; D. *Brinkhurstia* sp. 2, penial seta, ectal tip; E. *Brinkhurstia* sp. 2, penial seta, ental end; F. *Brinkhurstia* sp. 2, penial seta, ectal end. Scale bars equal 100 μm.

as many other sites. It was frequently associated with one or more species of *Pristinella*, including *P. jenkinae*, found only at two sandy sites, *P. notopora* and *P. longidentata*. Other than a single record for a *Dero* species, naidids were not found in fine, silty sediment samples.

The greatest taxonomic diversity was found in samples with associated aquatic vegetation, usually five or six species of opistocystid, naidid and tubificid oligochaetes. Sediment types in these locations were either sandy or mixed muds. The greatest diversity, 11 species, was found in a single sample from Barami-

ta which is located at a higher elevation than the other sites, in the upper reaches of the Barama River drainage system.

Although we have not noted any regular species associations by visual surveys of our data (Table 1), there are some co-occurrences that are of interest. In particular, the occurrence together of species of the same genus suggests very fine habitat partitioning, given that an ecological component of species divergence is accepted. This kind of sympatric distribution also raises questions about the evolutionary relationships of the co-occurring species – are they sister species or distant relatives – and speciation modes. Co-occurring congeners include: *Pristinella longidentata, Pristinella* sp. 1 and *Pristinella* sp. 2; *P. jenkinae* and *P. notopora*; *Pristina americana, Pristina* sp. 1, *P. aequiseta, P. biserrata, P. synclites, P. macrochaeta*; and *Dero indicus, Dero* sp. 1, *D. digitata* and *D. lodeni*. These records also indicate that very careful taxonomic studies, including morphometry, are needed to show that the species indeed can be and are consistently distinguished.

Species distributions

Among the species identified to an appropriate taxonomic level, a range of known distributions are displayed (Table 2) (Ercolini, 1969; Brinkhurst & Jamieson, 1971; Martinez Ansemil & Giani, 1982; Brinkhurst & Marchese, 1989; Righi & Hamonui, 1989; Gluzman, 1990a, 1990b; Coates *et al.*, in press). All the alluroidids have very limited known distributions, with three of the four recorded only from Guyana and the fourth from Argentina, Brazil and Guyana. Some of the enchytraeid genera identified also show a South American distribution with limited diversity in North America, Europe, and possibly northern Africa, i.e., *Aspidodrilus, Guaranidrilus* and *Tupidrilus*. A few of the species of naidids, opistocystids and tubificids have similar known distributions – *Dero borellii, D. lodeni, Pristina americana, Pristinella notopora, Pristinella sima, Slavina evelinae, Opistocysta funiculus, Bothrioneurum americanum, Paranidrilus descolei,* and *Tubifex siolii*. In addition to these, there are also five possible new species of naidid and two of tubificid for Guyana.

Another recurring distribution of taxa is in South America plus southeast Asia, occasionally in conjunction with North America and/or Africa (Grimm, 1987). This is seen in one of the enchytraeid genera, *Hemienchytraeus*, and in a few naidid species, *Dero indicus, D. palmata, D. sawayai, Pristina biserrata, P. macrochaeta,* and *P. synclites*.

The rest of the species are relatively widespread and include species that are thought to be ecologically tolerant and commonly introduced (Di Persia, 1980), such as *Limnodrilus hoffmeisteri, Aulodrilus pigueti, Nais elinguis, Dero digitata,* and *D. furcatus*.

Discussion and conclusions

The naidid fauna, at least, is very diverse and includes a number of poorly known and possibly new species. This gives the impression that freshwater habitats of western Guyana are 1) relatively undisturbed and 2) a remaining source of information about biodiversity of such systems in these and similar latitudes. The small number of cosmopolitan species is also an indication of the relatively undisturbed nature of the habitats investigated, not yet modified by introductions of tolerant exotics or highly disturbed by human activities. Similar habitats are rapidly disappearing throughout the world and Guyana presents an invaluable source of information on biodiversity and diversification processes among a major group of invertebrates, in a very important and rapidly deteriorating habitat type.

The associations of species with particular habitat types (Table 1) is, to a great extent, a reflection of the habitats sampled by us and not of the ecological range of the species themselves. Our records are generally consistent with literature on these associations (Learner *et al.*, 1978; Di Persia, 1980; Varela, 1990).

The freshwater oligochaete fauna is zoogeographically interesting, including many taxa with restricted and southern distributions. However, lack of cladistic hypotheses and broken distribution records for South and central American taxa limit biogeographic hypotheses.

The species found in these collections, from the different families, are identified to varying taxonomic ranks (Table 2); this impedes comparative statements on their zoogeography. Nonetheless, the naidid fauna is clearly an assemblage of southern, primarily neotropical, species (Table 2). Even at the generic level, some of the taxa show the same distribution. Identification of geographical and diversification patterns (Morrone, 1994; Lynch, 1989; Humphries & Parenti, 1986) await additional collections and completed identifications.

Acknowledgements

This research was assisted by an individual NSERC research grant #OGP0046464 to K.A. Coates and by grants from the Royal Ontario Museum Science Field Studies Committee. A grant from the Canadian Associates of the Bermuda Biological Station supported facilities costs during KAC residence at the Biological Station. Field work in Guyana was assisted by Youth Challenge International. Mr A. Strange assisted greatly with the production of photographic illustrations. R. Grimm confirmed the identity of *Uncinais uncinata*, J.W. Reynolds made preliminary examinations of the unidentified sparganophilid. E. Martinez Ansemil made many helpful suggestions and provided a number of references for distributional information. This is contribution #1407 of the Bermuda Biological Station for Research, Inc.

References

Brinkhurst, R. O. & B. G. M. Jamieson, 1971. Aquatic Oligochaeta of the World, Oliver and Boyd, Edinburgh, 860 pp.

Brinkhurst, R. O. & M. Marchese, 1987. A contribution to the taxonomy of the aquatic Oligochaeta (Haplotaxidae, Phreodrilidae, Tubificidae) of South America. Can. J. Zool. 65: 3154–3165.

Brinkhurst, R. O. & M. Marchese, 1989. Guide to the freshwater aquatic Oligochaeta of South and Central America. Coleccion CLIMAX 6. Asociacion Ciencias Naturales del Litoral. Santo Tome, Argentina, 179 pp.

Coates, K. A., 1990. Redescriptions of *Aspidodrilus* and *Pelmatodrilus*, enchytraeids (Annelida, Oligochaeta) ectocommensal on earthworms. Can. J. Zool. 68: 498–505.

Coates, K. A. & D. F. Stacey, 1994. Oligochaetes (Naididae, Tubificidae, Enchytraeidae and Alluroididae) of Guyana, Peru and Ecuador. Hydrobiologia 278: 79–84.

Coates, K. A., S. R. Gelder, J. Madill, J. W. Reynolds & M. J. Wetzel (in review). Common and Scientific Names of Aquatic Invertebrates from the United States and Canada: Clitellata (including earthworms) and Aphanoneura (Phylum Annelida). American Fisheries Society.

Di Persia, D. H., 1980. The aquatic Oligochaeta of Argentina: current status of knowledge. In R.O. Brinkhurst & D.G. Cook (eds), Aquatic Oligochaete Biology. Plenum Press, New York: 225–240.

Ercolini, A. 1969. Su alcuni Aeolosomatidae e Naididae della Somalia (Oligochaeta, Microdrili). Monit. Zool. Ital. 2: 9–36.

Gillespie, L. J. & V. A. Funk, 1993. Biodiversity and conservation in Guyana: A new centre for the study of biodiversity. Global Biodiversity 3: 7–11.

Gluzman, C., 1990a. Neuvos aportes al conocimiento de los oligoquetos acuáticos de Chile. Stud. Neotrop. Faun. Envir. 25: 89–92.

Gluzman, C., 1990b. A new species of the freshwater genus *Allonais* (Oligochaeta, Naididae) from among the roots of *Eichhornia crassipes* (Argentina). Stud. Neotrop. Faun. Envir. 25: 121–124.

Grimm, R., 1987. Contributions towards the taxonomy of the African Naididae (Oligochaeta). IV. Zoogeographical and taxonomical considerations on African Naididae. Hydrobiologia 155: 27–37.

Harman, W. J., R. O. Brinkhurst, & M. Marchese, 1988. A contribution to the taxonomy of the aquatic Oligochaeta (Naididae) of South America. Can. J. Zool. 66: 2233–2242.

Humphries, C. J. & L. R. Parenti, 1986. Cladistic Biogeography. Oxford University Press, Oxford, 98 pp.

Juget, J. & M. Lafont, 1994. Distribution of Oligochaeta in some lakes and pools of Bolivia. Hydrobiologia 278: 125–128.

Lynch, J. D., 1989. The gauge of speciation: on the frequencies and modes of speciation. In D. Otte & J. A. Endler (eds), Speciation and Its Consequences. Sinauer Associates, Sunderland, Massachussetts: 527–553.

Martinez Ansemil, E. & N. Giani. 1982. Contribución al concimiento del género *Pristina* (Oligochaeta, Naididae) en la Península Ibérica. Bol. R. Soc. Española Hist. Nat. 80: 249–260.

Martinez Ansemil, E. 1993. Etudes sur les oligochètes aquatiques des pays du poutour de la Mediterranée: taxonomie, phylogénie, biogéographie et écologie. Unpub. PhD thesis. Universite Paul Sabatier de Toulouse.

Morrone, J. J., 1994. On the identification of areas of endemism. Syst. Biol. 43: 438–441.

Nielsen, C. O. & B. Christensen, 1959. The Enchytraeidae: critical revision and taxonomy of European species. Studies on Enchytraeidae VII. Nat. Jutl. 8–9: 1–160.

Omodeo, P. 1995. *Kathrynella*, a new oligochaete genus from Guyana. Hydrobiologia 334: 11–15.

Righi, G & V. Hamonui. 1989. *Pristina longidentata* e a taxonomia das Naididae, Oligochaeta. Revue Brasil Biol. 49: 409–414.

Rosen, G., 1976. A vicariance model of Caribbean biogeography. Syst. Zool. 24: 431–464.

Sambugar, B. 1986. I Naididi Italiani (Oligochaeta). Unpub. Ph.D. Thesis. Universita' degli studi di Padova.

Varela, M. E., 1990. Notas taxonómicos y ecológicas sobre algunos oligoquetos dulcealcuicolas del nordeste Argentino. I. Naididae. Stud. Neotrop. Faun. Envir. 25: 223–233.

Hydrobiologia **334**: 31–36, 1996.
K. A. Coates, Trefor B. Reynoldson & Thomas B. Reynoldson (eds), Aquatic Oligochaete Biology VI.
© 1996 *Kluwer Academic Publishers.*

Species separation and identification in the Enchytraeidae (Oligochaeta, Annelida): combining morphology with general protein data

Rüdiger M. Schmelz
University of Osnabrück, FB 5, AG Spezielle Zoologie D-49069 Osnabrück, Germany

Key words: Enchytraeidae, Oligochaeta, taxonomy, general protein, morphology

Abstract

In the Enchytraeidae, species separation and identification is often problematic due to high morphological similarity of closely related species and considerable intraspecific variability of crucial characteristics. Immature specimens are almost undeterminable. To meet these difficulties, this paper recommends the consulting of general protein patterns as exhibited by non-specific silver-staining after isoelectric focusing. A method is presented which allows the successful inclusion of protein data in taxonomic studies and field surveys. Two examples from an investigation on *Fridericia* field populations show that, with the help of general protein pattern analysis, a clear taxonomic decision on the identity of morphologically aberrant forms and of juvenile specimens as well can be achieved with comparatively little expenditure of time. A combined use of morphological and protein data for taxonomic purposes is suggested.

Introduction

Enchytraeid diversity is high in many aquatic or terrestrial habitats (Healy, 1979; Juget, 1984; Nielsen & Christensen, 1959). They play an important role in the metabolism of soil nutrients (Didden, 1993) and prove to be highly suitable as environmental indicators because of their species richness and abundance (Graefe, 1993a, 1993b). Their significance for soil and sediment ecology and for bio-monitoring has encouraged extensive taxonomical work on this group in the past decades (see Nielsen & Christensen, *op. cit.*; Kasprzak, 1986; Healy, *loc. cit.*). In spite of these efforts, however, identification of single specimens often remains difficult and uncertain due to the fact that important characteristics can be seen clearly only in live specimens, and to taxonomic problems. Even widespread and frequent species are still far from being completely known (Graefe, pers. comm.). Interspecific uniformity of crucial characteristics in closely related species together with considerable intraspecific variability means that species boundaries are difficult to establish. Immature specimens are almost indeter-

minable since identification is mainly based on the study of the reproductive organs.

The use of protein data in addition to morphological examination may serve to clarify many of these taxonomically intricate situations. This paper presents a comparatively little time-consuming method which allows the use of protein patterns in the study of enchytraeid field populations. It is not based on the patterns of isozymes as are most of the protein-taxonomic investigations (see Hillis & Moritz, 1990) but on protein patterns as visualized by non-specific silver staining after isoelectric focusing in polyacrylamide gels (PAGIF). These patterns offer some advantages over isozyme analysis, especially when only few and small animals are available; a single electropherogram produces a large number of specifically distributed bands, so that even single individuals can be assigned to a species (Schmidt & Westheide, 1994) – whereas several isozyme assays with a larger number of animals must ordinarily be carried out to achieve such information. Moreover, the results obtained by isoelectric focusing are highly reproducible.

The method suggested here is demonstrated with two examples which represent problems commonly

32

encountered in field studies of Enchytraeidae. The first example deals with a doubtful species separation of a few morphologically variant specimens, the second with the identification of immature individuals.

1. Species separation: In a field survey of the enchytraeid species composition of an urban parkland in Osnabrück, Germany, two individuals of the enchytraeid genus *Fridericia* Michaelsen, 1889 were found, which differed slightly from the abundant *Fridericia galba* (Hoffmeister, 1834) but which could not reliably be separated from this species on the basis of morphology. *Fridericia galba* is known to be highly variable with respect to important characteristics such as the number of diverticula per spermatheca, which may vary from 2 to 8 (Nielsen & Christensen, 1959). The two individuals, named "Y" in the following account, had 2 diverticula per spermatheca, closely resembling those of *F. galba* (Fig. 1), but exhibited differences in the arrangement of chaetae and cutaneous glands. It was important to know, whether "Y" was a form of *F. galba* or a separate species, since an inclusion of "Y" into the latter species would amplify the already considerable morphological variability of *F. galba*. General protein distribution was analyzed in order to see whether protein patterns of *F.* "Y" agreed with those in *F. galba* or differed from them.

2. Species identification: In the same survey, an apparently new *Fridericia* species (named "Z" in the following account) was detected, in which, among other features, the coelomocytes were characterized by large refractile vesicles that surrounded the cells completely or were concentrated on one side (Fig. 2). All the other *Fridericia* species at the sampling site lacked this characteristic. Additionally, a considerable number of immature specimens of *Fridericia* also had these peculiar coelomocytes. Since the species assignment in enchytraeids is principally based on the study of the reproductive organs, it was not possible to know whether a decision to identify immatures of *F.* sp. "Z" by the presence of vesicle-surrounded coelomocytes was justified or not. Protein analysis was performed in order to see whether juvenile *Fridericia* exhibiting vesicle-framed coelomocytes agreed with mature *F.* sp. "Z" with respect to protein band patterns.

In both cases, individuals were first examined thoroughly for morphology and processed for general protein patterns afterwards.

Materials and methods

The number of *Fridericia* specimens investigated was 41 for *F. galba*, 2 for "Y", 12 for "Z" and 64 for juveniles of *Fridericia*, the latter being a mixed group of unknown species composition including specimens with and without vesicle-surrounded coelomocytes. The specimens were sampled in an urban parkland on the outskirts of Osnabrück, Germany, in May and June 1994. Soil columns of 10 cm depth were taken with the help of a soil corer, and extracted for enchytraeids in the laboratory according to Graefe (1984): ordinary hemisphere-shaped plastic tea strainers were filled with soil subsamples and placed onto water-filled plastic bowls, so that the soil was completely submerged. The water was at room temperature except in the hot season when extraction was carried out in a cooling room at 11° C. After 24 h, most of the enchytraeid worms had left the soil, entered the water body and fallen to the bottom of the bowl from where they could be picked up easily with a pipette and the help of a stereomicroscope.

The animals were examined alive in a drop of water between slide and coverglass with interference contrast microscopy. Mature individuals were identified to species, immatures were examined for key characteristics, such as chaetal patterns or coelomocyte shape. After morphological investigation, the worms were put back into water to let them defecate completely. Having emptied the gut, which could take 24 h or more, the worms were cut into aliquots of about equal biovolume of 0.25 mm^3. Each aliquot was put separately in an Eppendorf cap filled with 20 μl of distilled water and stored at $-64°$ C. No buffer was added. Protein electrophoresis was carried out following the protocol of Brockmeyer (1991) established for *Enchytraeus* species. Aliquots were thawed and homogenized with an electric drill (grinding attachment) at 3000 rpm for 1 s within the caps. Isoelectric focusing was performed in polyacrylamide gels ($0.5 \times 125 \times 265$ mm) on a horizontal system (LKB Multiphor II). The total amount of acrylamide (T) was 7.5%; the degree of crosslinkage (C) was 3%. The aqueous gel solution contained a mixture of 7% ampholyte solutions at pH 4–9 and pH 4–7, ratio 2 : 1. Polymerization was started by 0.05% (v/v) N,N,N',N'-tetramethylethylenediamine (TEMED) and 0.05% (v/v) ammonium persulfate (APS). Gels were cast on polyester foils (Gel BOND Film for PAGE, Serva) in glass cassettes and remained at room temperature for 1 h for polymerization. The electrode wicks were 9 cm apart.

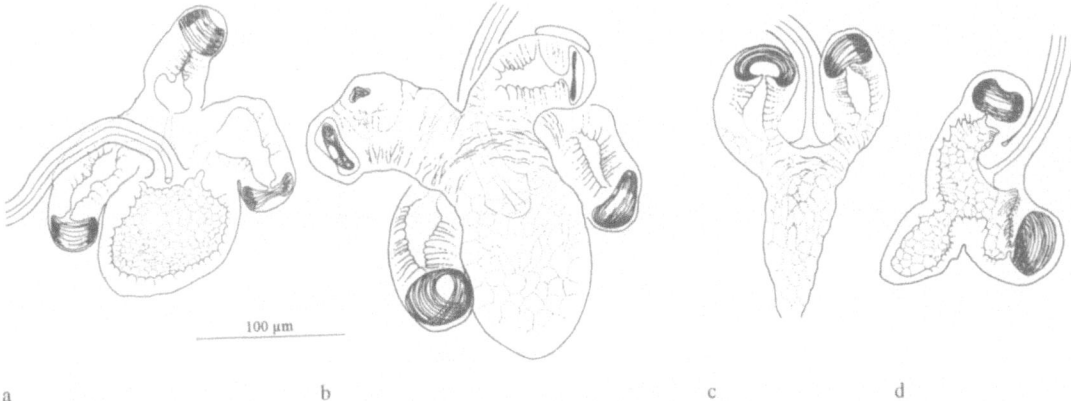

a   b   c   d

Fig. 1. Spermatheca of *Fridericia galba* (a, b) and *F.* form "Y" (c, d), as seen by interference contrast microscopy. In *F. galba*, a tridiverticulate and a four-diverticulate form is shown. Similarities: texture of inner (a, d) and outer (b, c) ampulla surface, a subchamber in each diverticulum devoid of sperm communicating with the ampullate lumen.

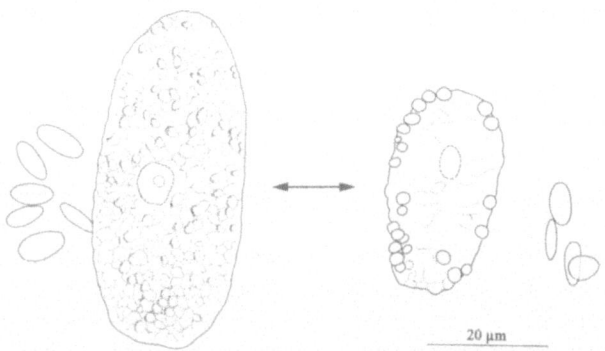

Fig. 2. Nucleate (large) and anucleate (small) coelomocytes of *Fridericia* sp. "Z" (right) and *F. galba* (left). The peculiar framing of nucleate cells by refractile vesicles in *F.* sp. "Z" did not occur in any other species at the sampling site, whereas the *F. galba* – type occurred in several other species as well.

The electrode solutions contained 0.025 M arginine, 0.025 M lysine, 2 M ethylenediamine for the cathode and 0.025 M aspartic acid and 0.025 M glutamic acid for the anode. Samples were applied in slots (7 × 1 mm). During focusing, the gels were cooled down to 10° C. The probes were first prefocused with 3 W constant power for 30 min. The focusing was continued with 5 W for another 2 h until the final total voltage reached about 900 V. The silver staining of proteins was performed according to the method of Heukeshoven & Dernick (1983).

In the first example (*F. galba* and form "Y"), protein patterns were analyzed by marking band positions which showed up distinctly for all mature specimens of *F. galba*. The resulting reference pattern was compared with the band pattern of specimens of *F.* form

"Y". In the second example (identification of juvenile *F.* sp. "Z"), protein patterns of all juvenile *Fridericia* were compared with the pattern found diagnostic for mature specimens of *F.* sp. "Z", to see whether juveniles with vesicle-framed coelomocytes shared the *F.* sp. "Z"-band pattern.

Results

The pattern analysis of silver stained proteins proves that *F.* form "Y" is not a morphological variant of *F. galba* but a different species. Fig. 3 shows the results of the comparison between 41 specimens of *F. galba* (Fig. 3, a–h) and the two individuals of form "Y" (Fig. 3, i-k). Although *F. galba* exhibits intraspecific

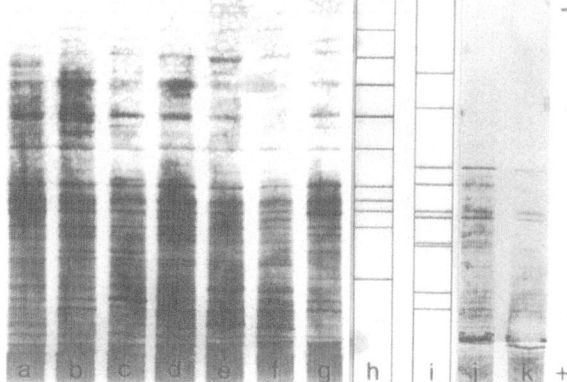

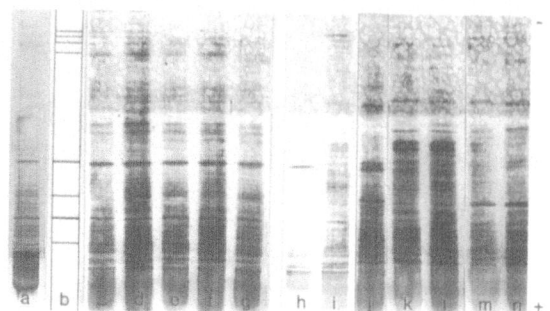

Fig. 3. Comparison of general protein patterns of *Fridericia galba* (a–h) and *F.* form "Y" (i–k) shows that the latter is not a variation of the former but a separate species. a–g: band patterns of 7 (out of 41) mature specimens of *F. galba*; h: band patterns found diagnostic for all 41 mature specimens of *F. galba* investigated; i: diagnostic band pattern of *F.* form "Y", derived from j and k; j, k: band patterns of the two specimens of *F.* form "Y". (PAGIF: pH 4–9, pH 4–7 stretched; separation distance 9 cm; top: cathode, bottom: anode).

Fig. 4. Comparison of general protein patterns of *Fridericia* sp. "Z" (a–b) and *F.* sp. juv. (c–n) show that coelomocyte shape is a reliable trait for identifying juvenile *F.* sp. "Z". a: band pattern of 1 (out of 12) mature specimen of *F.* sp. "Z"; b: band positions found diagnostic for all 12 mature specimens of *F.* sp. "Z" investigated. Specimens a and c–g had the "Z"-specific vesicle-framed coelomocytes, specimens h–n didn't. According to the band patterns a and c–h belong to *F.* sp. "Z": Thus, with the exception of h, juveniles with "Z"-specific band patterns also have "Z"-specific coelomocytes. The number of juveniles of *F.* sp. was 64. (PAGIF: pH 4–9, pH 4–7 stretched; separation distance 9 cm; top: cathode, bottom: anode).

variablity in the anodic area, a series of bands typical of the species can be ascertained. The two individuals of form "Y" do not share any of these diagnostic band positions, they exhibit a conspicuous pattern of their own.

Silver stained proteins also show that immature individuals with vesicle-framed coelomocytes belong to *F.* sp. "Z". Fig. 4 shows the results of the comparison between 12 mature specimens of *F.* sp. "Z" and 64 immature specimens of *Fridericia*. With one exception (Fig. 4h), all immatures with vesicle-framed coelomocytes share the band positions found characteristic for *F.* sp. "Z" (Fig. 4, a, c–g). Immature specimens with different coelomocytes, in turn, do not agree with mature specimens of *F.* sp. "Z" with regard to band patterns (Fig. 4, i–n). Only one single small specimen (Fig. 4, h) shares the band positions of *F.* sp. "Z", but does not show the "Z"-specific coelomocytes.

Discussion

Up to the present, there are only few examples where general protein patterns have been used for taxonomic purposes (see Brockmeyer, 1991; Carmona *et al.*, 1989; Schmidt & Westheide, 1994), whereas isozyme patterns are currently employed in nearly all groups of organisms, including aquatic oligochaetes (Anlauf, 1994; Christensen, 1980; Christensen & Jelnes, 1976; Milbrink & Nyman, 1973; Terhivuo *et al.*, 1994; Weider, 1992). This may be due to the fact that gener-

al protein patterns cannot be interpreted in respect of population genetics because the multi-locus coding of silver-stainable proteins makes it impossible to identify single bands as alleles of a specified gene locus. In cases, however, which do not concern genetics but rather the identification of single individuals, the use of general proteins is more suitable than the use of isozymes. In routine assays on food components general protein pattern analysis is currently employed for the quick identification of meat and plant varieties (Tinbergen & Olsman, 1976; Stegemann, 1984). Taxonomic investigations on *Enchytraeus* species (Brockmeyer, 1991) and on polychaete species of the family Nephtyidae (Schmidt & Westheide, 1994) have shown that a single general protein electropherogram is just as diagnostic as a series of zymograms.

In this study, the results obtained by protein analysis show that difficulties concerning species identification and species separation in field populations of Enchytraeidae can be resolved without too much expenditure of labour by simply comparing the general protein patterns of the respective specimens with one another. In both cases, the procedure is alike. Initially, a series of electropherograms is examined, belonging to mature specimens of the species of reference (*Fridericia galba* and *F.* sp. "Z", respectively). Distinct and prominent protein bands are outlined that show up in all electropherograms of one species, the resulting pattern thus representing a scheme which is diagnostic for the species. In doing this, the presence of intraspecific variation is eliminated and the complexity of the pherogram is reduced to a level that facilitates further comparisons. In a second step, the electropherograms of the specimens in question (specimens of *F*. form "Y" and juveniles of *F*. sp., respectively) are confronted with the diagnostic band pattern scheme of the reference species to see whether they are congruous with it or not.

In both examples presented a clear answer was obtained: *Fridericia* form "Y" is a species separate from *F. galba*, and juvenile *Fridericia* specimens exhibiting vesicle-framed coelomocytes can be identified as *F*. sp. Z". The only exception – see Fig. 4h – may be interpreted as a very small, perhaps newly hatched, specimen of *F*. sp. "Z", in which the characteristic vesicles are not yet formed.

This comparatively simple approach to the inclusion of protein data in taxonomic analyses has several advantages. The procedure is not very time-consuming; processing of the 119 animals, isoelectric focusing of protein homogenates and subsequent silver-staining took not more than three days of work. The comparison of patterns is carried out directly, without numerical analyses. Single individuals can be analyzed, since the electropherogram yields a whole pattern of specifically distributed bands. The results obtained are highly reproducible due to the technique of isoelectric focusing. The sensitive silver-staining allows the use of quantities down to 0.1 mm^3, which is about the biovolume of a worm measuring 3×0.2 mm. Another advantage is the combined use of morphology and proteins. The initial questions arose out of morphological studies, and general protein patterns were consulted in such a way that the answer was also given at the morphological level. In this way, routine species identification does not become dependant on obtaining protein data for every single specimen, which is of crucial importance in field studies dealing with large quantities of animals.

It may be objected that the outlining of a diagnostic band pattern scheme contains a strong subjective component, especially when all variations are deliberately eliminated. Species identity, however, of *Fridericia galba* and *F*. sp. "Z" was not established on the protein level but morphologically. Consequently, all variations at the protein level were interpreted as intra-specific and the selection of only those bands that showed in every specimen was justified. The treatment of protein data is dependant on the nature of the question.

One of our questions was: "Are the two forms "Y" identical with what we call, based on morphology, *F. galba*?" Another question such as "Is what we call *F. galba* really one species?" would require a completely different treatment of the same protein data.

Intraspecific variability may limit the use of protein data in the way suggested here. Carmona *et al.* (1989) found both seasonally and geographically determined variability in general protein patterns of a rotifer species. In our study, *F. galba* from a single population exhibits considerable variability of band patterns in the anodic area (Fig. 3), indicating a genetic diversity similar to what Christensen *et al.* (1992), using isozyme patterns, found for a Danish *Fridericia galba* field population. It was nonetheless possible to identify a group of prominent bands as species-specific for the sampling site. Moreover, Brockmeyer (1991) did not find any variability at all within monospecific laboratory cultures of *Enchytraeus* species, irrespective of age, body region or nutritional state of the individuals. Schmelz (1995) could also demonstrate that interspecific differences of general protein patterns between sympatric field populations of *Fridericia* species exceed those within the respective species.

The two examples presented in this paper concern sympatric populations only, and protein patterns have been used for comparatively limited questions within the framework of a field study. Protein data, however, may also be suitable to fix species boundaries as a whole, provided that the geographical variability of band patterns is low enough. Based on results obtained with laboratory cultures of *Enchytraeus* species, Westheide & Brockmeyer (1992) have suggested the establishment of a general protein index for each enchytraeid species, additional to the morphological species definition. Such an index would certainly help to answer many of the yet confusing taxonomic questions in the Enchytraeidae.

Acknowledgements

I am grateful to Prof W. Westheide (Osnabrück, Germany) for his support on this study. Further thanks go to Dr B. Healy (Dublin, Ireland) and Prof H. Stegemann (Braunschweig, Germany), who made useful comments on a previous version of the paper. I am also indebted to H. Schmidt (Osnabrück), who patiently introduced me to the technique of isoelectric focusing. This paper is part of the PhD project "Revision of *Fridericia* Michaelsen, 1889 (Enchytraeidae, Oligochaeta) with special respect to biochemical methods", which is funded by a dissertation fellowship from Cusanuswerk – Bischöfliche Studienstiftung, 53115 Bonn, Germany.

References

Anlauf, A., 1994. Some characteristics of genetic variants of *Tubifex tubifex* (Müller, 1774) (Oligochaeta: Tubificidae) in laboratory cultures. Hydrobiologia 278: 1–6.

Brockmeyer, V., 1991. Isozymes and general protein patterns for use in discrimination and identification of *Enchytraeus* species (Annelida, Oligochaeta). Z. zool. Syst. Evolut.-forsch. 29: 343–361.

Carmona, M. C., M. Serra & M. R. Miracle, 1989. Total protein analysis in rotifer populations. Biochem. Syst. Ecol. 17: 409–415.

Christensen, B., Hvilsom, M., and Pedersen, B. V., 1992. Genetic variation in coexisting sexual diploid and parthenogenetic triploid forms of *Fridericia galba* (Enchytraeidae, Oligochaeta) in a heterogeneous environment. Hereditas 117: 153–162.

Christensen, B., 1980. Constant differential distribution of genetic variants in polyploid parthenogenetic forms of *Lumbricillus lineatus* (Enchytraeidae). Hereditas 92: 193–198.

Christensen, B. & J. Jelnes, 1976. Sibling species in the oligochaete worm *Lumbricillus rivalis* by enzyme polymorphisms and breeding experiments. Hereditas 83: 237–244.

Didden, W. A. M., 1993. Ecology of Terrestrial Enchytraeidae. Pedobiologia 37: 2–29.

Graefe, U., 1984. Eine einfache Methode der Extraktion von Enchytraeiden aus Bodenproben. In H. Köhler (ed), Workshop zu Methoden der Mesofaunaerfassung, Univ. Bremen: 17.

Graefe, U., 1993a. Die Gliederung von Zersetzergesellschaften für die standortsökologische Ansprache. Mitteilgn. Dtsch. Bodenkundl. Gesellsch. 69: 95–98.

Graefe, U., 1993b. Veränderungen der Zersetzergesellschaften im Immissionsbereich eines Zementwerkes. Mitteilgn. Dtsch. Bodenkundl. Gesellsch. 72: 531–534.

Healy, B., 1979. Records of Enchytraeidae (Oligochaeta) in Ireland. J. Life Sci. R. Dubl. Soc. 1: 39–70.

Heukeshoven, J. & R. Dernick, 1983. Vereinfachte und schnelle Methode zur Silberfärbung von Proteinen in Polyacrylamidgelen: Bemerkungen zum Mechanismus der Silberfärbung. In B. J. Radola (ed), Elektrophorese Forum '83. München: 92–97.

Hillis, D. M. & C. Moritz (eds), 1990. Molecular Systematics. Sinauer Ass. Inc. Publ., Massachusetts, 588 pp.

Juget, J., 1984. Oligochaeta of the epigean and underground fauna of the alluvial plain of the French upper Rhône (biotypological trial). Hydrobiologia 115: 175–182.

Kasprzak, K., 1986. Skaposzczety Wodne i Glebowe, II. Panstwowe Wydawnictwo Naukowe, Warsaw: 366 pp.

Milbrink, G. & Nyman, L., 1973. On the protein taxonomy of aquatic oligochaetes. Zoon 1: 29–35.

Nielsen, C. O. & B. Christensen, 1959. The Enchytraeidae. Critical revision and taxonomy of European species. Nat. Jutl. 8–9: 1–160.

Schmelz, R., 1995. Separation of sympatric *Fridericia* species (Enchytraeidae, Oligochaeta) with isozyme and general protein patterns. In M. Bauer (ed), Newsletter on Enchytraeidae, 4: Proceedings of the Symposium on Enchytracidae. Vienna 1.12.1994. Institut für Zoologie, Universität für Bodenkultur, Vienna: 97–104.

Schmidt, H. & Westheide, W., 1994. Isozymes and general protein patterns as taxonomic markers in the taxon Nephtyidae (Annelida, Polychaeta). Mar. Biol. 119: 31–38.

Stegemann, H., 1984. Retrospect on 25 years of cultivar and species identification by protein patterns and prospect on developments. In: Biochemical Tests for Cultivar Identification. Proc. ISTA Symposium Cambridge (England) 1983, pp. 20-31.

Terhivuo, J., A. Saura & K. Hongell, 1994. Genetic and morphological variation in the parthenogenetic earthworm *Eiseniella tetraedra* (Sav.) (Oligochaeta: Lumbricidae) from South Finland and North Norway. Pedobiologia 38: 81–96.

Tinbergen, B. J. & Olsman, W. J., 1976. Isoelektrische Fokussierung als Technik zur Speziesidentifizierung in der Lebensmittelüberwachung. Fleischwirtschaft 10: 1495–1498.

Weider, L., 1992. Allozymic variation in tubificid oligochaetes from the Laurentian Great Lakes. Hydrobiologia 234: 79–85.

Westheide, W. & V. Brockmeyer, 1992. Suggestions for an index of enchytraeid species (Oligochaeta) based on general protein patterns. Z. zool. Syst. Evolut.-forsch. 30: 89–99.

Hydrobiologia **334**: 37–49, 1996.
K. A. Coates, Trefor B. Reynoldson & Thomas B. Reynoldson (eds), Aquatic Oligochaete Biology VI.
© 1996 *Kluwer Academic Publishers.*

Aquatic Oligochaeta in Italy, with special reference to Naididae

Andreina Di Chiara Paoletti[1] & Beatrice Sambugar[2]
[1]*Department of Biology, Università degli Studi di Milano, Via Celoria 26, 20133 Milano, Italy*
[2]*Centro Ricerche S.A.R., Via Anfiteatro 9, 37121 Verona, Italy*

Key words: Oligochaeta, Naididae, taxonomy, distribution, Italy

Abstract

We present a list of Italian freshwater and marine Oligochaeta in the families Lumbriculidae, Haplotaxidae, Tubificidae, Naididae, Propappidae, Criodrilidae, and Lumbricidae, representing 57 genera and 130 species. Published data reflect the incomplete knowledge of the Italian oligochaete fauna, restricted to certain geographical areas. Subterranean aquatic and marine fauna are of particular interest as these have been studied the least. We provide a comprehensive review of the Naididae including, for the first time, southern Italy and the islands of Sicily and Sardinia. The distribution of species is discussed and taxonomic problems arising from the morphological variability of Italian material are examined.

Introduction

The earliest research on aquatic Oligochaeta in Italy, at the turn of the century, focused on the fauna of Lake Garda and the River Adige (Garbini, 1898). Chinaglia (1912) was the first to catalogue the species. Subsequently, Sciacchitano (1934) published a number of studies, including the second catalogue of Italian species. The third catalogue was that of Brinkhurst (1963). Ercolini (1956) also published systematic and ecological data on the Naididae. More recently, with the development of ecological research, our knowledge of Oligochaeta has greatly improved. Important contributions include studies of the major prealpine lakes (Nocentini, 1963; Bonomi & Gerletti, 1967), as well as of the volcanic lakes of central Italy (Stella & Argenti, 1953; Nocentini, 1973; Bazzanti, 1981, 1983; Bazzanti & Seminara, 1987; Bazzanti *et al.*, 1988). Bonacina *et al.* (1992) published a review of data on the large Italian lakes.

Studies of river coenoses have focused on the Tevere (Stella, 1951) and Po (Paoletti & Sambugar, 1984), as well as the Adige and its basin (Sambugar, 1976; Centurioni & Sambugar, 1986).

Only a few studies have focused on the fauna of subterranean habitats, these include Juget & Dumnicka

(1986), Erséus & Dumnicka (1988), and Dumnicka (1990).

Published work on the marine oligochaete fauna of Italy were non-existent except for those on the genera *Phallodrilus* and *Limnodriloides* by Pierantoni (1902, 1903) and recent works, for instance, by Bonomi & Erséus (1984, 1985), Erséus (1987a, 1987b, 1987c), and Erséus & Bonomi (1987).

This paper contains an up-to-date list of freshwater, brackish water, and marine Oligochaeta found in Italy (the family Enchytraeidae is not considered herein). The report is based on data in the literature, as well as on data from a study of the Naididae for a Ph.D. Thesis (Sambugar, 1987).

Sampling sites, materials and methods

The geographical areas covered by this research (Sambugar, 1987) include Sicily and Sardinia, the center and south, and certain parts of the northeast of Italy. These areas were chosen because very little was known about naidids for some areas in the north, and research on this group in the south and on the islands was almost non-existent. Sampling sites are noted in Fig. 1. Samples collected for this study came from different freshwater

38

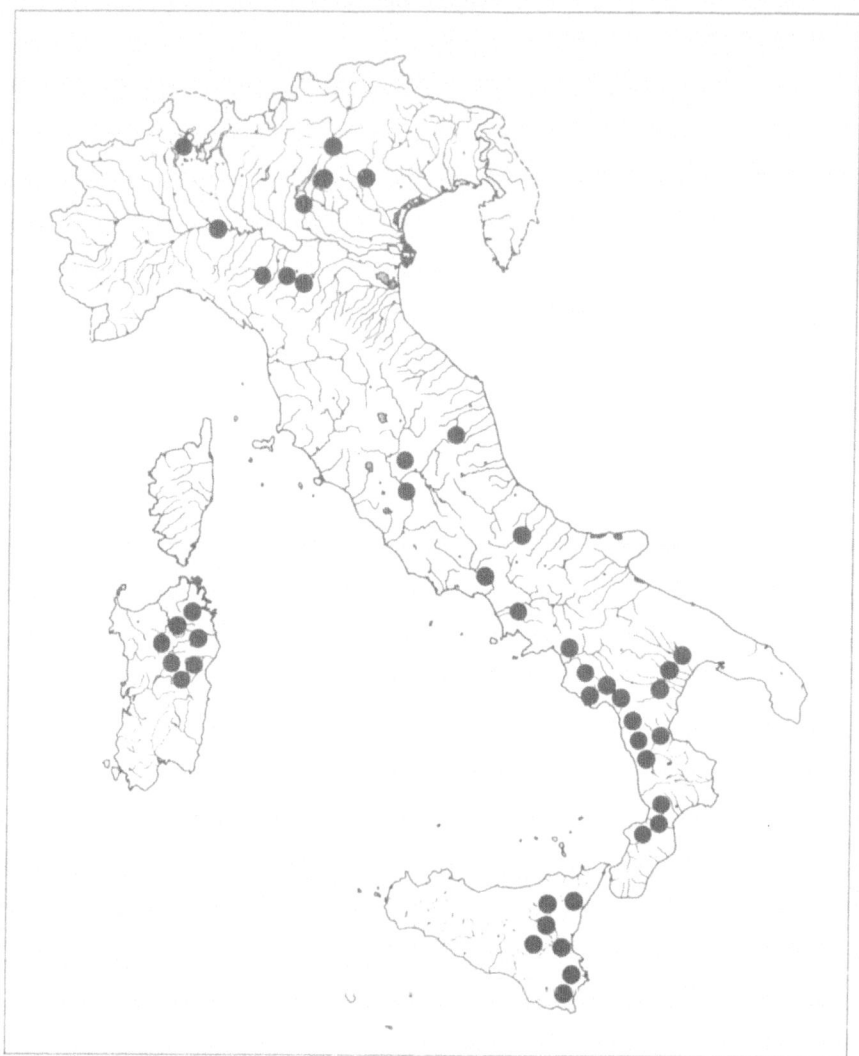

Fig. 1. Location of naidid sampling sites in Italy.
NORTH ITALY
Rivers: Adige, Isarco, Brenta, Po, Parma, Stirone, Secchia.
Lakes: Garda, Maggiore.
SOUTH ITALY
Rivers: Tevere, Sangro, Mergani, Liri, Volturno, Sele, Calore, Palistro, Mingardo, Bussento, Camastra, Basento, Bradano, Crati, Lao, Abatemarco, Corvino, Lamato, Pesipe, Angitola.
Lake: Piediluco.
SICILY
Rivers: Alcantara, Simeto, Ciane, Cassibile, Dittaino, Troina.
Lake: Biviere di Cesarò.
SARDINIA
Rivers: San Giovanni, Riu Badu, Posada, Tirso, Su Gologone.
Lakes: Coghinas, Posada.

habitats – springs, streams, rivers, lakes, pools, and upland cattle ponds.

Specimens collected in the field were fixed with 10% formalin and stored in 70% ethanol. All spec-

imens were mounted whole either as semipermanent mounts in 50% glycerol/50% lactic acid or as permanent mounts in Hydramount TM or Canada balsam. Drawings were completed using a Zeiss drawing tube

Table 1. Freshwater and marine Oligochaeta and geographic distribution in Italy: N = Northern Italy; S = central and southern Italy; Si = Sicily, Sa = Sardinia; FAO fishery zones: 3 = western basin, 4 = upper and middle-Adriatic, 5 = remaining basins.

Family Lumbriculidae

Bichaeta sanguinea Bretscher, 1900	N	S
Eclipidrilus lacustris (Verrill, 1871)	N	
Lumbriculus variegatus (Müller, 1774)	N	S
Rhynchelmis limosella Hoffmeister, 1843	N	
Stylodrilus aurantiacus (Pierantoni, 1904)		S
Stylodrilus brachystylus Hrabe, 1928		S
Stylodrilus heringianus Claparède, 1862	N	S
Stylodrilus lemani (Grube, 1879)	N	S
Trichodrilus allobrogum Claparède, 1862	N	S
Trichodrilus claparedei Hrabe, 1937		S
Trichodrilus leruthi Hrabe, 1937		S
Trichodrilus pragensis Vejdovsky, 1875	N	
Trichodrilus stammeri Hrabe, 1937	N	
Trichodrilus strandi Hrabe, 1936	N	

Family Haplotaxidae

Haplotaxis gordioides (Hartmann, 1821)	N	S

Family Tubificidae

Abyssidrilus cuspis (Erséus & Dumnicka, 1988)		S
Aulodrilus limnobius Bretscher, 1899	N	S
Aulodrilus pigueti Kowalewski, 1914	N	
Aulodrilus pluriseta (Piguet, 1906)	N	S
Branchiura sowerbyi Beddard, 1892	N	S
Bothrioneurum vejdovskyanum Stolc, 1888	N	S
Frearidrilus pescei (Dumnicka, 1980)		S
Haber monfalconensis (Hrabe, 1966)	N	
Haber speciosus (Hrabe, 1931)	N	
Haber zavreli (Hrabe, 1942)		S
Ilyodrilus templetoni (Southern, 1909)	N	
Limnodrilus claparedeianus Ratzel, 1868	N	S
Limnodrilus hoffmeisteri Claparède, 1862	N	S
Limnodrilus profundicola (Verrill, 1871)	N	S
Limnodrilus udekemianus Claparède, 1862	N	S
Monopylephorus limosus (Hatai, 1888)	N	
Potamothrix bavaricus (Öschmann, 1913)	N	
Potamothrix bedoti (Piguet, 1913)	N	
Potamothrix hammoniensis (Michaelsen, 1901)	N	S
Potamothrix heuscheri (Bretscher, 1900)	N	S
Potamothrix vejdovskyi (Hrabe, 1941)	N	
Psammoryctides albicola (Michaelsen, 1901)	N	S
Psammoryctides barbatus (Grube, 1861)	N	S
Rhyacodrilus coccineus (Vejdovsky, 1875)	N	S
Rhyacodrilus falciformis Bretscher, 1901	N	
Sketodrilus flabellisetosus (Hrabe, 1966)	N	
Spirosperma ferox (Eisen, 1879)	N	
Spirosperma velutinus (Grube, 1879)	N	S

Table 1. Continued

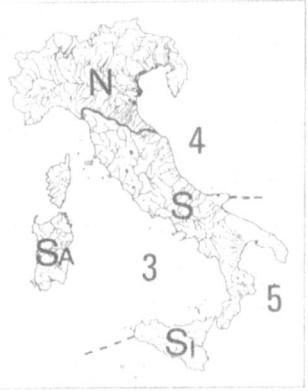

Tubifex blanchardi Vejdovsky, 1891	N	
Tubifex costatus (Claparède, 1863)	N	
Tubifex ignotus (Stolc, 1886)	N	S
Tubifex nerthus Michaelsen, 1908	N	
Tubifex newaensis (Michaelsen, 1903)	N	
Tubifex tubifex (Müller, 1774)	N	S
Tubificoides benedii (d'Udekem, 1855)	N	

Marine species

Adelodrilus pusillus Erséus, 1978	3
Aktedrilus cuneus Erséus, 1984	3, 4
Aktedrilus magnus Erséus, 1980	3
Aktedrilus mediterraneus (Erséus, 1980)	3
Aktedrilus monospermathecus Knöllner, 1935	4
Aktedrilus sardus Erséus, 1987	3
Bathydrilus adriaticus (Hrabe, 1971)	3
Coralliodrilus giacobbei Erséus, 1982	3
Coralliodrilus statutus Erséus, 1982	3
Coralliodrilus tyndariensis Erséus, 1982	3
Gianius densespectinis (Erséus, 1987)	3
Heterodrilus arenicolus Pierantoni, 1902	3
Heterodrilus subtilis (Pierantoni, 1917)	3
Inanidrilus bonomii Erséus, 1984	3
Limnodriloides agnes Hrabe, 1967	3
Limnodriloides appendiculatus Pierantoni, 1903	4, 5
Limnodriloides hrabetovae Erséus, 1987	4
Limnodriloides maslinicensis (Hrabe, 1971)	4
Limnodriloides pectinatus Pierantoni, 1904	3
Limnodriloides pierantonii (Hrabe, 1971)	3
Limnodriloides roseus Pierantoni, 1904	3
Phallodrilus parthenopaeus Pierantoni, 1902	3
Pirodrilus messanensis (Erséus, 1987)	3
Pectinodrilus rectisetosus (Erséus, 1979)	3
Tectidrilus pranzoi Erséus, 1987	3
Thalassodrilides gurwitschi (Hrabe, 1971)	3
Tubificoides vestibulatus Erséus & Bonomi, 1987	4, 5
Tubificoides swirencowi Jaroschenko, 1948	4

Table 1. Continued

Family Naididae		
Amphichaeta leydigii Tauber, 1879	S	
Amphichaeta sannio Kallstenius, 1892	S	
Chaetogaster cristallinus Vejdovsky, 1883	N S	
Chaetogaster diaphanus (Gruithuisen, 1828)	N S	Sa
Chaetogaster diastrophus (Gruithuisen, 1928)	N S	Si, Sa
Chaetogaster langi Bretscher, 1896	N S	
Chaetogaster limnaei Von Baer, 1827	N	Si
Chaetogaster parvus Pointner, 1914	N	
Dero (A.) furcata (Müller, 1773)	S	
Dero digitata (Müller, 1773)	N S	Sa
Dero dorsalis Ferronière, 1899		Sa
Dero nivea Aiyer, 1930		Sa
Dero obtusa d'Udekem, 1855	N S	Si, Sa
Haemonais waldvogeli Bretscher, 1900	N	
Homochaeta naidina Bretscher, 1896	N	
Homochaeta setosa (Moszynski, 1933)	N	
Nais alpina Sperber, 1948	N S	
Nais barbata Müller, 1773	N S	Si, Sa
Nais behningi Michaelsen, 1923	N	
Nais bretscheri Michaelsen, 1899	N S	
Nais christinae Kasprzak, 1973	N S	
Nais communis Piguet, 1906	N S	Si, Sa
Nais elinguis Müller, 1773	N S	
Nais pardalis Piguet, 1906	N S	Si, Sa
Nais pseudobtusa Piguet, 1909	N S	Si, Sa
Nais simplex Piguet, 1906	N S	
Nais stolci Hrabe,1981	S	
Nais variabilis Piguet, 1906	N S	Si, Sa
Ophidonais serpentina (Müller, 1773)	N S	
Paranais birsteini Sokolskaya, 1971		Si
Paranais frici Hrabe,1941	N S	
Paranais litoralis (Müller, 1784)	N	Si
Pristina aequiseta Bourne, 1891	N S	Si, Sa
Pristina longiseta Ehrenberg, 1828	S	Si, Sa
Pristinella amphibiotica (Lastockin, 1927)	N	
Pristinella bilobata (Bretscher, 1903)	N	Sa
Pristinella jenkinae (Stephenson, 1931)	N S	Si, Sa
Pristinella menoni (Aiyer, 1930)	N S	Sa
Pristinella notopora (Cernositov, 1937)	N	
Pristinella osborni (Walton, 1906)	S	
Pristinella rosea (Piguet, 1906)	N S	
Pristinella sima (Marcus, 1944)	S	
Slavina appendiculata (d'Udekem, 1855)	N S	Sa
Specaria josinae (Vejdovsky, 1883)	N S	
Stylaria lacustris (Linnaeus, 1767)	N S	
Uncinais uncinata (Örsted, 1842)	N S	
Vejdovskyella comata (Vejdovsky, 1883)	N S	
Vejdovskyella intermedia (Bretscher, 1896)	N	

Family Propappidae		
Propappus volki Michaelsen, 1922	N S	

Table 1. Continued

Family Criodrilidae		
Criodrilus lacuum Hoffmeister, 1845	N S	Si

Family Lumbricidae		
Eiseniella neapolitana (Örley, 1885)	N S	Si, Sa
Eiseniella tetraedra (Savigny, 1826)	N S	Si, Sa
Helodrilus patriarchalis (Rosa, 1893)	N	

attachment. Setae and other parts of the body were photographed with a Philips SEM505 scanning electron microscope. Specimens are kept in the personal collection of Beatrice Sambugar; additional material is deposited at the Museo Civico di Storia Naturale, Verona.

Results

Faunal survey

Lumbriculidae, Haplotaxidae, Tubificidae, Naididae, Propappidae, Criodrilidae and Lumbricidae of Italy are listed in Table 1. The list comprises 57 genera, totaling 130 species.

Some records for Italy still require confirmation as they are sporadic and often refer to only one specimen, from ecological studies. In particular these records are: *Eclipidrilus lacustris* (Centurioni & Sambugar, 1986), *Tubifex newaensis* (Ghetti *et al.*, 1977), *Tubifex nerthus* and *Potamothrix bedoti* (Bonacina *et al.*, 1992).

Naididae

The family Naididae is represented in Italy by 16 genera and 48 species. The inventory in Limnofauna Europaea (Brinkhurst, 1978) is supplemented by the addition of 14 species: *Chaetogaster parvus*, *Amphichaeta sannio*, *Paranais frici*, *P. litoralis*, *P. birsteini*, *Nais alpina*, *N. christinae*, *N. stolci*, *Vejdovskyella intermedia*, *Dero nivea*, *Pristinella jenkinae*, *P. sima*, *P. notopora*, and *P. osborni*. Another four species listed in Limnofauna Europaea (zone 4: Alps) are new to Italy, having been reported to occur only in the French and Swiss Alps. These species are *Haemonais waldvogeli*, *Amphichaeta leydigii*, *Nais behningi*, and *Dero dorsalis*.

In Italy, *Nais* is the most widespread genus, with marked prevalence over other genera in the north. Next in order of prevalence are *Chaetogaster*, *Pristina*, and *Pristinella*. The situation is different in the southern and island regions, where *Nais* is less common while *Pristina* and *Pristinella* are more frequently encountered. These are followed by *Dero* and *Chaetogaster*. *Pristina*, *Pristinella*, and *Dero* are characteristic of fauna from warmer climates, and thus increase in occurrence as one moves south. *Pristinella sima*, for example, in Europe is restricted to the south of Italy and Spain (Martinez-Ansemil & Giani, 1982). *Nais christinae*, a thermophilous species, is mostly found in the southern part of Italy. *Paranais birsteini* var. *maghrebensis* has been found in Sicily. This variety is known in Morocco, Algeria, and Tunisia (Martinez-Ansemil & Giani, 1987). The division of the species into two distinct varieties is thus confirmed – one having a Siculo-maghrebian range, the other occurring along the Pacific rim (Brinkhurst & Coates, 1985).

Another species not previously known outside Africa, *Dero (A.) africana*, has now been identified in Italy. Two specimens collected to date match the description provided by Grimm (1989) for both size and shape of setae. Some reservations must be expressed about this finding, given the limited material available and the difficulty of evaluating contracted branchial apparatus.

Variety *iorensis* of *N. bretscheri*, previously known only in the Middle East (Lebanon, Israel, Georgia) (Martinez-Ansemil & Giani, 1987; Martinez-Ansemil, 1993), has been found in the River Adige. Reports of *Nais stolci* in the Rivers Lao and Bussento are the first outside the Czech Republic. In the interstitial coenosis of rivers, Naididae are the dominant family. The presence of several *Pristina* and *Pristinella* species (*aequiseta*, *bilobata*, *jenkinae*, *menoni*, *notopora*) confirms the stygophilous character of these taxa, emphasizing the interactions between surface and groundwaters (Lafont *et al.*, 1992).

Other families

Haplotaxidae. The only representative of the Haplotaxidae in Italy is the cosmopolitan *Haplotaxis gordioides*.

Lumbriculidae. The family is represented in Italy by six genera and 14 species. Some of these species are widely distributed (*Lumbriculus variegatus*, *Stylodrilus heringianus*, *S. lemani*). Others are found in subterranean waters – particularly species of *Trichodrilus*, (*T. allobrogum*, *T. stammeri*, *T. strandi*, *T. claparedei* and *T. leruthi*).

Trichodrilus allobrogum is the most widely distributed species, present in both the northern and southern parts of Italy. This species also is reported elsewhere in Europe, in North America and North Africa (Algeria, Morocco, Tunisia). *Trichodrilus stammeri* is a species of uncertain determination; it was mentioned but not formally described by Hrabe (1937, 1942): it is reported to occur in Istria and Austria. *Trichodrilus claparedei* and *T. leruthi* recently were reported by Dumnicka (1990) from central Italian caves.

Tubificidae. Of the 35 freshwater tubificid species recorded from Italy, 19 are common worldwide. These 19 species are also widely distributed in Italian lakes and water courses. Known distributions of most of the remaning species consist of sporadic reports from only one site.

Of particular interest is the record from Italy of an Asiatic brackish water species, *Monopylephorus limosus*. A relatively abundant, stable population of this species has been reported in the heavily polluted River Lambro (Erséus & Paoletti, 1986).

The genus *Rhyacodrilus* is represented in Italy by its two most widely distributed species, *R. coccineus*, and *R. falciformis*. The latter – a primarily subterranean water species – is common in river sources (Juget, 1987). It was found indeed in a source of the Adige basin (Sambugar, unpublished). *Tubifex tubifex* and *T. blanchardi* have been found together in some water courses in the north of Italy (Paoletti & Rusconi, 1985). Negative laboratory testing for hybridization provides evidence that the latter is, indeed, a separate species rather than a form of *T. tubifex* (Paoletti, 1989). Nonetheless, widespread distribution of *T. tubifex* and possible wide variability in setae, which can be induced experimentally (Chapman & Brinkhurst, 1987), make it difficult to positively distinguish between these species. In this case, however, the species are sympatric in the River Lambro. As well as showing marked differences in size, *T. tubifex* has a higher number of segments and larger eggs and cocoons than *T. blanchardi*. Of the two, only *T. tubifex* can reproduce by parthenogenesis (Paoletti, 1989).

Of particular interest are the species reported in subterranean waters. Stammer (1932) examined the fauna at the mouth of the Timavo (Trieste), an underground karst river which comes to the surface a short distance from its mouth; he listed *Spirosperma benedeni*, *S. velutinus*, *Sketodrilus flabellisetosus*, and *Haber speciosus*. The identity of the last two species were subsequently confirmed by Hrabe (1966), who also reported *Haber monfalconensis* in caves and a small river in the same area. More recently, Dumnicka (1990) in a study on the subterranean water fauna at several sites in central Italy reports *Haber zavreli*, *Frearidrilus pescei*, and *Abyssidrilus cuspis*.

Marine tubificids known from Italy represent 13 genera and 26 species, of which most are reported only from the Tyrrhenian Sea (Erséus, 1987a, 1987b, 1987c).

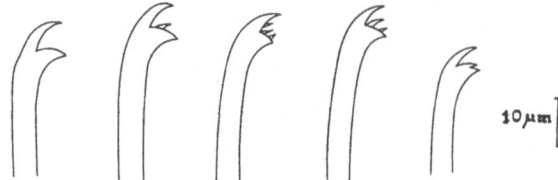

Fig. 2. *Chaetogaster diaphanus*. Setae of segment II.

Taxonomic remarks on Naididae

Since the publication of Sperber (1948), there has been increasing controversy regarding the taxonomy of many species in this family. The problems recur for the various fauna considered – China (Brinkhurst *et al.*, 1990), South America (Harman *et al.*, 1988), North America (Smith, 1985), and Africa (Grimm, 1987; Martinez-Ansemil, 1993). The taxonomy of the Naididae is heavily dependent upon the characteristics of the setae, which may vary considerably in response to ecological and other factors. The material collected in Italy shows considerable morphological variability: the main variables are presence/absence of intermediate teeth in bifid setae and presence/absence of giant setae.

Intermediate teeth. In specimens of the genus *Chaetogaster*, collected from the Su Gologone source in Sardinia, some setae of segment II have intermediate teeth (Fig. 2). In terms of other characteristics (size and shape of setae, body and prostomium shape), the specimens are identified as *C. diaphanus*. The pectinate setae (167 to 187 μm long) are shorter than the simple bifid setae (200 to 300 μm) suggesting that they are replacements. Variability of the setae in *Chaetogaster* was described by Hiltunen & Klemm (1980), who identified a reduced proximal tooth attached to the distal tooth in *C. setosus* (a species previously recognized as having only simple pointed setae). Specimens with bifid and simple pointed setae were described as a new species, *C. diversisetosus* (Sporka, 1983); probably this variation in setal form should be referred to the intraspecific variability of *C. setosus*.

In *Dero*, the presence of needle setae with intermediate teeth is confirmed in *Dero (A.) furcata* (Fig. 3), as described by Grimm (1987) in African specimens, and in *D. nivea* (Fig. 4), as observed by Smith (1985) and Grimm (1987) in specimens from America and Africa, respectively. Intermediate teeth are also reported here in the dorsal needle setae of *D. obtusa* (Fig. 5).

In *Nais*, the ventral setae of some specimens of *N. barbata* from the Alpone (a stream in the Adige basin), the River Angitola and Lake Coghinas have irregular intermediate teeth (Fig. 6). This characteristic is similar to that reported for *N. andina pectinata* by Martinez-Ansemil & Giani (1986) and Harman *et al.* (1988), in Bolivia and Peru, respectively. For *N. bretscheri*, the setae shown in Fig. 7 are a variation found in the *iorensis* variety.

Modified ventral setae. Enlarged and giant setae are found in *Pristina aequiseta*, *P. longiseta*, *Nais pardalis*, *N. bretscheri*, and *Vejdovskyella intermedia*. Loden & Harman (1980) considered *Pristina foreli* (devoid of giant setae) to be an ecomorph whose junior synonym is *P. aequiseta*. This synonymy has been the object of controversy (Dumnicka, 1986), but is now generally accepted (Harman *et al.*, 1988; Martinez-Ansemil, 1993). The material observed during the present study supports this synonymy. Recently formed zooids (in the chain of individuals) do not have enlarged setae but these are found in the first zooid.

Pristina leidyi and *P. longiseta* have been considered forms of a single species (Rodriguez, 1987; Collado & Martinez-Ansemil, 1991). The Italian material confirms that *P. longiseta* is a variable taxon comprising the two forms, *leidyi* and *longiseta*. 'Intermediate' specimens with variability in hair serration, in shape and length of the needle setae (range: 20 to 60 μm), and in thickness and length of the ventral setae in segments II and III have been observed. Given that newly formed zooids of *P. longiseta* are often devoid of lengthened hair setae in segment III, it could also be argued that *P. proboscidea* is a synonym of *P. longiseta*, a juvenile form with setae still poorly developed.

We often noticed that the second and subsequent zooids of individual chains of both *Nais pardalis* and *N. bretscheri* have 'normal' rather than enlarged and giant ventral setae. This is consistent with the view that enlarged setae appear in older specimens and that what has often been considered two species are in fact two

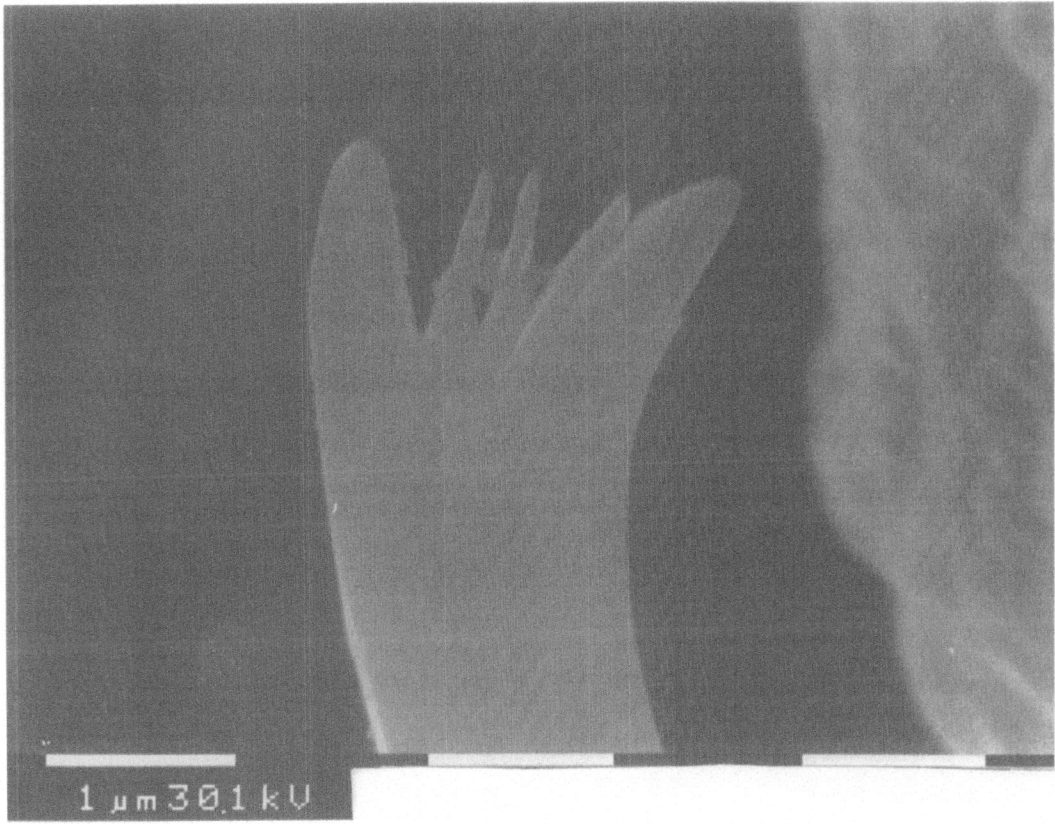

Fig. 3. Dero (A.) furcata. Needle seta.

developmental stages of the same species, as is also the case with *Pristina.*

Length of the setae. To solve the problem of setal form variability, other characteristics must be considered. Following his research into the measurements and reciprocal ratios of the three types of setae (ventral – anterior and posterior; needle setae; hair setae), Grimm (1986, 1987) suggested using the length of the posterior ventral setae. To compare the measurements and ratios of African Naididae with those of the Italian population, we used only species found in significant numbers (≥ 10) to determine mean lengths (Table 2). Our data confirm many of the findings reported in the African population. The standard deviation of the mean lengths equals an average of 8.5 to 10.7% for ventral and needle setae; and of 13.4% for hair setae. Similar values were reported by Grimm (1986): <10% for the ventral and needle setae, and about 15% for the hair setae. Grimm also reported a mean length ratio of 1.1 (± 0.2) between the needle and posterior ventral setae. For the Italian specimens, the ratio was 1.2 (± 0.2).

A comparison of the mean values and standard deviations (Table 2) within the species in the genera, show that, for some genera (*Nais, Pristina,* and *Pristinella*) mean values of the various species are too similar to be used as a differentiating characteristic. In other cases, measurements can differentiate species (*Dero dorsalis* and *D. digitata, Chaetogaster diastrophus* and *C. langi, Paranais frici* and *P. birsteini*).

In conclusion, data resulting from the measurement of setae may permit the separation of some species which otherwise would not be easily distinguished.

Discussion

The oligochaete data in the literature indicate, at best, a partial knowledge of Italian fauna. In many instances, findings are not supported by any morphological, biological, or ecological data. In most cases, fauna are listed in ecological reports; and few taxonomic studies have been published. Fewer species are reported to

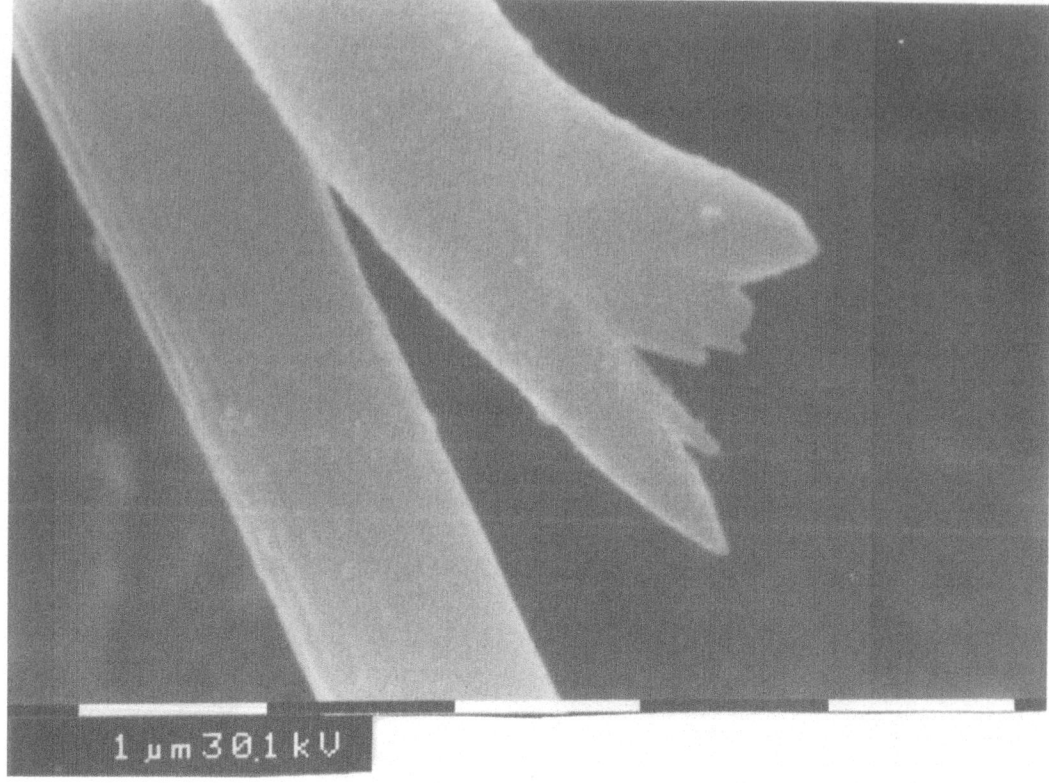

Fig. 4. Dero nivea. Needle seta.

occur in the southern part of Italy than in the north. These results are more a reflection of the limited number of studies that have been conducted and published than of the actual distribution of species in southern Italy. The only exception is seen in the distribution of species in the family Naididae, for which the data may be considered complete, for they are the result of a survey which involved observation of more than 10 000 specimens from a variety of freshwater habitats throughout Italy.

Naididae

The Italian fauna (48 species) is comparable with that of other European countries, in terms of both composition and quantity. Martinez-Ansemil (1993) reported 42 species for the Iberian Peninsula (all, but *Chaetogaster setosus*, present in Italy) and 45 for France (of which only *Chaetogaster setosus*, *Piguetiella blanci*, and *Ripistes parasita* were not reported from Italy).

Data in the literature list 60 species for Europe. Those not found in Italy include *Nais bihorensis*, *Pristinella arcaliae*, *Bratislavia elegans*, *Chaetogaster krasnopolskiae*, *C. setosus*, *C. diversisetosus*, *Paranais botniensis*, *P. multisetosa*, *P. simplex*, *Arcteonais lomondi*, *Ripistes parasita*, and *Piguetiella blanci*. These species are mostly rare (like *Piguetiella blanci* or *Chaetogaster setosus*), confined to particular habitats (*Ripistes parasita*), or found only in a restricted area (*Paranais simplex* and *P. multisetosa* two species only from the Caspian Sea). Similarly, a survey of fauna in other Mediterranean countries (Morocco, Algeria, Tunisia, Egypt, Lebanon, Syria, Turkey) does not suggest major differences in the known distribution of Naididae. This similarity is attributed to the numbers of species reported for North Africa (22) and the Near East (30) (Martinez-Ansemil, 1993).

The only species from North Africa not found in Italy is the tropical *Allonais paraguayensis*; neareastern species not yet collected in Italy include *Pristinella arcaliae* and *Stephensoniana trivandrana*. The large number of species common to the European and Mediterranean countries is evidence of the wide eco-

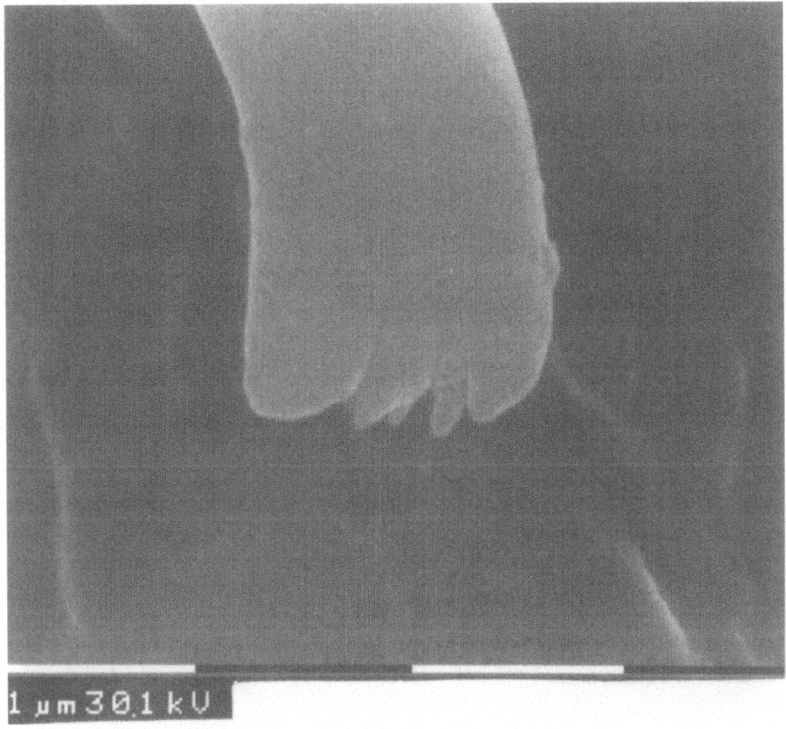

Fig. 5. Dero obtusa. Needle seta.

logical range in which naidids occur. Many species are cosmopolitan, or at least widely distributed.

One feature of the naidid fauna, related to the geographical position of Italy, is the presence of taxa common to other Mediterranean areas such as the Middle East (*Nais bretscheri* var. *iorensis*) and North Africa (*Paranais birsteini* var. *maghrebensis*, *Dero (A.) africana*). Another finding is the difference in the northern and southern areas in Italy. *Nais* was identified as the prevalent genus in the north, while *Pristina*, *Pristinella*, and *Dero* were more commonly distributed in the south. From a taxonomical point of view, morphological differences observed in the setae of *Chaetogaster*, *Nais*, and *Dero* confirm that intermediate teeth are frequent in species previously described as having only bifid setae. The authors consider that pectination is possible in all furcate needle setae – and probably also in the ventral setae – of Naididae. These modifications may be related to ecological factors, though further investigation is required to allow a definitive evaluation of the question. Many of the modifications concerned are, for example, present only in some specimens from the samples studied – sometimes only in some bundles from a given specimen. The presence of enlarged and giant setae in species of *Pristina* and *Nais* also seems related to ecological factors or to the age of the specimens. Measurement of the setae in series of individuals from different sites, however, confirms that setal lengths can be a useful parameter for identification of some species.

The authors consider that problems raised by morphological variability in the Naididae, as highlighted by research all over the world, cannot be resolved by study of fixed material alone. An essential useful complement is *in vivo* observation of variability according to age and environmental conditions.

Other families

Data on species in other oligochaete families are too fragmented, both geographically and temporally, to support conclusions regarding their presence and distribution. To date, available information allows us to indicate few characteristics of the Italian fauna.

Haplotaxidae. The only representative of the Haplotaxidae in Italy is the cosmopolitan *Haplotaxis gordioides*, although 6 other species in this family are known to occur elsewhere in Europe. These species

46

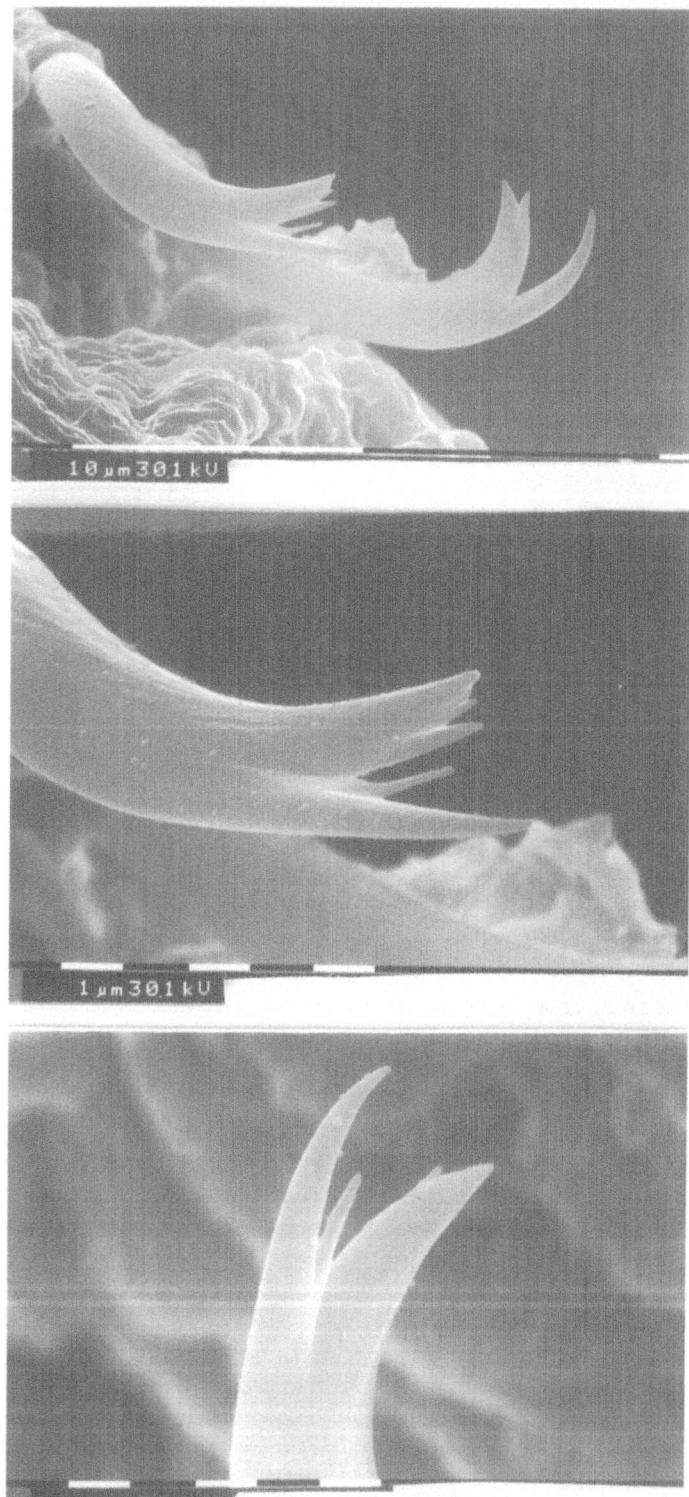

Fig. 6. Nais barbata. Ventral setae.

Table 2. Mean values and standard deviations of setal lengths of Italian Naididae. Abbreviations: AVS = anterior ventral setae; PVS = posterior ventral setae; NS = needle setae; HS = hair setae; m= mean setal length in μm; s = standard deviation; s% = standard deviation expressed in percent of the mean value.

Species	AVS			PVS			NS			HS		
	m	s	s%	m	s	s%	m	s	s%	m	s	s%
Chaetogaster diastrophus	91	12	13	72	9.3	13						
C. langi	78	6	7.7	70	6.6	9.4						
C. parvus	56	4.2	7.5	42	4.5	11						
C. diaphanus	215	27	12	177	24	13						
C. cristallinus	137	14	10	107	8	7.5						
Dero dorsalis	138	15	9	134	7	5	116	4.4	4	258	28	11
D. digitata	118	9	8	85	7.7	9	78	9.3	12	216	24	11
D. obtusa	95	13	14	73	3	4	63	6.5	10	158	24	15
D. nivea	106	12	11	63	3.5	5.5	55	4	7.3	141	10	7
D. (Aulophorus) furcata	70	6	9	63	5	8	53	8	15	160	15	9
Nais christinae	107	12	11	92	7.6	8	66	5.6	8.5	196	33	17
N. bretscheri	98	11	11	79	9.7	12	60	4.8	8	124	31	25
N. group *comm.-var.*	70	10	14	70	8.2	12	56	6	11			
N. pardalis	89	7.6	8.5	69	6.4	9	58	6.3	11	119	11	9
N. elinguis	91	10	11	88	8.4	9.5	78	10	13	238	25	10
N. group *barb.-pseudob.*	92	16	17	75	15	20	80	19	24	202	26	13
N. simplex	91	7	8	76	5	6.6	76	5	6.6	180	22	12
N. alpina	92	7.2	8	82	5	6	60	7	12	176	19	11
N. behningi	104	14	13	85	9	10	58	10	17	176	22	12
Ophidonais serpentina	170	15	8.8	145	9.6	6.6	164	7.5	4.5			
Paranais frici	102	9	9	85	8.5	10	85	8.5	10			
P. birsteini	69	2.3	3	66	2.5	4	66	2.5	4			
Pristina aequiseta	49	8	16	47	4.5	9.6	45	3.4	7.5	180	31	17
P. longiseta	68	6.5	10	54	2.2	4	42	9.3	22	177	37	21
Pristinella jenkinae	51	6.8	13	55	4.2	7.6	59	8	13	212	27	13
P. menoni	36	4	11	38	3.5	9.2	52	6	11	175	15	8.6
P. osborni	39	2.6	6.7	39	1.5	3.8	32	2.8	8.7	94	10	11
Slavina appendiculata	117	7	6	106	7	6.6	63	7.6	12	315	37	12
Stylaria lacustris	196	11	5.6	196	11	5.6	98	10	10	582	136	23
Uncinais uncinata	128	14	11	116	12	10	102	10	10			
m (s%)	10			8.5			10.7			13.4		

are restricted to subterranean waters; their apparent absence in Italy can be attributed to a lack of data for the fauna of this habitat.

Lumbriculidae. Of particular interest is the presence of 4 species of *Trichodrilus* in subterranean waters. *Trichodrilus* includes mostly psychrostenothermic species, known from European subterranean waters (Juget & Dumnicka, 1986; Rodriguez & Giani, 1994). The distribution of each species is limited and discontinuous, and geographical isolation may account for the diversity of the genus (Juget & Dumnicka, 1986). No fewer than eight species have been reported from subterranean habitats in southwestern Europe (Martinez-Ansemil, 1993).

Tubificidae. For this family too, the species of subterranean waters are most interesting. The strictly stygobiont species (*Haber zavreli, Frearidrilus pescei* and *Abyssidrilus cuspis*) illustrate the often endemic or at least discontinuous distribution of subterranean fauna. *Haber zavreli*, for example, is known in the Czech Republic and southern Poland.

The genus *Rhyacodrilus* is represented by its two most widely distributed species: *R. coccineus*, and *R. falciformis*. A number of other species of *Rhya-*

48

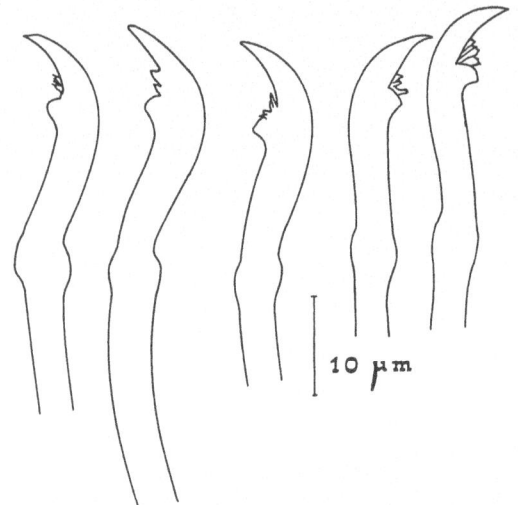

Fig. 7. Nais bretscheri var. *iorensis*. Setae of segments VIII and IX.

codrilus are known primarily from long-established lakes or subterranean waters in southwestern Europe (Martinez-Ansemil, 1993). These species probably require stable abiotic conditions. Once again there has been insufficient research in Italy on subterranean waters.

For the marine Tubificidae, reported distributions in the three coastal basins is heavily influenced by the concentration of sampling sites – with almost all species reported from the Tyrrhenian Sea.

Acknowledgments

We are grateful to Dr C. Erséus for his help in preparing the list of marine Tubificidae. Thanks to Prof B. Baccetti (University of Siena) for granting us access to scanning electron microscopy facilities. The authors are also indebted to the curators of the Museo Civico di Storia Naturale, Verona, for kindly making their premises available, and to Prof S. Ruffo for his helpful advice and suggestions.

References

Bazzanti, M., 1981. Survey of the macrobenthic community in an area of Lake Bracciano (Italy). Boll. zool. 48: 295–303.

Bazzanti, M., 1983. Composition and diversity of the macrobenthic community in the polluted Lake Nemi (Central Italy) 1979–1980. Acta Oecol. 4: 211–220.

Bazzanti, M. & M. Seminara, 1987. Profundal macrobenthos in a polluted lake. Depth distribution and its relationship with biological indices for water quality assessment. Acta Oecol. 8: 15–26.

Bazzanti, M., M. Seminara & C. Tamorri, 1988. Macrobenthos profondo del lago artificiale di Campotosto (Abruzzo, Italia Centrale). Riv. Idrobiol. 27: 161–175.

Bonacina, C., G. Bonomi & A. Pasteris, 1992. Some remarks on the macrobenthos community of the profundal zone of the large Italian lakes. Mem. Ist. ital. Idrobiol. 50: 79–106.

Bonomi, G. & C. Erséus, 1984. A taxonomic and faunistic survey of the marine Tubificidae and Enchytraeidae (Oligochaeta) of Italy. Introduction and preliminary results. Hydrobiologia 115: 207–210.

Bonomi, G. & C. Erséus, 1985. Marine Oligochaeta of Italy: a preliminary survey. Nova Thalassia 6: 713–714.

Bonomi, G. & M. Gerletti, 1967. Il lago d'Iseo: primo quadro limnologico generale (termica, chimica, plancton e benton profondo). Mem. Ist. ital. Idrobiol. 22: 149–175.

Brinkhurst, R. O., 1963. The aquatic Oligochaeta recorded from Lake Maggiore with notes on the species known from Italy. Mem. Ist. ital. Idrobiol. 16: 137–150.

Brinkhurst, R. O. 1978. Oligochaeta. In G. Illies (ed), Limnofauna Europea. A. Fisher Verlag, Stuttgart: 139–144.

Brinkhurst, R. O. & K. A. Coates, 1985. The genus *Paranais* (Oligochaeta: Naididae) in North America. Proc. biol. Soc. Wash. 98: 303–313.

Brinkhurst, R. O., Q. Sang & L. Liang, 1990. The aquatic Oligochaeta from the People's Republic of China. Can. J. Zool. 68: 901–916.

Centurioni, C. & B. Sambugar, 1986. Gli Oligocheti. In M. G. Braioni & S. Ruffo (eds), Ricerche sulla qualità delle acque dell'Adige. Mem. Mus. Civ. St. Nat. Verona 1: 125–151.

Chapman, P. M. & R. O. Brinkhurst, 1987. Hair today, gone tomorrow: induced chaetal changes in tubificid oligochaetes. Hydrobiologia 155: 45–55.

Chinaglia, L., 1912. Catalogo sinonimico degli Oligocheti d'Italia. Boll. Mus. zool. Anat. comp. Torino 2: 71–76.

Collado, R. & E. Martinez-Ansemil, 1991. Les Oligochètes aquatiques de la Peninsule Ibérique: premières données sur le Portugal. Bull. Soc. Hist. nat. Toulouse 127: 57–69.

Dumnicka, E., 1986. Naididae (Oligochaeta) from subterranean waters of West Indian Islands. Distribution, taxonomic remarks and description of a new species. Bijdr. Dierk. 56: 267–281.

Dumnicka, E., 1990. Oligochaetes from subterranean waters of Italy and Greece. Mem. Biospéléol. 18: 233–236.

Ercolini, A., 1956. Prime note sistematiche e biologiche sopra i Naididi del Piemonte. Boll. Mus. Zool. Anat. comp. Torino 5: 143–163.

Erséus, C., 1987a. Marine Limnodriloidinae (Oligochaeta, Tubificidae) from Italy, with description of two new species. Boll. Zool. 54: 159–164.

Erséus, C., 1987b. Taxonomic revision of the marine interstitial genus *Aktedrilus* (Oligochaeta, Tubificidae), with description of three new species. Stygologia 3: 108–124.

Erséus, C., 1987c. Seven new marine species of *Phallodrilus* (Oligochaeta, Tubificidae) from various parts of Europe, and a re-examination of the type species *P. parthenopaeus* Pierantoni. J. nat. Hist. 21: 915–931.

Erséus, C. & G. Bonomi, 1987. A new species of *Tubificoides* (Oligochaeta, Tubificidae) from the Adriatic Sea. Boll. Zool. 54: 165–168.

Erséus, C. & E. Dumnicka, 1988. A new *Phallodrilus* (Oligochaeta, Tubificidae) from subterranean waters in Central Italy. Stygologia 4: 116–120.

Erséus, C. & A. Paoletti, 1986. An italian record of the aquatic oligochaete *Monopylephorus limosus* (Tubificidae), previously known only from Japan and China. Boll. Zool. 53: 115–118.

Garbini, A., 1898. Fauna. In L. Sormani Moretti (ed), La provincia di Verona, Verona: 314–315.

Ghetti, P. F., A. Canuti, G. Bonazzi, G. Canuti & M. De Crecchio, 1977. I corsi d'acqua della città di Cremona: caratteristiche chimiche, biologiche e prospettive urbanistiche. At. Parm. Acta Nat. 13: 263–275.

Grimm, R., 1986. Beiträge zur Systematik der afrikanischen Naididae (Oligochaeta). III. Untersuchungen zur qualitativen und quantitativen Chaetotaxonomie der Naididae. Mitt. hamb. zool. Mus. Inst. 83: 101–115.

Grimm, R., 1987. Contributions towards the taxonomy of the african Naididae (Oligochaeta). IV. Zoogeographical and taxonomical consideration of African Naididae. Hydrobiologia 155: 27–37.

Grimm, R., 1989. Beiträge zur Systematik der afrikanischen Naididae (Oligochaeta). VII. Naidinae (teil 2). Mitt. hamb. zool. Mus. Inst. 86: 107–126.

Harman, W. J., R. O. Brinkhurst & M. R. Marchese, 1988. A contribution to the taxonomy of the aquatic Oligochaeta (Naididae) of South America. Can. J. Zool. 66: 2233–2242.

Hiltunen, J. K. & D. J. Klemm, 1980. A guide to the Naididae (Annelida, Oligochaeta) of North America. Great Lakes Fish. Lab. EPA-600/4-80 – 031 N.S. (Environmental Protection Agency, Cincinnati), 48 pp.

Hrabe, S., 1937. Contribution a l'étude du genre *Trichodrilus* (Oligochaeta, Lumbriculidae) et description de deux espèces nouvelles. Bull. Mus. r. Hist. Nat. Belg. 13: 1–23.

Hrabe, S., 1942. Zur Kenntnis der Brunnen und Quellen fauna aus Slovenien. Sb. Přír. Klubu Brně 24: 23–30.

Hrabe, S., 1966. New or insufficiently known species of the family Tubificidae. Spisy přír. Fac. Univ., Brně 470: 57–77.

Juget, J., 1987. Contribution to the study of Rhyacodrilinae (Tubificidae, Oligochaeta), with description of two new stygobiont species from the alluvional plain of the French upper Rhône, *Rhyacodrilus amphigenus*, sp.n. and *Rhizodriloides phreaticola*, g.n., sp.n. Hydrobiologia 155: 107–118.

Juget, J. & E. Dumnicka, 1986. Oligochaeta (incl. Aphanoneura) des eaux souterraines continentales. In Botosaneanu (ed), Stygofauna mundi – E. J. Brill, Leiden: 234–244.

Lafont, M., A. Durbec & C. Ille, 1992. Oligochaete worms as biological describers of the interactions between surface and groundwaters: a first synthesis. Regulated Rivers: Research & Management 7: 65–73.

Loden, M. S. & W. J. Harman, 1980. Ecophenotypic variation in setae of Naididae (Oligochaeta). In R. O. Brinkhurst and D. G. Cook (eds), Aquatic Oligochaete Biology. Plenum Press, N.Y.: 33–39.

Martinez-Ansemil, E., 1993. Études sur les Oligochètes aquatiques des pays du pourtour de la Méditerranée: taxonomie, phylogenie, biogeographie et ecologie. Ph.D. Université Paul Sabatier Toulouse, France, 616 pp.

Martinez-Ansemil, E. & N. Giani, 1982. Contribution al conocimento del genero Pristina (Oligochaeta, Naididae) en la Peninsula Iberica. Bol. R. Soc. Esp. Hist. nat. (Biol.) 80: 249–260.

Martinez-Ansemil, E. & N. Giani, 1986. Algunos oligoquetos acuáticos de Bolivia. Oecol. aquatica. 8: 107–115.

Martinez-Ansemil, E. & N. Giani, 1987. The distribution of aquatic oligochaetes in the south and eastern Mediterranean area. Hydrobiologia 155: 293–303.

Nocentini, A., 1963. La fauna macrobentonica litorale del Lago Maggiore. Mem. Ist. ital. Idrobiol. 16: 189–274.

Nocentini, A. M., 1973. La fauna macrobentonica litorale e sublitorale dei laghi di Bolsena, Bracciano e Vico. Mem. Ist. ital. Idrobiol. 30: 97–148.

Paoletti, A., 1989. Cohort cultures of *Tubifex tubifex* forms. Hydrobiologia 180: 143–150.

Paoletti, A., & C. Rusconi, 1985. Preliminary data on the biology of a tubificid community in a small stream. S. It. E. Atti 5: 277–280.

Paoletti, A. & B. Sambugar, 1984. Oligochaeta of the middle Po River (Italy): principal component analysis of the benthic data. Hydrobiologia 115: 145–152.

Pierantoni, U., 1902. Due nuovi generi di oligocheti marini rinvenuti nel Golfo di Napoli. Boll. Soc. nat. Napoli 16: 113–117.

Pierantoni, U., 1903. Altri nuovi Oligocheti del Golfo di Napoli (*Limnodriloides* n. gen.). II. Note sui Tubificidae. Boll. Soc. Nat. Napoli 17: 185–192.

Rodriguez, P., 1987. The variability of setae of *Pristina longiseta* Ehrenberg (Oligochaeta, Naididae). Hydrobiologia 155: 39–43.

Rodriguez, P. & N. Giani, 1994. A preliminary review of the taxonomic characters used for the systematics of the genus *Trichodrilus* Claparède (Oligochaeta, Lumbriculidae). Hydrobiologia 278: 35–51.

Sambugar, B., 1976. Oligocheti. In: U. Ferrarese & B. Sambugar, Ricerche sulla fauna interstiziale iporreica dell'Adige in relazione allo stato di inquinamento del fiume. Riv. Idrobiol. 15: 87–91.

Sambugar, B., 1987. I Naididi italiani (Oligochaeta). Ph.D. Thesis, Università di Padova, Italy, 246 pp.

Sciacchitano, I., 1934. Sulla distribuzione geografica degli Oligocheti in Italia. Arch. Zool. ital. 20: 1–31.

Smith, M. E., 1985. Setal morphology and its intraspecific variation in *Dero digitata* and *Dero nivea* (Oligochaeta: Naididae). Trans. am. microsc. Soc. 104: 45–51.

Sperber, C., 1948. A taxonomical study of the Naididae. Zool. Bidr. Upps. 28: 1–296.

Sporka, F., 1983. *Chaetogaster diversisetosus* sp.n. a new species of Naididae (Oligochaeta) from Czechoslovakia. Vest. cs. Spolec. Zool. 47: 137–139.

Stammer, H. J., 1932. Die Fauna des Timavo. Zool. Jb. (Syst.) 63: 521–656.

Stella, E., 1951. Studio zoologico preliminare sulle zoocenosi del Tevere. Verh. int. Ver. Limnol. 11: 383–391.

Stella, E. & G. Argenti, 1953. Il lago di Albano. III. Le società bentoniche profonde. Boll. Pesca, Pisc. Idrobiol. 8: 154–173.

Hydrobiologia **334**: 51–62, 1996.
K. A. Coates, Trefor B. Reynoldson & Thomas B. Reynoldson (eds), Aquatic Oligochaete Biology VI.
© 1996 *Kluwer Academic Publishers.*

The distribution of Oligochaeta on an exposed rocky shore in southeast Ireland

Brenda Healy
Department of Zoology, University College, Belfield, Dublin 4, Republic of Ireland

Key words: Enchytraeidae, intertidal, rock, wave-exposed, crevices, mats

Abstract

The distribution of oligochaetes was studied on a wave-exposed, granite shore at Carnsore Point, County Wexford. Habitats sampled were crevices and shallow surface cracks in rock, *Lichina pygmaea* turf, a *Mytilus edulis* bed, barnacles, *Laurencia pennatifida* turf, *Corallina officinalis* turf and pools. *Lichina*, mussels, barnacles greater than 8 mm in height, *Laurencia* and *Corallina* were sampled quantitatively.

Three species of Tubificidae and nine species of Enchytraeidae were recorded including two new species described from this locality and five which are undescribed. Oligochaetes were present in all the habitats sampled but were rare in pools and numbers were low on barnacle-covered rock except where barnacles were more than 8 mm in height. The number of species increased with decreasing tide level and was highest in lower shore crevices. Highest densities were recorded from *Corallina* turf and in barnacles >8 mm. *Lumbricillus semifuscus* was the most widespread species, occurring in crevices and mats at all tide levels; other species had more restricted distributions. *Grania* species were confined to *Corallina* turf. Only three species of oligochaete were recorded from cracks where they were almost the only fauna present. Species diversity and density were influenced by physical structure of the habitat, particularly the amount of retained sediment.

The oligochaetes are members of a rich cryptofauna in habitats which provide them with organic matter and moisture and protect them from environmental extremes and wave damage and from predators during immersion. Reasons for the scarcity of oligochaete records from exposed rocky shores and the high proportion of new species in this study are discussed.

Introduction

This paper describes the distribution of oligochaetes on wave-beaten rocks where the only habitats which provide adequate protection are the mats and incrustations of larger organisms or the cracks and crevices of the rock itself. The worms constitute a significant proportion of a species-rich cryptofauna consisting of small animals, many of meiofaunal dimensions, occupying secondary space.

Studies of intertidal oligochaetes have mainly been concerned with species inhabiting sediments or tidal debris and there are relatively few records from rocky shores. In most accounts of faunal studies of the rocky intertidal, the shore is only briefly described and the habitats where oligochaetes have been found, usually under stones, among fucoid algae, or in driftweed at high water mark, indicate fairly sheltered conditions. In more wave-exposed situations, stones, fucoid algae and wrack beds are rare or absent and most small, mobile animals live cryptically in algal turf, sessile macrofauna or crevices or are confined to pools. A few oligochaete species records from the rocky intertidal refer specifically to certain cryptohabitats, e.g., named algal species (Elmhirst & Stephensen, 1926; Stephenson, 1932; Colman, 1940; McGrath, 1975; Erséus, 1976, 1987; Shurova, 1977), lichens (Colman, 1940), mussels (Erséus, 1976, 1990), and barnacles (Erséus, 1993). In most general studies of cryptofauna, however, oligochaetes are either not mentioned (Glynne-Williams & Hobart, 1952; Morton, 1954; Chapman, 1955; Sloane *et al.*,1961; Kensler, 1964a;

Reish, 1964; Kensler & Crisp, 1965; Dommasnes, 1969; Hicks, 1971; Sarma & Ganapati, 1972; Reimer, 1976; Raffaelli, 1978; Myers & Southgate, 1980; Choat & Kingett, 1982; Tsuchiya & Nishihira, 1986) or not identified (Delamare-Deboutteville & Bougis, 1951; Wieser, 1952; Gorvett, 1958; de Marguia & Seed, 1987; Tsuchiya & Retière, 1992). This is thus the first study of rocky shore cryptofauna focused on oligochaetes.

The locality of the present study has been the subject of an intensive baseline survey (unpublished) and a number of investigations of selected faunal groups and species; much background information is therefore available. An account of marine Oligochaeta in SE Wexford (Healy, 1979) includes records from this shore but some habitats were not examined and others were only superficially searched.

Site description

The investigation was carried out at Carnsore Point, County Wexford (Irish Grid Reference T 130035), a rocky headland in the SE of Ireland consisting of about 0.5 km of exposed bedrock which gives way westwards to a gravel barrier and northeastwards to a more sheltered boulder shore. The rock is a red, Precambrian granite, a strongly porphyritic, two-mica formation with phenocrysts up to 5 cm in length of microline-microperthite (feldspar) which produce a rough, irregular surface (Baker, 1955). On the lower half of the shore, however, scouring by sand and grazing by gastropods have made the surface smoother. The S-SW aspect of this section of the coast exposes it to the prevailing W and SW winds and a long wave fetch and both physical and biological features of the shore are strongly affected by wave action. Any crevices which appear may be relatively short lived as violent storms frequently tear off sizeable portions of rock. Patches of mussels and barnacles may also be detached by storm waves, although the predatory activities of birds and fish may also be responsible for the bare patches which appear among mussels and barnacles.

On the biological exposure scale of Ballantine (1961), which grades British shores on a scale of 1–8, the site corresponds approximately to grade 3 (exposed) which is characterised by wide lichen belts in the splash zone, very reduced fucoid zones, represented on the study site by scattered plants of *Pelvetia canaliculata* (L.) Dene. & Thur. and a few patches of *Fucus vesiculosus* f. *linearis* L. in sheltered places, distinct zones of *Lichina pygmaea* (Lightf.) and coralline algae, and an abundance of barnacles, limpets and mussels. The corresponding grade on the scale of Dalby *et al.* (1978), devised for use on Norwegian coasts, is between 2 and 3. Tides are semidiurnal and the mean spring tidal range is 3.6 m. The salinity of inshore water at the sampling site in 1976–78 was 29–34‰.

In the following description, and throughout this paper, tide levels are abbreviated as follows: EHWS - extreme high water of spring tides, MHWS - mean high water of spring tides, MHWN - mean high water of neap tides, MTL - mean tide level, MLWN - mean low water of neap tides, HLWS - highest of low water spring tides. The shore profile (Fig. 1) presents a zonation pattern typical of exposed shores in Ireland and SW Britain. The following zones are readily distinguished at Carnsore but there are some local variations:

1. *Upper lichen zone.* Grey and orange lichen belts below the level of flowering plants.
2. *Verrucaria zone.* A wide, black lichen belt, incorporating the upper limit of *Littorina* spp.
3. *Bare zone.* A narrow zone of bare rock without conspicuous life except for some discoloration by cyanobacteria.
4. *Upper barnacle zone.* Low-density barnacles starting at around MHWS, with scattered plants of *Pelvetia*.
5. *Lichina zone.* Dense barnacles and numerous limpets with scattered patches of *Lichina* and a few plants of *Fucus vesiculosus* f. *linearis,* extending from about MHWN to MTL. Width of zone 9–17 m.
6. *Mussel zone.* A mosaic mussel bed around MTL with barnacles and limpets between, and isolated patches of *Laurencia pennatifida* (Huds.) Lamour near the lower edge of the bed. Width of zone about 4 m.
7. *Lower barnacle zone.* A dynamic zone in which barnacles and mussels are subject to predation pressure and densities vary from year to year. Patches of *Laurencia* occur, especially on vertical surfaces, and *Corallina officinalis* L. is present in depressions and drainage channels. Width of zone 5–9 m.
8. *Corallina zone.* A dense turf of *Corallina*, extending from just above HLWS into the sublittoral, with scattered plants of *Mastocarpus stellatus* (Stackh.) Guiry and *Himanthalia elongata* (L.) S.F. Gray and, at a lower level, some *Alaria esculenta* (L.) Grev. and *Laminaria digitata* (Huds.) Lamour.

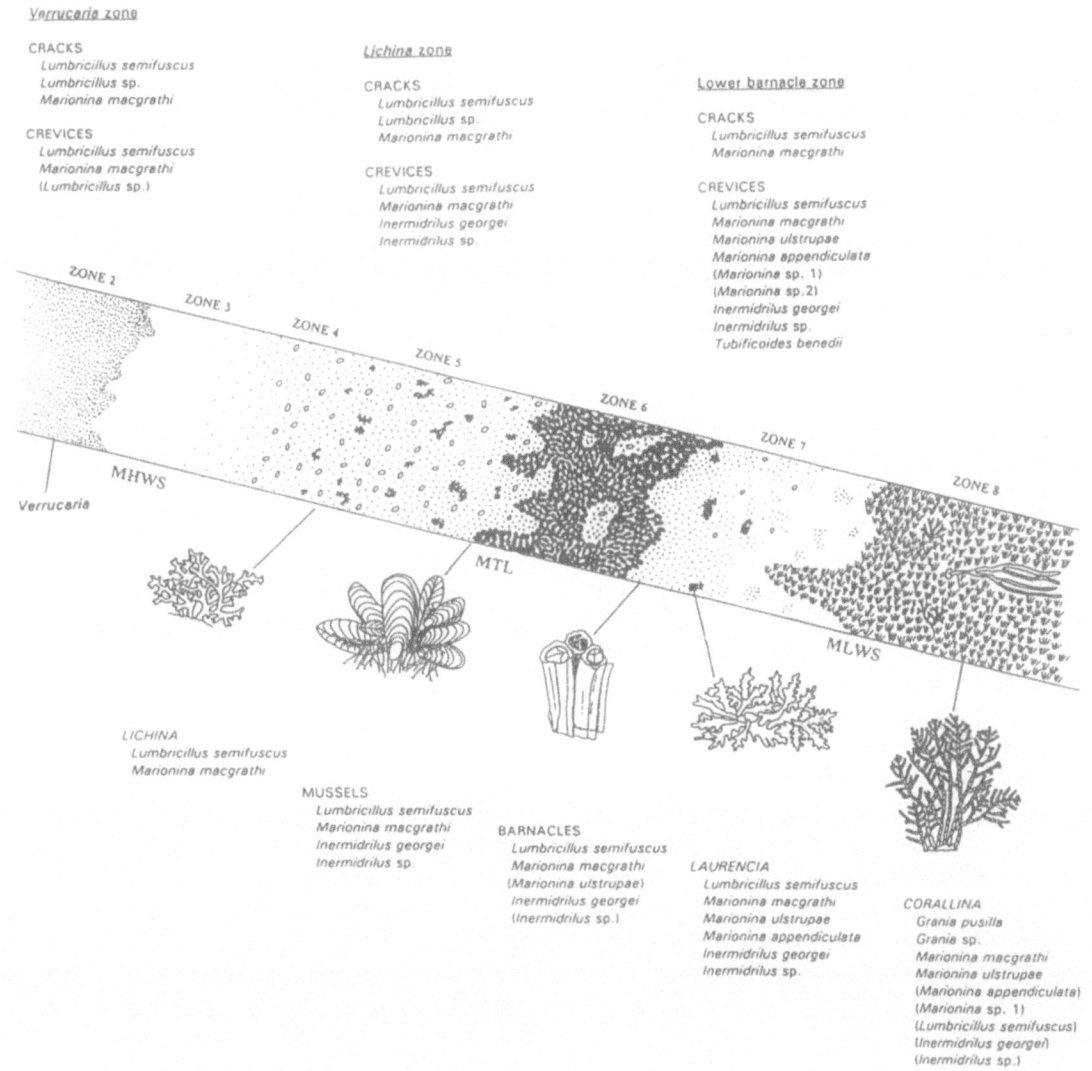

Verrucaria zone

CRACKS
Lumbricillus semifuscus
Lumbricillus sp.
Marionina macgrathi

CREVICES
Lumbricillus semifuscus
Marionina macgrathi
(Lumbricillus sp.)

Lichina zone

CRACKS
Lumbricillus semifuscus
Lumbricillus sp.
Marionina macgrathi

CREVICES
Lumbricillus semifuscus
Marionina macgrathi
Inermidrilus georgei
Inermidrilus sp.

Lower barnacle zone

CRACKS
Lumbricillus semifuscus
Marionina macgrathi

CREVICES
Lumbricillus semifuscus
Marionina macgrathi
Marionina ulstrupae
Marionina appendiculata
(Marionina sp. 1)
(Marionina sp.2)
Inermidrilus georgei
Inermidrilus sp.
Tubificoides benedii

ZONE 2 ZONE 3 ZONE 4 ZONE 5 ZONE 6 ZONE 7 ZONE 8

MHWS

Verrucaria

MTL

MLWS

LICHINA
Lumbricillus semifuscus
Marionina macgrathi

MUSSELS
Lumbricillus semifuscus
Marionina macgrathi
Inermidrilus georgei
Inermidrilus sp.

BARNACLES
Lumbricillus semifuscus
Marionina macgrathi
(Marionina ulstrupae)
Inermidrilus georgei
(Inermidrilus sp.)

LAURENCIA
Lumbricillus semifuscus
Marionina macgrathi
Marionina ulstrupae
Marionina appendiculata
Inermidrilus georgei
Inermidrilus sp.

CORALLINA
Grania pusilla
Grania sp.
Marionina macgrathi
Marionina ulstrupae
(Marionina appendiculata)
(Marionina sp. 1)
(Lumbricillus semifuscus)
(Inermidrilus georgei)
(Inermidrilus sp.)

Fig. 1. Distribution of oligochaetes on the exposed shore at Carnsore Point. Zones 2–8 (not to scale) are: Zone 2 - *Verrucaria* zone; Zone 3 - bare zone; Zone 4 - upper barnacle zone with limpets; Zone 5 - *Lichina* zone with barnacles and limpets; Zone 6 - mussel zone with patches of barnacles and limpets; Zone 7 - lower barnacle zone with limpets and sparse barnacles (tall barnacles sampled at a different site); Zone 8 - *Corallina* turf with scattered *Macrocarpus* and *Himanthalia*. Further details in text under Site Description. Species in brackets were uncommon in this habitat.

Zone 1 corresponds to the maritime zone (supralittoral zone), Zones 2 and 3 to the littoral fringe (supralittoral fringe, splash zone), and zones 3–8 lie within the eulittoral zone (midlittoral zone) (Lewis, 1961). In this paper, zones 4–5 are referred to as the upper shore, zone 6 as the middle shore, and zones 7–8 as the lower shore.

Description of habitats

The habitats investigated fall into three categories: rock, living plant or animal mats and pools.

Rock

(a) *Crevices.* The non-foliaceous rock develops few crevices but some were present at all levels. They were at least 2 mm wide at the opening, extending for 10–60 cm into the rock at various angles but mostly horizontally. Conditions within the crevices were very variable depending on their width and orientation and the amount and particle size of the sediment and organic matter trapped within them. Some were well-drained and the accumulated sediment a brown or rust-coloured silt or gravelly silt while others were partially waterlogged and the sediments anaerobic and blackened. In all crevices there was a gradient of environmental conditions from the outside inwards with a corresponding zonation of fauna, the majority of species being concentrated in an outer zone. Oligochaetes were among the few groups inhabiting the sediments of the inner zones. The diverse faunal community was similar to that described for crevices in Britain (Morton, 1954; Glynne-Williams & Hobart, 1952; Kensler & Crisp, 1965), and Norway (Kensler, 1964a).

(b) *Surface cracks* were present wherever the rock surface was rough, i.e. they were rare on the lower shore. They were particularly numerous near water-filled depressions in the *Verrucaria* and bare zones where the rock could sometimes be crumbled by hand. The cracks were less than 0.5 mm in width at the surface and extended into the rock both vertically and horizontally, usually to about 1–1.5 cm. They are the result of weathering processes which differentially dissolve feldspar and could be detected by prising off projecting feldspar crystals. The cracks were generally lined with films of green algae or cyanobacteria; wider cracks also contained a little silt and fine organic detritus.

Living plant or animal mats

The term 'mat' is used throughout this paper to include algal and lichen turfs and the incrustations of barnacles and mussels.

(a) *Lichina turf.* Tufts of *Lichina*, 5–15 cm across, 5–10 mm high when dry, somewhat taller when wet, formed a patchy zone about 12 m wide. The zone is exposed to the air for at least 6 hours at each low tide and *Lichina* is subject to desiccation. The hygroscopic tissues prevent the plants becoming brittle, however, and their close growth at the base holds moisture and some sediment forming a retaining network for small animals, especially *Lasaea rubra* (Mont.), *Littorina saxatilis* (Olivi), *Melaraphe neritoides* (Mont.), *Hyale nilssoni* (Rathke), *Campecopaea hirsuta* (Mont.), *Ligia oceanica* (L.), mites, and enchytraeids.

(b) *Mussel beds.* A mosaic-like mussel bed, approximately 5–8 m in width, occurred just below the *Lichina* zone. The bed is known to date from 1986, a previous similar bed having been washed away during storms. Juveniles settling below the main bed are subject to predation and are relatively short-lived; small rosettes, a few cms across are present here in most years but were absent during the sampling period. The mussels, a mixed population of *Mytilus edulis* L., *M. galloprovincialis* (Lam.) and hybrids (Gosling & Wilkins, 1981), were of various sizes (maximum 30 mm), packed tightly, one or two deep, and attached to the rock and each other by a strong network of byssus threads which traps sediment and organic matter and remains wet throughout tidal emersion. Moisture retention is further enhanced by growths of various algae on the mussels in some areas, particularly *Porphyra umbilicalis* (L.) J. Ag., *Ceramium* sp. and *Enteromorpha* sp. Dominant members of the infauna were *Cirratulus cirratus* (Müll.), errant polychaetes, *Idotea pelagica* Leach, *Parasinelobus chevreuxi* (Dollf.), *Lasaea rubra*, *Hydrogamasus littoralis* G. & R. Can., and nemerteans.

(c) *Barnacles.* Three species were present with overlapping distributions giving three main zones: an upper band of relatively low density *Chthamalus montagui* Southw. above MHWN, a middle zone of mixed *C. stellatus* (Poli), *C. montagui*, and *Semibalanus balanoides* (L.) extending to MTL or just above MLWN, and a lower zone of *C. stellatus* and *S. balanoides* with *C. stellatus* decreasing seawards to disappear at about MLWS. Most individuals did not exceed 5 mm in height but patches of taller specimens of *C. stellatus* and *S. balanoides*, reaching 16 mm, occurred in certain areas in the mussel and lower barnacle zones. Tall barnacles were frequent in some areas but were rare at the main sampling site. They were frequently encrusted with short growths of red, green or brown algae and films of cyanobacteria, and their skeletal plates were often perforated by endolithic algae, lichens or sponges. They were thus probably several years old. Empty barnacles shells were present at all levels but were especially common on the lower shore where *Nucella lapillus* (L.) the principal predator, was most abundant.

Cryptofauna were scarce among barnacles less than 7 mm high but were abundant and diverse among taller specimens where small animals lived in spaces between individual barnacles, in grooves between the skeletal plates and in empty shells. Dominant taxa were oligochaetes, mites, larvae of Chironomidae, *Lasaea rubra, Fabricia sabella,* juvenile tanaids and *Idotea,* and small amphipods.

(d) *Laurencia turf. Laurencia* occurred between the middle of the mussel zone and the upper edge of *Corallina* turf. Erect plants grew on vertical and sloping surfaces but a few patches of more prostrate plants, up to 20 cm across, occurred on more horizontal rock providing a moist, protective layer in which some sediment accumulated. The mats contained a rich fauna consisting of elements of both the mussel and *Corallina* communities.

(e) *Corallina turf.* A dense turf of *Corallina officinalis* covered horizontal or gently sloping rock from HLWS into the sublittoral and also occurred in shallow pools and drainage channels on the lower shore. Most parts of the turf are never uncovered for more than one to three hours. Plants reach a maximum height of 20 mm, their branches forming a close mesh which traps mineral and organic particles and provides a shelter for small animals which prevents their being washed out to sea. The sediment layer reached 10 mm in depth and consisted mainly of fine sand which was light or dark grey indicating oxygen depletion. It harboured a diverse fauna of small invertebrates, especially *Lasaea rubra,* gastropods, sabellids, amphipods, tanaids, harpacticoids, and enchytraeids.

Pools

Most pools were fringed with *Corallina* or *Ulva* and/or *Enteromorpha* and the deeper ones contained varying amounts of *Fucus serratus* (L.), *Halidrys siliquosa* (L.) Lyngb. and *L. digitata.* Small amounts of sediment were sometimes present at the bottom of the deeper pools and some sediment was trapped by compact growths of *Corallina* or *Ulva/Enteromorpha.* On the upper shore and in the splash zone, fragments of decaying algae were sometimes present, creating anaerobic conditions, recognised by a black discoloration, when sufficiently abundant. Upper shore pools are subject to wide fluctuations in temperature and salinity and those in the splash zone may dry out in summer.

Materials and methods

Sampling

The main investigation was carried out in August and late September–October 1994 on a section of the shore, approximately 50 m in width, about 150 m east of the southernmost point of the coast, called here The Point. One series of samples was taken at The Point. Qualitative samples for species presence were collected from rock crevices and surface cracks, *Lichina,* mussels, barnacles, *Laurencia* patches, *Corallina* turf below HLWS, and pools. Surface mats were sampled quantitatively in late September-October.

Rock crevices were sampled at three levels: the *Verrucaria* zone, the *Lichina* zone, and the lower barnacle zone between the mussel bed and the *Corallina* turf. Crevices were forced open with a cold chisel and hammer and sediment scraped or brushed from both surfaces into water which was then screened on a 500 μm sieve. In some crevices, the inner and outer zones were analysed separately.

Surface cracks were sampled in the same three zones by prising off small portions of rock up to 5 cm across which were kept for laboratory examination. Oligochaetes on remaining surfaces were removed with forceps.

Surface mats. For qualitative studies, portions of mats were scraped from rock, together with any underlying barnacles, and washed through a 500 μm sieve to remove fine sediment. For quantitative estimates, ten replicate samples of 6.25 cm^{-2} (*Lichina,* barnacles > 8 mm, *Laurencia,* and *Corallina*) or 25 cm^{-2} (mussels) were taken from the centre of patches (avoiding edges) and treated in the same way using a 250 μm sieve.

Pools were sampled qualitatively in three zones as for rock. The dominant algae (*F. serratus, H. siliquosa, Enteromorpha, Ulva* and *Corallina*) were washed vigorously in a bucket and the residue screened on a 500 μm sieve. Bottom sediments were sampled in the same way, each sample being washed several times.

Laboratory procedures

All samples were sorted at ×8–10 magnification, in seawater, while the worms were still alive. Most mature

56

individuals could be identified. Permanent mounts were made of all mature specimens of *Marionina* and *Grania* so that their identification could be verified by comparing them with reference material collected from the same location for taxonomic studies (Healy, 1996). Immature specimens could be distinguished from their congeners by differences in setal shape (*Grania*), the number of pharyngeal glands and shape of the coelomocytes (*Lumbricillus*) or glands on the body wall (*Inermidrilus*). Immature *Marionina* species could not be distinguished with confidence.

Results

Species distribution

Oligochaetes were present at all tide levels and in all the habitats sampled. Twelve species are recorded, only five of which were previously known. A species list, with known ecological distributions, is given in Table 1. It includes three Tubificidae and nine Enchytraeidae. Two enchytraeids were species new to science and are described elsewhere (Healy, 1996), one tubificid and four enchytraeids remain undescribed.

Species distribution was correlated with both tide level and habitat with no more than nine species occurring in any habitat (Fig. 1, Table 2). The greatest number of species (nine) occurred in lower shore crevices and the fewest (two) in *Lichina*. *Lumbricillus semifuscus* was recorded from all tide levels but was rare below LWN. Other species were more restricted in their tidal range: *Lumbricillus* sp. was only taken from the upper shore and splash zone, *Marionina macgrathi* from the upper and middle shore, *Inermidrilus georgei*, *Inermidrilus* sp., *Marionina ulstrupae*, *Marionina* sp. 1 and *Marionina appendiculata* from the middle and lower shore, *Marionina* sp. 2 and *Tubificoides benedii* from the lower middle shore, and *Grania pusilla* and *Grania* sp., only from below HLWS.

Only three species, *Lumbricillus* sp., *M. macgrathi* and juvenile *L. semifuscus* were present in cracks where they were practically the only animals seen. *Lumbricillus* sp. was almost entirely confined to this habitat, only rarely occurring in crevices. Ten species were taken from crevices, three of which were absent from mats. *Lumbricillus semifuscus* was the dominant oligochaete in all crevices. Cocoons of *L. semifuscus* were frequent in crevices; confirmation of their identity was obtained from laboratory cultures. In the *Lichina* zone, where crevices were all well drained, it was joined by *M.*

macgrathi and *Inermidrilus* sp. while below the mussels, where crevices presented a wider range of conditions and were generally wetter, the faunal composition was more variable. *Tubificoides benedii* occurred only in crevices where the sediment was wet and partially blackened; *Marionina* sp. 2 was found only twice.

Nine species were recorded from mats with the number of species increasing downshore from two in *Lichina* to eight in *Corallina*. Three of the *Corallina*-associated species, *L. semifuscus*, *M. appendiculata* and *Marionina* sp. 1, were only found in August and were not present in the later, quantitative samples. The only species present in mats but absent from crevices were *G. pusilla* and *G.* sp. which were confined to *Corallina* turf.

Observations made during earlier investigations of mat and crevice fauna point to a seasonal variation in species distributions with a downshore shift in summer and an upshore shift in winter. More species were present in *Corallina* in summer than in winter, while *L. semifuscus* was rare in the *Verrucaria* zone in summer but extended above it into the yellow lichen zone in winter.

Oligochaetes were generally rare in pools although *G. pusilla* was frequent in some shallow pools below the mussel zone where it occurred in *Corallina* with deep, sandy sediment. *Lumbricillus semifuscus* and *M. macgrathi* were present in low numbers among algae and in bottom sediments of pools on the middle and lower shore. No oligochaetes were taken from algae or sediments in upper shore pools or in *Corallina* turf which fringed the pools on the middle shore.

Species density in mats

Oligochaete density was lowest in *Lichina* (216 100 cm^{-2}) and highest in tall barnacles (562 · 100 cm^{-2}) and *Corallina* (699 · 100 cm^{-2}) (Table 2). The more frequent species reached their maximum densities in different habitats: *I. georgei* in mussels, *Inermidrilus* sp. in *Laurencia*, *M. macgrathi* and *L. semifuscus* in tall barnacles, and *M. ulstrupae* in *Corallina*. *Grania* spp. were confined to *Corallina*. *Lumbricillus semifuscus*, which had the widest tidal range, was dominant in *Lichina* (79%) but formed a progressively smaller proportion of the oligochaetes downshore to disappear in *Corallina*. The population structure of *L. semifuscus* varied between habitats with the highest proportion of mature individuals (42%) in *Lichina*, 24% and 20% respectively in barnacles and *Laurencia*, very few in mussels and none in cracks.

Table 1. Species present on the exposed shore at Carnsore and other habitats in which they are known to occur.

Tubificidae	
Inermidrilus georgei (Erséus, 1987)	Intertidal (?). *Corallina* in rock pool
Inermidrilus sp. indet.	
Tubificoides benedii (d'Udekem, 1885)	Intertidal, organic sediments
Enchytraeidae	
Lumbricillus semifuscus (Claparède, 1861)	Intertidal; rocky, fucoid shores
Lumbricillus sp. indet.	
Grania pusilla Erséus, 1974	
Grania sp.	
Marionina appendiculata Nielsen &	
Christensen, 1959	Intertidal, mainly muddy habitats
Marionina macgrathi Healy, 1996	
Marionina ulstrupae Healy, 1996	
Marionina sp. 1	
Marionina sp. 2	

Table 2. Densities of oligochaete species in mats in September-October 1994 (ind. 100 $cm^{-2} \pm$ S.D., n=10). Barnacles (>8 mm only) sampled at a different site.

	Lichina	Mussels	Barnacles	*Laurencia*	*Corallina*
Lumbricillus semifuscus	171±224	169±152	203±375	66±147	0
Marionina macgrathi	26±49	0	93±90	30±35	0
Inermidrilus georgei	0	109±126	50±108	40±104	11±21
Inermidrilus sp.	0	0.4±1.2	14±35	3±4	5±11
Marionina ulstrupae	0	0	0	14±14	194±184
Marionina appendiculata	0	0	0	3±4	0
Grania pusilla	0	0	0	0	200±153
Grania sp.	0	0	0	0	19±29
Marionina sp. 1	0	0	0	0	2±7
Marionina immature	19±29	8.2±5.4	202±147	198±213	246±214
Total ind. 100 cm^{-2}	216±248	287±240	562±531	345±521	699±485
% cover of habitat	10	35	5	2	60
ind. m^{-2} in zone	2,160	10,045	2,631*	708	41,940

* all sizes of barnacles

Oligochaete density for each of the main zones (*Lichina*, mussels, lower barnacles and *Corallina*) was calculated using an estimate of cover of the habitat investigated in the zone, assuming insignificant numbers in other habitats, chiefly barnacle-covered rocks in which the oligochaete density was < 5100 cm^{-2} (Table 2). In the case of tall barnacles, only a very rough estimate can be given because their size frequency could not be determined with any confidence without destructive sampling. Barnacle studies were in any case carried out at a different site from that of other mats where tall barnacles were absent. The overall density was approximately 2,160 m^{-2} in the *Lichina* zone, 10,045 m^{-2} in the mussel zone, 41,900 m^{-2} in the *Corallina* zone, and 2,600 m^{-2} in the zone between the mussels and *Corallina* turf (comprising short and tall barnacles and *Laurencia*).

Both of the species occurring in *Lichina* were highly aggregated (SD> mean), probably as a result of progressive desiccation of the turf during emersion. On a dry, sunny day in August the worms were seen to have formed tight balls at the base of a few moist fronds. *Inermidrilus* spp. were also highly aggregated in all the mats in which they occurred.

58

In mussels, oligochaetes were less numerous in samples containing six or more specimens of *Cirratulus cirratus* which secrete a sticky mucus. Tube-forming tanaids and sabellid polychaetes, on the other hand, tended to increase the amount of sediment accumulated, especially among barnacles and *Laurencia*, and seemed to favour oligochaetes. *Marionina ulstrupae* and *I. georgei* were frequently seen among or inside empty, sand-encrusted tubes.

Discussion

Mats and crevices are important habitats on wave exposed shores where they increase faunal diversity by providing a range of safe environments which protect the infauna from wave shock and displacement by water currents, retain organic matter which provides food, buffer fluctuations of environmental conditions, maintain a moist microclimate, and provide a refuge from possible predators, especially fish, which invade the shore during tidal immersion. These habitats may occupy only a small proportion of the available rock but the cryptofauna is species- rich and densities can be high (Wieser, 1952; Gorvett, 1958; Reimer, 1976; Suchanek, 1985, 1992; Gibbons & Griffiths, 1988). The communities have received much less attention than the macrofauna occupying primary space, however, and information on faunal composition and species interactions is scarce for west European shores.

The different habitat and tide level preferences exhibited by the oligochaete species recorded in this investigation (Fig. 1, Table 2) imply a range of specialisations for distinctly different sets of conditions. Two factors, tide level and the physical structure of the substrate, are of primary importance because they influence other conditions relevant for survival of the fauna. The importance of tide level is well documented and many authors have noted critical levels which have more or less universal application (Colman, 1933; Morton, 1940; Evans, 1947; Glynne-Williams & Hobart, 1952). For oligochaetes, the most relevant level is probably around MHWN, which marks the highest point at which regular tidal flooding can occur and which corresponds to a transition zone where terrestrial and aquatic faunas overlap (Glynne-Williams & Hobart, 1952; Morton, 1954). On exposed shores, however, this limit is raised and partly obscured by wave surge and splash. The physical structure of the habitat may be important in a variety of ways. For example, the width and orientation of a crevice deter-

mine the extent of water penetration, grain size of deposits and size of the inhabitants; the size, degree of branching and closeness of algal fronds and byssus threads determine the amount and nature of sediment which can be held, moisture retention, the force of water currents and the ability of an animal to cling or entwine and thus avoid being swept away; the height of barnacles determines the length of grooves which can be occupied and the microclimate and amount of organic matter within them.

Cracks and crevices provide an effective buffer against environmental extremes allowing inhabitants to penetrate higher up the shore than on exposed rock (Glynne-Williams & Hobart, 1952). They are relatively short-lived features which develop as a result of weathering processes, gradually enlarging until a portion of rock finally breaks away. During evolution of a crevice, there is an ecological succession of the fauna in response to changes in environmental conditions such as the particle size of retained sediment, degree of water penetration, and the capacity to retain air during immersion (Glynne-Williams & Hobart, 1952; Morton, 1954). The successional changes are reflected in the gradient of faunal composition and environmental conditions within crevices, with the inner regions containing fine sediments corresponding to early stages and the wetter, outer regions with coarser sediments representing later stages. The persistence of air pockets during immersion, reported by several workers, allows colonisation by air-breathing invertebrates, even at low tide levels, thus there can be an overlap of terrestrial and aquatic species. Of the species recorded in this study, only *Lumbricillus sp.* (and possibly *Marionina* sp. 2 which was only found in two crevices) can be described as an obligatory crevice species (Kensler, 1964a; Kensler & Crisp, 1965). Together with *M. macgrathi* and juvenile *L. semifuscus*, it is among the first colonisers of developing crevices, appearing in narrow cracks containing little other life except encrusting green algae and cyanobacteria. *Lumbricillus semifuscus,* a common inhabitant of crevices at all levels, is probably the next oligochaete to colonise. Other crevice species are essentially 'aquatic' since they do not occur above the *Lichina* zone and are also present in mats where they are subject to regular wetting or, in the case of *T. benedii*, occur in other wet marine habitats, but their preference for the inner zone of crevices suggests that most, if not all, characterise early stages in crevice formation.

Oligochaetes also occur in intertidal rock crevices at several localities near Dublin where they are particularly abundant in slate. They are not mentioned, however, in any of the accounts of crevice fauna of the northeast Atlantic from a variety of rock types, including mica-schist and slate, in Britain (Glynne-Williams & Hobart, 1952; Morton, 1954; Kensler & Crisp, 1965), Norway (Kensler, 1964a), the Iberian Peninsula and North Africa (Kensler, 1965) and the Mediterranean (Kensler, 1964b). Environmental conditions, particularly on the upper shore, would appear to be ideal for enchytraeids and it seems probable that oligochaetes have been overlooked, or not listed, in at least some of these studies.

Mussel beds are rich in organic matter derived from attached algae, detritus trapped by the byssus matrix and the faeces and pseudofaeces produced by the mussels (Suchanek, 1985). The mussels provide a large surface area for attachment of sessile organisms, including algae, as well as retaining coarse sediment and broken shells. A wide diversity of organisms lives within this framework, particularly in the thicker beds or patches of older mussels which contain more sediment, fragments and byssus threads (Tsuchiya & Nishihira, 1986). Species richness and abundance are also related to patch size (Tsuchiya & Retière, 1992). More than 300 taxa were recorded from structurally complex beds of *Mytilus californianus* on exposed shores of the Pacific coast of North America, making the system one of the most diverse temperate communities known to date (Suchanek, 1992). Unlike other mussels, however, *M. edulis* has the ability to clean fouling organisms from its shell (Thiesen, 1972) and encrusting species are therefore less frequent and the physical structure of the beds less complex that those of *M. californianus*. At Carnsore, the amount of sediment retained in mussels was less than in algal turf and the average size of inhabitants was larger. This may explain why oligochaete density was lower than in turfs. The activities of larger carnivores such as errant polychaetes and nemerteans, which were common in the mussel samples, and the abundant mite *Hydrogamasus littoralis*, which was observed attacking oligochaetes, may also limit populations. The few records of oligochaetes from mussel beds suggest that they are not a common component of this habitat; for example they were rare ($<1.100\,\mathrm{cm}^{-2}$) in patches of *M. edulis* on the north coast of France (Tsuchiya & Retière, 1992). Erséus (1976) recorded *Paranais litoralis* (Müll.), *Lumbricillus semifuscus, L. scoticus* and *L. pumilis* from among byssus on the coast of Iceland but the mussels were subtidal. *Phallodrilus albanensis* Erséus, 1990, *Bathydrilus edwardsi* Erséus, 1984, and *Limnodriloides agnes* Hrabe, 1967

were recorded from intertidal mussels in Western Australia (Erséus, 1990).

The mobile fauna associated with barnacles has mainly been studied in empty shells which provide the most obvious refuge, especially for small gastropods (e.g. Raffaelli, 1978) and few authors refer to the meiofauna which live on opercular plates or in grooves between the lateral plates. Gorvett (1958) recorded a faunal density of $7,024 \cdot 100\,\mathrm{cm}^{-2}$ among *S. balanoides* on an exposed Scottish shore of which 4,224 were mites and 220 were oligochaetes. In the present study, a higher oligochaete density of $561 \cdot 100\,\mathrm{cm}^{-2}$ is explained by the selective sampling of tall barnacles with long vertical grooves which are of an appropriate size and shape for small worms. On some shores, there are extensive mats of tall barnacles which can grow to even greater heights and among these the faunal community could be even more abundant and diverse. However, oligochaetes were rare or absent from barnacles on two North Wales shores, even among densely packed, elongated individuals (de Marguia & Seed, 1987), thus factors other than crevice size and shape may be limiting on some shores.

Algal turfs have received more attention than other cryptic habitats of the intertidal. Most authors stress the relationship between the size range, density and composition of the faunal community and the structural complexity of the turf, which is generally summarised and quantified as a surface:volume ratio (Gibbons, 1991) but the concept can be enlarged to include frond size and stiffness, size range of interstices, surface texture and sediment-trapping properties (Myers & Southgate, 1980). Invertebrates are both more abundant and more diverse among compact algae with a complex branching structure. The degree of complexity is seen as particularly important as a measure of the effectiveness of turfs as refuges from fish predation (Coull & Wells, 1983; Dean & Connell, 1987; Gibbons, 1988a, 1988b). The importance of structure, as distinct from tide level, as the prime factor determining species richness and abundance has been tested using both artificial substrates (Myers & Southgate, 1980; Gibbons, 1988b) and an algal species with a wide tidal range (Wieser, 1952). These studies show that structural complexity mainly affects faunal abundance and diversity indirectly through its influence on sediment accumulation. At Carnsore, oligochaetes were most abundant in *Corallina* turf which contained the most sediment. Here, and in *Laurencia*, they may live entirely within the sediment layer but the width, rigidity and texture of fronds may also be important, affecting the ability of individuals to entwine and so avoid displacement by water currents or foraging fish. Tide level appears to be the principal factor determining the species composition of turfs at Carnsore but narrow, stiff fronds, close growth and depth of sediment are probably also responsible for the high density in *Corallina*.

Among the surface mats investigated, *Lichina* turf is the most demanding habitat for oligochaetes because its rather open structure and high level on the shore subject its inhabitants to wide fluctuations in moisture content, temperature and salinity. *Lumbricillus semifuscus* reaches high densities in *Lichina* turf, with a high proportion of mature individuals. Field observations in summer showed a marked tendency to aggregate in warm, dry weather and many individuals may ultimately succumb to desiccation as the *Lichina* tufts dry out. This section of the *L. semifuscus* population may be severely depleted unless individuals find refuge in nearby crevices or wet depressions. In other kinds of mat, water, of more constant salinity, is always present and other factors must limit species distribution and population density. Saturated deposits appear to be necessary for *Grania* spp. on this shore, as elsewhere; members of the genus are mainly subtidal and rarely occur above the lower intertidal zone (Coates, 1990; Coates & Stacey, 1996). *Marionina ulstrupae*, which is restricted to the lower shore, may also require long periods of immersion.

Exposed rocky shores are highly dynamic ecosystems, the perennial communities of sessile organisms being subject to frequent disturbance, especially during storms. The result is a mosaic of patches of different sizes, ages and life-spans. Among the habitats studied during this investigation, many, including crevices, were present as small patches, no more than 20–30 cm across, and generally surrounded by relatively hostile rock and barnacles. Even mussel beds and *Corallina* turf, which appear to be long-lived, more or less continuous habitats, are actually mosaics of seral stages produced by successive, small scale disturbances (Sousa, 1984). Only *Lichina* patches are known to be stable and persistent (Boney, 1961). The mobile inhabitants of these patches need a set of behavioural responses which enable them not only to remain within the range of conditions to which they are adapted, but also to colonise new patches following altered conditions or destruction of the present one. Almost nothing is known about either the behavioural responses or passive dispersal of rocky shore oligochaetes but the clear habitat preferences demonstrated by most

species during this study suggest that comparatively little movement of individuals takes place between habitats. Wieser (1952) found no difference in infaunal composition of the compact alga *Gelidium* at high and low tide but obvious differences in fauna of larger algae with a more open structure. On exposed shores, therefore, where oligochaetes live in complex habitats which provide good protection, they may be unaffected by wave action unless their habitat is damaged. This does not preclude the possibility of active migrations during emersion or calm conditions, especially when conditions become unfavourable, for example, the moisture-seeking behaviour of *L. semifuscus* in *Lichina*. Marine oligochaetes in general are known to tolerate wide variations in environmental conditions (Giere & Pfannkuche, 1982) and could thus easily survive in suboptimal conditions during dispersal. The age segregation of the wide ranging *L. semifuscus*, which reaches its highest density in the high-risk *Lichina* turf, and the presence of low densities of some species in pools, which could act as temporary refuges, suggests that some movement of individuals outside their preferred habitats, either active or passive, does occur.

The scarcity of oligochaete records from rocky shores, emphasised by the presence at Carnsore of two species described from the locality and five undescribed species, may have several explanations. Taxonomists have generally focused on habitats where oligochaetes are known to be numerous and may have been deterred from sampling exposed shores by their inhospitable appearance. It seems probable that in many general studies, oligochaetes have been overlooked or ignored by workers unfamiliar with the group or that they were absent or rare in samples due to the methods employed; for example, only a small proportion might be dislodged from compact algae by rinsing, even after narcotisation which itself can kill fragile oligochaetes. The dimensions of most marine oligochaetes have precluded them from consideration in studies of phytal meiofauna, defined as organisms passing through a 1 mm sieve (Choat & Kingett, 1982; Gibbons, 1988a, 1988b), yet they would be considered too small to be included in macrofaunal studies. Thus, the absence of oligochaetes from faunal lists is not a reliable indication of their absence from the habitats investigated and they are probably more widespread and make a greater contribution to cryptofaunal communities than is currently realised. The available evidence, however, points to a large scale patchiness in oligochaete distributions. Numbers were low in several cryptofaunal studies (Colman, 1940;

Edgar, 1983; de Marguia & Seed, 1987; Tsuchiya & Retière, 1992) but were comparable to those at Carnsore or 'numerous' at others (Delamare Deboutteville & Bougis, 1951; Wieser, 1952; Gorvett, 1958). Wieser reported oligochaetes from the compact alga *Gelidium* in southern England (1952) but not from *Corallina* in the Mediterranean (1959). Some of these differences may be due to variations in turbidity of the sea which carries more suspended matter on exposed shores than on more sheltered ones (Kensler, 1964b) or to temporal variations in populations, either seasonal or noncyclical, related to climatic factors. Oligochaetes are usually rare among intertidal algae in warm climates (e.g., Healy & Coates, 1996) where high temperatures and oxygen deficiency may become intolerable during emersion. Even on temperate shores, climatic factors, particularly air temperature and relative humidity, may prove to be important in limiting oligochaete populations, either in the long or short term.

Acknowledgements

I am grateful to Dr David McGrath for drawing my attention to the presence of oligochaetes in rock cracks and for his contributions to the baseline study which led to this investigation. Thanks are also due to Geoffrey Oliver and Michael Mullen for help with field work.

References

Baker, J. W., 1955. Pre-Cambrian rocks in Co. Wexford. Geol. Mag. 2: 63–68.

Ballantine, W. J., 1961. A biologically-defined exposure scale for the comparative description of rocky shores. Field. Stud. 1: 1–19.

Boney, A. D., 1961. A note on the intertidal lichen *Lichina pygmaea*. J. mar. biol. Ass. U.K. 41: 123–126.

Chapman, G., 1955. Aspects of the fauna and flora of the Azores. VI. The density of animal life in the coralline alga zone. Ann. Mag. nat. Hist. ser. 12, 8: 801–805.

Choat, J. H. & P. D. Kingett, 1982. The influence of fish predation on the abundance cycles of an algal turf invertebrate fauna. Oecologia (Berl.) 54: 88–95.

Coates, K. A., 1990. Marine Enchytraeidae (Oligochaeta, Annelida) of the Albany area, Western Australia. In F.E. Wells, D.I. Walker, H. Kirkman & R. Lethbridge (eds), Proceedings of the Third International Marine Biological Workshop: The Marine Flora and Fauna of Albany, Western Australia. Western Australian Museum: 13–41.

Coates, K. A. & D. F. Stacey, 1996. Enchytraeids (Oligochaeta: Annelida) of the lower shore and shallow subtidal of Darwin Harbour, Northern Territory, Australia. In R. Hanley & G. Caswell (eds), Proceedings of the Sixth International Marine Biological Workshop: The Marine Flora and Fauna of Darwin Harbour, Northern Territory, Australia. (In press.)

Colman, J., 1933. The nature of intertidal zonation of plants and animals. J. mar. biol. Ass. U.K. 18: 435–476.

Colman, J., 1940. On the faunas inhabiting intertidal seaweeds. J. mar. biol. Ass. U.K. 24: 129–183.

Coull, B. C. & J. B. J. Wells, 1983. Refuges from fish predation: experiments with phytal meiofauna from the rocky intertidal. Ecology 64: 1599–1609.

Dalby, D. H., E. B. Cowell, W. J. Syratt & J. H. Crothers, 1978. An exposure scale for marine shores in western Norway. J. mar. biol. Ass. U.K. 58: 975–996.

Dean, R. L. & J. H. Connell, 1987. Marine invertebrates in an algal succession. III. Mechanisms linking habitat complexity with diversity. J. exp. mar. Biol. Ecol. 109: 249–273.

Delamare-Deboutteville, C. & P. Bougis, 1951. Recherches sur le trottoir d'algues calcaires effectuées à Banyuls pendant le stage d'été 1950. Vie Milieu 2: 161–181.

de Marguia, A. M. & R. Seed, 1987. Some observations on the occurrence and vertical distribution of mites (Arachnida: Acari) and other epifaunal associates of intertidal barnacles on two contrasted rocky shores in North Wales. Cah. Biol. mar. 28: 381–388.

Dommasnes, A., 1969. On the fauna of *Corallina officinalis* L. in western Norway. Sarsia 38: 71–86.

Edgar, G.J., 1983. The ecology of south-east Tasmanian phytal communities. I. Spatial organisation on a local scale. J. exp. mar. Biol. Ecol. 70: 129–157.

Elmhirst, R. & J. Stephenson, 1926. On *Lumbricillus scoticus* n. sp. J. mar. biol. Ass. U.K. 14: 469–473.

Erséus, C., 1976. Littoral Oligochaeta (Annelida) from Eyjafjördur, North Coast of Iceland. Zool. Scr. 5: 5–11.

Erséus, C., 1987. Seven new marine species of *Phallodrilus* (Oligochaeta: Tubificidae) from various parts of Europe, and a re-examination of the type species *P. parthenopaeus* Pierantoni. J. nat. Hist. 21: 915–931.

Erséus, C., 1990. The marine Tubificidae and Naididae (Oligochaeta) of south-western Australia. In F.E. Wells, D.I. Walker, H. Kirkman & R. Lethbridge (eds), Proceedings of the Third International Marine Biological Workshop: The Marine Flora and Fauna of Albany, Western Australia. Western Australian Museum: 43–88.

Erséus, C., 1993. The marine Tubificidae (Oligochaeta) of Rottnest Island, Western Australia. In F.E. Wells, D.I. Walker, H. Kirkman & R. Lethbridge (eds), Proceedings of the Fifth International Marine Biological Workshop: The Marine Fauna and Flora of Rottnest Island, Western Australia. Western Australian Museum: 331–390.

Evans, R. G., 1947. The intertidal ecology of Cardigan Bay. J. Ecol. 34: 273–309.

Gibbons, M. J., 1988a. Impact of predation by juvenile *Clinus superciliosus* on phytal meiofauna: are fish important as predators? Mar. Ecol. Prog. Ser. 45: 13–22.

Gibbons, M. J., 1988b. The impact of sediment accumulations, relative habitat complexity and elevation on rocky shore meiofauna. J. exp. mar. Biol. Ecol. 122: 225–241.

Gibbons, M. J., 1991. Rocky shore meiofauna: a brief overview. Trans. r. Soc. S. Afr. 47: 595–603.

Gibbons, M. J. & C. L. Griffiths, 1988. An improved quantitative method for estimating intertidal meiofaunal standing stock on an exposed rocky shore. S. Afr. J. mar. Sci. 6: 55–58.

Giere, O. & O. Pfannkuche, 1982. Biology and ecology of marine Oligochaeta, a review. Oceanogr. mar. Biol. ann. Rev. 20: 173–308.

Glynne-Williams & J. Hobart, 1952. Studies on the crevice fauna of a selected shore in Anglesey. Proc. zool. Soc. Lond. 122: 797–825.

Gorvett, H., 1958. Animal life on wave-beaten rocks. Nature 182: 1652–1653.

62

Gosling, E. M. & N. P. Wilkins, 1981. Ecological genetics of the mussels *Mytilus edulis* and *M. galloprovincialis* on Irish coasts. Mar. Ecol. Prog. Ser. 4: 221–227.

Healy, B., 1979. Marine fauna of County Wexford. −1. Littoral and brackishwater Oligochaeta. Ir. Nat. J. 19: 418–422.

Healy, B., 1996. New species of *Marionina* (Oligochaeta: Enchytraeidae) from a wave-exposed shore in SE Ireland. (In press.)

Healy, B. & K. A. Coates, 1996. Enchytraeids (Oligochaeta: Annelida) of the mid and upper intertidal of Darwin Harbour, Northern Territory, Australia. In R. Hanley & G. Caswell (eds), Proceedings of the Sixth International Marine Biological Workshop: The Marine Flora and Fauna of Darwin Harbour, Northern Territory, Australia. (In press.)

Hicks, G. R. F., 1971. Checklist and ecological notes on the fauna associated with some littoral corallinacae algae. Bull. nat. Sci. Victoria Univ. biol. Soc. 2: 47–58.

Kensler, C. B., 1964a. The crevice habitat in western Norway. Sarsia 17: 21–33.

Kensler, C. B., 1964b. The Mediterranean crevice habitat. Vie Milieu 15: 947–977.

Kensler, C. B., 1965. Distribution of crevice species along the Iberian Peninsula and northwest Africa. Vie Milieu 16: 851–887.

Kensler, C. B. & D. J. Crisp, 1965. The colonisation of artificial crevices by marine invertebrates. J. Anim. Ecol. 34: 507–516.

Lewis, J. R., 1961. The littoral zone on rocky shores. Oikos 12: 280–301.

McGrath, D., 1975. Notes on some Irish marine littoral and freshwater Oligochaeta (Annelida). Ir. Nat. J. 18:216–218.

Morton, J. E., 1954. The crevice fauna of the upper intertidal zone at Wembury. J. mar. biol. Ass. U.K. 33: 187–224.

Myers, A. A. & T. Southgate, 1980. Artificial substrates as a means of monitoring rocky shore cryptofauna. J. mar. biol. Ass. U.K. 60: 963–975.

Raffaelli, D. G., 1978. Factors affecting the population structure of *Littorina neglecta* Bean. J. moll. Stud. 44: 223–230.

Reimer, A. A., 1976. Description of a *Tetraclita stalactifera panamensis* community on a rocky intertidal Pacific shore of Panama. Mar. Biol. 35: 225–238.

Reish, D. J., 1964. Studies on the *Mytilus edulis* community in Alamitos Bay, California: II. Population variations and discussion of the associated organisms. Veliger 6: 202–207.

Sarma, A. L. N. & P. N. Ganapati, 1972. Faunal associations of algae in the intertidal region of Visakhapatnam. Proc. ind. Acad. Sci. 38: 380–396.

Sousa, W. P., 1984. Intertidal mosaics: patch size, propagule availability and spatially variable pattern of succession. Ecology 65: 1918–1935.

Shurova, N. M., 1977. New species of littoral oligochaetes of the genus *Lumbricillus* (Oligochaeta). (In Russian.) Biologiya Morya 1: 57–62.

Sloane, J. F., R. Bassindale, E. Davenport, F. J. Ebling & J. A. Kitching, 1961. The ecology of Lough Ine. IX. The flora and fauna associated with undergrowth-forming algae in the rapids area. J. Ecol. 49: 353–368.

Stephenson, J., 1932. Oligochaeta from Australia, North Carolina, and other parts of the world. Proc. zool. Soc. Lond. (1932): 899–941.

Suchanek, T. H., 1985. Mussels and their role in structuring rocky shore communities. In P. G. Moore & R. Seed (eds), The Ecology of Rocky Coasts. Hodder & Stoughton: 70–96.

Suchanek, T. H., 1992. Extreme biodiversity in the marine environment: mussel bed communities of *Mytilus californianus*. Northwest env. J. 8: 150–152.

Theisen, B. F., 1972. Shell cleaning and deposit feeding in *Mytilus edulis* L. (Bivalvia). Ophelia 10: 49–55.

Tsuchiya, M. & M. Nishihira, 1986. Islands of *Mytilus edulis* as a habitat for small intertidal animals: effect of *Mytilus* age structure on the composition of the associated fauna and community organization. Mar. Biol. Prog. Ser. 31: 171–178.

Tsuchiya, M. & C. Retière, 1992. Zonation of intertidal organisms and community structure of small animals associated with patches of the mussel *Mytilus edulis* L. along the rocky coast of Dinard, Brittany, France. Bull. Coll. Sci. Univ. Ryukyus 54: 47–81.

Wieser, W., 1952. Investigations on the microfauna inhabiting seaweeds on rocky coasts. IV. Studies on the vertical distribution of the fauna inhabiting seaweeds below the Plymouth Laboratory. J. mar. biol. Ass. U.K. 31: 145–174.

Wieser, W., 1959. Zur Ökologie der Fauna mariner Algen mit besonderer Berücksichtigung des Mittelmeeres. Int. Rev. ges. Hydrobiol. 44: 137–179.

Hydrobiologia **334**: 63–72, 1996.
K. A. Coates, Trefor B. Reynoldson & Thomas B. Reynoldson (eds), Aquatic Oligochaete Biology VI.
© 1996 *Kluwer Academic Publishers.*

63

Oligochaeta and Aphanoneura in ancient lakes: a review

Patrick Martin
Institut royal des Sciences naturelles de Belgique (I.R.Sc.N.B.), Section Biologie des Eaux Douces
(Royal Belgian Institute of Natural Sciences, Freshwater Biology), 29 rue Vautier, B-1040 Brussels, Belgium

Key words: aquatic Oligochaeta, ancient lakes, zoogeography, diversity, speciation, endemicity

Abstract

By their antiquity, history, rarity, great depth in many instances and the presence of highly diverse faunas with many endemics, ancient lakes constitute ecosystems of a special nature, clearly apart from the large majority of extant lakes. While the fauna of these lakes is becoming better and better known for various animals groups, the Oligochaeta are still poorly known. Tubificidae and Naididae are found in each ancient lake. On the other hand, some families are restricted to only one lake, such as Aeolosomatidae and Proppapidae in Lake Baikal or Eudrilidae and Ocnerodrilidae (megadriles) in Lake Tanganyika, but such a distribution is probably due to a lack of knowledge or sampling biases. All ancient lakes have an endemic oligochaete fauna except Lake Kinneret (Israël). The oldest, Lake Baikal (20–25 Ma), holds the most abundant and diverse oligochaete fauna, in which species flocks are even recognizable or suspected. In contrast, the oligochaete fauna of the slightly younger Lake Tanganyika is very scarce. This is partly due to an obvious lack of studies, as the oligochaete fauna of other great African lakes is virtually unknown, but this might be the result of an environment in these lakes less favourable to oligochaetes. Some factors likely to interact with speciation in oligochaetes are discussed but nothing can be concluded to date. A recent interest in African great lakes revealed a more diverse oligochaete fauna than previously assumed but a better study of this fauna is still badly needed.

Introduction

There are about 10 000 lakes larger than 1 km^2 on earth; only ten or so can be considered as ancient (Gorthner, 1994). Most lakes date back to the last glaciations, as a consequence of the retraction of continental ice sheets in northern Europe and America. They are not older than 20 000 years and most will disappear during the next 100 000 years due to infilling by sediments.

In contrast, ancient lakes are outstandingly older, by 2 to 3 orders of magnitude (1 to 25 Ma; Martens *et al.*, 1994). Most of them have a tectonic origin and owe their longevity to peculiar processes which counterbalance their infilling. These are either subsidence along border faults as in graben-type lakes such as Baikal, Tanganyika, Malawi/Nyasa, Ohrid (Hutchinson, 1957; Martin, 1994; Coulter, 1994a; Ribbink, 1994a; Salemaa,1994) or tectonic movements associated with rift valleys such as Lake Victoria and Lake

Kinneret (Greenwood, 1994; Gophen & Nishri, 1994) or with orogenesis (Lake Titicaca; Dejoux, 1994). As ancient lakes, Lake Biwa and Kinneret are borderline cases.The former originated from inland basins but was completely filled up with sediments twice during its history (Nakajima & Nakai, 1994). Similarly, the precursor of Lake Kinneret, Lake Lisan, dried up some 18 000 to 11 000 years ago (Gophen & Nishri, 1994). However, in both cases, some aquatic environment persisted since the formation of their basin, 4 and 20 million years ago, respectively.

In addition to their antiquity and rarity, ancient lakes share common characteristics making them ecosystems of a special nature, clearly apart from the large majority of extant lakes. Some ancient lakes are the deepest lakes in the world, which is scarcely surprising given the tectonic origin of most of them. Lakes Baikal, Tanganyika and Malawi constitute the most famous examples, being the three deepest of the planet's lakes (1637, 1470 and 785 m, respectively; Martin 1994; Coulter, 1994a; Ribbink,1994a).

Due to their persistence through millions of years, all ancient lakes have suffered wide vicissitudes, such as periods of desiccation associated with dramatic fluctuations of surface level, orogenesis or glaciation events, depending on the lake (Coulter, 1994b; Martens et al., 1994).

Lastly, nearly all ancient lakes harbour a highly diverse fauna, with many endemics and even species flocks. The gammarid amphipods are a well-known example in Lake Baikal (259 species, 98% of which are endemic; Kamaltynov, 1993), while African great lakes are famous for their cichlid fishes (of a total of 172 species in Lake Tanganyika, 167 species are endemic; Coulter, 1991).

From these qualities, it is evident that ancient lakes are of the greatest interest for a biologist. While the fauna of these lakes is becoming better and better known for various animal groups (see Martens et al., 1994), the Oligochaeta are still poorly known, aside from a few exceptions (e.g., Lake Baikal; Snimschikova & Akinshina, 1994). The present paper is an attempt to fill this deficiency.

Oligochaete fauna

Among all extant ancient lakes, only eight have been well studied, at least for one animal group. These are the lakes Baikal, Ohrid, Kinneret, Biwa, Titicaca, Tanganyika, Malawi/Nyasa and Victoria (Fig. 1).

As far as the oligochaetes are concerned, a recent interest in African Great Lakes has led to the discovery of new species in lakes Tanganyika and Malawi/Nyasa (Martin & Brinkhurst, 1994; Martin & Giani, 1994) but Lake Victoria still remains completely unknown. A tentative list of species of oligochaetes in ancient lakes can, however, be drawn up (Table 1*), synthesized at the family and genus levels (Table 2) for ease of comparison. This suggests the following comments on zoogeography, diversity, endemism and species flocks.

---

* One significant publication appeared since this manuscript was prepared, which adds a series of new species to Lake Biwa, gives a description of *Rhyacodrilus hiemalis*, and brings the total number of species and genera to 34 (one endemic) and 23, respectively: Ohtaka, A. & M. Nishino, 1995. Studies on the Aquatic Oligochaete Fauna in Lake Biwa, Central Japan. I. Checklist with Taxonomic Remarks. Jap. J. Limnol. 56: 167–182.

Zoogeography

At the family level, the distribution of oligochaetes is in good accord with what is known of these families (Brinkhurst, 1971, 1982).

Tubificidae and Naididae are cosmopolitan and are found in all lakes. Lumbriculidae is largely restricted to the northern hemisphere (except two species introduced to the southern hemisphere) and is present only in Lakes Baikal and Ohrid, as expected. In the former lake, it even shows an extensive radiation (see further).

Some families are restricted to only one lake, such as Aeolosomatidae (Aphanoneura) and Proppapidae in Lake Baikal, Biwadrilidae in Lake Biwa and Alluroididae, Eudrilidae and Ocnerodrilidae in Lake Tanganyika. Among the megadriles, Biwadrilidae is an endemic family to Japan. But the absence of Alluroididae, Eudrilidae, Ocnerodrilidae in several African great lakes and Proppapidae from Lake Ohrid is likely due to a lack of knowledge. Similarly, Aeolosomatidae are probably present in other ancient lakes, according to their cosmopolitan distribution (Van der Land, 1971), but these tiny worms are easily overlooked in ordinary benthic samples, or are lost through nets and screens.

The Phreodrilidae is an interesting family because of its probable Gondwanian origin (Brinkhurst & Jamieson, 1971; Martin & Brinkhurst, 1994). Until recently, records of the species were mainly restricted to the southern hemisphere with one notable exception in Sri Lanka. The recent discovery of new species, one in Morocco (Martin & Giani, 1996), another in the surroundings of Lake Titicaca (Juget & Lafont, 1994) gave more support to a pan-Gondwanian origin of the family. In African great lakes, the Phreodrilidae were represented until recently by only one species (*Insulodrilus tanganyikae* Brinkhurst, 1970, in Lake Tanganyika). New species have been discovered now in Lake Tanganyika as well as in Lake Malawi (Martin & Brinkhurst, 1994; Martin & Giani, 1995b).

Diversity and endemism

A spectacular diversity (defined here as the number of species) is found in Lake Baikal (152 species but 179 if the 27 named but undescribed potential species listed by Snimschikova & Akinshina, 1994, are included) while all other ancient lakes have less than 25 known species (Table 1). Similarly, Lake Baikal is the richest in terms of endemic species, while Lake Biwa and Lake Titicaca have only one endemic and Lake Kinneret none. In a recent publication, Lake Biwa has been said

65

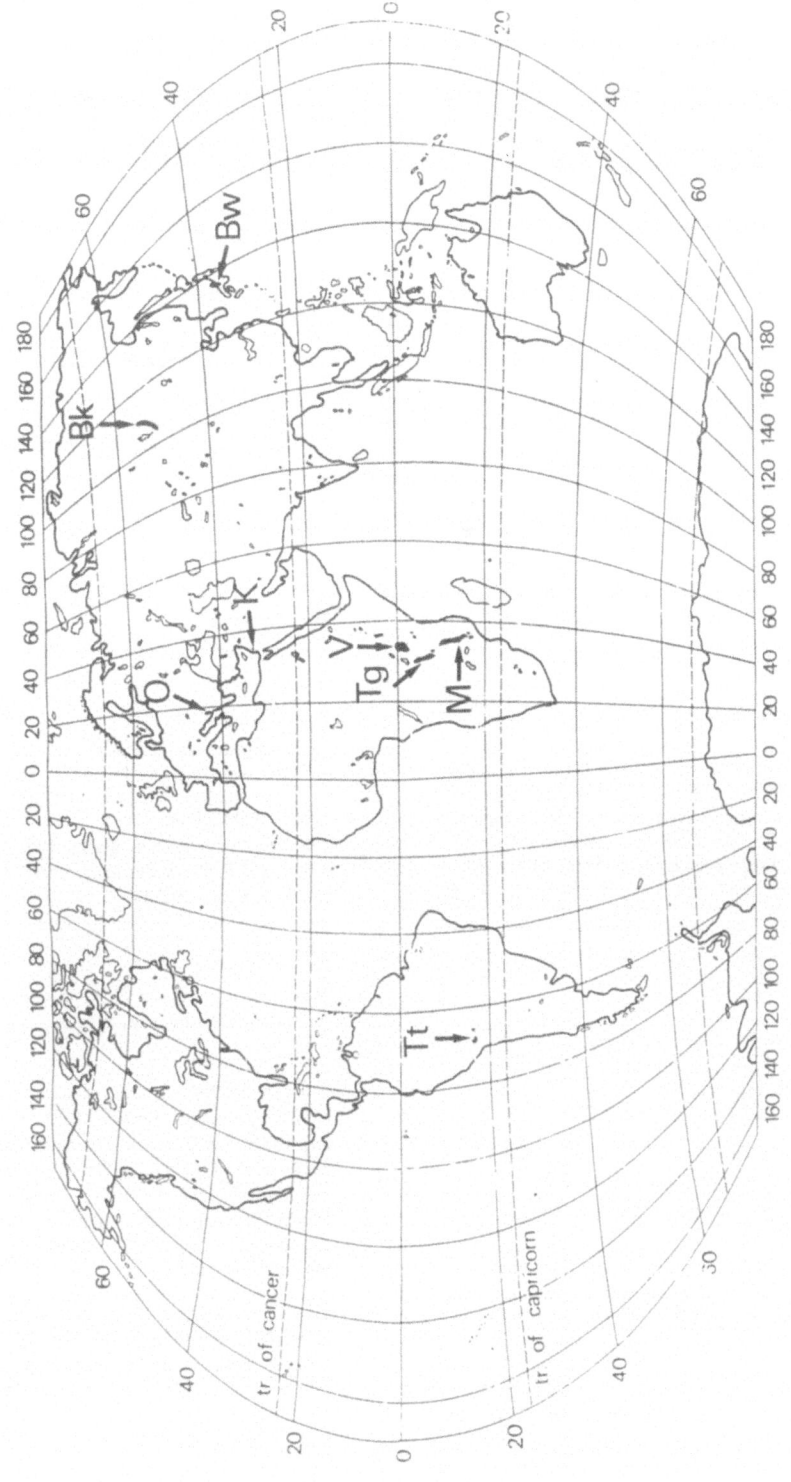

Fig. 1. Map showing the geographical position of extant ancient lakes considered in this review. Bk = Baikal; Bw = Biwa; K = Kinneret; O = Ohrid; M = Malawi (Nyasa); Tg = Tanganyika; Tt = Titicaca; V = Victoria.

Table 1. Continued..

Table 1. List of species of Oligochaeta in ancient lakes. The sign * symbolises endemic species.

BAIKAL: a list of species has been recently given by Snimschikova & Akinshina, 1994. It must be used with caution, however, due to numerous species, subspecies, or even genera (e.g. *Hrabeus*) still *nom. nud.* to date. 179 species, sp. *nom. nud.* incl.; 152 sp., 111 endemics, without sp. *nom. nud.*.

BIWA (Mori & Miura, 1990; Ohtaka, 1993, 1994).

Fam. Biwadrilidae
Biwadrilus bathybates (Steph., 1917).

Fam. Lumbriculidae
Lumbriculus sp.

Fam. Naididae
Amphichaeta sp. *Arcteonais lomondi* (Mart., 1907). *Chaetogaster limnaei* Baer, 1827; *diastrophus* (Gruith., 1928); *diaphanus* (Gruith., 1821). *Dero* spp. *Nais barbata* (Müll., 1774); *bretscheri* Mich., 1899; *communis* Piguet, 1907; *pardalis* Piguet, 1909. *Paranais* sp. *Pristina aequiseta* Bourne, 1891. *Slavina appendiculata* Vejd., 1883. *Stylaria fossularis* Leidy, 1852.

Fam. Tubificidae
Aulodrilus pigueti Kow., 1914. *Bothrioneurum vejdovskyanum* (Stolč, 1886). *Branchiura sowerbyi* Bedd., 1892. *Embolocephalus yamaguchii** (Brink. *et* Jamie., 1971). *Limnodrilus claparedianus* Ratzel, 1869; *grandisetosus* Nomura, 1932; *hoffmeisteri* Clap., 1862; *profundicola* (Verr., 1871). *Rhyacodrilus* sp. 1 (*hiemalis* Ohtaka, *nom. nud.*); sp. 2. *Tubifex* sp.

KINNERET (adapted from Serruya, 1978).

Fam. Criodrilidae
Criodrilus lacuum Hoffm., 1845

Fam. Haplotaxidae
Haplotaxis gordioides (Hartm., 1821).

Fam. Naididae
Dero digitata (Müll., 1774). *Stephensoniana trivandrana* (Aiyer, 1926).

Fam. Tubificidae
Limnodrilus hoffmeisteri Clap., 1862. *Potamothrix bavaricus* (Osch., 1913); *hammoniensis* (Mich., 1901); *heuscheri* (Bret., 1900). *Psammoryctides albicola* (Mich., 1901).

MALAWI/NYASA (Beddard, 1908; Martin & Giani, 1995a; (1) Martin, *unpubl. data*).

Fam. Naididae
Dero sp. *Nais* sp. *Pristina longiseta* Ehrbg., 1828. *Homochaeta* sp.[1]. *Stephensoniana* sp.[1]

Fam. Phreodrilidae
Insulodrilus sp.[1]

Fam. Tubificidae
Aulodrilus sp.[1]. *Branchiura sowerbyi*[1] Bedd., 1892. *Epirodrilus mammosus* Martin *et* Giani 1995a.

OHRID (adapted from Stanković, 1960).

Fam. Criodrilidae
Criodrilus lacuum Hoffm., 1845; *ochridensis** Geo.,1950.

Fam. Enchytraeidae
Fridericia bisetosa (Levinsen, 1884).

Fam. Haplotaxidae
Haplotaxis dubius (Hrabě, 1931).

Fam. Lumbriculidae
Lamprodrilus pygmaeus Mich., 1902. *Rhynchelmis komáreki* Hrabě, 1927. *Stylodrilus leucocephalus** Hrabě, 1931; *parvus* (Hrabě *et* Čern., 1927).

Fam. Naididae
Nais variabilis Piguet, 1907; *pseudobtusa* Piguet, 1909. *Stylaria lacustris* (Linnaeus, 1767).

Fam. Tubificidae
Haber speciosus (Hrabě, 1931). *Limnodrilus hoffmeisteri* Clap., 1862; *udekemianus* Clap., 1862. *Potamothrix hammoniensis* (Mich., 1901); *isochaetus* (Hrabě, 1931); *ochridanus** (Hrabě, 1931). *Psammoryctides deserticola* Hrabě, 1950. *Rhyacodrilus punctatus** Hrabě, 1931. *Spirosperma tenuis* (Hrabě, 1931); *stankovici** (Hrabě, 1931). *Tubifex tubifex* (Müll., 1774).

TANGANYIKA (Martin & Brinkhurst, 1994; Martin & Giani, 1995a, 1995b; (1) Martin, *unpubl. data*).

Fam. Alluroididae
*Alluroides tanganyikae** Lauzanne, 1968.

Continued on p. 67

Table 1. Continued..

Fam. Eudrilidae

*Metschaina tanganyikae** Bedd., 1906. *Stuhlmannia inermis** Bedd., 1906.

Fam. Naididae

Dero digitata (Müll., 1773); *pectinata* Aiyer, 1929. *Homochaeta* sp.[1]. *Nais* sp.[1]. *Pristinella jenkinae*[1] (Steph., 1931).

Fam. Ocnerodrilidae

*Ocnerodrilus cunningtoni** Bedd., 1906. *Pygmaeodrilus grawerti** Mich., 1901.

Fam. Phreodrilidae

*Insulodrilus genitalisetifera** Martin *et* Brinkhurst, 1994; *martensi** Martin *et* Giani, 1995b; *tanganyikae** (Brinkhurst, 1970).

Fam. Tubificidae

*Epirodrilus tanganyikae** Martin *et* Giani, 1995a. *Psammoryctides* sp.[1]. *Rhizodrilus* sp.[1]

TITICACA (Gavrilov, 1978; Lafont & Juget, 1992; Juget & Lafont, 1994).

Fam. Enchytraeidae

Enchytraeus buchholzii Vejd., 1879. *Hemienchytraeus stephensoni* (Cogn., 1927). *Lumbricillus lineatus* (Müll.,1774).

Fam. Haplotaxidae

Metataxis americanus sp. dub. (Čern., 1939).

Fam. Naididae

Dero obtusa Udek., 1855; *sawayai* Mar., 1943. *Nais andina* Čern., 1939; *pardalis* Piguet, 1909; *variabilis* Piguet, 1907. *Pristina leidyi* F. Sm., 1896.

Fam. Tubificidae

Bothrioneurum americanum Bedd., 1894. *Epirodrilus antipodum** Čern., 1939. *Isochaetides lacustris* Čern., 1939. *Limnodrilus udekemianus* Clap., 1862; *hoffmeisteri* Clap., 1862. *Potamothrix bavaricus* (Osch., 1913); *hammoniensis* (Mich., 1901); *heuscheri* (Bret., 1900). *Rhyacodrilus* sp. *Tubifex ignotus* (Stolč, 1886).

---

to hold one endemic genus and species, *Kawamuria japonica* Stephenson 1917 (Mori & Miura, 1990), but this species was previously considered synonymous with *Branchiura sowerbyi* Beddard 1892 (Brinkhurst & Jamieson, 1971).

Lake Baikal harbours 12 endemic genera. No other lake has endemic genera. As they are approximately 2–4 Ma old (Table 2), we can consider that this period of time is not sufficient for the formation of supra-specific differences. Surprisingly enough, Lake Tanganyika, which has a similar age as Lake Baikal, has no endemic genus but it seems obvious that our knowledge concerning its oligochaete fauna is incomplete.

Endemism results either from the survival of an old fauna which has long since become extinct in the surrounding areas ('palaeoendemics' or relict species; Mayr, 1947), or from the arising of a new fauna by intra lacustrine speciation ('neoendemics'; Martens *et al.*, 1994). The *Baikalodrilus* group (Tubificidae) is probably a good example of neoendemicity as this genus is restricted to Baikal and because a novel phylogenetic analysis of the Tubificinae clearly showed its recent origin within the oligochaete complex with papillate body wall (Brinkhurst, 1991; Snimschikova & Timm, 1992). In contrast, *Epirodrilus* species in ancient lakes might be considered as palaeoendemic because the genus is reported from the Palaeartic, Neotropical and Ethiopian regions (Martin & Giani, 1994). No phylogenies at the species level are, however, available for the latter genus so that this assumption cannot be further substantiated thus far.

It should be noted that endemism is not restricted to ancient lakes. Lake Tahoe (Sierra Nevada, USA) has two endemic species, *Spirosperma beetoni* Brinkhurst 1965 and *Rhyacodrilus brevidentatus* Brinkhurst 1965 (Brinkhurst, pers. comm.) in spite of its youth (it lies in a tectonic trough 10 to 30×10^6 years old but the origin of the lake itself goes back to the last glaciations 10 000 years ago; Goldman, 1993). This suggests that speciation may be fairly rapid in oligochaetes. The possibility of being relict species cannot be discarded however, as they might well have been present in the original basin long before the present lake. Accordingly, speciation rates are only given here as a rough indication as they may not be calculated based on this single lake.

Species flocks

One of the most interesting features of ancient lakes is the presence of species flocks. Three principal criteria have been proposed to define a species flock, namely monophyly, endemism and speciosity (Coulter, 1991: 282–285; Martens *et al.*, 1994). The first two criteria are self-explanatory while speciosity has been defined as 'a disproportionate abundance of closely related species within a geographically circumscribed area' (Ribbink, 1984). The latter criterion helps to highlight those groups of species that are undergoing an unusually high rate of speciation (Coulter, 1991). In practice, however, a strict application of the three criteria is difficult and they should be interpreted broadly (Martens *et al.*, 1994). Species flocks are well-known in cichlid fishes of ancient African lakes and in gammarids of Lakes Baikal or Ohrid. For oligochaetes, species flocks are found only in Lake Baikal. To date, three families have at least one genus with large numbers of endemic species: Naididae (*Chaetogaster*, 9 species), Tubificidae (*Baikalodrilus*, 20 species, but many more suspected; Snimschikova & Timm, 1992), Lumbriculidae (*Lamprodrilus*, 20 species; Snimschikova & Akinshina, 1994). Among the tubificids, *Isochaetides* contains only five species but at least seven others are mentioned as *nom. nud.* by Snimschikova & Akinshina (1994). If these species were confirmed as genuine species in the future, the genus could be considered as having produced a species flock.

Factors affecting the evolution of oligochaete diversity in ancient lakes

Each lake is a combination of unique features (depth, heterogeneity, age, isolation, climatic history, founding stocks; Coulter, 1994b) having interacted with the evolution of their fauna and diversity. However, some constants can be distinguished. Diversity is the net result of a balance between a positive term (invasion and speciation) and a negative one (extinction). An attempt to understand oligochaete diversity in ancient lakes requires the examination of both terms.

During their history, all ancient lakes have been subjected to profound changes in their topography, as a result of tectonic and climatic modifications (orogenetic movements, periods of desiccation, etc). The communication with hydrological systems formerly separated from the lakes are probably responsible for 'invasions' by foreign elements. An example is given by Lake Baikal, where many stages of invasions have been distinguished during its history (Mazepova, 1990).

On the other hand, such profound changes were sometimes the cause of 'extinctions'. The occurrence of mass extinction phenomena in Lake Baikal is a matter of current dispute in Russian literature (Mazepova, 1990; Snimschikova & Timm, 1992). In contrast, catastrophic extinction is well-documented for Lake-

Turkana (= Lake Rudolf, one of the four largest lakes in East Africa) which has a relatively poor extant but a rich fossil fauna due to a period of complete desiccation (Coulter, 1994b). With respect to the oligochaete fauna, the disappearance of the precursors of lakes Kinneret and Biwa, during the history of these lakes, is probably responsible for their low extant diversity and absence of endemics in the former (Table 2).

While there is no doubt that multiple invasions can play an important role in diversity, at least as founding stocks, 'speciation' is probably the main source of diversity in a lake, especially as regards the lakes Baikal, Tanganyika and Malawi, the fauna of which is deemed to have originated largely by intralacustrine speciation (Fryer, 1991).

If the rate of speciation is much shorter than the life span of a species, then species may gradually accumulate. This assumption seems reasonable according to the rate of speciation known for some groups (200 to 1000 years for cichlid fishes; Owen *et al.*, 1990 – even if this is an extreme example of rapid speciation, it gives an order of magnitude), and to an estimated average duration of fossil species of about 4 Ma (Raup, 1991). A 30 Ma life span of an ostracod species has even been reported (Witte *et al.*, 1992). As a result, the older a lake is, the more diverse fauna it harbours. This may be true of Lake Baikal which holds the most abundant and diverse oligochaete fauna. The age of a lake is, however, not a sufficient explanation on its own (Ribbink, 1994b) and in some instances, diversity may be scarcer than expected due to some local confounding circumstance that makes a lake unique.

For example, the presence of only one endemic species in Lake Titicaca (*Epirodrilus antipodum* Cernosvitov, 1939) is surprising, given the age of the lake (3 Ma) which is similar to that of Lake Ohrid (2–3 Ma) for which five endemics are known to date. No convincing explanation has been offered so far. Two special features of Lake Titicaca, assumed to be unfavourable to oligochaetes, have been invoked, namely its eutrophic status associated with a permanent oxygen deficiency below the depth of 100 m on the one hand and a high electrical conductivity of the waters on the other hand, (Lafont & Juget, 1992).

A comparison between Lake Baikal and Lake Tanganyika

Lake Tanganyika is perhaps a good illustration of the age problem as, while being just younger than Lake Baikal, it has a very limited oligochaete diversity. This

Table 2. List of Oligochaeta and Aphanoneura in ancient lakes, at the family and genus levels, including the number of species for each genus. Values in brackets indicate the number of endemic species within the genus. The open circles in front of some generic names indicate that the genus is endemic. Numbers of endemic species given for Lake Tanganyika and Malawi include species not yet described but identified as valid sp. n. These are also minimal values as some species whose genus is mentioned here for the first time have not been identified so far. 1 : Snimschikova & Akinshina (1994); 2 : Stankovic (1960); 3 : Serruya (1978); 4 : Mori & Miura (1990), Ohtaka (1993, 1994); 5 : Gavrilov (1978), Lafont & Juget (1992), Juget & Lafont (1994); 6 : Coulter (1991), Martin & Brinkhurst (1994), Martin & Giani (1995a, 1995b), Martin (unpubl. data); 7 : Beddard (1908), Martin & Giani (1995a), Martin (unpubl. data).

Family	Genus	Baikal[1]	Ohrid[2]	Kinneret[3]	Biwa[4]	Titicaca[5]	Tanganyika[6]	Malawi[7]
Aeolosomatidae	*Aeolosoma*	(1) 1						
Alluroididae	*Alluroides*						1	
Biwadrilidae	*Biwadrilus*				1			
Criodrilidae	*Criodrilus*		(1) 2	1				
Enchytraeidae	*Enchytraeus*	(1) 1				1		
	Fridericia		1					
	Hemienchytraeus					1		
	Lumbricillus					1		
	Marionina	(1) 1						
	Mesenchytraeus	(1) 1						
Eudrilidae	*Metschaina*						(1) 1	
	Stuhlmannia						(1) 1	
Haplotaxidae	*Haplotaxis*	(1) 2	1	1				
	Metataxis					1		
Lumbriculidae	° *Agriodrilus*	(1) 1						
	Lamprodrilus	(20) 22	1					
	Lumbriculus	1						
	° *Pseudolycodrilus*	(1) 1						
	° *Pseudorhynchelmis*	(1) 1						
	Rhynchelmis	(2) 3	1					
	Stylodrilus	(6) 7	(1) 2					
	Styloscolex	(6) 7						
	° *Teleuscolex*	(4) 4						
Naididae	*Amphichaeta*	(2) 3						
	Chaetogaster	(9) 10			1			
	Dero	1		1		2	2	1
	Homochaeta						1	(1) 1
	Nais	(6) 13	2		1	3	1	1
	° *Neonais*	(1) 1						
	Pristina					1		1
	Pristinella						1	
	Ripistes	1						
	Slavina	1						
	Specaria	1						
	Stephensoniana			1				(1) 1
	Stylaria	2	1		1			
	Uncinais	2						
	Vejdovskyella	(1) 2						
Ocnerodrilidae	*Ocnerodrilus*						(1) 1	
	Pygmaeodrilus						(1) 1	
Phreodrilidae	*Insulodrilus*						(3) 3	(1) 1
Propappidae	*Propappus*	(1) 2						

Continued on p. 70

Table 2. Continued.

Family	Genus	Baikal[1]	Ohrid[2]	Kinneret[3]	Biwa[4]	Titicaca[5]	Tanganyika[6]	Malawi[7]
Tubificidae	*Aulodrilus*							1
	° *Baikalodrilus*	(20) 20						
	Bothrioneurum				1	1		
	Branchiura				1			1
	Embolocephalus	1			(1) 1			
	Epirodrilus					(1) 1	(1) 1	(1) 1
	Haber	(1) 1	1					
	Ilyodrilus	1						
	Isochaetides	(3) 5				1		
	Limnodrilus	2	2	1	4	2		
	° *Lycodrilides*	(1) 1						
	° *Lycodrilus*	(3) 3						
	° *Lymphachaeta*	(1) 1						
	° *Pararhyacodrilus*	(1) 1						
	Potamothrix	1	(1) 3	3		3		
	Psammoryctides	1	2	1			1	
	Rhizodrilus						(1) 1	
	° *Rhyacodriloides*	(1) 1						
	Rhyacodrilus	(3) 6	(1) 1		1	1		
	Spirosperma	2	(1) 2					
	° *Svetlovia*	(3) 3						
	Tasserkidrilus	(3) 3						
	Teneridrilus	(2) 3						
	Tubifex	(5) 6	1		1	2		
TOTAL	Species	152	23	9	13	21	16	9
	Endemic species	111	5	0	1	1	>9	>4
	Genera	45	15	7	10	13	13	9
	Families	7	6	4	3	4	6	3
Age of the lake	Ma	20-25	2-3	0.018 (20)	0.3 (4)	3	20	>2

is all the more surprising because both lakes are not only of a similar age but they share features that could lead to species diversification. For instance, both have a very great depth (1680 m and 1470 m for Lake Baikal and Lake Tanganyika, respectively) which was probably a protection against catastrophic events such as periods of desiccation or glaciations (Coulter, 1994a; Snimschikova & Timm, 1992). For this reason, Lake Tanganyika is considered as the only great African lake, having the longest continuous history (Coulter, 1994a).

It has been suggested above that the apparently very poor extant oligochaete diversity in Lake Tanganyika seems to be due to a lack of studies, as recent new interest in great African lakes gave indications of a more diverse oligochaete fauna than previously assumed (Martin & Brinkhurst, 1994; Martin & Giani, 1995a, b; Martin, unpubl. data; Table 1). Nevertheless, it is not impossible that Lake Tanganyika and other great African lakes are a less favourable environment to oligochaetes. The scarcity of oxygen availability in the tropics has been suggested by Martin & Brinkhurst (1994) as the main reason.

Lake Baikal is indeed oxygenated to the deepest point (Weiss *et al.*, 1991; Martin, 1994) while Lake Tanganyika is virtually anoxic below 100 to 240 m (Coulter & Spigel, 1991; Coulter, 1994b). As a result, Lake Baikal * has about 3.9 times more potentially

---

* It can be easily estimated that the area of oxygenated bottom is only 8150 km^2 in Lake Tanganyika versus 31 500 km^2 in Lake Baikal given that (1) their depth is very low compared with their breadth ($\pm$ 3% only) so that their surface area can be considered similar to their bottom area, and (2) the area of oxygenated sediments is estimated to a maximum of 20% in Lake Tanganyika (Tiercelin & Mondeguer, 1991).

oxygenated, and habitable, sediments than Lake Tanganyika. The influence of this theoretically available area on diversity in Lake Baikal is nevertheless queried for oligochaetes, even if a genuine abyssal fauna is well known for other groups (e.g. ostracods; Mazepova, 1994). Most tubificids indeed seem to have a eurybathic distribution (Snimschikova & Akinshina, 1994) and while lumbriculids are said to be rather stenobathic (Snimschikova & Akinshina, ibid.), this claim should be tempered as no revision of this family for Lake Baikal has been made in the last 32 years (Isossimov, 1962).

Oxygen measurements in the sediment by means of microelectrodes have shown that the sediment is virtually anoxic below the first millimetre(s) in Lake Tanganyika (Goddeeris et al., unpubl.data) and Malawi (Martin et al., 1993a; and unpubl. data) while it can be oxygenated deeper than 50 mm in Lake Baikal (Martin et al., 1993b; and unpubl. data). Preliminary data indicate that oligochaetes are restricted to the uppermost centimetres in great African lakes (Martens et al., unpubl. data) while they can be found deeper than 10 cm in Lake Baikal (Martin et al., 1993c), suggesting the importance of oxygen in the vertical distribution of these animals and in their potential habitats.

Considering oxygen as a restrictive factor, it is worthy of note that, among oligochaetes found in lakes Tanganyika and Malawi, some are periphytic naidids (Homochaeta, Nais and Pristina), others are characterized by the possession of gills or respiratory tail (Dero, Branchiura, Aulodrilus), still others (Insulodrilus, Epirodrilus, Rhizodrilus) are found in sandy, well-oxygenated, sediments.

From this comparison between the oligochaete diversity in Lake Baikal and in Lake Tanganyika, it is concluded that the ancient African great lakes are still virtually unknown and that a better study of this fauna is badly needed. This review would like to be an appeal for such a study. Lake Tanganyika might reveal as diverse an oligochaete fauna as the one in Lake Baikal.

References

Beddard, F. E., 1908. The oligochaetous fauna of Lake Birket el Qurun and Lake Nyassa. Nature 77 (2009): 608.

Brinkhurst, R. O., 1971. Distribution and ecology. 1. Microdriles. In R. O. Brinkhurst & B. G. Jamieson (eds), Aquatic Oligochaeta in the World. Oliver and Boyd, Edinburgh: 104–146.

Brinkhurst, R. O., 1982. Oligochaeta. In S. P. Parker (ed), Synopsis and Classification of Living Organisms. Mc Graw-Hill Book Company, Inc., New York, Vol. 2: 50–61.

Brinkhurst, R. O., 1991. A phylogenetic analysis of the Tubificinae (Oligochaeta, Tubificidae). Can. J. Zool. 69: 392–397.

Brooks, J. L., 1950. Speciation in ancient lakes. Quart. Rev. Biol. 25: 30–60, 131–176.

Coulter, G., 1991. Lake Tanganyika and its life. Natural History Museum Publications, Oxford Univ. Press, London, 354 pp.

Coulter, G. W. & R. H. Spigel, 1991. Hydrodynamics. In G. W. Coulter (ed), Lake Tanganyika and its life. Natural History Museum Publications, Oxford Univ. Press, London: 49–75.

Coulter, G., 1994a. Lake Tanganyika. In K. Martens, B. Goddeeris & G. Coulter (eds), Speciation in Ancient Lakes, Arch. Hydrobiol. Beih. Ergebn. Limnol. 44: 13–18.

Coulter, G., 1994b. Speciation and Fluctuating Environments, with Reference to Ancient East African Lakes. In K. Martens, B. Goddeeris & G. Coulter (eds), Speciation in Ancient Lakes, Arch. Hydrobiol. Beih. Ergebn. Limnol. 44: 127–137.

Dejoux, C., 1994. Lake Titicaca. In K. Martens, B. Goddeeris & G. Coulter (eds), Speciation in Ancient Lakes, Arch. Hydrobiol. Beih. Ergebn. Limnol. 44: 35–42.

Fryer, G., 1991. Comparative aspects of adaptive radiation and speciation in Lake Baikal and the great rift lakes of Africa. Hydrobiologia 211: 137–146.

Gavrilov, K., 1978. Oligochaeta. In S. H. Hurlbert (ed), Aquatic biota of southern South America being a compilation of Taxonomic Bibliographies for the fauna and flora of inland waters of Southern South America. San Diego State University, San Diego: 99–121.

Goldman, C. R., 1993. The conservation of two large lakes: Tahoe and Baikal. Verh. int. Verein. Limnol. 25: 388–391.

Gophen, M. & A. Nishri, 1994. Lake Kinneret. In K. Martens, B. Goddeeris & G. Coulter (eds), Speciation in Ancient Lakes, Arch. Hydrobiol. Beih. Ergebn. Limnol. 44: 65–71.

Gorthner, A., 1994. What is an ancient lake? In K. Martens, B. Goddeeris & G. Coulter (eds), Speciation in Ancient Lakes, Arch. Hydrobiol. Beih. Ergebn. Limnol. 44: 97–100.

Greenwood, P. H., 1994. Lake Victoria. In K. Martens, B. Goddeeris & G. Coulter (eds), Speciation in Ancient Lakes, Arch. Hydrobiol. Beih. Ergebn. Limnol. 44: 19–26.

Hutchinson, G. E., 1957. A treatise on limnology. Volume I, Geography, Physics, and Chemistry. J. Wiley & Sons, New York, 1015 pp.

Isossimov, V. V., 1962. Oligochaetes worms of the family Lumbriculidae of Lake Baikal. Proc. Limnol. Instit. Acad. Sci. U.S.S.R., Siberian Sect., 1: 3–126.

Juget, J. & M. Lafont, 1994. Distribution of Oligochaeta in some lakes and pools of Bolivia. Hydrobiologia, 278: 125–127.

Kamaltynov, R. M., 1993. On the present state of amphipod systematics. Hydrobiol. J. 26: 82–92.

Lafont, M. & J. Juget, 1992. The benthic oligochaetes. In C. Dejoux & A. Iltis (eds), Lake Titicaca. A Synthesis of Limnological Knowledge. Monographiae Biologicae, 68, Kluwer Academic Publishers, Dordrecht: 302–306.

Martens, K., G. Coulter & B. Goddeeris, 1994. Speciation in Ancient Lakes – 40 years after BROOKS. In K. Martens, B. Goddeeris & G. Coulter (eds), Speciation in Ancient Lakes, Arch. Hydrobiol. Beih. Ergebn. Limnol. 44: 75–96.

Martin, P., B. Goddeeris & K. Martens, 1993a. Sediment oxygen distribution in ancient lakes. Verh. int. Ver. Limnol. 25: 793–794.

Martin, P., B. Goddeeris & K. Martens, 1993b. Oxygen concentration profiles in soft sediment of Lake Baikal (Russia) near the Selenga Delta. Freshwat. Biol. 29: 343–349.

72

Martin, P., B. Goddeeris & K. Martens, 1993c. Spatial distribution of Oligochaetes in Lake Baikal (Siberia-Russia). Hydrobiologia 278: 151–156.

Martin, P., 1994. Lake Baikal. In K. Martens, B. Goddeeris & G. Coulter (eds), Speciation in Ancient Lakes, Arch. Hydrobiol. Beih. Ergebn. Limnol. 44: 3–11.

Martin, P. & R. O. Brinkhurst, 1994. A new species of *Insulodrilus* (Oligochaeta, Phreodrilidae) from Lake Tanganyika (East Africa) with notes on the oligochaete fauna of the lake. Arch. Hydrobiol. 130: 249–256.

Martin, P. & N. Giani, 1995a. Two new species of *Epirodrilus* (Oligochaeta, Tubificidae) from Lake Nyasa and Tanganyika (East Africa), with redescription of *E. slovenicus* and *E. michaelseni*. Zool. Scr., 24: 13–19.

Martin, P. & N. Giani, 1995b. *Insulodrilus martensi*, a new species of the Phreodrilidae (Oligochaeta) from Lake Tanganyika (East-Africa). Ann. Limnol., 31: 5–8.

Martin, P. & N. Giani, 1996. A new species of Phreodrilidae (Oligochaeta), *Astacopsidrilus naceri* sp. n., from Morocco (North Africa) with notes on the biogeography of the family. Can. J. Zool., *in press*.

Mayr, E., 1947. Ecological factors in speciation. Evolution 1: 263–288.

Mazepova, G., 1990. Rakushkovye ratchki (Ostracoda) Baikala (=Ostracods (Ostracoda) of Lake Baikal). Nauk. Sib. Otdel. Akad. Nauk. SSSR, Novosibirsk, 470 pp.

Mazepova, G., 1994. On comparative aspects of ostracod diversity in the Baikalian fauna. In K. Martens, B. Goddeeris & G. Coulter (eds), Speciation in Ancient Lakes, Arch. Hydrobiol. Beih. Ergebn. Limnol. 44: 197–202.

Mori, S. & T. Miura, 1990. List of plant and animal species living in Lake Biwa (corrected third edition). Mem. Fac. Sci. Kyoto Univ. (Ser. Biol.) 40: 13–32.

Nakajima, T. & K. Nakai, 1994. Lake Biwa. In K. Martens, B. Goddeeris & G. Coulter (eds), Speciation in Ancient Lakes, Arch. Hydrobiol. Beih. Ergebn. Limnol. 44: 43–54.

Ohtaka, A., 1993. Oligochaeta. In M. Nishino (ed), Handbooks of Zoobenthos in Lake Biwa III. Porifera, Platyhelminthes, Annelida, Tentaculata and Crustacea. Lake Biwa Research Institute, Otsu, Japan: 18–41.

Ohtaka, A., 1994. Redescription of *Embolocephalus yamaguchii* (Brinkhurst, 1971), comb. nov. (Oligochaeta,Tubificidae). Proc. Japan. Soc. Syst. Zool. 52: 34–42.

Owen, R. B., R. Crossley, T. C. Johnson, D. Tweddle, I. Kornfield, S. Davison, D. H. Eccles & D. E. Engstrom, 1990. Major low levels of Lake Malawi and their implications for speciation rates in cichlid fishes. Proc. r. Soc. Lond. 240: 519–553.

Raup, D. M., 1991. Extinction: bad genes or bad luck? New Scientist 14 Sep.: 46–49.

Ribbink, A. J., 1984. Is the species flock concept tenable? In A. A. Echelle & I. Kornfield (eds), Evolution of fish species flocks. Univ. Maine, Orono Press: 21–25.

Ribbink, A. J., 1994a. Lake Malawi. In K. Martens, B. Goddeeris & G. Coulter (eds), Speciation in Ancient Lakes, Arch. Hydrobiol. Beih. Ergebn. Limnol. 44: 27–33.

Ribbink, A. J., 1994b. Rates and modes of speciation. In K. Martens, B. Goddeeris & G. Coulter (eds), Speciation in Ancient Lakes, Arch. Hydrobiol. Beih. Ergebn. Limnol. 44: 505–508.

Salemaa, H., 1994. Lake Ohrid. In K. Martens, B. Goddeeris & G. Coulter (eds), Speciation in Ancient Lakes, Arch. Hydrobiol. Beih. Ergebn. Limnol. 44: 55–64.

Serruya, C., 1978. Lake Kinneret. Monographiae Biologicae, 32, Junk, The Hague, 501 pp.

Snimschikova, L. N. & T. Timm, 1992. Review of the genus *Baikalodrilus* Holmquist, 1978 (Oligochaeta, Tubificidae). Bull. Inst. r. Sci. nat Belg., Biologie 62: 53–85.

Snimschikova, L. N. & T. W. Akinshina, 1994. Oligochaete fauna of Lake Baikal. Hydrobiologia 278: 27–34.

Stanković, S., 1960. The balkan Lake Ohrid and its living world. In F. S. Bodenheimer & W. W. Weisbach (eds), Monographiae Biologicae, IX, Junk, The Hague, 357 pp.

Tiercelin, J.-J. & A. Mondeguer, 1991. The geology of the Tanganyika Trough. In G. W. Coulter (ed), Lake Tanganyika and its life. Natural History Museum Publications, Oxford Univ. Press, London: 7–48.

Van der Land, J., 1971. Family Aeolosomatidae. In R. O. Brinkhurst & B. G. Jamieson (eds), Aquatic Oligochaeta in the World. Oliver and Boyd, Edinburgh: 665–695.

Weiss, R. F., E. C. Carmack & V. M. Koropalov, 1991. Deep-water renewal and biological production in Lake Baikal. Nature 349: 665–669.

Witte, L., Th. Lissenberg & H. Schuurman, 1992. Ostracods from the Albian/Cenomanian boundary in the Achterhoek area (eastern part of the Netherlands). Scr. Geol. 102: 33–84.

Hydrobiologia **334**: 73–83, 1996.
K. A. Coates, Trefor B. Reynoldson & Thomas B. Reynoldson (eds), Aquatic Oligochaete Biology VI.
© 1996 *Kluwer Academic Publishers.*

Distribution patterns of aquatic oligochaetes inhabiting watercourses in the Northwestern Iberian Peninsula

Enrique Martínez-Ansemil & Rut Collado
Departamento de Bioloxía Animal e Bioloxía Vexetal, Facultade de Ciencias, Universidade da Coruña. Campus da Zapateira s/n, 15071 A Coruña, Spain

Key words: aquatic Oligochaeta, running waters, distribution patterns, ordination

Abstract

Distribution patterns of aquatic oligochaete assemblages, inhabiting largely unpolluted watercourses, in the Northwestern Iberian Peninsula, and their relationships with chemical and physiographical characteristics were analyzed by means of multivariate analyses. Qualitative and quantitative samples from 47 stations were obtained seasonally during 1983/84 and 1988/90. The variables included in this study were altitude, order number, distance from the origin, mean width, mean depth, substrate, current velocity, oxygen content, pH, conductivity, alkalinity and concentration of different ions.

From a chemical point of view, major changes in the structure of oligochaete assemblages were related to water mineralization. The correlation between faunal distribution and environmental variables showed a great increase when physiographic parameters were included in the analyses. Substrate type and current velocity are the principal variables explaining the community structure.

Introduction

Most of the recent ecological studies on the distribution of oligochaetes in watercourses from Europe deal with polluted waters, and pollutants appear to be the major factor explaining the composition and the structure of the communities (e.g. Kasprzak, 1976; Dumnicka & Pasternak, 1978; Uzunov, 1982; Giani, 1984).

Some papers consider the influence of other environmental variables (Korn, 1963; Wachs, 1967; Timm, 1970; Kasprzak, 1976; Kasprzak & Szczesny, 1976; Dumnicka, 1978, 1994; Dumnicka & Pasternak, 1978; Giani & Martínez-Ansemil, 1981; Schwank, 1982a, 1982b; Uzunov, 1982; Rodríguez, 1984; Centurioni & Sambugar, 1986; Dumnicka & Kownacki, 1988; Lafont, 1989; Martínez-Ansemil, 1990). However, most of them refer to a particular watercourse or watershed and their results are ordinated without a joint analysis of species and environmental variables.

The aim of this study is to analyze, by means of multivariate analyses, the main patterns of oligochaete distribution in relation to environmental variables, in natural unpolluted watercourses of the NW Iberian Peninsula (west corner of Europe).

Study area

The study area (Galicia and N Portugal) (Fig. 1) is in the Hesperian Massif and shows great lithological diversity. Except for some isolated areas, where sediment material and scarce limestone and dolomite outcrops are found, the lithology is dominated by metamorphic rocks: granite, schist, slate, quartzite, gneiss, migmatite, etc. On the whole, the lithology of the study area is dominated by hard materials of low solubility.

The climate of the study area has a strong oceanic influence. The prevailing westerly winds extend the influence from the Atlantic through the entire region, and, although the average yearly rainfall varies, most of the study area reaches precipitation values of over 1000 mm.

The topography of this area, together with the climate, determines the presence of an extensive hydro-

74

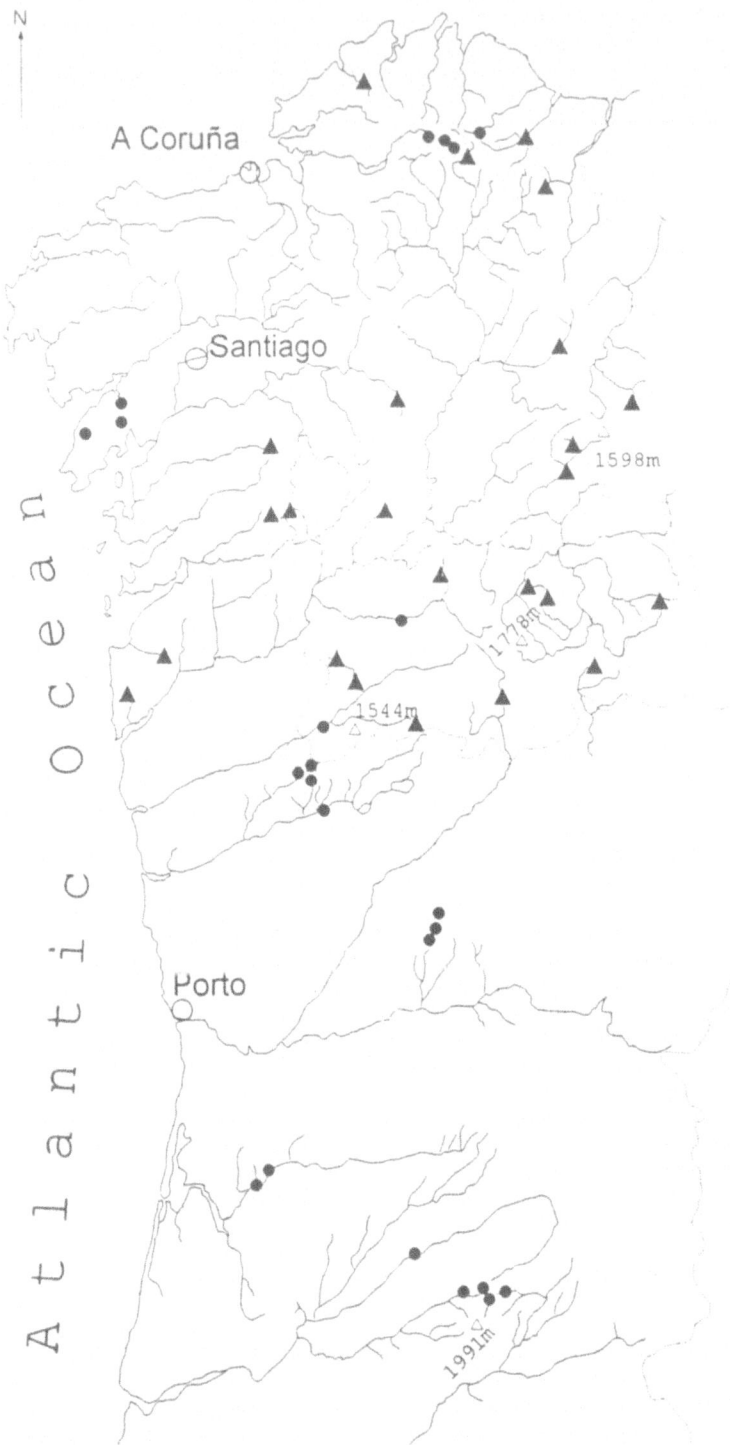

Fig. 1. The study area (Galicia and N. Portugal). ▲, stations sampled in 1983/84 (studied in more detail); •, stations sampled in 1988/89 or 1989/90.

75

Table 1. Mean, minimum and maximum values of the environmental variables analyzed in the sampling stations. *, only considered in the 25 ▲ stations (see Fig. 1).

VARIABLES	CODES	MEAN	MIN.–MAX.
*Order number	ON	2.2	1–4
*Distance from origin (km)	OD	5.85	1.2–111.4
Mean width (m)	W	4.94	0.7–20
Mean depth (m)	DEPTH	0.3	0.07–0.60
Altitude (m)	ALT	666	20–1,540
Temperature (°C)	T	12.08	1.5–26.0
O_2(mg l^{-1})	O	10.22	6.7–14.5
pH	PH	6.22	3.2–9.5
Conductivity (μS cm^{-1})	COND	40.94	9–210
*Ca^{2+} (mg l^{-1})	CA	4.50	0.2–39.5
*Mg^{2+} (mg l^{-1})	MG	1.08	0.1–5.0
*K^+ (mg l^{-1})	K	0.37	0.1–1.1
*Na^+ (mg l^{-1})	NA	2.99	0.7–8.8
*Cl^- (mg l^{-1})	CL	5.73	1.5–15.7
*Si (mg l^{-1})	SI	2.75	0.10–6.8
*SO_4^{2-} (mg l^{-1})	SO4	3.73	0.4–22
*PO_4^{2-} (μg l^{-1})	PO4	8.02	0.6–35.9
*NO_3^- (μg l^{-1})	NO3	0.23	0.01–0.85
*Alkalinity (μeq l^{-1})	ALC	117.8	2–1,068

and preserved in 70% ethanol until identified to the species level. All specimens were mounted whole in glycerine for an initial identification. When necessary for accurate identifications, the worms were stained in haematoxylin and dissected. Some specimens of each species were mounted in Canada balsam and kept in the authors' collection.

For all sampling stations, several physiographical variables (altitude and approximate mean depth and width of watercourse taking into account the seasonal variations) were recorded, and the most general chemical parameters (oxygen content, conductivity, pH) were measured seasonally; more complete analyses (Membiela *et al.*, 1991) were carried out seasonally in the 25 stations sampled during 1984/85 (Fig. 1▲). (See Table 1 for summarized data). Temperature was excluded from the analysis due to its great daily variability. For the chemical analytical techniques used see Membiela *et al.* (1991).

For each sample we recorded the substratum (pebbles [P], gravel [G], sand [S], mud [MU], macrophytes [MA], moss [MO], plant debris [D]) and current velocity [VEL] (from 0 to 4: very slow, slow, moderate, fast, very fast).

Data analysis

By means of different multivariate analyses using the CANOCO program (Ter Braak, 1987) V. 3.12 we analyzed the faunistic and environmental data in order to determine the main patterns of distribution of the oligochaete fauna in the study area, and their relation to the environmental variables. The list of species is given in Table 2.

For the analyses carried out on quantitative species data at the station level, the unidentified immature individuals for each family (e.g. immature tubificids with or without hair setae, immature lumbriculids...) were proportionately assigned to the identified species with mature individuals. As is well known, identification to the species level can be made from immature and/or mature specimens, depending on the family. This method, already used by Martínez-Ansemil (1990), is an estimation carried out from many samples collected along the year, which allow us to integrate quantitative data from different families in a joint analysis. The species abundance was log-transformed ($\ln (y + 1)$).

For the analyses carried out on quantitative species data at the sample level, we did not consider the unidentified immature individuals and, after a logarithmic

graphic network, particularly dense in Galicia. The highest alititudes for the whole network are around 2000 m. There is a clear relationship between rainfall patterns and the flow regime of the watercourses, although in some cases there is a snow-melt component. For more complete information see De Brum Ferreira (1981), Ribeiro *et al.* (1988), Membiela *et al.* (1991).

Materials and methods

Sampling

Forty-seven sampling stations were selected in different, natural, unpolluted watercourses distributed throughout the study area (Fig. 1).

Two to six qualitative or quantitative oligochaete samples were taken seasonally in each station in 1983/84 (25 stations: Fig. 1▲) and 1988/90 (22 stations: Fig. 1●). A Surber sampler (pore size: 200 μm, sampler area: 1/10m^2) was used to take samples in bottoms with pebbles, gravel or vegetation, and a core sampler (diameter: 8 cm) for soft bottoms. The samples were sieved, and fixed in formalin (4%). The oligochaetes were sorted out under a stereomicroscope,

Table 2. Principal oligochaete species from the study area. IND, total number of individuals caught; ST, number of stations where the species were present (for 44 sampling stations); +, #, x, species included respectively in the analyses over the 44 stations, the 25 ▲ stations and the 172 quantitative samples (see text).

SPECIES	CODES	IND.	ST.
+ # x *Lumbriculus variegatus* (Müller, 1774)	LUMVAR	344	16
+ # x *Stylodrilus heringianus* Claparède, 1862	STYHER	1867	33
+ # x *S. parvus* (Hrabe & Cernosvitov, 1927)	STYPAR	984	21
+ # x *Nais alpina* Sperber, 1948	NAIALP	7217	43
+ # # *N. communis* Piguet, 1906	NAICOM	75	6
+ # x *N. pardalis* Piguet, 1906	NAIPAR	478	14
+ # x *N. variabilis* Piguet, 1906	NAIVAR	597	20
+ # # *Chaetogaster crystallinus* Vejdovsky, 1883	CHACRY	14	6
+ # # *C. diastrophus* (Gruithuisen, 1828)	CHADIA	99	7
+ # x *Pristina aequiseta* Bourne, 1891	PRIAEQ	475	28
+ # # *Pristinella* cf. *bilobata* (Bretscher, 1903)	PRIBIL	76	7
+ # # *P. osborni* (Walton, 1906)	PRIOSB	13	6
+ # x *P.* gr. *rosea*	PRIROS	87	14
+ # # *Stylaria lacustris* (Linné, 1767)	STYLAC	28	5
+ # x *Vejdovskyella comata* (Vejdovsky, 1883)	VEJCOM	300	18
+ # x *Slavina appendiculata* (d'Udekem, 1855)	SLAAPP	157	18
+ # x *Specaria josinae* (Vejdovsky, 1883)	SPEJOS	212	11
+ # x *Tubifex tubifex* (Müller, 1774)	TUBTUB	1112	6
+ # x *T. ignotus* (Stolc, 1886)	TUBIGN	333	8
+ # # *Limnodrilus hoffmeisteri* Claparède, 1862	LIMHOF	336	4
+ # x *Spirosperma (E.) velutinus* (Grube, 1879)	SPIVEL	92	7
+ # *Protuberodrilus tourenqui* Gi. & Mar.-Ans., 1979	PROTOU	33	5
# # # *Rhyacodrilus falciformis* Bretscher, 1901	RHYFAL	19	3
+ # x *Cognettia cognettii* (Issel, 1905)	COGCOG	452	27
+ # # *C. glandulosa* (Michaelsen, 1888)	COGGLA	15	5
+ # x *C. sphagnetorum* (Vejdovsky, 1877)	COGSPH	180	15
+ # # *Cognettia* sp.	COGSP	237	25
+ # # *Marionina argentea* (Michaelsen, 1889)	MARARG	10	5
# # # *Fridericia bisetosa* (Levinsen, 1884)	FRIBIS	33	3
+ # x *F. perrieri* (Vejdovsky, 1877)	FRIPER	104	8
+ # # *Achaeta eiseni* Vejdovsky, 1877	ACHEIS	5	4
+ # # *Achaeta* spp.	ACHSPP	133	17
+ # x *Cernosvitoviella atrata* (Bretscher, 1903)	CERATR	192	22
+ # # *C.* cf. *tatrensis* (Kowalewski, 1916)	CERTAT	120	5
+ # x *C. palustris* Healy, 1979	CERPAL	223	11
+ # # *Mesenchytraeus lusitanicus* Coll., Mar.-Ans. & Gi., 1993	MESLUS	19	4
+ # x *Eiseniella tetraedra* (Savigny, 1826)	EISTET	57	7

transformation of the data, we scaled the abundance of each species by dividing its abundance in each sample by its maximum abundance (0–1 scale).

For the analyses considering the chemical variability in the 25 ▲ stations, only the mean values for each parameter at each sampling station were included. Although chemical and faunistic samples were taken during different periods, previous results show that spatial variability is much more important than temporal (Membiela *et al.*, 1991).

The analyses that were included in this study are:

– Two correspondence analyses (CA) (for qualitative and quantitative data) applied to 44 sampling stations and 35 species (Table 2, +), after omitting the species present in three stations or less and the stations with three species or less.

Table 3. Some statistical values from the partial CCA analyses carried out on different variables: substrate-current velocity, conductivity, pH, O_2.

Total inertia: 4.996	Subs-Vel	Con	pH	O_2
Eigenvalue (axis 1)	0.147	0.067	0.078	0.035
Species-environment correlation (axis 1)	0.592	0.514	0.539	0.364
% variance of species data (axis 1)	3.4	1.7	1.9	0.9
% variance of spec-env. relation (axis 1)	33.8	100	100	100
Sum of all unconstr. eigenv. (after fitting covariables)	4.388	4.019	4.030	3.987
Sum of all canonical eigenv. (after fitting covariables)	0.436	0.067	0.078	0.035
P-value of Monte Carlo test (first canonical axis)	0.02	0.01	0.01	0.17

Table 4. Scores and tolerances for the species with the highest scores in partial CCA using conductivity or pH as variables.

	Sc.	Tol.
Conductivity		
EISTET	0.68	1.27
SPIVEL	0.65	0.78
NAIPAR	0.63	0.72
COGSPH	−0.48	0.85
pH		
NAIPAR	0.90	0.55
VEJCOM	0.69	1.16
SPEJOS	0.61	1.36
FRIPER	−0.53	0.44
CERPAL	−0.41	1.12

– A principal component analysis (PCA) carried out with the correlation matrix of the chemical variables of the 25 ▲ stations (Table 1), in order to reduce the multidimensional information of such variables to a series of independent principal components that would be included in subsequent canonical analyses.

– A canonical correspondence analysis (CCA) applied to the 25 ▲ stations and 33 species (Table 2, #) (after omitting the species present in two stations or less), including as environmental variables the physiographical ones (Table 1) and the scores in the first and second axes extracted from the PCA (PC1 and PC2). A partial CCA on the same data, taking PC1 and PC2 as covariables, was also carried out.

– Several partial CCA applied to the 172 quantitative samples and 20 species (Table 2, ×), after omitting the species present in nine samples or less. The parameters included as environmental variables were: oxygen content, pH, conductivity, substratum (as a nominal variable), current velocity, sample depth, altitude and mean width of the sampling station. The variables considered for the different partial CCA were: (a) substratum and current velocity, (b) conductivity, (c) pH, (d) oxygen content, including the other variables as covariables. The statistical significance of the canonical axes was assessed by Monte Carlo randomization tests.

For details about the ordination methods used, the reader is referred to the extensive account in Jongman *et al.* (1987). These methods have been succesfully used by many authors in similar studies (e.g. Verdonschot, 1990; Walker *et al.*, 1991; Kingston *et al.*, 1992).

Results

General patterns of spatial distribution

The CA carried out with the quantitative data of the 44 stations, shows that species have broad distributions over the study area, with only 13.7% of the variance explained by the first axis and 10.3% by the second (Fig. 2). Moreover, the results obtained by using only qualitative data are very similar (cumulative percentage of variance for the first two axes = 21.3).

The large scale environmental factors have little effect in the distribution of the oligochaetes in the unpolluted watercourses of the NW Iberian Peninsula. Nevertheless, the distribution of species along the first axis reflects a 'physiographical distribution' of the stations, from the small streams or the upper course of the rivers (left side) to the lower course of the rivers (right side). Although the fact that stations were located in different catchment areas, the distribution of the species along the first axis could be compared with a longitudinal distribution in a single watercourse. On the whole, the aquatic Enchytraeidae are characteristic of the upper courses, the Lumbriculidae of the

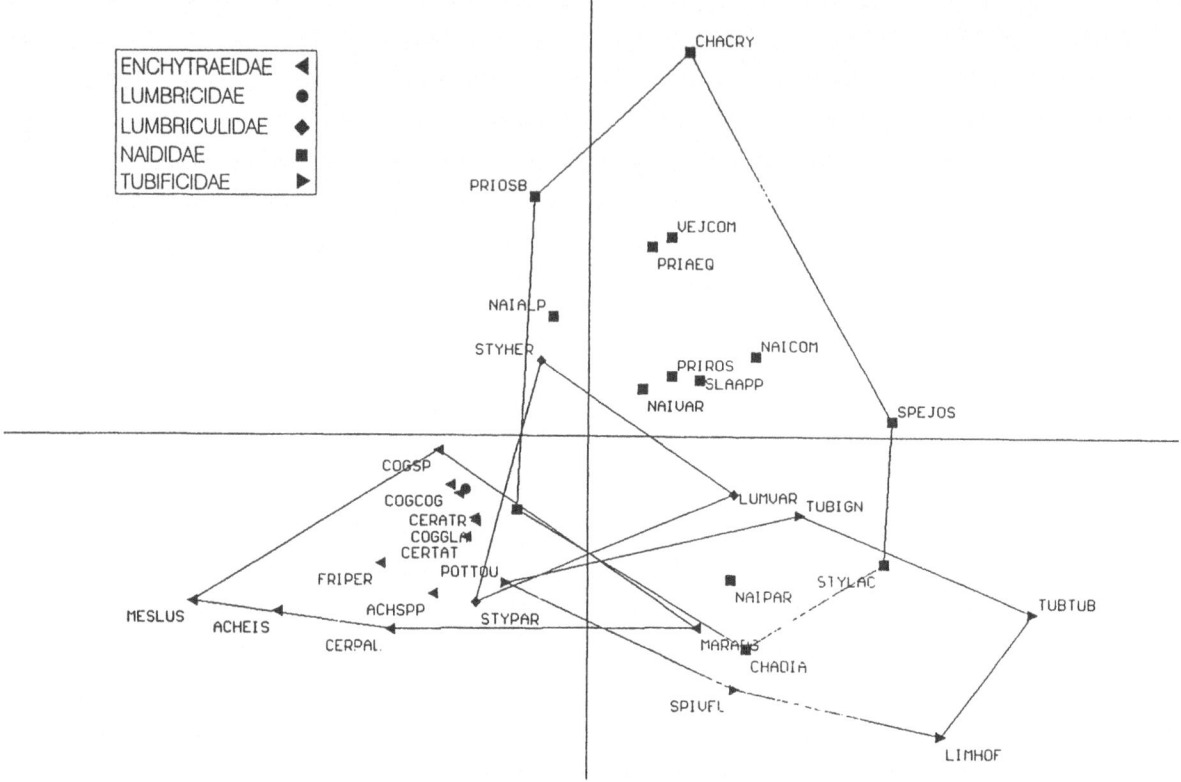

Fig. 2. CA diagram for the species scores for axes 1 and 2, based on oligochaete species abundance from the 44 sampling stations. Lines join the species showing the extreme values for each family, giving a picture of its distribution in the plot. (See codes in Table 2.)

upper-middle, the Naididae of the lower-middle and the Tubificidae of the lower courses.

Major factors influencing the species distribution throughout the study area

In order to evaluate the factors influencing the distribution of the oligochaetes in the natural watercourses of the NW Iberian Peninsula we analyzed the 25 ▲ stations.

The results of the PCA carried out on the correlation matrix of the chemical variables are shown in Fig. 3. PC1 (% variance = 46.6) mainly reflects the degree of water mineralization, and PC2 (% variance = 24.5) the content in two different kind of ions, from softer to harder and more productive waters.

The CCA carried out using as environmental variables PC1, PC2 and the physiographical characteristics, explains 28.9% of the variance in the species-environment relation for the first axis and 23.5% for the second (Fig. 4). Correlations between the chemical (associated with canonical axis 2) and physiographi-

cal variables (axis 1) were very low, with the exceptions of PC1-ALT ($r = -0.72$) and PC2-ALT ($r - 0.60$). The partial CCA using PC1 and PC2 as covariables shows a very similar ordination pattern in relation to the physiographical characteristics. W, ON, OD and DEPTH, all strongly correlated, show a positive correlation with the Tubificidae (lower courses) and a negative one (upper courses) with the Lumbricidae (*Eiseniella tetraedra*) and all the Enchytraeidae except *Cernosvitoviella* cf. *tatrensis* and *Marionina argentea*. The Lumbriculidae *Stylodrilus heringianus* and *Stylodrilus parvus* are more frequent in the upper-middle courses and *Lumbriculus variegatus* in the lower-middle ones. The Naididae are largely widespread, *Stylaria lacustris*, with *Pristinella* cf. *bilobata* and *Specaria josinae* being the most characteristic species of the lower courses and *Pristinella osborni* of the upper ones.

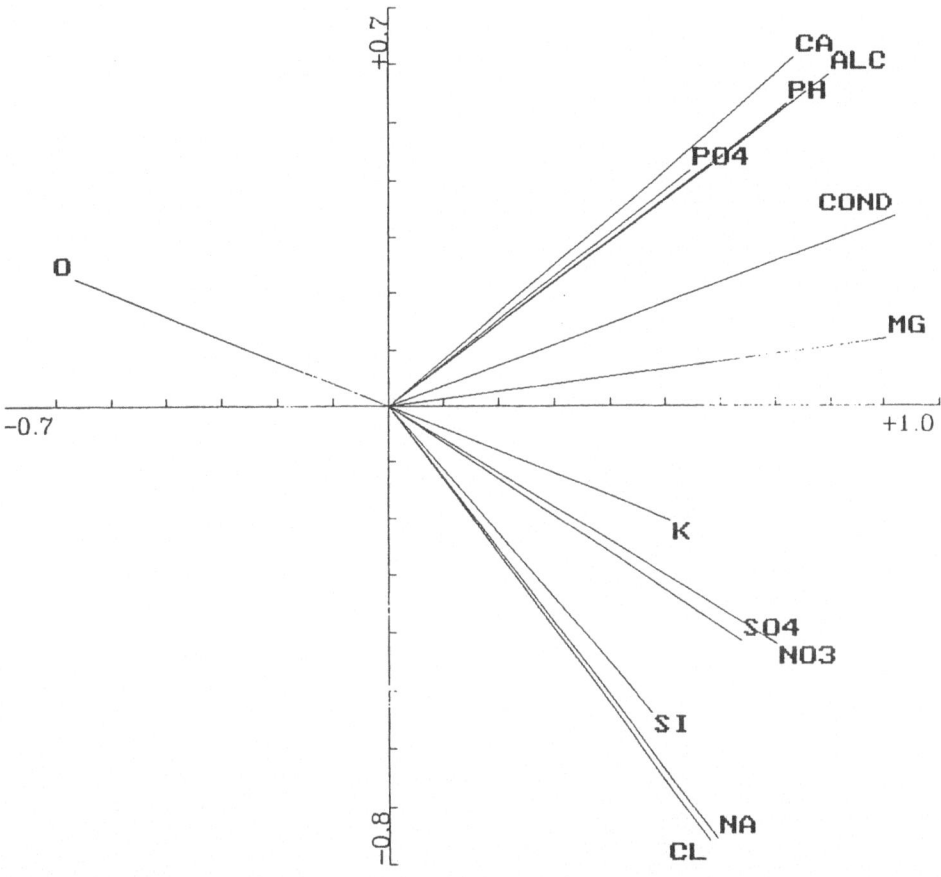

Fig. 3. PCA diagram for axes 1 and 2 carried out with the correlation matrix of the mean chemical values from the 25 ▲ stations. (See stations in Fig. 1 and codes in Table 1).

Principal factors influencing the composition and structure of the communities at the sample level

Table 3 summarizes the principal results of the different CCAs carried out on the 172 quantitative samples in order to determine the influence of the substratum and current velocity, mineralization, pH and oxygen content over the species distribution.

Substratum and current velocity. The first axis (% variance of the species-environment relation = 33.8) (Fig. 5) appears to be related to the current velocity and the associated substrate: mineral with greatest granulometry (P, G) and attached vegetation (MA, MO) are opposed to D and mineral substrate of minor granulometry (S, MU). The third axis (% variance = 13.8) separates the mineral and the vegetal substrata. We cannot attribute any ecological meaning to the second axis, which explains 22.7% of the variance.

Figs 6 and 7 show the species scores and tolerances (root mean squared deviation) for the first and third axes respectively. Although most of the species show a wide range in regard to substrate and current velocity, *Tubifex tubifex, S. josinae, Pristinella* gr. *rosea, Vejdovskyella comata* and *Tubifex ignotus* are more characteristic of the lentic habitats and *Fridericia perrieri, Nais alpina, Cognettia sphagnetorum, E. tetraedra* and *Cognettia cognettii* of the lotic ones; *S. parvus, T. ignotus, Nais pardalis, T. tubifex, Pristina aequiseta, E. tetraedra* and *F. perrieri* are species more representative of mineral substrates and *Nais variabilis* and *V. comata* from vegetal substrates.

Mineralization, pH and oxygen content. Conductivity, as a measure of the degree of mineralization, as well as pH, significantly affect ($p = 0.01$) species distribution, whereas oxygen content does not have a significant effect. Table 4 shows the highest species scores related to conductivity and pH.

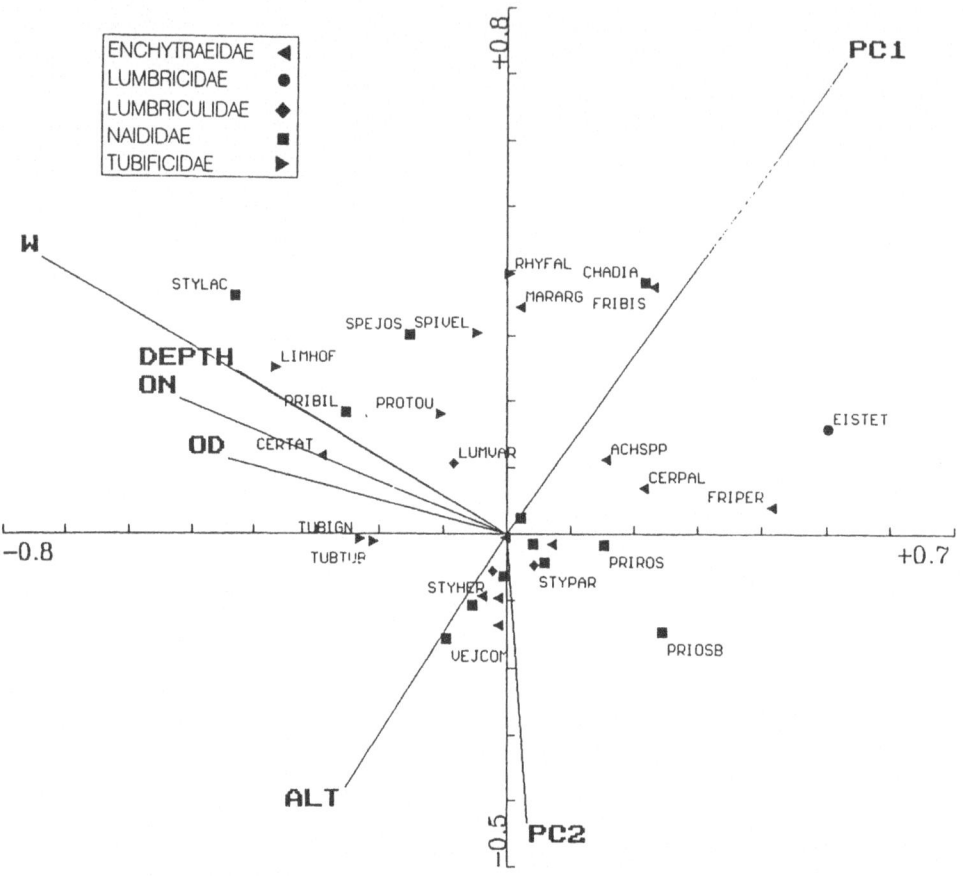

Fig. 4. CCA diagram for axes 1 and 2, based on oligochaete species abundance and physiographical and chemical variables (values for axes 1 and 2 from the PCA: PC1, PC2) from the 25 ▲ stations. (See stations in Fig. 1 and codes in Table 1).

Discussion

Most of the oligochaete species found in the unpolluted watercourses of the NW Iberian Peninsula exist over a wide range of environmental conditions, showing a large spatial distribution. Nevertheless, we can discern some general trends at the family level: the common lumbricid, *E. tetraedra*, and most of the enchytraeids inhabit the upper courses, the lumbriculids *S. heringianus* and *S. parvus* are characteristic species of the upper-middle courses, the tubificids and the lumbriculid *L. variegatus* are representative of the lower-middle courses, and the naidids, although largely widespread, are more characteristic of the middle and lower courses. Such general patterns of distribution can also be deduced from several other papers (Kasprzak & Szczesny, 1976; Giani & Lavandier, 1977; Schwank, 1982a; Vagner, 1987; Dumnicka, 1994).

Substratum and current velocity appear to be the major factors explaining the spatial distribution of the species in the study area. Wachs (1967), Kasprzak & Szczesny (1976), Giani & Martínez-Ansemil (1981) and Martínez-Ansemil (1990), studying watercourses with insignificant pollution and moderated thermic and chemical conditions, state that substratum and current velocity are the principal factors influencing the longitudinal distribution of oligochaetes. The role of these factors in the spatial distribution of oligochaetes also has been emphasized by Korn (1963), Fomenko (1972), Learner *et al.* (1978), Dumnicka (1978, 1982, 1994), Giani & Martínez-Ansemil (1981), Schwank (1982a, 1985), Uzunov (1982), Dumnicka & Kownacki (1988), Collado & Martínez-Ansemil (1991).

Although our study concerns many different stations from different watersheds, we could state that changes in the community structure along a nat-

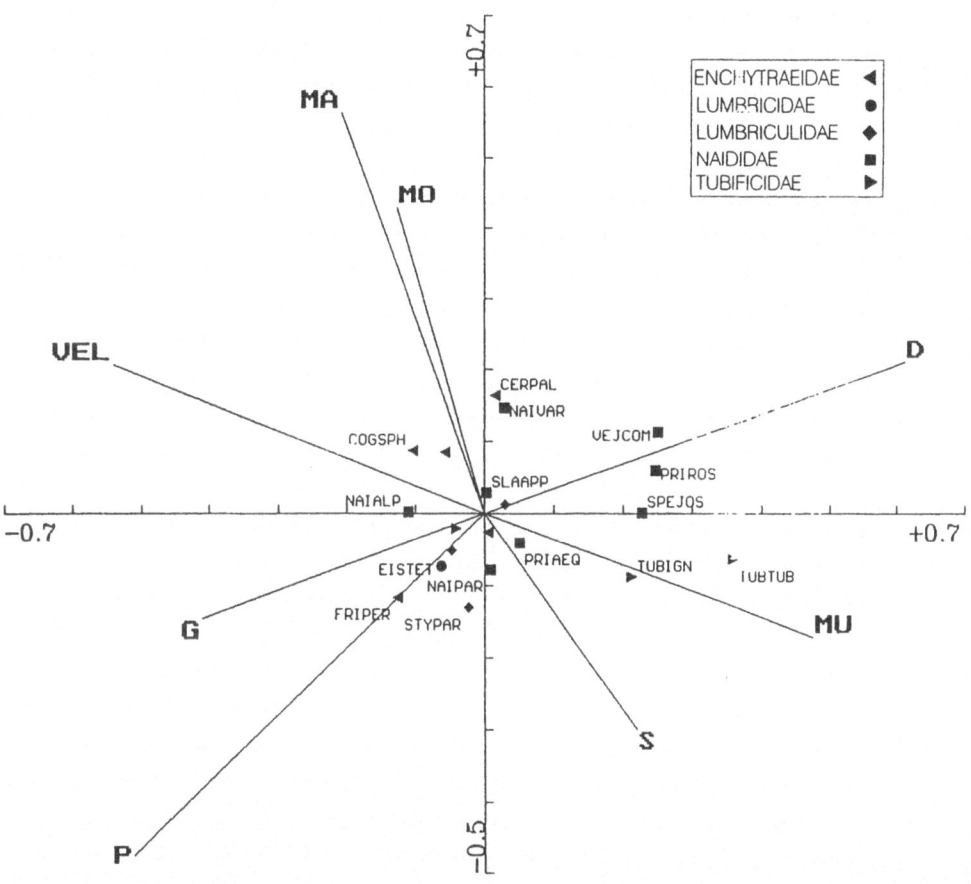

Fig. 5. CCA diagram for axes 1 and 3, based on oligochaete species abundance and substrate and current velocity as environmental variables in 172 samples. (See species codes in Table 2 and variables codes in the text: sampling).

ural watercourse mainly reflect the succession of the biotope mosaïcs that characterize each locality.

But, are the substratum and current velocity really the main factors directly related to oligochaete distribution? If we analyze the oligochaete substrate and current velocity preferences in our watercourses and others from different geographical regions (e.g. Berg, 1948, river Susaa; Korn, 1963, upper Danube; Wachs, 1967, river Fulda, Kasprzak, 1976, river Pradnik; Schwank, 1982a, 1985, several mountain watercourses in Germany; Lafont, 1989, several watercourses from Central and North France; Centurioni & Sambugar, 1986, Adige bassin) we observe that many species show different preferences for substratum and current velocity in different areas, considerably enlarging the ranges of their distribution with regard to these factors. So, it is likely that other less well-known factors, such as food, may play a more important role in the oligochaete distribution, as can be expected by

the results of studies on the feeding habits and diet of some species (Brinkhurst, 1974; Learner *et al.*, 1978; Moore, 1979; Frenzel, 1981; McMurtry *et al.*, 1983; Smith & Kaster, 1986; Lazim & Learner, 1987).

Wachs (1964), Rodríguez (1984) and Giani & Martínez-Ansemil (1981) found appreciable changes in the oligochaete community structure of rivers that could be related to important changes in water mineralization. We also found a possible relation within oligochaete distribution and water mineralization in spite of the relatively narrow range of our conductivity values. However as for the substratum and current velocity, water mineralization must have an even smaller direct effect than appears in our analyses. In fact, many species of oligochaetes inhabiting poorly mineralized freshwaters, have also been found in brackish waters (see for example Laakso, 1969; Särkkä, 1969; Pfannkuche, 1977).

82

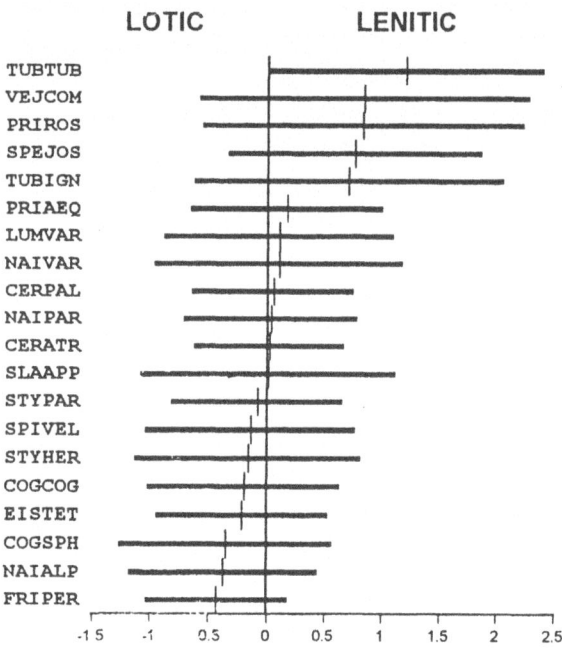

Fig. 6. Species scores and tolerances for axis 1 from the CCA based on oligochaete species abundance and substrate and current velocity as variables in 172 samples. (See codes in Table 2).

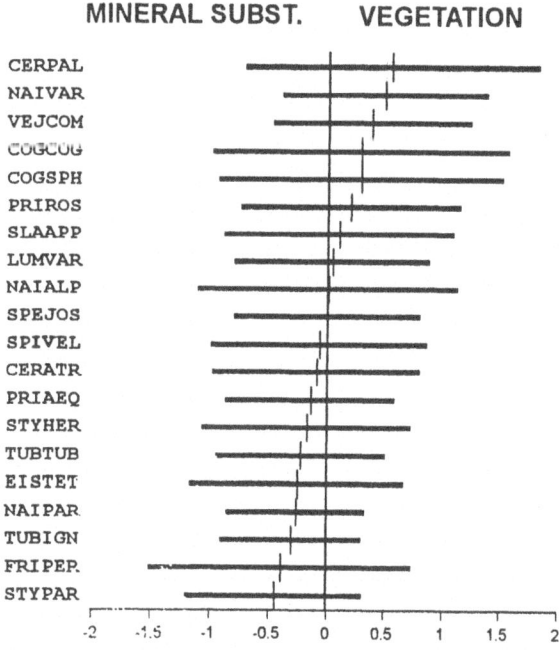

Fig. 7. Species scores and tolerances for axis 3 from the CCA based on oligochaete species abundance and substrate and current velocity as variables in 172 samples. (See codes in Table 2).

Most of the species inhabiting very acid waters in our watercourses, also inhabit basic waters in other areas, leading us to the same critical conclusion about the role of such a variable on the spatial distribution of oligochaetes.

Acknowledgments

We are much indebted to Dr Juan Freire for his assistance in the statistical treatments.

References

Berg, K., 1948. Biological studies on the River Susaa. Folia limnol. scand. 4: 1–318.

Brinkhurst, R. O., 1974. Factors mediating interspecific aggregation of tubificid oligochaetes. J. Fish. Res. Bd Can. 31: 460–462.

Centurioni, C. & B. Sambugar, 1986. I macroinvertebrati del bacino idrografico dell'Adige. I. Gli Oligocheti. In M. G. Braioni & S. Ruffo (eds), Ricerche sulla qualità delle acque dell'Adige. Mem. Mus. Civ. St. Nat. Verona (II Ser.), Sez. Biologica 6: 125–151.

Collado, R. & E. Martínez-Ansemil, 1991. Les Oligochètes aquatiques de la Péninsule Ibérique: premières données sur le Portugal. Bull. Soc. Hist. nat. Toulouse 127: 57–69.

De Brum Ferreira, D., 1981. Carte géomorphologique du Portugal. Mem. Centro Estudos Geog. 6: 1–53.

Dumnicka, E., 1978. Communities of oligochaetes (Oligochaeta) of the River Nida and its tributaries. Acta hydrobiol. 20: 117–141.

Dumnicka, E., 1982. Stream ecosystems in mountain grassland (West Carpathians). 9. Oligochaeta. Acta hydrobiol. 24: 391–398.

Dumnicka, E., 1994. Communities of oligochaetes in mountain streams of Poland. Hydrobiologia/Dev. Hydrobiol. 278: 107–110.

Dumnicka, E. & A. Kownacki, 1988. A regulated river ecosystem in a polluted section of the Upper Vistula. Acta hydrobiol. 30: 81–97.

Dumnicka, E. & K. Pasternak, 1978. The influence of physico-chemical properties of water and bottom sediments in the River Nida on the distribution and numbers of Oligochaeta. Acta hydrobiol. 20: 215–232.

Fomenko, N. V., 1972. Ecological groups of Oligochaeta worms in the Dnieper basin. In Aquatic Oligochaeta worms. Taxonomy, ecology and faunistic studies in the USSR. Amerind Publish., New Delhi (1980) (translated from Russian): 105–118.

Frenzel, P., 1981. Untersuchungen zur Ökologie der Naididae des Bodensees. Die Nische von *Chaetogaster* und *Amphichaeta*. Arch. hydrobiol. 91: 45–55.

Giani, N., 1984. Le Riou Mort, affluent du Lot, pollué par métaux lourds. IV. Etude des Oligochètes. Ann. Limnol. 20: 167- 181.

Giani, N. & P. Lavandier, 1977. Les Oligochètes du torrent d'Estaragne (Pyrénées centrales). Bull. Soc. Hist. nat. Toulouse 113: 234–243.

Giani, N. & E. Martínez-Ansemil, 1981. Contribution à la connaissance des Oligochètes aquatiques du bassin de l'Argens. Ann. Limnol. 17: 121–141.

Jongman, R. H. G., C. J. F. ter Braak & O. F. R. van Tongeren, 1987. Data Analysis in Community and Landscape Ecology. Pudic, Wageningen, 299 pp.

Kasprzak, K., 1976. Materials to the fauna of Oligochaeta of the Ojców National Park and its vicinity – the Pradnik-Bialucha stream. Acta hydrobiol. 18: 277–289.

Kasprzak, K. & B. Szczesny, 1976. Oligochaetes (Oligochaeta) of the River Raba. Acta hydrobiol. 18: 75–87.

Kingston, J. C., H. J. B. Birks, A. J. Uutala, B. F. Cumming & J. P. Smol, 1992. Assessing trends in fishery resources and lake water aluminum from paleolimnological analyses of siliceous algae. Can. J. Fish. aquat. Sci. 49: 116–127.

Korn, H., 1963. Studien zur Ökologie der Oligochaeten in der oberen Donau unter Berücksichtigung der Abwassereinwirkungen. Arch. Hydrobiol. 27: 131–182.

Laakso, M., 1969. Oligochaeta from brackish water near Tvärminne, south-west Finland. Ann. Zool. Fennici 6: 98–111.

Lafont, M., 1989. Contribution à la gestion des eaux continentales: Utilisasion des Oligochètes comme descripteurs de l'état biologique et du degré de pollution des eaux et des sédiments. Thèse d'Etat, Univ. Claude Bernard- Lyon I, 311 pp.

Lazim, M. N. & M. A. Learner, 1987. The influence of sediment composition and leaf litter on the distribution of tubificid worms (Oligochaeta). Oecologia (Berlin) 72: 131–136.

Learner, M. A., G. Lochhead & B. D. Hughes, 1978. A review of the biology of British Naididae (Oligochaeta) with emphasis on the lotic environment. Freshwat. Biol. 8: 357–375.

Martínez-Ansemil, E., 1990. Etude biologique et écologique sur les Oligochètes aquatiques de la rivière Tambre et ses milieux associés (Galice, Espagne). Ann. Limnol. 26: 131–151.

McMurtry, M. J., D. J. Rapport & K. E. Chua, 1983. Substrate selection of tubificid oligochaetes. Can. J. Fish. aquat. Sci. 40: 1639–1646.

Membiela, P., C. Montes & E. Martínez-Ansemil, 1991. Características hidroquímicas de los ríos de Galicia (NW Península Ibérica). Limnetica 7: 163–174.

Moore, J. W., 1979. Influence of food availability and other factors on the composition, structure and density on a subartic population of benthic invertebrates. Hydrobiologia 62: 215–223.

Pfannkuche, O., 1977. Ökologische und systematische untersuchungen an naidomorphen Oligochaeten brackiger und limnischer biotope. Dissertation, Univ. Hamburg, 138 pp.

Ribeiro, O., H. Lautensach & S. Daveau, 1988. Geografia de Portugal. II. O ritmo climático e a paisagem. J. Sá da Costa, Lisboa, 623 pp.

Rodríguez, P., 1984. Los Oligoquetos acuáticos del río Nervión (Vizcaya, España): Resultados faunísticos generales. Limnetica 1: 169–178.

Särkkä, J., 1969. The bottom fauna at the mouth of the river Kokemäenjoki, southwestern Finland. Ann. Zool. Fennici 6: 275–288.

Schwank, P., 1982a. Turbellarien, Oligochaeten und Archianneliden des Breitenbachs und anderer oberhessischer Mittelgebirgsbäche. IV. Allgemeine Grundlagen der Verbreitung von Turbellarien und Oligochaeten in Fliessgewässern. Arch. Hydrobiol. 62: 254–290.

Schwank, P., 1982b. Turbellarien, Oligochaeten und Archianneliden des Breitenbachs und anderer oberhessischer Mittelgebirgsbäche. III. Die Taxozönosen der Turbellarien und Oligochaeten in Fliessgewässern – eine synökologische Gliederung. Arch. Hydrobiol. 62: 191–253.

Schwank, P., 1985. Differentiation of the coenoses of helminthes and annelida in exposed lotic microhabitats in mountain streams. Arch. Hydrobiol. 103: 535–543.

Smith, M. E. & J. L. Kaster, 1986. Feeding habits and dietary overlap of Naididae (Oligochaeta) from a bog stream. Hydrobiologia 137: 193–201.

Ter Braak, C. J. F., 1987. CANOCO – A FORTRAN program for canonical community ordination by [partial] [detrended] [canonical] correspondence analysis, principal component analysis and redundancy analysis. TNO Institute of Applied Computer Science, Wageningen, the Netherlands.

Timm, T., 1970. On the fauna of the Estonian Oligochaeta. Pedobiologia 10: 52–78.

Uzunov, J., 1982. Statistical assessment of the significance of both bottom substrata and saprobity for the distribution of aquatic Oligochaeta in rivers. Limnologica (Berlin) 14: 353–361.

Vagner, D., 1987. A contribution to knowledge of fauna of Oligochaeta (Annelida, Clitellata) of the river Una. Biosistematika 13: 45–61.

Verdonschot, P. F. M., 1990. Ecological characterization of surface waters in the province of Overijssel (The Netherlands). Thesis. University of Wageningen, 255 pp.

Wachs, B., 1964. Beitrag zur Oligochaeten-Fauna eines schiffbaren Flusses. Z. angew. Zool. 51: 179–192.

Wachs, B., 1967. Die Oligochaeten-Fauna der Fliessgewässer unter besonderer Berücksichtigung der Beziehungen zwischen der Tubificiden-Besiedlung und dem Substrat. Arch. Hydrobiol. 63: 310–386.

Walker, I. R., J. P. Smol, D. R. Engstrom & H. J. B. Birks, 1991. An assessment of Chironomidae as quantitative indicators of past climatic change. Can. J. Fish. aquat. Sci. 48: 975–987.

Hydrobiologia **334**: 85–88, 1996.
K. A. Coates, Trefor B. Reynoldson & Thomas B. Reynoldson (eds), Aquatic Oligochaete Biology VI.
© 1996 *Kluwer Academic Publishers.*

The effect of water movement on the distribution of Oligochaeta in lakes

T. D. Slepukhina
Institute of Lake Research, Russian Academy of Sciences, ull. Sevastjanova 9, St. Petersburg, 196199 Russia

Key words: Oligochaeta, large lakes, wind and wave effects

Abstract

The Oligochaeta are more exposed to disturbance under conditions of intense water movement than other invertebrates, particularly when the substrate is unstable. Furthermore, low mobility of bottom water, which can lead to excessive accumulation of organic matter in sediments, and to the formation of 'sapropel', can also be limiting to the distribution of oligochaetes. Organic contamination in conditions of low water dynamics can result in the absence of Oligochaeta through oxygen depletion. Dynamic water movement can stimulate the mass development of Oligochaeta in organically polluted areas.

Introduction

Mass water movement effects the distribution and development of bottom habitats in lakes through a number of processes. Variation in water movement near the bottom sediment can form heterogeneous sediment distributions. The concentration of oxygen, the water temperature and other physico-chemical variables near the sediment-water interface are also affected by patterns of water movement. Finally, extremely rapid water movement may be physically detrimental to benthic organisms.

In large lakes wave activity is the dominant type of water movement in the shallow areas in deep ones and of the entire area of shallow ones. The effect of wave action on bottom invertebrates has been noted in the literature, but there has been little quantitative evaluation. Although Giziński (1974) established a correlation between the quantity of benthos and the morphometry of lakes, and introduced the term 'mixing index'.

Our work on Kubenskoye Lake (Raspopov *et al.*, 1976) showed a quantitative dependance between certain hydrodynamic indices and the density and biomass of benthic populations. We have postulated that as bottom habitats develop, over months or even years, a correlation between attributes of the benthic community and wave action will be found on the basis of long

term data on average and maximal wind velocities and wave measurments. This concept has been discussed by Håkanson (1977), Maitland (1979), Barton & Carter (1982) and Bailey (1988).

Materials and methods

These investigations took place on the large lakes of north-west Russia: Kubenskoye (surface area 417 km^2), Vozhe (surface area 418 km^2), Beloje (surface area 1280 km^2), Onega (surface area 9690 km^2) and Ladoga (surface area 17 670 km^2).

Samples of zoobenthos were obtained using an Ekman grab (two sample-units per station). Invertebrates sampled from macrophytes, were collected by washing the leaves and stems of the plants.

To analyse the influence of hydrodynamic conditions we used the Russian 'Building Standards and Rules' (1983). We calculated the horizontal component of orbital maximal velocity of water movement near the bottom (V_{bm}) by:

$$V_{bm} = \pm \frac{h}{\sqrt{\frac{21}{\pi g} \cdot sh\frac{4\pi}{\lambda} \cdot H}} m \cdot sec^{-1}, \quad (1)$$

where h = height of waves; λ = length of waves; H = water depth; g = acceleration due to gravity; sh = hyperbolic sine.

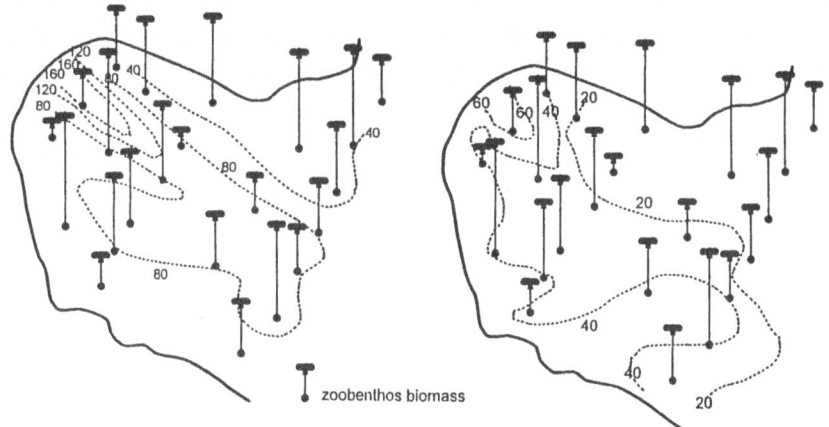

Fig. 1. The biomass of zoobenthos (vertical lines g.m^{-2}) in Pensguba and Pmax (μH.sm^{-1}) under two wind speeds from the SE 17 m.sec^{-1} and 7 m.sec^{-1}.

The value of orbital velocity of water movement was variable at each point. Each wave, moving through some point, changes its direction twice after stopping, when the velocity was equal to zero. In our work we used its maximal value under the most frequent and under maximal wind conditions (Raspopov *et al.*, 1988).

Wave pressure on zoobenthos was determined as $P_{\max}$ – maximal resultant pressure on the length unit of horizontal cylindrical streamlined obstacle with diameter of 0.01 m, situated on the bottom and defined by:

$$P_{\max} = P_x^2 + P_z^2; P_x = \gamma \pi d \frac{h^2}{\lambda} \cdot E_x; P_z = -\frac{g}{5} P_x.$$

(2)

where E_x = velocity coefficient, depending on the depth of the water; d = diameter of zoobenthos; γ = density of water; P_x = horizontal pressure on zoobenthos; P_z = vertical pressure on zoobenthos; $P_{\max}$ = maximal resulting pressure.

All the values were calculated, however, they were checked experimentally.

Results and discussion

On exposed shallow shores of large lakes a variety of wind velocity and direction the wave action affects oligochaetes at each point in shallow water communities, even though the intensity of $P_{\max}$ and V_{bm} varies under different wind conditions (Fig.1). Oligochaetes,

relative to other invertebrates, were more sensitive to disturbance under excessive water movements and particularly when the substrate was unstable. In exposed habitats, with winds greater than 7 m.s^{-1} (the mean wind velocity on Ladoga and Onega lakes) and depths, where V_{bm} was equal to or more than 1.8 m.s^{-1}, oligochaetes were absent from sand and gravel substrates. However, in stony habitats a number of worm species were observed. The stability of the substrate was a critical factor in determining the effect of rapid water movement.

Low water mobility near the bottom, which leads to the accumulation of organic matter in sediments, and the formation of sapropel, can also be limiting to oligochaetes. This is shown in data collected in 1981–1982 from Glubokaja Bay (Lake Onega), which is protected from winds and waves from the open lake. North west winds of 17 m.s^{-1} from the direction of the lake reduced V_{bm} and $P_{\max}$ along the bay and in the southern end of the bay values were close to zero (Fig. 2). The main substrate in the southern part of the bay was mud with a high (24 – 57%) organic content (Raspopov *et al.*, 1988, 1990). The density, biomass and number of species of oligochaetes decreased in the stagnant part of the bay; at most sites oligochaetes were absent or only *Limnodrilus hoffmeisteri* Clap. and *Potamothrix hammoniensis* (Mich) occurred (Table 1).

Our work on several large lakes has shown, that the density and biomass of invertebrates on the stems and leaves of macrophytes were lower than the density and

Table 1. Density (ind. m^{-2}), biomass (mg m^{-2}), number of oligochaete species (sp.) and species diversity (div) for all invertebrates for 37 sites in Glubokaja Bay, Onega Lake.

Site	Den.	Biom.	Sp.	Div
1	2	20	0	0.69
2	0	0	0	0.44
3	120	720	1	0.52
4	0	0	0	0.85
5	0	0	0	0.49
6	0	0	0	0
7	0	0	0	0.5
8	0	0	0	0.2
9	0	0	0	0.2
10	0	0	0	0.62
11	65	390	1	0.84
12	130	940	2	0.26
13	80	360	2	0.69
14	20	10	1	0.48
15	0	0	0	1.07
16	0	0	0	1.07
17	0	0	0	0.16
18	0	0	0	0
19	0	0	0	0.93
20	0	0	0	0.53
21	20	20	1	0.98
22	0	0	0	0.81
23	0	0	0	1.04
24	0	0	0	1.23
25	180	340	1	1.8
26	0	0	0	1.04
27	0	0	0	0.97
28	40	20	1	1.41
29	600	1680	4	1.61
30	200	440	2	1.78
31	20	80	1	1.92
32	680	2280	8	2.53
34	1055	2775	8	2.48
35	130	280	4	1.37
36	120	220	1	1.35
37	280	380	3	1.15

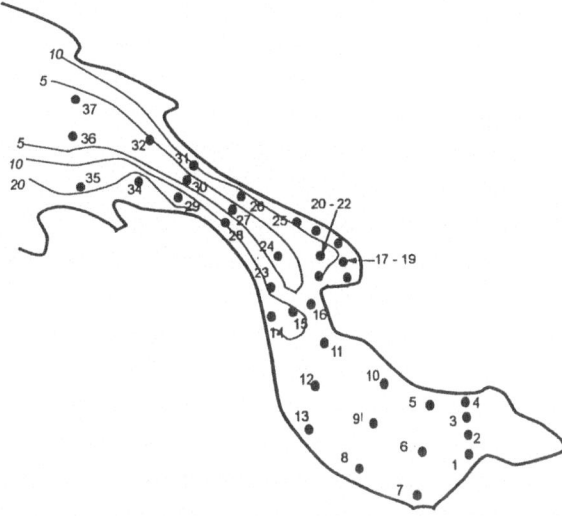

Fig. 2. Sampling site (●) and P_{max} under NW wind 17 m sec^{-1} in Glubokaja Bay, Lake Onega.

Organic contamination in areas with low water movement results in the absence of oligochaetes in these areas because of low oxygen. In Lake Ladoga such situations occur in isolated bays and bottom depressions. Up until 1986 in Shchuchij Bay of Lake Ladoga and in a 20–30 m deep depression near Priosersk there was a 'dead zone' where no macroinvertebrates were observed. Between Shchuchij Bay and the depression there was a large shallow area, covered with stones and pebbles, where intense wave action and currents resulted in an oxygenated bottom layer. Several oligochaetes species were observed in this area including *Lamprodrilus isoporus* Mich, *Uncinais uncinata* Oersted, *Nais barbata* Mull. The intense water movement stimulated the mass development of oligochaetes in such organically polluted areas (Slepukhina, 1984).

Conclusion

The distribution of oligochaetes in lakes depends on many different factors, most of them regulated by hydrodynamics. Thus, the structure and development of oligochaete communities is directly related to water movement. Exsessive dynamic action and stagnation are equally detrimental to these animals.

biomass in bottom sediment in exposed habitats and the reverse in protected habitats (Table 2). In exposed habitats the low numbers of animals on macrophytes was due to the direct action of waves (washing off organisms) and the reduced periphyton on the macrophyte surface, presumably also an effect of wave action (Richkova, 1978, 1987). In protected habitats, low water movement results in oxygen deficiency in sediment habitats, and reduced benthic populations.

88

Table 2. Density (ind. m^{-2}) and biomass (g m^{-2}) of invertebrates on stems of *Phragmites* and in bottom sediment in weed beds in exposed and protected zones in four large Russian lakes.

Lake	Zones	Periphyton biomass	Stems den.	Biom.	Bottom den.	Biom.
Ladoga	Exposed	0.03–0.14	200	0.5	1500–2100	5.0–12.0
	Protected	2.0–12.0	450–10990	0.5–7.7	0–100	0–0.5
Onega	Exposed	0.05–0.4	100	0.1	200–1800	11.0–12.0
	Protected	5.0	400–10000	0.5–10.0	0–100	0–0.5
Kubenskoje	Exposed	0	72	0.04	2100	6.3
	Protected	0.68	15740	2.5	0–100	0–0.5
Vozhe	Exposed	0.14	2400	0.7	1040	5.9
	Protected	3.12	8300	2.0	2600	2.0

References

Bailey, R. C., 1988. Correlations between species richness and exposure: Freshwater molluscs and macrophytes. Hydrobiologia 162, 2: 183–191.

Barton, D. R. & J. C. Carter, 1982. Shallow water epilithic invertebrate communities of eastern Georgian Bay, Ontario, in relation to exposure to wave action. Can. J. Zool. 60, 5: 984–933.

Giziński, A., 1974. Typologia faunistiezna eutroficznych jezior Połnocnej Polski. Toruň, 75 s.

Håkanson, L.,1977. The influence of wind, fetch and water depth on distribution of sediments in lake Vännern, Sweden Can. J. Earth Sci. 14, 3: 397–412.

Maitland, P. S., 1979. The distribution of zoobenthos and sediments in Loch Leven, Kinross, Scotland. Arch. Hydrobiol. 85, 1: 98–125

Raspopov, I. M., T. D. Slepukhina, F. F. Vorontzov & M. A. Richkova, 1976. Rol dinamiki vod v formirovanii biocenosov litorali oz. Kubenskogo. The role of water dynamics in formation of littoral biocenoses in Kubenkoye Lake. Abstracts of communications of the 3rd All union Congress of Hydrobiological society. Riga, 2: 235–236 [In Russian.]

Raspopov, I. M., T. D. Slepukhina, F. F. Vorontzov & O. H. Dotzenko, 1988. Wave effects on the bottom biocenoses in the Onega Lake bays. 1988. Arch. Hydrobiol. 112, 1; 115–124.

Raspopov, I. M., F. F. Vorontzov, T. D. Slepukhina, O. H. Dotzenko & M. A. Richkova, 1990. Rol volneniya v formirovanii biocenosov bentosa bolshikh ozer. Role of the water roughness in the benthos biocenoses forming in the great lakes. Nauka. Leningrad, 114 p. [In Russian.]

Richkova, M. A., 1978. Perifyton ozer Vozhe i Lacha. V kn. Gidrobiologia ozer Vozhe i Lacha. Periphyton of Vozhe and Lacha lakes. In Hydrobiology of Vozhe and Lacha lakes. Nauka, Leningrad: 28–33. [In Russian.]

Richkova, M. A., 1987. Perifiton ozera i ego produktsia. V kn. Sovremennoye sostoyanie ekosistemy Ladozhskogo ozera. Periphyton of the lake and its productivity. In Modern state of Ladoga Lake ecosystem. Nauka. Leningrad: 116–119. [In Russian.]

Slepukhina, T. D., 1984. Comparison of different methods of water quality evaluation by means of oligochaetes. Hydrobiologia 115: 183–186.

'Stroitelnye normy i pravila'. 'Building Standards and Rules', 1983. Moscow, 38 pp. [In Russian.]

Hydrobiologia **334**: 89–95, 1996.
K. A. Coates, Trefor B. Reynoldson & Thomas B. Reynoldson (eds), Aquatic Oligochaete Biology VI.
© 1996 *Kluwer Academic Publishers.*

Oligochaeta of Lake Taimyr: a preliminary survey

Tarmo Timm

Institute of Zoology and Botany (Estonian Academy of Sciences), Võrtsjärv Limnological Station. EE–2454 Rannu, Tartumaa, Estonia

Key words: Oligochaeta, large lakes, Siberia, cryofauna

Abstract

Lake Taimyr in Siberia is northernmost among the world's large lakes: 73°40′–75°20′ N, 99–106° E. The lake area is up to 4650 km^2 in summer, with a maximum depth of 26 m and a mean depth of only 2.8 m. The ice-free period lasts about three months. The water level sinks 5.5–6 m during winter, so that 85% of the bottom surface is frozen into ice for some time and subjected to negative temperatures, probably down to -20 °C. In artificially melted sediment samples, 75–92% of animals survived. The average summer biomass of zoobenthos is about 1 g m^{-2} wet weight, a half of this being formed by Oligochaeta. Altogether 76 samples with 3742 oligochaete specimens collected by V. N. Grëze in 1943–1944 were studied. At least 14 taxa of Tubificidae, Lumbriculidae, and Enchytraeidae were found in the lake, and some more enchytraeids in an adjacent river. Many immature animals could not be identified to species. Naididae were completely lacking probably due to the absence of macrovegetation. The shallow freezing zone is inhabited mostly by *Alexandrovia ringulata*. The profundal fauna is dominated by *Lamprodrilus isoporus*, *Stylodrilus* sp., and *Isochaetides* sp.

Introduction

From papers on Oligochaeta collected during the Arctic expeditions (e.g., Eisen, 1879; Čejka, 1914; Smith & Welch, 1919), one can conclude that the freshwater and terrestrial northern oligochaete fauna consists mostly of Enchytraeidae, with a few species of Lumbricidae and Lumbriculidae. The northernmost lake studied in North America (Char Lake at 75° N) is inhabited by only two enchytraeid species (Nurminen, 1973).

However, Lake Taimyr in North Siberia supports a much more diverse oligochaete fauna. It was investigated in 1943–1944 by an expedition of the Siberian Branch of the All-Union Research Institute of Lake and River Fisheries, led by V. N. Grëze (1947a, 1947b, 1957). The oligochaetes collected on that expedition are studied further here. In spite of severe environmental conditions, Lake Taimyr shelters a diverse oligochaete fauna made up mainly of Tubificidae and Lumbriculidae, with a smaller complement of enchytraeids.

In this paper, the general composition of the oligochaete fauna and its zonal distribution in the lake are dealt with. A taxonomic scrutiny of the separate species is in progress and will be published elsewhere.

Habitat description

Lake Taimyr lies in the middle of the Taimyr Peninsula in North Siberia, Russia, 73°40′–75°20′ N and 99–106° E (Atlas SSSR, 1962), on average 6 m above sea level (Haack Grosser Weltatlas, 1968), between the Byrranga Mountains in the north and a tundra-covered lowland in the south. The Lake is northernmost among the world's large lakes. The Upper Taimyra River is its biggest inflow, the outflowing Lower Taimyra River runs into the Kara Sea (Fig. 1). The catchment area of the lake occupies 80 000 km^2; it was partially flooded by a late marine transgression (Grëze, 1947a). Due to the continuing rise of the land, the average water level is decreasing at a rate of about one meter each century (Grëze, 1947a, 1957).

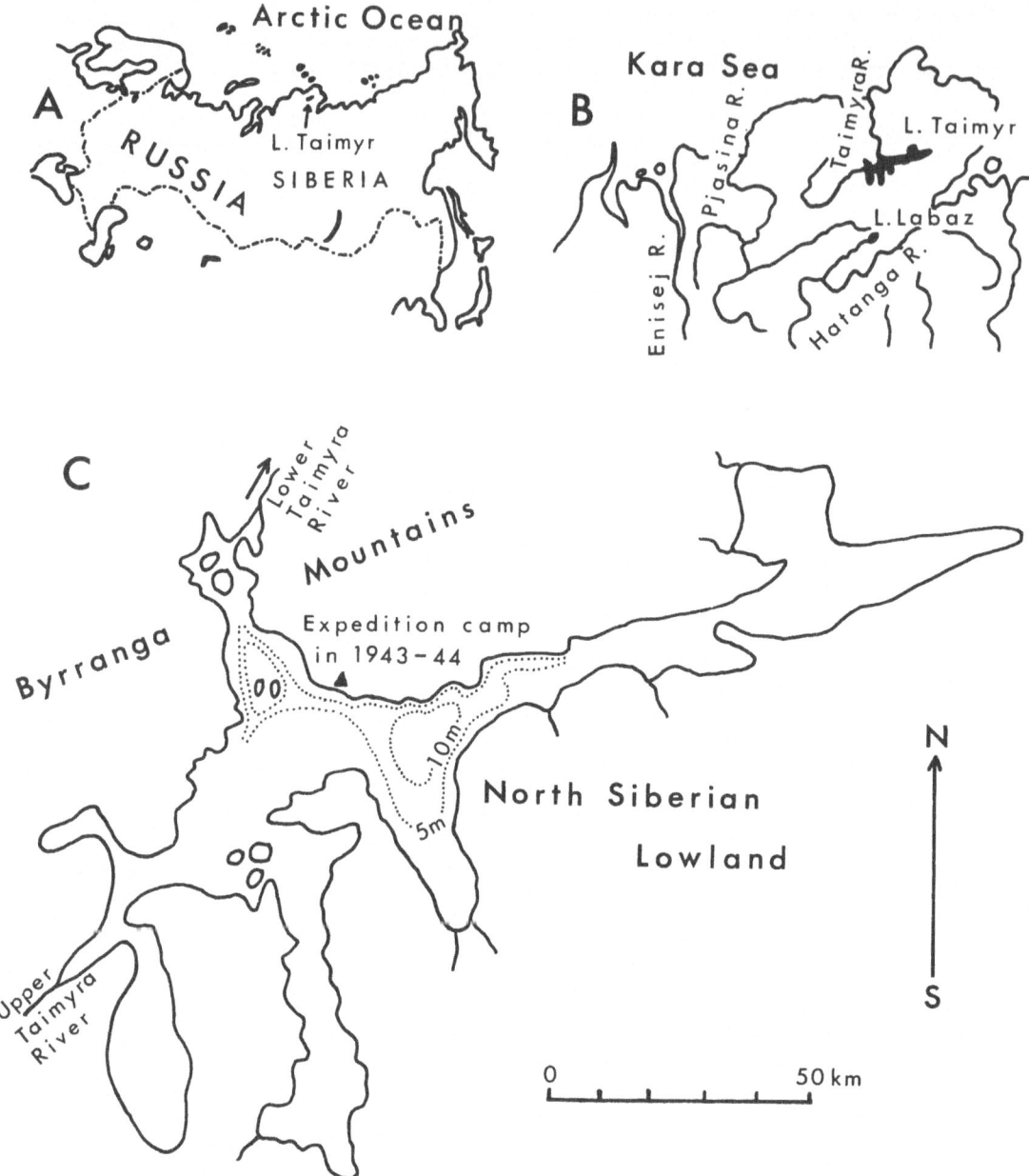

Fig. 1. Location and schematic map of Lake Taimyr. A – Russia; B – Taimyr Peninsula; C – Lake Taimyr.

The surface area of Lake Taimyr reaches 4650 km² in early July when snow-melt water accumulates, decreasing to 4000 km² in autumn. The maximum depth in the high water period reaches 26 m, with an average depth of only 2.8 m. For eighty percent of the lake surface area the water depth is less than 4 m. The lake is covered with ice about 2 m thick for at least nine months of a year. The last ice masses thaw in July, then freezing starts again in mid-September; thus the fully ice-free period lasts for not more than two months. The water level continues to fall through the winter. By spring, 75% of the water volume has discharged, and the water level has dropped by 5.5–6 m. In autumn, the shallows freeze to the bottom and the descending ice surface becomes concave in the deeper areas (Fig. 2). Surface sediments of 85% of the bottom,

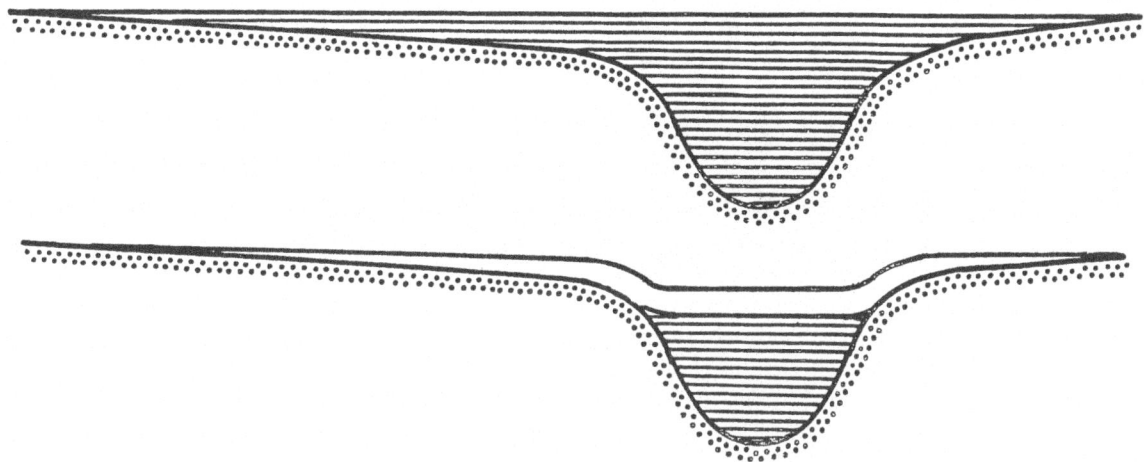

Fig. 2. Scheme of the annual changes of the water and ice level in Lake Taimyr: above – situation in summer; below – situation in late winter; stippled, sediments; lined, water; unshaded, ice.

down to summer depths of up to 7–8 m, freeze solid in winter.

Temperatures of −10.0 ° to −11.4 °C were measured in May and June in the sediment, under an ice sheet 1 m thick; presumably temperatures can be about −17 ° to −20 °C in December. The lake is homothermic in summer. Surface water temperatures were 4.3 ° to 12.3 °C in July 1943 and 0.1 ° to 8.8 °C in July 1944. There were only 500 positive degree-days in the water of one bay, in 1943 (Grëze, 1947a, 1947b).

Water transparency was 0.25 m at flood times but 1.2–1.5 m in August 1943. The water was soft, 1.44 German degrees, dH ° (equal 4.03 International degrees of hardness, or 4.03 meq l^{-1}), with a CaO content of 9.60 mg l^{-1}, MgO 3.47 mg l^{-1}, SiO$_2$ 1.60 mg l^{-1}, SO$_4^{2-}$ 5.35 mg l^{-1}, CO$_3^{2-}$ 10.2 mg l^{-1}; no Fe or P ions were found. The winter oxygen saturation of the near-bottom water was high at shallow depths, gradually decreasing to 15% at a depth of 20 m (Grëze, 1947a).

The lake is devoid of any macrovegetation and, consequently, of phytophilous animal groups. Common 'glacial relict' crustaceans, some marine tintinnoideans, and some possible Baikalian elements [*Manayunkia baicalensis* Nusbaum (presently regarded as a synonym of *M. speciosa*) (Polychaeta), *Hislopia placoides* Korotneff (Entoprocta), species of *Echinogammarus* (Amphipoda)] are characteristic of the Lake Taimyr fauna (Grëze, 1947a). The summer plankton biomass (mostly of diatoms and rotifers) was 1.7–5.0 g

m^{-3} in 1943–1944, which corresponds to the 'polar eutrophic' lake type (Grëze, 1957).

Five benthal zones were distinguished by Grëze (1957):

(1) Sand at summer depths of 0–2.5 m, freezing on an average for 290 days a year. The bottom fauna was poor, consisting mainly of the chironomid genus *Orthocladius*, with an average abundance of 378 ind m^{-2} and wet biomass 0.246 g m^{-2}.

(2) Muddy sand at depths of 2.5–5 m, freezing for about 250 days. Zoobenthos (562 ind m^{-2}, 1.139 g m^{-2}) was dominated by Oligochaeta.

(3) Mud at depths of 5–15 m, usually not freezing. Zoobenthos was the richest here (2,403 ind m^{-2}, 4.608 g m^{-2}), dominated by Oligochaeta and Amphipoda.

(4) Mud in the deepest profundal, at depths over 15 m, subjected to a moderate oxygen deficiency in winter. Zoobenthos (900 ind m^{-2}, 0.550 g m^{-2}) consisted almost exclusively of Oligochaeta.

(5) Stony bottom washed by currents, at depths of 3–4 m, rarely up to 9 m, mostly freezing in winter. Zoobenthos was diverse and abundant in deeper, non-freezing places only. The average numbers of zoobenthic organisms were low (188 ind m^{-2}, 0.585 g m^{-2}). Oligochaeta and Amphipoda dominated.

The average wet biomass for the whole lake was about 1 g m^{-2} in the summer of 1943 and 1944, a half of this being formed by Oligochaeta. About 38% of the number of animals observed in the lake in summer must be frozen into ice together with the sediment in winter, with most of them (mainly Chironomidae, Oligochaeta, and Nematoda) surviving there (Grëze, 1947b, 1957).

Material and methods

The collection of oligochaetes from Lake Taimyr comprised 74 vials containing 3,612 specimens; specimens from the Gorbita River (upper course of the Taimyra River) were in 2 vials containing 130 specimens. All the specimens had been preserved and were stored in formalin. The samples were collected at 33 stations on the lake from July 1943 to July 1944, and at two stations on the Gorbita River in April 1943. Not all stations can be relocated on the Lake since no sampling diary was available. Several stations were sampled on a number of dates, others only once. Winter sampling was done from ice holes.

A Petersen grab with an area of 250 cm^2 served as the main sampling instrument. A sample commonly consisted of two hauls. On a hard bottom, a semi-quantitative 'Grëze drag' was trailed for a distance. Solid rectangular pieces with a surface area of 500 cm^2 were cut out of the frozen sediment. The samples were washed on a silk sieve No 26 (26 threads per cm) and hand-sorted (Grëze, 1947b, 1957).

Oligochaetes were studied here as whole mounts in glycerine. Permanent whole mounts in Canada balsam were made as voucher specimens of every taxon. Sagittal and transverse serial sections 5 μm thick, stained after Mallory or Hauser, were made of some mature individuals. Most of the taxa have so far been identified to a generic or family level only.

Results

In Lake Taimyr, 14 taxa belonging to the families Tubificidae (7 taxa), Lumbriculidae (3), and Enchytraeidae (4) were observed. Five more enchytraeid taxa were found only in the Gorbita River. Over all samples, Tubificidae formed 76.7% of the individuals, Lumbriculidae 19.5%, and Enchytraeidae only 3.8%.

Tubificidae

Alexandrovia ringulata (Sokol'skaja, 1961) was the most common and abundant tubificid species, with 2,341 individuals out of the total 3,742. It was found in 40 samples, including one from the river. Shallows

covered with 2–5 m of water in summer and frozen in winter are its main habitat. Only a few individuals were found in deeper locations that do not freeze at any time. *Alexandrovia ringulata* is usually characterized by an armour of adhering silt particles. Many scurfy individuals, with peeling armour, were observed while many others appeared entirely naked and transparent. The proportion of the latter was greater among small worms and in cooler seasons. Among the numerous mature individuals, clitellate specimens were found in July, August, and October, post-reproductive specimens, with a naked cuticle in the clitellar region, were found only once, in May. Mature individuals always formed a minority of the specimens in a sample. Cocoons, identifiable as those of *A. ringulata*, were in collections from September and November.

Limnodrilus profundicola (Verrill, 1871) was represented by a single mature specimen with penial sheaths. Some immature specimens might belong to this species but they are difficult to distinguish from the young of *Isochaetides* sp., following.

Isochaetides sp. is Limnodrilus-like in habitus and setae but often has spermathecal setae when mature. A dissected specimen revealed a tube-like atrium with a subapical entrance of the vas deferens, and a very short and thin, ring-shaped penial sheath as in the Baikalian tubificids, *Isochaetides excavatus* Hrabě, 1982, *Isochaetides eximius* Semernoj, 1982, *Tubifex bazikalovae* Čekanovskaja, 1975, and *Tubifex crassiseptus* Semernoj, 1982.

Isochaetides sp. occurred in 33 samples, possibly mixed with immatures of *L. profundicola* and, in one of these samples, with the single mature specimen of the latter, taken from different depth zones of the lake and also from the river. Combined these two taxa were represented by 496 individuals. Clitellate specimens of *Isochaetides* sp. formed a small part of the population in June–August and one specimen was found in January.

Tasserkidrilus sp., 21 specimens including two maturing ones (in summer samples), was identified in ten samples from various depths below the freezing zone. The shape of the penial sheath resembles that of *Tasserkidrilus kessleri* (Hrabě, 1962) and *Tasserkidrilus acapillatus* (Finogenova, 1972). Several individuals have a very few hair setae while others have none.

Baikalodrilus sp. is a papillate tubificid with hair setae, with single-pointed and bifid ventral setae, but lacking spermathecal setae and with an organic 'crystal' in the atrium. Eight specimens in three samples were collected below the freezing zone. Clitellate worms and distinctive cocoons were found in samples taken in April and May.

Two tubificids of indeterminate genera were each represented by two juveniles. Immature specimens of the first species have two bifid setae per bundle, with an enlarged and bent upper tooth; specimens of the second have commonplace hair and bifid setae.

Lumbriculidae

Lamprodrilus isoporus Michaelsen, 1901 was the most abundant lumbriculid species, with 535 specimens in 42 lacustrine samples. It was found in different depth zones including some shallow stations, but not in samples of frozen sediment. Maturing and clitellate individuals were observed from October to June, new-born worms and empty cocoons in May and June. Only immature worms belonging to two different size groups occurred in the ice-free period.

Stylodrilus sp. is a small, delicate lumbriculid with bifid setae, male pores in X, and one pair of preclitellate spermathecae. One hundred eighty-seven specimens were found in 25 samples, mostly samples from moderate depths but below the freezing zone. About one-half of the specimens identified were maturing or clitellate individuals, they occurred in samples from all seasons.

A third lumbriculid species remains unidentified as all eight specimens were immature. Representatives of this taxon were found in five samples taken at moderate depths. It is distinguished from the two other species reported here by paired single-pointed setae that are notably enlarged and strongly curved in anterior ventral bundles. It was observed at five stations of moderate depths.

Enchytraeidae

Four enchytraeid taxa occurred in the lacustrine material. *Marionina* sp., a small worm with one straight seta per bundle (some bundles entirely lacking), was represented by seven immature individuals from two profundal samples taken in June. Two maturing specimens of *Cernosvitoviella* sp. were found in two shallow-water samples taken in September. The third taxon, of an indeterminate genus, with 2–3 (rarely 4) straight or slightly curved setae per bundle, was the most abundant. Twenty-five individuals, among them several mature ones, were identified from 11 samples taken in different seasons. These were mainly shallow-water

samples and included samples of frozen sediments. The fourth taxon, a single juvenile worm with fanlike setal bundles of the Lumbricillus-type, was also found in frozen mud.

Five different enchytraeid taxa occurred in the samples taken from the Gorbita River. One of them (eight specimens) has setal bundles of the Lumbricillus-type. Two other taxa (94 and one specimen, respectively) have bundles of the Henlea-type, with straight setae that are shorter in the middle of the bundle. A single immature specimen of a fourth taxon has extremely unequal setae within a bundle and has a thick cuticle. It is probably a species of *Fridericia*. A fifth taxon (three small immature specimens) has bisetate bundles.

Discussion

The material from the Gorbita River corresponds to the traditional conception of Arctic Oligochaeta being dominated by terrestrial or semi-aquatic enchytraeids, with a few cold-tolerant tubificids (species of *Limnodrilus* or *Isochaetides*, *Alexandrovia ringulata*, *Tasserkidrilus* sp.).

Lake Taimyr reveals a more usual northern lacustrine fauna, not very rich in species, but relatively diverse and clearly zoned. The lack of Naididae is unusual for a lake but is likely due to the absence of macrovegetation. Instead of a plant-covered littoral, there is an extensive zone flooded only in summer; however, it does not dry out when the water level sinks, as it remains under the ice sheet. Here the bottom sediment freezes together with its animal life and reaches very low temperatures in mid-winter, probably −20 °C. This 'cryolittoral' is inhabited by a depauperate oligochaete community dominated by *A. ringulata*. Some enchytraeids occur here, together with occasional individuals of other tubificids and lumbriculids.

Deeper areas cannot be divided into zones such as the sublittoral and profundal on the basis of oligochaetes since oxygen does not limit their depth distribution in this lake. Even at the deepest stations (11.5–18.5 m), *Isochaetides* sp., *Tasserkidrilus* sp., Tubificidae gen. sp. with bisetate bundles, *Lamprodrilus isoporus*, *Stylodrilus* sp., *Marionina* sp., and a single individual of *A. ringulata* were found in April–June 1944.

Grëze (1947b) performed experiments of thawing the frozen bottom sediments of Lake Taimyr. Of the Chironomidae, Nematoda and unidentified Oligochaeta (presumably *A. ringulata*) in the sedi-

94

ments, 75–92% revived. The proportion of revivable individuals may be even greater in nature where the warming is more gradual. Semernoj (1971) made analogous experiments in the Ivano-Arahlej Lakes, east of Lake Baikal, where the littoral benthos is encased for 6–7 months in the frozen sediment, at −2 ° to −11 °C. *Rhyacodrilus coccineus* (Vejdovský, 1875), *Rhyacodrilus* sp., *Tubifex tubifex* (Müller, 1774), and *Lumbriculus variegatus* (Müller, 1774) were curled in special cysts but unidentified Enchytraeidae were free in the sediments. All came to life when thawed slowly. *Stylaria fossularis* Leidy, 1852, *Nais simplex* Piguet, 1906, *Nais pseudobtusa* Piguet, 1906, *Nais communis* Piguet, 1906, *Chaetogaster diastrophus* (Gruithuisen, 1828), *Chaetogaster diaphanus* (Gruithuisen, 1828), and *Aeolosoma hemprichi* Ehrenberg, 1828 (Annelida: Aphanoneura) spent their cold diapause as cocoons only. Either emergence from cocoons was observed, or hatched young were found some days later. Holmquist (1973) presented a discussion on animal (including oligochaetes) life in shallow permafrost lakes. Danell (1981) and Olsson (1981) studied the hibernation of *L. variegatus* and unidentified tubificids in the frozen bottom sediment of some Swedish waterbodies where −2 °C was measured in the sediment by Olsson (*loc. cit*). *Lumbriculus variegatus* was also found in an encysted state.

Semernoj (1971), referring to Rey (1959), claimed −20 °C to be a critical temperature below which ice crystals are always formed in cells during slow freezing. This could be why many lake-dwelling species of oligochaetes are not found in the freezing zone of Lake Taimyr whereas they are in the more southern Ivano-Arahlej Lakes.

Alexandrovia ringulata, which was dominant in the cryolittoral, may be more frost-tolerant than the other oligochaetes. *Alexandrovia ringulata* is a truly northern species distributed only from Karelia to Alaska (Popčenko, 1988; Zaloznyj, 1984; Sokol'skaja, 1961; Holmquist, 1974). The holarctic species *L. profundicola* is the most cold hardy among its congeners. *Lamprodrilus isoporus* is the only widely distributed species of a genus that is diverse in Lake Baikal, occurring throughout northern Europe.

The oligochaete fauna of Lake Taimyr differs from those of neighbouring lakes. In the lakes of the Putorana Mountains between 67–69° N (Zinov'ev, 1981; unpublished data by V. P. Semernoj), as well as in Lake Labaz, 72° N, (unpublished data by V. P. Semernoj), *Peloscolex velutinus* (Grube, 1879) and *T. tubifex* dominated, whereas both are apparently absent from

Lake Taimyr. The identification of *P. velutinus* from these lakes is uncertain subsequent to the revision of the genus and redescription of the species by Holmquist (1979) and cannot be verified as the original material is lost.

Grēze (1947a) emphasized the probable connection of Lake Taimyr with Lake Baikal through the Enisej and Pjasina Rivers, in the recent past. Presumably, Baikalian elements will be discovered in the Lake Taimyr oligochaete fauna, *Baikalodrilus* sp. is the first candidate. Both lakes are permanently cool – the first due to its huge volume, the second due to its northern location. Unlike Lake Baikal, Lake Taimyr is young and rapidly shrinking as a result of land uplift. The present, deeper sediments that remain ice-free represent only a small relict of a larger and deeper lake, and probably the bottom fauna is also relatively depauperate.

Different reproductive periods occur among the Lake Taimyr oligochaetes. *Alexandrovia ringulata* and *Isochaetides* sp. are summer-breeders like most northern tubificids. As summer conditions are of short duration in Lake Taimyr, it is expected that cocoons would appear in autumn, under the new ice cover. The cocoons and hatched young may freeze for many months, thus extending the life cycle to several years. In contrast, the cold-stenotherm species *L. isoporus* shows the same reproduction pattern as described by Timm (1970) for populations in Lake Peipsi, which is much warmer: maturing and egg-laying under the ice cover only, with all cohorts immature in summer. Apparently, the same pattern is characteristic of *Baikalodrilus* sp. The ecological significance of these reproductive tactics in the consistently cool Lake Taimyr is unknown. No clear seasonality was observed in the maturation of *Stylodrilus* sp..

Acknowledgments

I express my cordial gratitude to the Zoological Institute (Russian Academy of Sciences) in Saint-Petersburg, and especially to Dr Nonna P. Finogenova for preserving this valuable material for many years and entrusting it to me for identification. The research became possible partly by the allotment of Grant No LD5 000 from the International Science Foundation. Mrs Ester Jaigma kindly checked my English text.

References

Atlas SSSR, 1962. GUGK, Moskva. 186 pp. [In Russian.]

Čejka, B., 1914. Die Oligochaeten der Russischen in den Jahren 1900–1903 unternommenen Nordpolarexpedition. IV. Verzeichnis der während der Expedition gefundenen Oligochaeten-Arten. Mém. Acad. Imp. Sci. 29(9): 25–32.

Danell, K., 1981. Overwintering of invertebrates in a shallow northern Swedish lake. Int. Revue ges. Hydrobiol. 66: 837–845.

Eisen, G., 1879. On the Oligochaeta collected during the Swedish expeditions to the Arctic regions in the years 1870, 1875 and 1876. K. svenska Vetensk. Akad. Handl. 15(7): 1–49.

Grëze, V. N., 1947a. Taimyrskoe ozero. Izvestija Vsesojuznogo geografičeskogo obščestva 79(3): 289–302. [Lake Taimyr. In Russian.]

Grëze, V. N., 1947b. Anabioz zoobentosa Taimyrskogo ozera i ego produktivnost'. Zool. Žurnal 26: 3–8. [Anabiosis and productivity of the zoobenthos in Lake Taimyr. In Russian.]

Grëze, V. N., 1957. Osnovnye čerty gidrobiologii ozera Taimyr. Trudy Vsesojuznogo gidrobiologičeskogo obščestva 8: 183–218. [The main features of hydrobiology of Lake Taimyr. In Russian.]

Haack Grosser Weltatlas, 1968. VEB Hermann Haack, Gotha/Leipzig, 480 pp.

Holmquist, C., 1973. Some Arctic limnology and the hibernation of invertebrates and some fishes in sub-zero temperatures. Arch. Hydrobiol. 72: 49–70.

Holmquist, C., 1974. On *Alexandrovia onegensis* Hrabě from Alaska, with a revision of the Telmatodrilinae (Oligochaeta, Tubificidae). Zool. Jb. Syst. Ökol. Geogr. Tiere 101: 249–268.

Holmquist, C., 1979. Revision of the genus *Peloscolex* (Oligochaeta, Tubificidae). Zool. Scr. 8: 37–60.

Nurminen, M., 1973. Enchytraeidae (Oligochaeta) from the Arctic archipelago of Canada. Ann. zool. fenn. 10: 403–411.

Olsson, T., 1981. Overwintering of benthic macroinvertebrates in ice and frozen sediment in a North Swedish river. Holarct. Ecol. 4: 161–166.

Popčenko, V. I., 1988. Vodnye maloščetinkovye červi (Oligochaeta limicola) Severa Evropy. Nauka, Leningrad, 287 pp. [Aquatic Oligochaeta of northern Europe. In Russian.]

Rey, L., 1959. Conservation de la vie par le froid. Paris.

Semernoj, V. P., 1971. Zimovka vodnyh oligohet v promerzajuščem grunte. Biologija vnutrennih vod, Informacionnyj bjulleten' 9: 29–32. [Overwintering of aquatic oligochaetes in the frozen sediment. In Russian.]

Smith, F. & P. Welch, 1919. Oligochaeta collected by the Canadian Arctic Expedition 1913–1918. Rep. Can. Arct. Exped. 1913–18. 9A: 3–19.

Sokol'skaja, N. L., 1961. Materialy po faune presnovodnyh maloščetinkovyh červej Kamčatki. Bjulleten' Moskovskogo obščestva ispytatelej prirody, biol. 66: 54–68. [Materials on the fauna of freshwater oligochaetes of Kamchatka. In Russian.]

Timm, T., 1970. On the fauna of the Estonian Oligochaeta. Pedobiologia 10: 52–78.

Zaloznyj, N. A., 1984. Rol' oligohet i pijavok v èkosistemah vodoëmov Zapadnoj Sibiri. In Biologičeskie resursy vnutrennih vodoëmov Sibiri i Dal'nego Vostoka. Nauka, Moskva: 124–143. [The role of oligochaetes and leeches in the aquatic ecosystems of West Siberia. In Russian.]

Zinov'ev, V. N., 1981. Zoobentos oz. Ajan. In Ozera Severo-Zapada Sibirskoj platformy, Novosibirsk, Trudy Limnologičeskogo instituta 33: 135–142. [Zoobenthos of Lake Ajan. In Russian.]

Hydrobiologia **334**: 97–101, 1996.
K. A. Coates, Trefor B. Reynoldson & Thomas B. Reynoldson (eds), Aquatic Oligochaete Biology VI.
© 1996 *Kluwer Academic Publishers.*

Influence of grain size on the distribution of tubificid oligochaete species

Georg Sauter & Hans Güde
Institut für Seenforschung, Untere Seestr. 81, D-88085 Langenargen, Germany

Key words: Tubificidae, clay and silt content, distribution of the species

Abstract

The grain size fraction less than 63 μm is an important environmental factor affecting the distribution and abundance of many of the tubificid species of Lake Constance. In substrates with a heterogeneous grain size where tubificids can be selective many species occur almost exclusively in their particle size preference range, which is highly species specific.

Introduction

The distribution of the tubificid species in lakes is highly variable (Probst, 1988), particularly in the littoral zone which is spatially heterogeneous in a small spatial scale (Sauter, 1995a). In the profundal zone of many eutrophic lakes oxygen is a limiting factor for most species (Volpers & Neumann, 1992) and pollution can also result in reduced numbers of species. (Chapman & Brinkhurst, 1984). Zahner (1981) showed that in Lake Constance the grain size composition of the substrate also influenced the distribution of species. Finally, the quantity and quality of available food influences the distribution and abundance of oligochaete species (Moore, 1979).

To conduct ecological assessments of communities from their habitat characteristics it is important to know the environmental requirements of individual species and to be able to correlate their occurrence with quantifiable environmental factors. These results describe the role of grain size composition on the occurrence of 14 tubificid and one lumbriculid species.

Materials and methods

To identify correlations between habitat structure and species distribution samples were taken from Lake Constance from 1992 to 1994. Lake Constance is a large prealpine lake which is divided into two parts, the shallow 'lower lake' and the deeper 'upper lake' which covers an area of 476 km^2 and at its deepest point is 252 m deep. The sampling grid covered the whole benthic area of the upper lake. Approximately 1200 cores were taken, either from the surface with a corer or, in the littoral zone, by diver. Each core was inserted into the sediment to a depth of 15 cm. In addition to the identification of all tubificids, including the immature animals, grain size analysis was conducted on each of the cores.

Determination of species

All tubificids were determined by microscope from their setal and habitus characteristics using the methods described by Sauter (1995b). The determination was checked by the penis sheaths or the spermathecal chaetae (if existing).

Determination of the grain size composition

The determination of the grain size fraction smaller than 63 μm was determined by wet sieving. The dry weight of the clay and silt fraction relative to the total dry weight of the sample is expressed as a percentage.

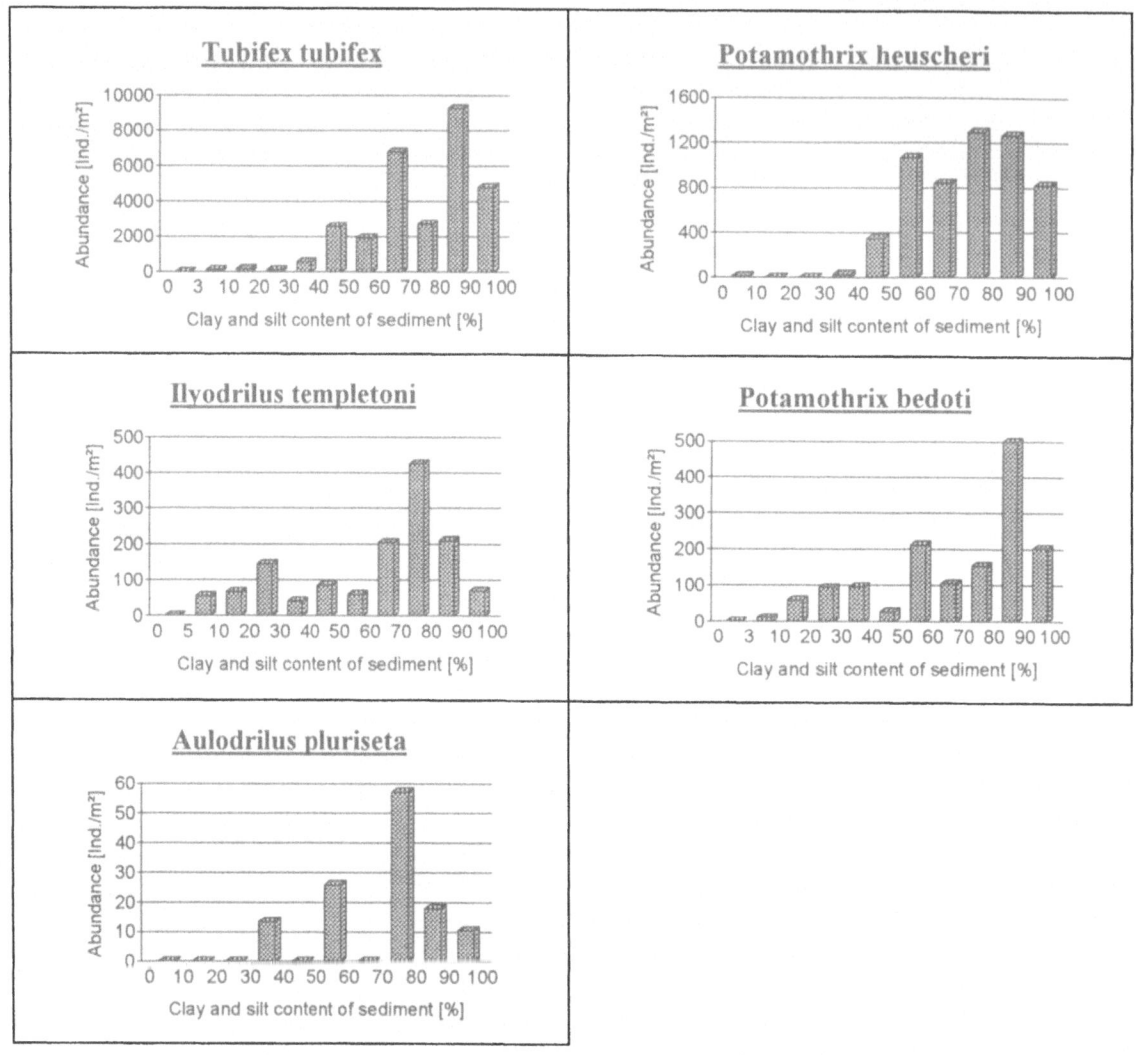

Figure 1. Abundance of species in relation to the clay and silt content of the sediment – species preferring fine grained substrates with a high clay and silt content. The total number of the samples n 1200, so that every column is in accordance with the mean value of n 110 (120) samples.

Results

Of the 14 tubificid species described in Lake Constance (Steinlechner, 1987), five species show a preference for substrates with a high clay and silt content. These are *Tubifex tubifex*, *Potamothrix heuscheri*, *Potamothrix bedoti*, *Ilyodrilus templetoni* and *Aulodrilus pluriseta* (Fig. 1).

Three species *Potamothrix moldaviensis*, *Limnodrilus claparedeianus* and *Tubifex ignotus* (Fig. 2) are more likely to occur in sandy substrates, having a clay and silt content of less than 10%. *Psammoryctides barbatus* also shows a preference for more sandy substrates.

The species *Limnodrilus udekemianus* and *Spirosperma ferox* are most abundant in sediment with a medium clay and silt content (Fig. 3).

Limnodrilus hoffmeisteri, *Limnodrilus profundicola* and *Potamothrix hammoniensis* occur quite frequently over a wide range of the clay and silt contents, so that the grain size is of little importance to these three species (Fig. 4).

At the family level tubificids are less abundant in sandy substrates while the opposite is true of the Chironomidae (Fig. 4).

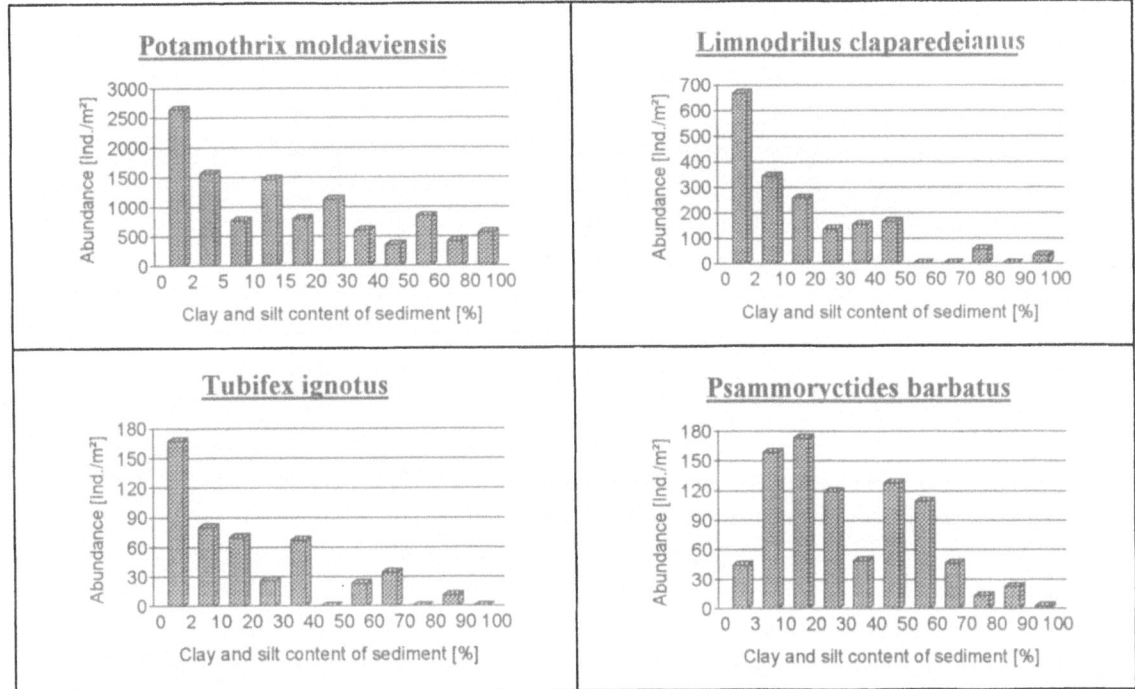

Figure 2. Abundance of species in relation to the clay and silt content of the sediment – species preferring sandy substrates with a lower clay and silt content. The total number of the samples n 1200, so that every column is in accordance with the mean value of n 110 (120) samples.

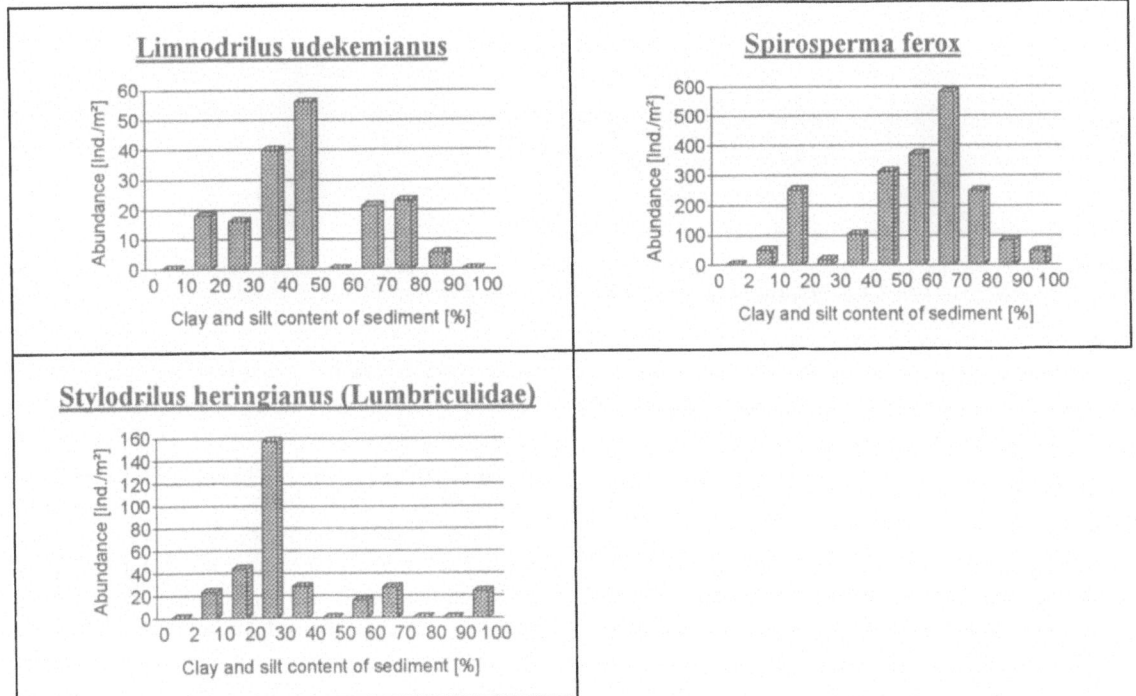

Figure 3. Abundance of species in relation to the clay and silt content of the sediment – species preferring substrates with a medium clay and silt content. The total number of the samples n 1200, so that every column is in accordance with the mean value of n 110 (120) samples.

100

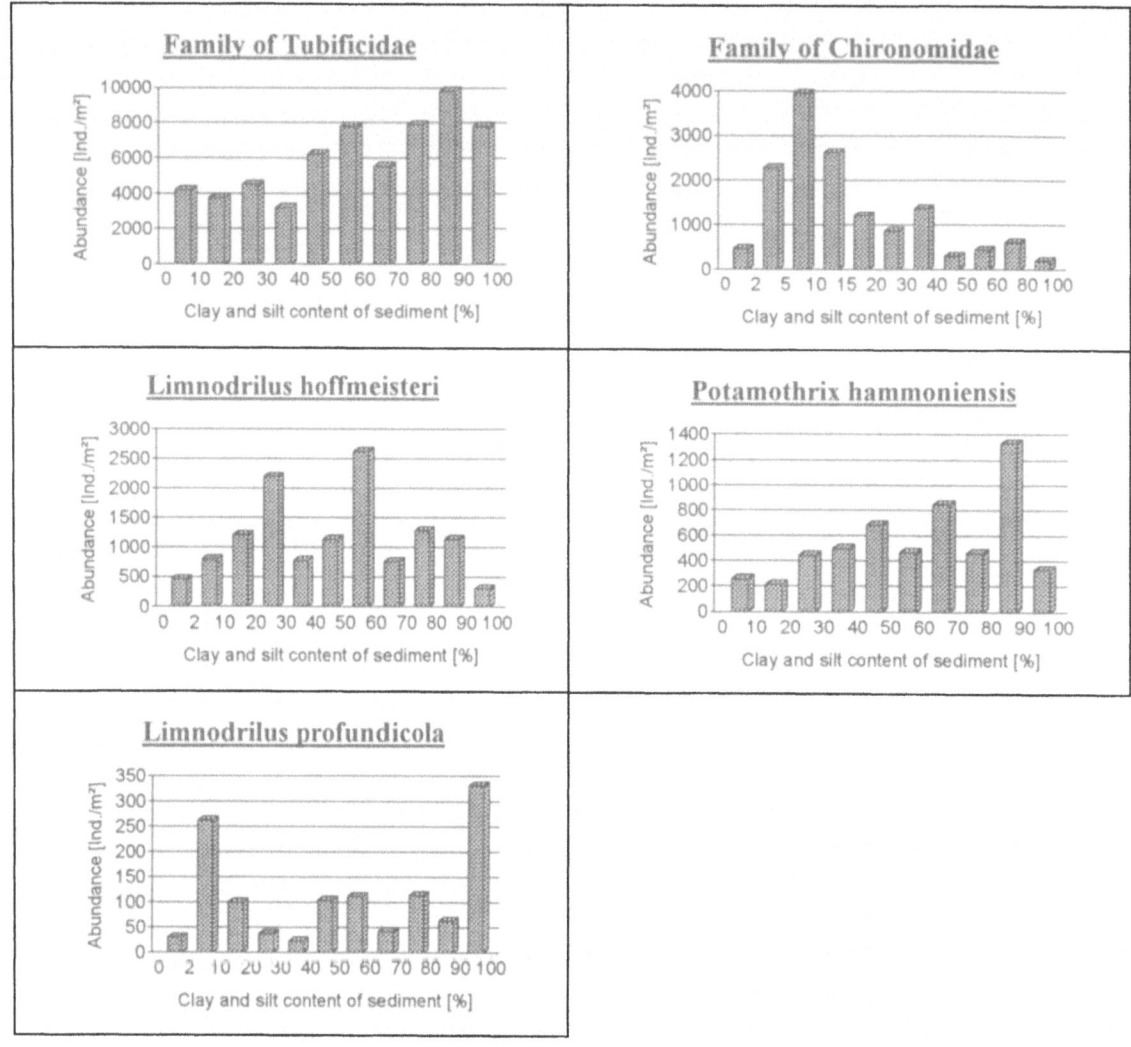

Figure 4. Abundance of species in relation to the clay and silt content of the sediment – non selective species and family preferences. The total number of the samples n 1200, so that every column is in accordance with the mean value of n 110 (120) samples.

The lumbriculid species *Stylodrilus heringianus* shows a clear preference for a clay and silt content between 20% and 30%.

Discussion

For the majority of tubificid species the clay and silt content of the substrate is an important factor effecting their distribution and abundance. However, the occurrence of a certain species is not determined by a single environmental factor. This was evident from the variable tubificid abundance in samples with the same grain size composition. However, the large number of samples allowed us to develop patterns in abundance related to grain size.

In the littoral zone, with more heterogeneous substrates exposing individual tubificid worms to a greater choice of grain size, species were more selective (or restricted) in the range of grain size in which they occurred compared to the profundal zone. For example, *Limnodrilus claparedeianus* occurs only in sandy substrates with a clay and silt content of up to 50% in the littoral zone, while *Potamothrix heuscheri*, in the littoral zone, is restricted to substrates with a clay and silt content of over 60%. In Lake Constance, the abundance of the tubificid worms is with 3780 ind m^{-2}, lower in the littoral than in the profundal zone, where the mean abundance is 6750 ind m^{-2}.

Benthic algae (Moore, 1979) and bacteria (Wavre & Brinkhurst, 1971) are the basic food sources of tubificids. Moore (1979) found that algae formed 65%, by volume, of the gut contents of littoral zone populations. It is likely that food supply is a decisive factor influencing the abundance of species, if there are no other environmental factors, for example the grain size composition or the oxygen content, restricting their occurrence. Oxygen is also an important factor limiting the occurrence of many species in the profundal zone of eutrophic lakes (Volpers & Neumann, 1992). The upper lake of Lake Constance is mesotrophic and it is not likely oxygen is a limiting factor, as in the deep profundal zone the oxygen content of the water is more than 5 mg l^{-1} throughout the year.

We consider that these results can be used to evaluate the effects of a local clay and silt content on the occurrence of the different tubificid species. This will assist in the use of the species as indicators of complex environmental conditions or the effects of specific environmental impacts.

References

Chapman, P. M. & R. O. Brinkhurst, 1984. Lethal and sublethal tolerances of aquatic oligochaetes with reference to their use as a biotic index of pollution. Hydrobiologia 115: 139–144.

Jones, J. G. & B. M. Simon, 1977. Increased sensitivity in the measurement of ATP in freshwater samples with a comment on the adverse effect of membran filtrations. Freshwat. Biol. 7: 253–260.

Moore, J. W., 1979. Influence of food availability and other factors on the composition, structure and density on a subarctic population of benthic invertebrates. Hydrobiologia 62: 215–223.

Probst, L., 1988. Die Oligochaeten im Bodensee als Indikatoren für die Belastung des Seebodens (1972 bis 1978). Int. Gewässerschutzkomm. Bodensee, Ber. 38: 69 pp.

Sauter, G., 1995a. Reaktion der Tubificiden auf unterschiedliche Habitatbedingungen im Bodenseelitoral. In DGL, Erweiterte Zusammenfassung der Jahrestagung 1994 in Hamburg. H. Kaltenmeier Söhne, Krefeld-Hüls.: 354–358.

Sauter, G., 1995b. Bestimmungsschlüssel für die in Deutschland verbreiteten Arten der Familie Tubificidae mit besonderer Berücksichtigung von nicht geschlechtsreifen Tieren. Lauterbornia 23 (ISSN 0935-333X), 52 pp.

Steinlechner, R., 1987. Identification of immature tubificids (Oligochaeta) of Lake Constance and its influence on the evaluation of species distribution. Hydrobiologia 155: 57–63.

Volpers, M. & D. Neumann, 1992. Unterschiedliche Überlebensfähigkeiten von Tubificiden unter natürlichen und simulierten Profundalbedingungen eutropher Seen. Verh. dt. zool. Ges. 85: 50.

Wavre, M. & R. O. Brinkhurst, 1971. Interactions between some tubificid oligochaetes and bacteria found in the sediments of Toronto harbour, Ontario. J. Fish. Res. Bd Can. 28: 335–341.

Zahner, R., 1981. Zum biologischen Zustand des Seebodens des Bodensees in den Jahren 1972 bis 1978. Int. Gewässerschutzkomm. Bodensee, Ber. 25: 289 pp.

Hydrobiologia **334**: 103–114, 1996.
K. A. Coates, Trefor B. Reynoldson & Thomas B. Reynoldson (eds), Aquatic Oligochaete Biology VI.
© 1996 *Kluwer Academic Publishers.*

Oligochaetes in the southern basin of the Venetian Lagoon: community composition, species abundance and biomass

Sandra Casellato
Dipartimento di Biologia, Università di Padova, Via Trieste 75, I-35121, Padova, Italy

Key words: Oligochaeta, Venetian Lagoon, composition, distribution

Abstract

The oligochaete community of the Chioggia Lagoon (southern basin of the Venetian Lagoon) has been studied from May 1992 to May 1994, at ten fixed sampling stations, distributed across the lagoon. Numerous data regarding composition, abundance, biomass and the breeding periods of the species in the community have been obtained. Nine tubificid species, recently reported for the Mediterranean and Ponto-Caspian areas have been collected. Among them only three were abundant everywhere, except in the innermost part of the lagoon, and throughout the year: *Tubificoides vestibulatus*, *T. swirencowi* and *Limnodriloides maslinicensis*. Only one enchytraeid genus, *Grania*, has been found restricted to an area near the connection with the sea. The values for the Shannon Diversity Index for the community never exceeded 0.8 and were mainly between 0.4 and 0.7. The most interesting result was the decrease of abundance and biomass, proceeding from the sea mouth (about 6000 ind m^{-2}) towards the innermost part of the lagoon (less than 100 ind. m^{-2}). Correlations are suggested with the different grain size of the sediment.

Introduction

Little is known about the marine and estuarine oligochaete faunas of Italy, despite the many recent studies on benthic invertebrates which have been done. The aquatic oligochaetes have the reputation of being a 'difficult' group, particularly marine species which have a low morphological diversity. Quite often surveys in marine and brackish water environments have included in the faunal list a particular taxonomic category: 'unidentified Oligochaeta' without further details.

Owing to the numerous studies particularly of H. R. Baker, R. O. Brinkhurst, K. A. Coates, B. Healy and C. Erséus, to date hundreds of new species of Tubificidae and Enchytraeidae have been described all over the world, and, even though we are still far from a full knowledge of the marine oligochaete diversity, we have many new criteria for revising generic and specific features than were used in the past (Brinkhurst, 1982; Brinkhurst & Jamieson, 1971).

More recently, a few studies have provided the first accounts of the presence and distribution of oligochaete species along the Tyrrhennian, Adriatic and Jonian coasts (Bonomi & Erséus, 1984; Erséus 1982a; Erséus, 1987a; Erséus & Bonomi, 1987) and in estuarine and lagoon areas of the Northern Adriatic coast (Casellato, 1994; Casellato & Poja 1984).

To help fill this gap, the benthic oligochaete community of the Chioggia Lagoon (southern basin of the Venice Lagoon) has been studied for two years, from May 1992 to May 1994. Thus, numerous data regarding composition, species abundance, biomass and dispersion patterns of individuals in the different areas have been obtained.

Investigated area and material and methods

The study area of the Chioggia Lagoon (110 km^2) is shown in Fig. 1, and comprises the southern basin of the Venetian Lagoon. It is connected with the Adriatic Sea through the 'harbour channel' of Chioggia. There is a good water exchange, thus the salinity differences between the sea and the lagoon water are low throughout the year. The terrestrial freshwater contribution is small, and there is no river flow directly into the lagoon. Thus, salinity on average is 31–32 ppt, ranging from 20 ppt to 36–37 ppt. over the year.

Water temperature shows a large annual range of about 24 °C; and the minimum temperature has been recorded in January–February at 6 °C and the maximum in July–August (about 30 °C).

The water depth in the lagoon varies from 0.7 to 3–4 m with a few deeper areas reaching 12–15 m. Often there is a thick layer of aquatic plants (*Zostera*) or macroalgae (particularly *Enteromorpha* spp. and *Ulva rigida*) present on the bottom.

In the first year of this study (May 1992, March 1993), seasonal sampling was done with a Van Veen grab (0.037 m² sampling area) with 3 replicate samples taken at each of the ten stations (Fig. 1), located at the edges of the channels. The water depth at the sampling stations varied from 1.5 to 2.5 m. The samples collected were sieved in the field using a 500 μm mesh sieve and preserved in 4% formalin. Worms were sorted and identified in the laboratory, under a light microscope. Some have been mounted in polyvinyl lactophenol, others fixed in glutaraldehyde for scanning electron microscopy, and the remainder preserved in alcohol (70%) or Bouin fluid, for further taxonomic investigation. The granulometric composition of the sediment at each station has been analyzed, using the Buchanan (1984) method and the organic content was determined according to the methods described by El Wakel & Riley (1956).

The dry weight of the collected worms was determined by drying at 105 °C for 24 h, this was done for two samples having a large number of individuals (>100), from each station and on each sampling occasion.

In the second study year (April 1993–May 1994) only qualitative data were collected to check the breeding periods for each species and to provide more material for further investigation of specific taxonomic problems.

Results

Community composition, species distribution, abundance and biomass

Nine tubificid species and one enchytraeid belonging to the genus *Grania*, (species not yet classified), were found in this study (Table 1). Of the tubificids only three species (*Tubificoides vestibulatus, T. swirencowi*

Table 1. Oligochaete species found in Chioggia Lagoon from May 1992 to May 1994

Tubificidae

Tubificoides vestibulatus Erséus and Bonomi, 1987
T. swirencowi Jaroschenko, 1948
Aktedrilus cuneus Erséus, 1984
A. monospermathecus Knoller, 1935
Limnodriloides appendiculatus Pierantoni, 1903
L. maslinicensis (Hrabê) 1971
L. hrabetovae Erséus, 1987
L. agnes Hrabê, 1967
Thalassodrilides gurwitschi (Hrabê) 1971

Enchytraeidae

Grania sp.

and *Limnodriloides maslinicensis*) were abundant and widely distributed, except in the innermost part of the lagoon.

Tubificoides vestibulatus, first described from Fano and Bari (Adriatic Sea) by Erséus & Bonomi (1987), is the most abundant species. It reaches the highest density in the area near the sea mouth (site 10) (see Fig. 2A), where the bottom is mostly sandy (70–80%) (Fig. 3A). This species breeds from March to July. Some slight morphological differences from specimens described from Fano and Bari have just been discussed in a previous paper (Casellato, 1994a).

Tubificoides swirencowi was first reported by Jaroshenko (1948) from the Black Sea and later recorded by Hrabê (1964, 1965) from the same area. More recently it has been noted by Moroz (1994) in the Dnieper estuary system. This species was widely distributed in the Chioggia Lagoon but was more numerous in the central part of the lagoon (sites 5, 9) (see Fig. 2A) where the sediment is sandy-muddy (see Fig. 3A). It is often associated with *T. vestibulatus*, but is never as abundant as this latter species. Mature specimens have been found from March to September.

Limnodriloides maslinicensis is abundant in the area near the sea mouth (see Fig. 2A), particularly from September to November and in this period many mature specimens, with spirally twisted spermatozoans in their spermathecae, were found. This species was first described by Hrabê (1971) from Jugoslavia, found at low depth, in sandy subtidal sediment. In the study area it is more abundant in sandy-muddy sediment, and it is almost completely absent from other parts of the lagoon.

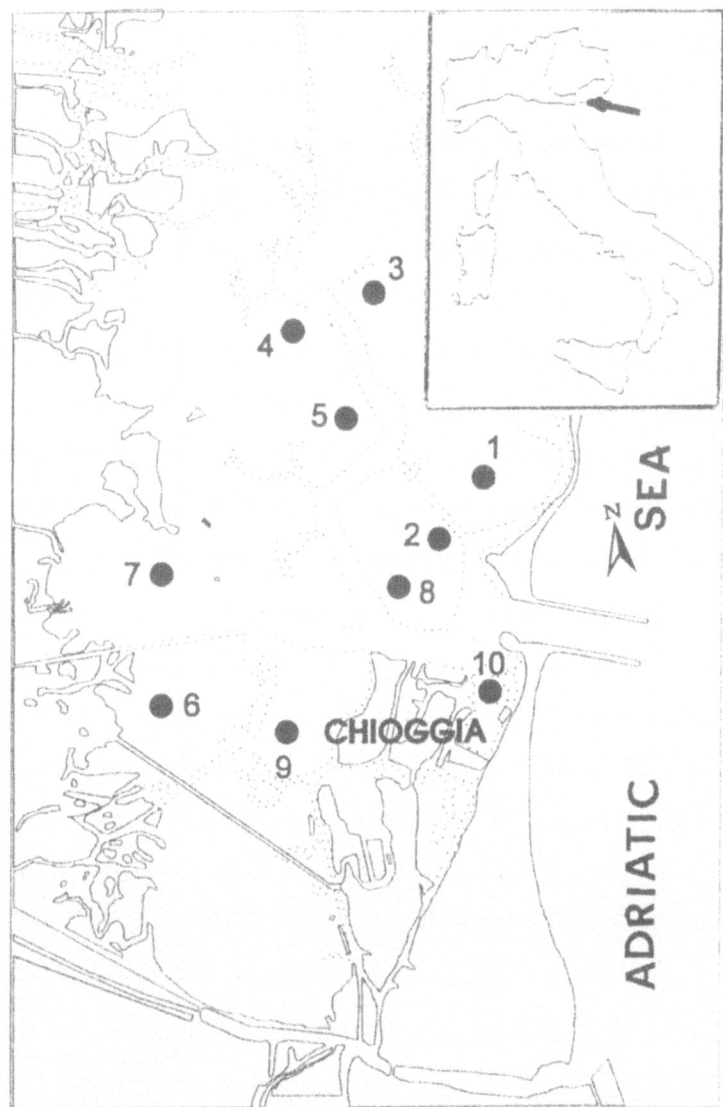

Fig. 1. Chioggia Lagoon: sampling stations.

Limnodriloides hrabetovae (Fig. 2A), *Limnodriloides agnes* and *Limnodriloides appendiculatus* (Fig. 2B) are less abundant than the previous congeneric species, particularly *L. agnes*, which was collected only at station 2, in November 1992, at low densities (5–10 ind. m^{-2}). No mature specimens were found during the study, thus this species could not be identified with certainty and is not reported quantitatively. *Limnodriloides hrabetovae* and *L. appendiculatus* could not easily be distinguished from external features (chaetae or other morphological aspects), thus internal genital characteristics were used for all specimens. The former species is more abundant than the latter in all areas, and attains densities of about 500 ind m^{-2} in the central part of the lagoon. Mature specimens of these two species, with oocytes in their ovisacs, were only found in July 1992.

Aktedrilus cuneus is more numerous than *A. monospermathecus*, particularly in the spring and in the area near the sea mouth (see Fig. 2B). *Thalassodrilides gurwitschi* was more abundant at stations 7, 8 and 10 (see Fig. 2B), particularly in September and

106

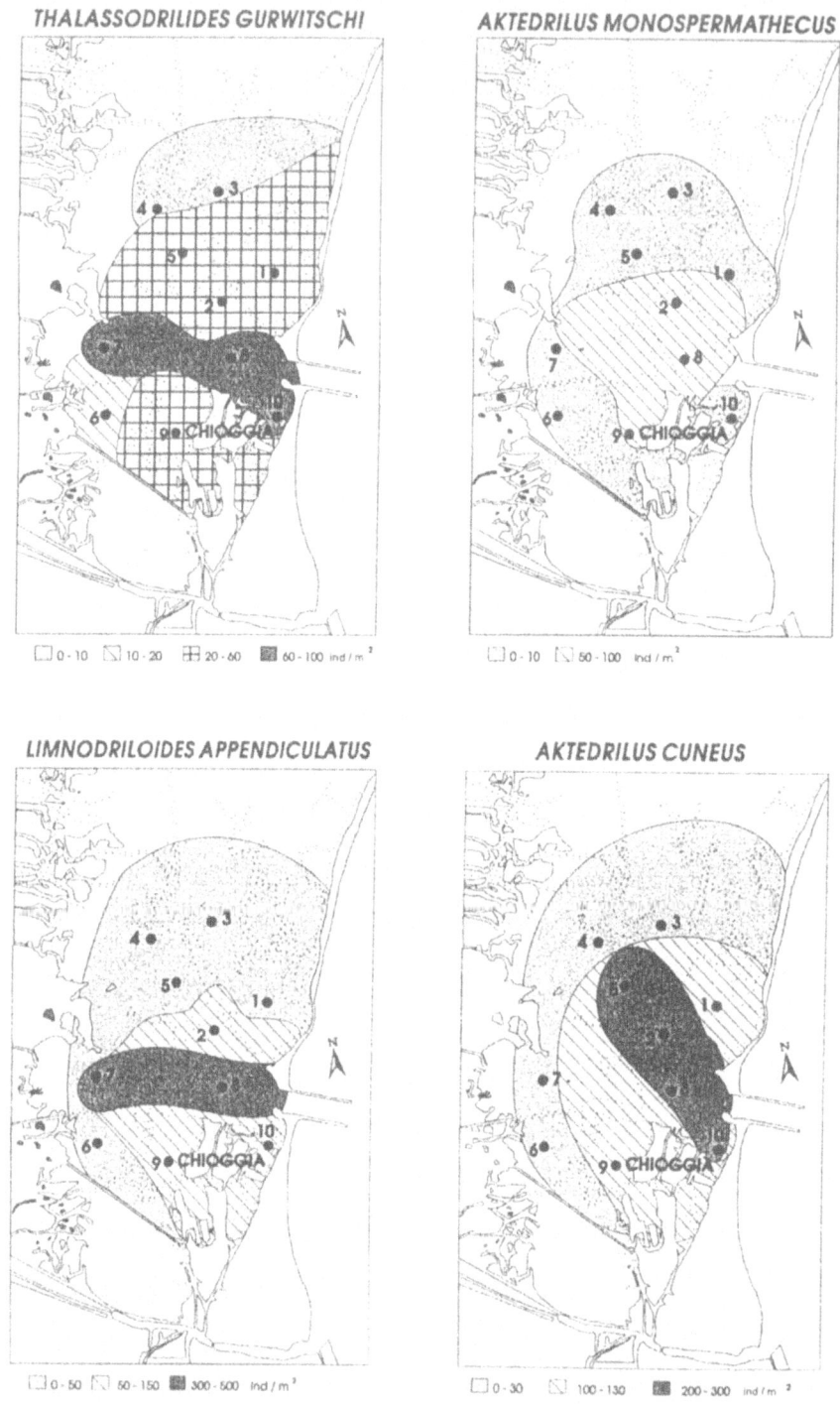

Fig. 2 (a). Mean annual (1992–93) density (ind m^{-2}) of different oligochaete species in the sampled areas.

107

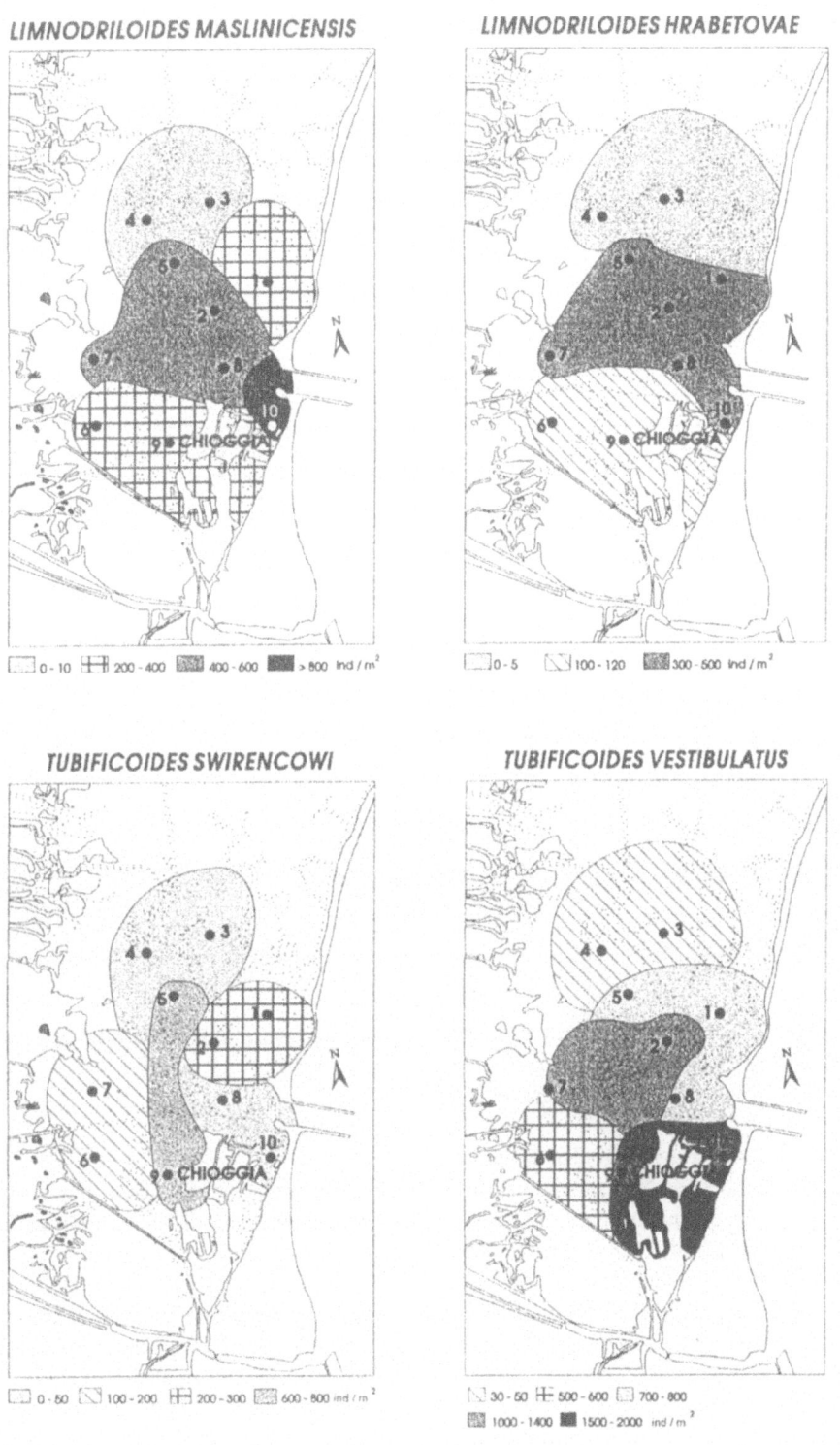

Fig. 2 (b). Mean annual (1992–93) density (ind m^{-2}) of different oligochaete species in the sampled areas.

108

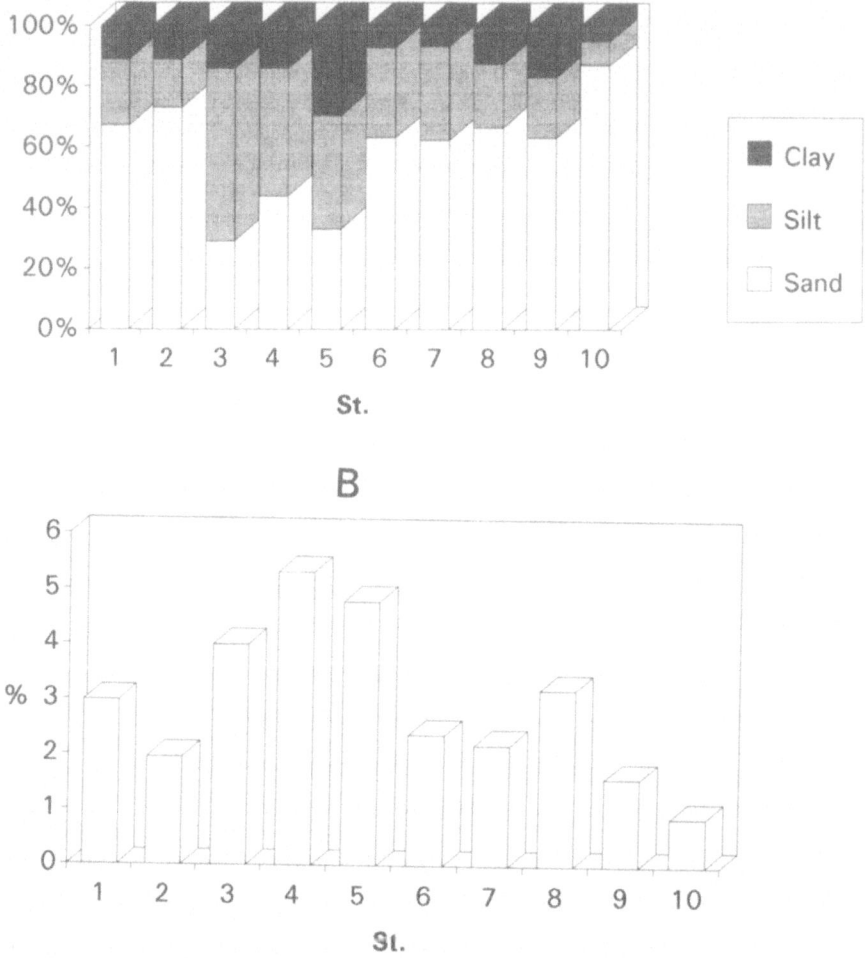

Fig. 3. Mean annual granulometric composition (A) and organic content (B) of sediment in Chioggia Lagoon.

November, at other periods of the year it was rare everywhere. Mature specimens of this species, with a well developed male apparatus were collected from May to July.

The highest values for total oligochaete density were observed in September and November, at most of the stations (see Fig. 4A); during this period the 'juveniles' are particularly abundant. They contribute more to the total density than to the total biomass (Fig. 4B), the highest biomass, was recorded in May, when the mature specimens are more abundant than the 'juveniles'.

Overall, the lowest densities and biomass were found in the innermost part of the lagoon (stations 3, 4), and values decrease as one proceeds from the area near

the sea to the innermost northern part of the lagoon (see Fig. 5).

Species diversity

The Shannon Index values calculated for each station for different periods of the year are consistently low. The values never exceeded 0.8, and more often lay between 0.4 and 0.7 (Fig. 6). Both community 'richness' and 'eveness' are low; most of the individuals belonging to a few species. The lowest values of the Shannon Index were at sites 3 and 4; in this area, primarily a muddy sediment having the highest organic matter content (Fig. 3B) the oligochaete component of the benthic fauna is extremely poor. A quite different situation is observed for the other benthic groups, par-

A

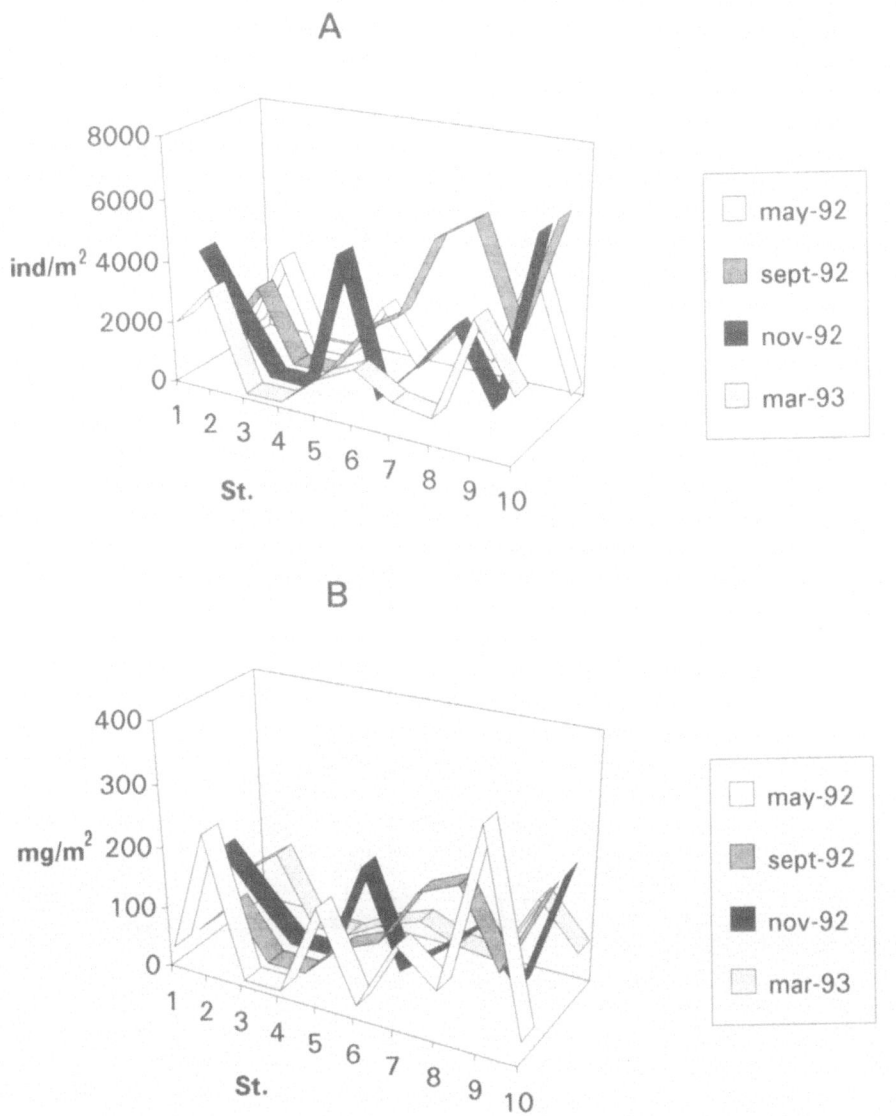

B

Fig. 4. Variation of density (A) and biomass (B) in the oligochaete fauna in different seasons.

ticularly the polychaetes and crustaceans (Casellato *et al.*, 1995).

Discussion and conclusions

The oligochaete species collected from the Chioggia Lagoon in 1992 and 1993 are very few compared to other benthic groups such as polychaetes (65), crustaceans (55) and molluscs (39), in the same habitat (Casellato *et al.*, 1995). However, their total density often represented more than 70% of the whole benthic community, particularly in autumn and spring. Unfortunately, very few data are available in the literature for a comparison with other similar environments.

Casellato (1994) in a recent ten year survey, exploring the brackish and estuarine zones of the Po Delta area, found even lower species diversity in areas other than the Chioggia. Lagoon. However, Moroz (1994) studying the oligochaete fauna of the Dnieper-Bug estuary system, reported a high species diversity (89 taxa) in those environments. Although, it is important

110

OLIGOCHAETA

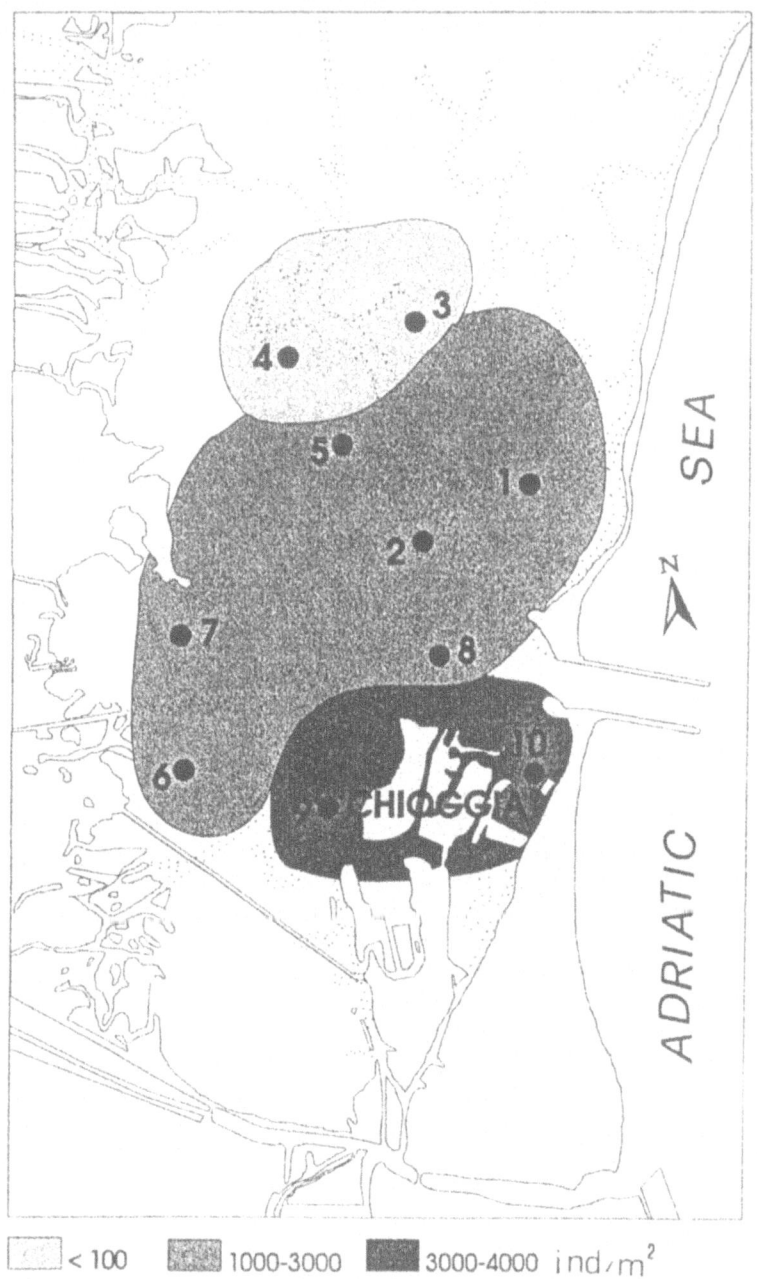

Fig. 5. Total mean annual oligochaete density (ind. m^{-2}) in different areas of Chioggia Lagoon.

to note that the area investigated by Moroz included several separate water bodies, with different hydrological and hydrochemical characteristics, thus providing a higher habitat heterogeneity when compared with the homogeneity of the lagoon environment studied in the northern Adriatic coast.

In contrast to the few studies of brackish environments, many studies have been conducted on the composition and distribution of the oligochaete fauna of

OLIGOCHAETA

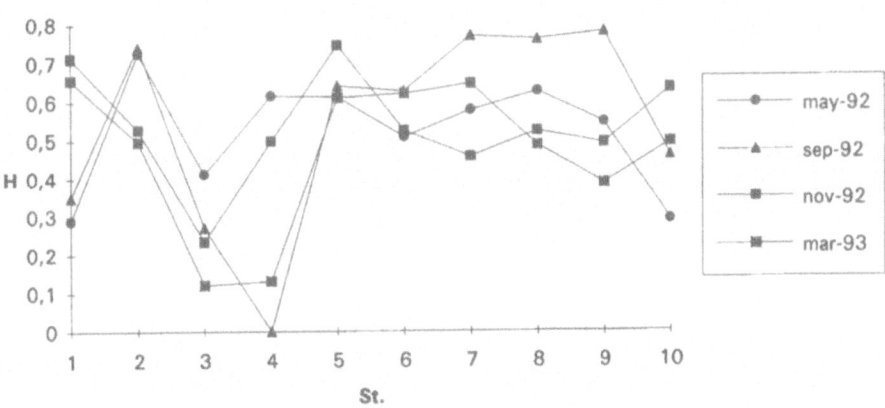

Fig. 6. Shannon diversity index values (H) for oligochaete communities at different stations in Chioggia Lagoon.

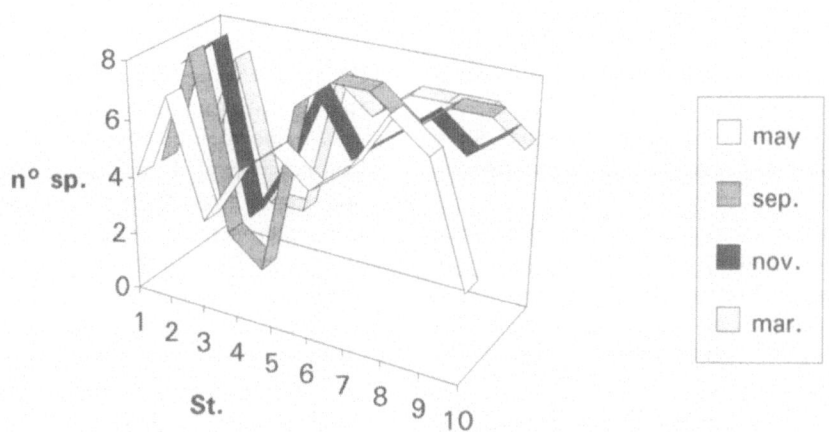

Fig. 7. Number of species at different stations in Chioggia Lagoon throughout the year.

intertidal and subtidal waters of the marine coasts all over the world. Baker (1984) compared the Tubificidae and Enchytraeidae fauna of intertidal shallow water of Pacific and Atlantic coasts of the United States and showed the west coast had a higher species diversity than the east one. In a recent studies (Erséus, 1990; Diaz & Erséus, 1994) the total number of species of marine Tubificidae in the barrier reef ecosystems of Central America have been shown to exceed that of the freshwater environments. These authors collected 52 species in the reef area, none of the species was dominant, thus indicating habitat stability and high niche specialization.

The tubificid species collected by Bonomi & Erséus (1984) were not as numerous along the Italian coasts, but these data represent a preliminary survey and have to be confirmed with more numerous samples in space and time. Even if preliminary, the data from that survey was particularly interesting and showed the species diversity of the Tyrrhenian coast to be higher than that of the Adriatic, or the Ligurian or the Ionian coasts.

All the marine tubificid species reported in that preliminary study and others describing the Adriatic Sea (Bonomi & Erséus, 1984; Erséus, 1981; Erséus, 1987a; Erséus, 1987b; Erséus & Bonomi, 1987; Hrabê, 1971, 1973, 1975) have been found in the Chioggia Lagoon, but here only three species were abundant year round: *Tubificoides vestibulatus. Limnodriloides maslinicensis* and *T. swirencowi*. The distribution of the first species was, until now, limited to the Italian Adriatic area (Erséus & Bonomi, 1987). The second species was previously recorded only in Jugoslavia (Hrabê, 1971) and from the South Adriatic coast (Erséus, 1987a), while the third species is widely distributed in the Ponto-Caspian area (Jaroshenko, 1948; Finogenova, 1972a, 1972b; Moroz, 1994). The other congeneric species of *Limnodriloides maslinicensis: L. hrabetovae, L. agnes* and *L. appendiculatus* have been reported recently in the Mediterranean area: the first species in the Adriatic only (Jugoslavia and Italy) (Erséus, 1982, 1987a), the second one has also been reported from the Bulgarian coast of the Black Sea (Erséus, 1982b) and the last has been recorded in the Tyrrhenian Sea (Pierantoni, 1903; Bolt & Schverin, 1928; Erséus, 1982b; Erséus, 1987a), in the Adriatic (Erséus, 1987a) and the Arabian Gulf coast of Saudi Arabia (Erséus, 1985).

Both *Aktedrilus cuneus* and *A. monospermathecus* have a wider distribution: the first has been reported from Galapagos (Erséus, 1984a), southern China (Erséus, 1992), Saudi Arabia and whole Mediterranean area (Erséus, 1987b, 1989; Delamare-Deboutville, 1954); the second species has been reported from the Baltic, North, Mediterranean and Caspian Seas (Erséus, 1980, 1987b, 1989; Knoller, 1935). The two species were often found associated in the lagoon areas explored, but the second species is much less abundant than the first.

Thalassodrilides gurwitschi is widely distributed: in Europe along the Mediterranean and the Black Sea (Hrabê, 1971; Erséus, 1981), in America (Brinkhurst & Baker, 1979); in Puerto Rico, Florida. Panama and Caribbean area (Righi & Kanner, 1979) in Saudi Arabia (Erséus, 1984b), in Australia (Erséus, 1993) and in China (Erséus, 1984c). Moreover, the species has been found in sites with highly fluctuating salinities (up to 42 ppt) (Erséus, 1981) Moreover, the literature

recorded a high tolerance for this species, but it is not the most abundant and widely distributed species in the environment studied.

The distribution of the oligochaete species in the different lagoon areas is an interesting aspect of this research. There are no water salinity, temperature, or oxygen gradients in this lagoon (Andreoli *et al.*, pers. comm.) however, the oligochaete species are unequally distributed. Higher species diversity is observed at the sites near the sea, and decreases toward the innermost part of the lagoon (Fig. 7). The total density also decreases along the transect from the sea to the innermost part of the lagoon (Fig. 5). It is not possible to attribute this to hydrochemical differences, but it is reasonable to attribute this to the sediment, particularly its granulometric composition, rather than its organic matter content. In fact, the enrichment of organic matter in the sediment in the innermost part of the lagoon might suggest the existence of a better habitat for these oligochaetes, which often occur in areas of organic pollution. However, this is not the case in the Chioggia Lagoon where the innermost sites (3, 4 and 5), which have a higher proportion of mud and organic matter content have the lowest density of oligochaetes year round.

One hypothesis for explaining this phenomenon is the granulometric composition of the sediment. The species occurring in the Chioggia Lagoon come from the marine subtidal coasts, where sand is the dominant component. In the lagoon this may also be the preferred habitat, as the sediment is similar in composition to that of the marine coastal areas. This explanation is supported by the fact that in the other estuarine and brackish areas explored on the Po Delta, the sediment was primarily silt and mud, and there also the oligochaete fauna was extremely poor.

A different explanation can be hypothesized for *Tubificoides swirencowi*; it is unrecorded from the marine coastal environment of the Adriatic Sea, until now. Furthermore, it is abundant in other brackish habitats in the Po Delta area and in the Chioggia Lagoon where, unlike other species, it was more numerous in the central area of the lagoon, where the sediment is equally muddy and sandy. Moroz (1994) suggested that it was an estuarine species and in these habitats the sediment is usually variable and sand is not a dominant component.

In conclusion, it is not possible to provide a comprehensive explanation of the distribution and abundance of the oligochaete community without information on the complete chemical characteristics of sediment. In addition, it is necessary to undertake more complete surveys along the Adriatic Sea coast, at several sampling sites year round, to collect quantitative and qualitative data.

Acknowledgments

This study was supported by a financial grant of the Ministero della Marina Mercantile. I am indebted to Prof. R. Brunetti for the analysis of sediment.

References

Baker, H. R., 1984. Diversity and zoogeography of marine Tubificidae (Annelida, Oligochaeta) with notes on variation in widespread species. Hydrobiologia 115: 191–196.

Bolt, W. & M. Schverin, 1928. Mitteilung uber Oligochaetes der Familie Tubificidae. Zool. Anz. 75: 143–151.

Bonomi, G, & C. Erséus, 1984. A taxonomic and faunistic survey of the marine Tubificidae and Enchytraeidae (Oligochaeta) of Italy. Introduction and preliminary results. Hydrobiologia 115: 207–210.

Brinkhurst, R. O., 1982. British and other marine and estuarine oligochaetes. Cambridge University Press, 127.

Brinkhurst, R. O. & H. R. Baker, 1979. A review of the marine Tubificidae (Oligochaeta) of North America. Can. J. Zool. 67: 1553–1569.

Brinkhurst, R. O. & B. G. M. Jamieson, 1971. Aquatic Oligochaeta of the World. Oliver & Boyd, Edinburgh, 861.

Buchanan, J. B., 1984. Sediment analysis. In: N. Holme & A. D. McIntyre (eds), Methods for the Study of Marine Benthos, II° ed. Blackwell Sc. Pub., Oxford: 41–65.

Casellato, S., 1994. Oligochaete fauna of estuarine areas and lagoons on the northern Adriatic coast (Italy). Boll. Zool. 61: 261–269.

Casellato, S. & R. Poja, 1984. Ecology of tubificids in the lower reaches of the rivers Adige and Brenta. Boll. Zool. 51: 339–352.

Casellato, S., F. Cauero & B. Santagiuliane, 1995. Composizione e distribuzione dei popolementi macrobenthomei nelle Laguna di Chioggie. Atti. S. Jt. E. 16: 57–60.

Delamare-Deboutville, C., 1954. Eaux souterraines littorales de la côte catalane française (mise au point faunistique). Vie Milieu 5: 408–451.

Diaz, R. J. & C. Erséus, 1994. Habitat preferences and species association of shallow-water marine Tubificidae (Oligochaeta) from the barrier reef ecosystem of Belize, Central America. Hydrobiologia 279: 93–105.

El Wakeel, S. K. & J. P. Riley, 1956. The determimation of organic carbon in marine muds. J. Cons. Intern. expl. Mar. 22: 180–183.

Erséus, C., 1980. Taxonomic studies on the marine genera Aktedrilus Knollner and Bacescuella Hrabê (Oligochaeta, Tubificidae), with descriptions of seven new species. Zoologica Scripta 9: 97–111.

Erséus, C., 1981. Taxonomy of the marine genus Thalassodrilides (Oligochaeta: Tubificidae). Trans. am. Microsc. Soc. 100: 333–344.

Erséus, C., 1982a. Three new species of the marine genus Coralliodrilus (Oligochaeta, Tubificidae) from Italy. Boll. Zool. 49: 241–247.

Erséus, C., 1982b. Taxonomic revision of the marine genus Limnodriloides (Oligochaeta: Tubificidae). Verh. naturwiss. Ver. Hamburg 25: 207–277.

Erséus, C., 1984a. Intertidal fauna of Galapagos XXXII: Tubificidae (Annelida Oligochaeta). Microfauna Marina 1: 1291–1298.

Erséus, C., 1984b. Annelida of Saudi Arabia marine Tubificidae (Oligochaeta) of the Arabian Gulf Coast of Saudi Arabia. Fauna of Saudi Arabia 6: 131–154.

Erséus, C., 1984c. The marine Tubificidae (Oligochaeta) of Hong Kong and Southern China. Asian Mar. Biol. 1: 135–175.

Erséus, C., 1985. Marine Tubificidae (Oligochaeta) of the Arabian gulf coast of Saudi arabia (Part 2). Fauna of Saudi Arabia 7: 59–63.

Erséus, C., 1987a. Marine Limnodriloidinae (Oligochaeta, Tubificidae) from Italy, with description of three new species. Boll. Zool. 54: 159–164.

Erséus, C., 1987b. Taxonomic revision of the marine intertidal genus Aktedrilus (Oligochaeta, Tubificidae), with description of three new species. Stylogia 3: 107–124.

Erséus, C., 1989. Marine Tubificidae (Oligochaeta) of the Arabian Gulf Coast of Saudi Arabia (Part 5). Fauna of Saudi Arabia 10: 11–19.

Erséus, C., 1990 The marine Tubificidae (Oligochaeta) of the barrier reef ecosystem at Carrie Bow Cay, Belize, and other parts of the Caribbean Sea, with description of twenty-seventh new species and revision of Heterodrilus, Thalassodrilides and Smitsonidrilus. Zool. Scr. 19: 243–303.

Erséus, C., 1992. Hong Kong's marine Oligochaeta: a supplement. In B. Morton (ed), The Marine Flora and Fauna of Hong Kong and Southern China III, Proc. 4th Int. Marine Biol. Workshop. Hong Kong University Press: 157–179.

Erséus, C., 1993. The marine Tubificidae (Oligochaeta) of Rottnest Island, Western Australia. In F. E. Wells, D. I. Walker, H. Kirkman & R. Lethbridge (eds), The Marine Flora and Fauna of Rottnest Island, Western Australia, Western Australian Museum: 332–389.

Erséus, C. & G. Bonomi, 1987. A new species of Tubificoides (Oligochaeta, Tubificidae) from the Adriatic Sea Boll. Zool. 54: 165–168.

Finogenova, N. P., 1972a. Oligochaetes in Ponto-Caspian brackish water basins. In G. M. Belyayev et al. (eds), Aquatic Oligochaeta, Moscow, Nauka: 65–74. [In Russian.]

Finogenova, N. P., 1972b. New species of Oligochaeta from the Dnieper-Bug Liman and the Black Sea with the revision of some species. Trudy Zool. Inst. Acad. Nauk SSSR. Leningrad 52: 97–116. [In Russian.]

Hrabê, S., 1964. On Peloscolex svirenkoi (Jaroshenko) and some other species of the genus Peloscolex. Spisy prir. Fac. Univ. Brno 450: 101–112.

Hrabê, S., 1965. New or insufficiently known species of the family Tubificidae. Spisy prir. Fac. Univ. Brno 57–77.

Hrabê, S., 1971. On new marine Tubificidae of the Adriatic Sea. Scr. Fac. Sci. Nat. Brno, 1: 215–226.

Hrabê, S., 1973. On a collection of Oligochaetes from various parts of Jugoslavia. Biol. Vest. 21: 39–50.

Hrabê, S., 1975. Second contribution to the knowledge of marine Tubificidae (Oligochaeta) from the Adriatic Sea. Sez. Vest. csl. Spol. Zool. 39: 111–119.

Knollner, F. N., 1935. Okologische und systematische Untersuchungen uber litorale und marine Oligochaten der Kieler Bucht. Zool. Jahb. (Syst.) 66: 425–512.

Jaroshenko, M. F., 1948. Oligochaetes of the Dnieper-Bug Liman. Nauc Zap. Modlav. nauc. issl. Bazy Acad. Nauk. SSSR 1: 57–68. [In Russian.]

114

Moroz, T. G., 1994. Aquatic Oligochaeta of the Dnieper-Bug estuary system. Hydrobiologia 279: 133–138.

Pierantoni, U., 1903. Altri nuovi oligocheti del Golfo di Napoli (*Limnodriloides* n. gen.) Boll. Soc. Nat. Napoli 17: 185–192.

Righi, G. & E. Kanner, 1979. Marine Oligochaeta (Tubificidae and Enchytraeidae) from the Caribbean Sea. Stud. Fauna Curacao 58: 44–68.

Hydrobiologia **334**: 115–123, 1996.
K. A. Coates, Trefor B. Reynoldson & Thomas B. Reynoldson (eds), Aquatic Oligochaete Biology VI.
© 1996 *Kluwer Academic Publishers.*

Seasonal dynamics of aufwuchs Naididae (Oligochaeta) on *Phragmites australis* in a eutrophic lake

Boris Löhlein
Ecosystem Research Centre, University of Kiel, Schauenburgerstr. 112, D-24118 Kiel, Germany

Key words: Naididae, Oligochaeta, population dynamics, seasonal succession, aufwuchs, periphyton, biomass

Abstract

The Oligochaeta and Aphanoneura in the aufwuchs on *Phragmites australis* in a eutrophic hardwater lake were studied at two sites over a period of one year, in order to elucidate the structure and dynamics of this assemblage. The naidids *Chaetogaster diastrophus*, *Nais* spp., and *Stylaria lacustris* dominated the assemblage at any season. At both sites these taxa showed the same distinct pattern of successive population maxima in spring and summer: *Chaetogaster diastrophus* reached its peak density first, followed by *Nais* species, and eventually by *Stylaria lacustris*. Differences in temporal dynamics between sites were small apart from a second *Stylaria* maximum which was only observed at one site. Total naidid densities reached peak values of 3.8 individuals per cm^2 reed stem surface area. With mean individual biomass of 2.2 μg dry mass for *Chaetogaster diastrophus*, 13.3 μg for *Nais*, and 86 μg for *Stylaria lacustris*, respectively, maximum total naidid biomass on reed stems was 44 μg dry mass per cm^2. The biomass peak occurred later than that of total naidid density because in summer larger naidids dominated the assemblage. The observed succession appears to be consistent with seasonal changes in periphytic algal communities on the reed stems.

Introduction

Research focusing on the littoral zone of lakes faces a rich life, both in numbers and species, inhabiting a wider variety of habitats than in any other part of the lake. This is appealing from a faunistical point of view, but rather discouraging for those who seek understanding of the littoral from a system perspective. We need information, however, on the functioning of littoral zones if we want to understand both the processes within lakes and exchanges with their surroundings.

Aufwuchs is characteristic of the littoral zone. In the present study, I prefer the term aufwuchs over periphyton to stress that this community includes 'dependent non-sessile forms' (Brown & Austin, 1971). Aufwuchs is hence defined as the community of animals, plants, and microorganisms living on, but not penetrating, submerged surfaces (Brown & Austin, 1971). In this community, algae, invertebrate grazers, and predators are the most prominent components of the food web, with the Naididae being one of the numerically most important (e.g. Beckett *et al.*, 1992; Botts & Cowell, 1993; Cattaneo, 1983), yet often neglected (Beckett *et al.*, 1992), grazer groups. Aufwuchs biota are known to exhibit strong seasonal fluctuations, both in subtropical (Botts & Cowell, 1993) and temperate (e.g. Kairesalo, 1984; McElhone, 1978; U. Müller, 1994) waters. The precise temporal sequence of these fluctuations and the factors that control them are not yet fully understood. Studies on seasonal development of aufwuchs communities which include identification of oligochaetes beyond the family level are scarce, but 'consistent seasonal changes in plant-dwelling invertebrate communities' are thought 'to be a general phenomenon in temperate lakes' (Beckett *et al.*, 1992). Thus, the primary aim of this paper is to describe the population and biomass dynamics of aufwuchs Naididae in a eutrophic lake over a period of one year, in order to serve as baseline information for future studies on food web dynamics.

Table 1. Physical and chemical water parameters of Lake Belau at the sampling sites in 1993.

Parameter	North Basin			South Basin		
	Mean	Min	Max	Mean	Min	Max
Temperature [°C]	9.3	0.4	18.4	9.3	1.0[a]	18.2
Conductivity [μS cm^{-1}]	359	307	393	358	317	393
pH	8.2	7.2	9.6	8.3	7.2	9.7
Alkalinity [mmol l^{-1}]	2.3	1.1	2.8	2.4	1.3	2.8
Oxygen Concentration [mg l^{-1}]	11.0	7.1	16.0	10.0	5.5	13.1
Oxygen Saturation [%]	93	60	129	89	56	124
NH_4^+-N[mg l^{-1}]	0.14	0.01	0.70	0.11	0.01	0.55
NO_3^--N [mg l^{-1}]	0.46	<0.001	1.05	0.40	<0.001	1.30
Total P [mg l^{-1}]	0.05	0.01	0.11	0.06	0.03	0.10

[a] No measurements during ice cover.

Study sites

The study was carried out in Lake Belau, a eutrophic hardwater lake of the 'Bornhöved Lakes Region' situated in a rural area in northern Germany (54° 06′ N, 10° 15′ E). Lake Belau has a surface area of 1.13 km, a maximum depth of 26 m, and a mean depth of 9 m (H. E. Müller, 1981). It consists of two basins: the main, north basin with a stable stratification during summer and the small and shallow (<2 m), polymictic south basin. A reed belt consisting primarily of *Phragmites australis* (Cav.) Trin. ex Steud., occasionally interspersed with isolated stands of *Typha angustifolia* L. and *Schoenoplectus lacustris* L. (U. Müller, 1994), covers about 65% of the 5.4 km shore line.

One sampling site was chosen on the western shores of both the north and south basins. Both sites were located on the lakeward border of homogenous *Phragmites* belts 10–20 m wide and with stem densities varying from 60 to 90 per m^2 in the north basin and from 80 to 120 per m^2 in the south basin (B. Schieferstein, pers. com.). Physical and chemical water parameters recorded at each sampling occasion were similar at the two sites (Table 1).

Materials and methods

Sampling started on 12 January 1993 and continued until 25 January 1994. Samples were taken at biweekly intervals by collecting four *Phragmites* stems at each station, except on four dates in winter when ice cover made sampling of the south basin site impossible. Young stems which had not yet reached the water surface were excluded from sampling. Reed stems were clipped off just above the sediment and carefully taken out of the water. The portion below the water level was cut into pieces 10 cm in length, which were individually placed in test tubes filled with lake water, and returned to the laboratory in a cool-box. This sampling procedure is assumed to be sufficiently accurate to document densities of aufwuchs naidids on *Phragmites* stems as Downing (1986) found no significant difference in estimated oligochaete densities between enclosing and simple non-enclosing sampling techniques.

In the laboratory, the animals associated with each stem section were extracted by vigorous agitation of the test tubes on a Vortex mixer. The rinse water was collected, and the procedure repeated ten times or more, depending on the density and type of algal growth. Preliminary experiments had shown that this protocol resulted in quantitative extraction of oligochaetes from aufwuchs samples. The sample volume was reduced by sieving through a 30 μm mesh screen. Oligochaetes and aphanoneurans were identified alive to family (Tubificidae, Enchytraeidae), genera (*Aeolosoma*, *Nais*, *Pristinella*) or species level and counted under a dissecting microscope at 12× magnification. On each sampling date, all tubificids and 10–20 individuals of *Nais* were fixed in 4% formalin for later identification to species level after mounting in polyvinyl-lactophenol. The surface area of each stem section was determined by measuring its length and diameter at both ends and assuming a conical shape. Densities are expressed regardless of depth distribution as mean (*n* = 4 stems) numbers of ind. cm^{-2} of reed stem surface area.

The individual biomass of *Chaetogaster diastrophus*, *Nais* spp., and *Stylaria lacustris* was determined

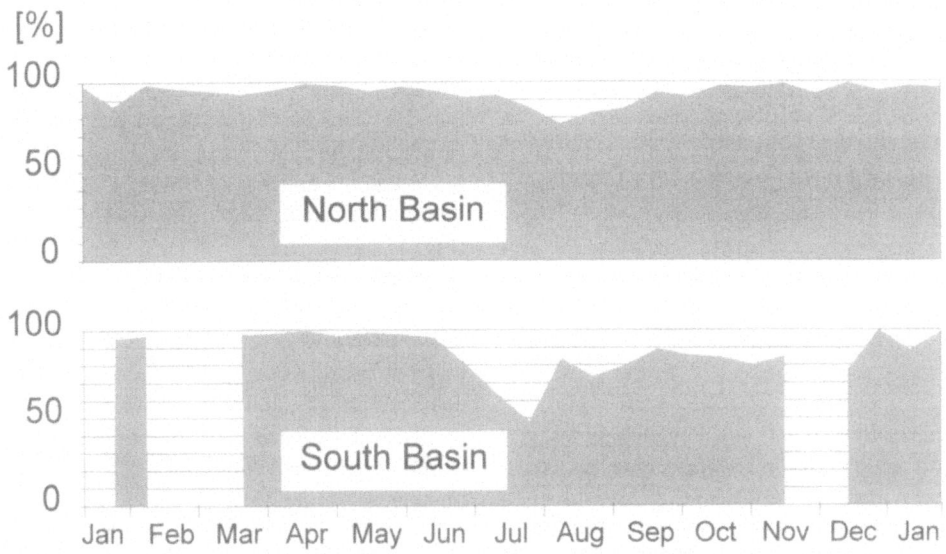

Fig. 1. Relative proportion of the dominant naidid taxa (*Chaetogaster diastrophus*, *Nais* spp. and *Stylaria lacustris*) in total aufwuchs oligochaete and aphanoneuran numbers over 1993. Gaps indicate ice cover.

Table 2. Oligochaete and aphanoneuran taxa found in aufwuchs on *Phragmites australis* in Lake Belau and their relative proportion in the community (dominance).

Family	Species	Dominance [%]	
		Family	Genus/species
Naididae		96.2	
	Chaetogaster cristallinus Vejdovsky		
	Chaetogaster diaphanus (Gruithuisen)		
	Chaetogaster diastrophus (Gruithuisen)		50.5
	Chaetogaster limnaei von Baer		
	Nais		31.1
	N. barbata Müller		
	N. pseudobtusa Piguet		
	N. variabilis Piguet		
	Pristina aequiseta Bourne		
	Pristina longiseta Ehrenberg		
	Pristinella sp.		
	Slavina appendiculata (d'Udekem)		
	Stylaria lacustris (L.)		11.3
Tubificidae		0.3	
	Psammoryctides albicola (Michaelsen)		
	Psammoryctides barbata (Grube)		
	immature specimens with hairs		
	immature specimens without hairs		
Enchytraeidae		0.8	
Aeolosomatidae		2.7	
	Aeolosoma spp.		
		100%	92.9%

on 5–11 dates in spring 1994. Depending on the size and number of specimens available, 3–42 individuals of each taxon were placed together on a tared aluminium chip, dried at 60 °C and weighed to the nearest μg on a Sartorius 4503 balance. Density data were converted to biomass by multiplying the number of individuals by the respective mean individual biomass of each taxon.

Results

Aufwuchs community

A total of 18 taxa of Oligochaeta and Aphanoneura was found to be associated with *Phragmites* stems in Lake Belau. The most abundant members of this assemblage belonged to the Naididae. They made up more than 96% of all individuals, with *Chaetogaster diastrophus*, *Nais pseudobtusa*, *N. variabilis* and *Stylaria lacustris* being the dominant species (Table 2). *Pristina longiseta* was the fifth most numerous oligochaete in the samples. Apart from *C. diastrophus*, the genus *Chaetogaster* was represented by *C. cristallinus*, *C. diaphanus* and *C. limnaei*, the latter inhabiting *Acroloxus lacustris* and *Bithynia* spp. present in the aufwuchs samples. Amongst the *Nais* specimens identified to species level, *N. pseudobtusa* and *N. variabilis* clearly dominated while *N. barbata* contributed little to the total *Nais* density. *Pristina aequiseta*, *Pristinella* sp. and *Slavina appendiculata* were rare, occurring on < 6% of all reed stems. Tubificidae, mostly small immature individuals, were found occasionally on the basal portion of stems. The few specimens that could be identified were *Psammoryctides albicola* and *P. barbata*. Likewise, Enchytraeidae were found only rarely and almost exclusively (80.4% of all individuals) in samples containing *Phragmites* leaf sheaths. Aeolosomatidae accounted for 2.7% of the assemblage (Table 2).

Seasonality

Chaetogaster diastrophus, *Nais* spp., and *Stylaria lacustris* dominated the aufwuchs oligochaete and aphanoneuran assemblage at any season. In terms of numbers their proportion dropped below 50 % only once at the south basin site (Fig. 1); in terms of biomass, however, their proportion always exceeded 80%.

All naidids and aphanoneurans attained maximum densities in spring or summer. *Chaetogaster cristallinus* and *C. diaphanus* reached their population peaks

in spring, but were never as abundant as *C. diastrophus*. *Pristina longiseta* showed a weak summer population peak not exceeding 0.03 ind. cm^{-2} reed stem surface area. Peak population densities of Aeolosomatidae were recorded in spring at the north and in summer at the south basin site. The latter amounted to 0.2 ind. cm^{-2} and was one cause of the low proportion the dominant taxa had in the assemblage at this time (Fig. 1).

At both sites, the dominant naidid taxa exhibited a distinct successional pattern over the year: *Chaetogaster diastrophus* reached its peak density in the period from February to April; the maximum of *Nais* spp. occurred in May, and was eventually followed by that of *Stylaria lacustris* in June and July (Fig. 2). For *C. diastrophus* and *Nais*, peak densities were approximately 2.0 (1.8–2.5) ind. cm^{-2}. Maximum *Stylaria* densities were lower, reaching only 0.5 ind. cm^{-2} at the north, and 0.3 at the south basin site. Overlapping periods of high densities of *C. diastrophus* and *Nais* species resulted in total naidid density maxima of 3.8 ind. cm^{-2} at the south, but only 2.5 ind. cm^{-2} at the north basin site. Population declines after these maxima were rapid, in particular for *C. diastrophus* and *Nais* at the south basin site (Fig. 2). During late summer and autumn, total naidid densities remained around 0.3 ind. cm^{-2} at the north basin site, but were on most dates less than 0.1 ind. cm^{-2} at the south basin site.

The mean individual biomass of the dominant naidid taxa was 2.2 μg dry mass for *C. diastrophus*, 13.3 μg for *Nais*, and 86 μg for *S. lacustris*, respectively (Table 3). These values are in accordance with data derived from the literature (Finogenova, 1984; Lochhead & Learner, 1983; Schönborn, 1984) if dry mass is assumed to be 12–15% of the wet mass. Although the individual biomass of naidids was highly variable the three taxa clearly belong to distinctly different size-classes. No seasonal trends in individual biomass were apparent in any taxon during spring 1994.

The seasonal development of total naidid biomass (Fig. 3) differed in timing from changes in numbers (Fig. 2) because of the differences in individual biomass. Specimens of *S. lacustris* weighed about 40 times as much as those of *C. diastrophus* and 6.4 times as much as those of *Nais* (Table 3). As a result, the maximum of total naidid biomass, compared to that of total naidid density, was shifted towards the summer when successively larger species dominated the assemblage. Maximum total naidid biomass amounted to 42 μg dry mass per cm^2 in July at the north, and

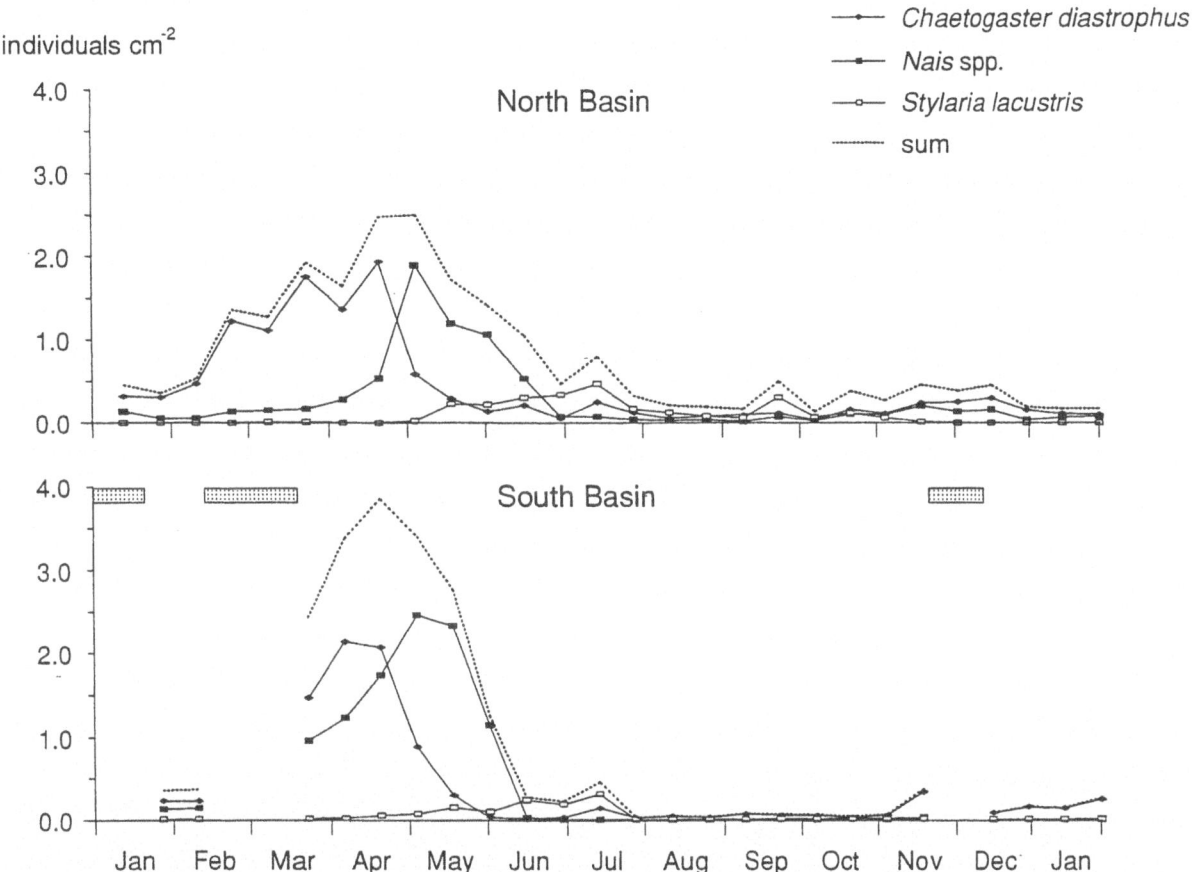

Fig. 2. Seasonal development of densities of dominant naidid taxa in aufwuchs on *Phragmites australis* in Lake Belau over 1993. Bars indicate ice cover.

44 µg per cm² in May at the south basin site. With decreasing numbers of *S. lacustris* from mid July on, total biomass was markedly reduced, in the south basin to almost zero (Fig. 3). An autumn maximum of total naidid biomass caused by a second population peak of *S. lacustris* in September was observed at the north, but not at the south basin site.

In spite of these differences, annual means of population densities and biomasses of the dominant naidids were similar at the two sites (Table 4), even though *Nais* were relatively more important at the south, and *Stylaria* at the north basin site, respectively.

Population growth of all dominant naidid taxa was caused by asexual reproduction. During the sampling period no sexually mature *Chaetogaster diastrophus* were observed. In October at a water temperature of 9 °C, 31% of *Nais* were mature, and 76% of the population of *S. lacustris* matured in October and November

at a temperature of 5 °C. *Chaetogaster diastrophus* and *Nais* persisted during the winter, reproducing asexually. In contrast, no specimens of *S. lacustris* were found in December and January. The first individuals occurring in early spring were very small, indicating that they had just hatched, having overwintered as cocoons.

Discussion

Aufwuchs community

The definition of aufwuchs as a community living on, but not penetrating solid submerged surfaces raises the question whether all oligochaete taxa found in the present study can be regarded as members of this community. Tubificids have been shown to leave the sediment and invade decaying leaf litter (Chauvet *et al.*,

120

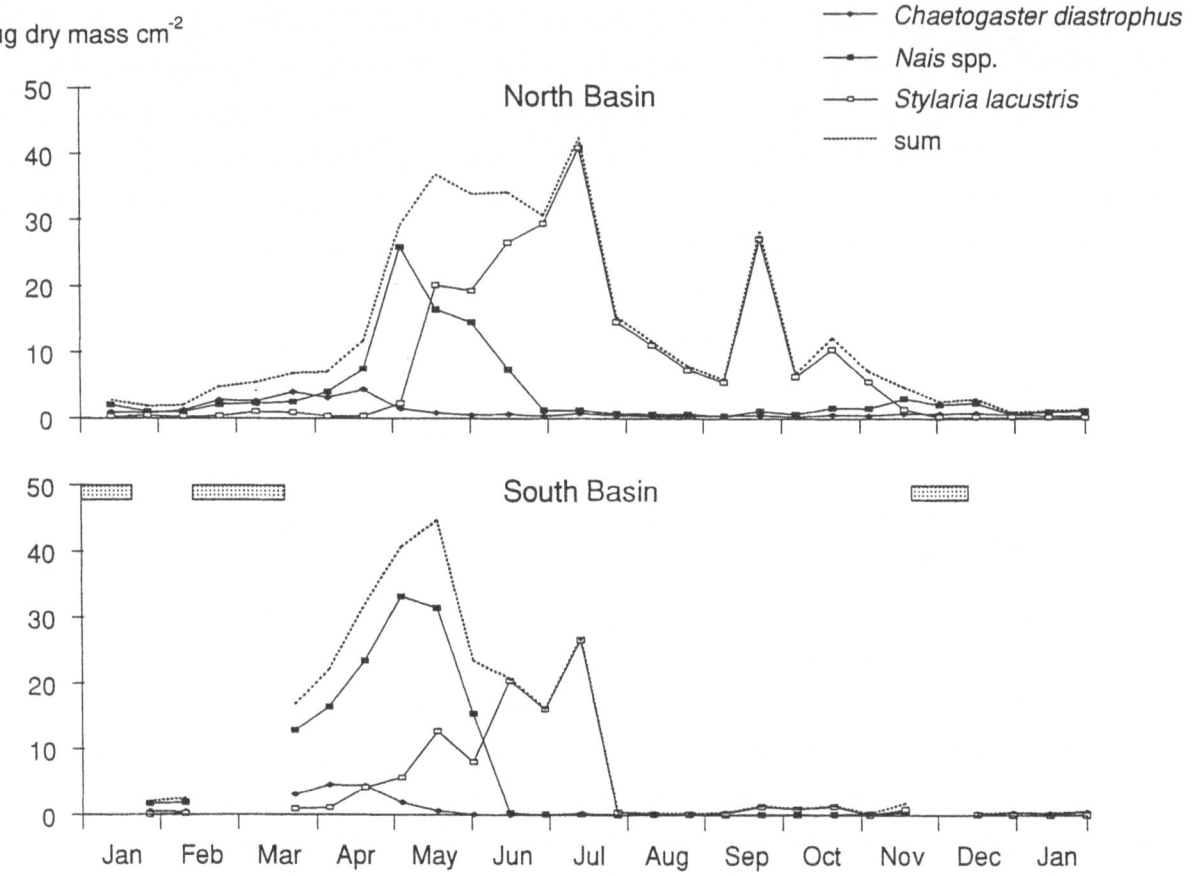

Fig. 3. Seasonal biomass development of dominant naidid taxa in aufwuchs on *Phragmites australis* in Lake Belau over 1993. Bars indicate ice cover.

Table 3. Individual biomass [µg dry mass] of dominant naidid taxa in aufwuchs in Lake Belau. CoV: coefficient of variation [%], n: number of independent measurements, i: mean number of individuals per measurement.

Species	Mean	Range	CoV	n	i
Chaetogaster diastrophus	2.2	1.3–2.9	29	6	12
Nais spp.	13.3	4.8–31.3	56	13	16
Stylaria lacustris	86	51–109	16	30	15

Table 4. Mean annual population density and biomass of dominant naidid taxa in aufwuchs on *Phragmites australis* in Lake Belau. *C.d.*: *Chaetogaster diastrophus*, *S.l.*: *Stylaria lacustris*.

Site	Density [individuals cm^{-2}]				Biomass [µg dry mass cm^{-2}]			
	C.d.	*Nais*	*S.l.*	Total	*C.d.*	*Nais*	*S.l.*	Total
North basin	0.45	0.28	0.10	0.83	0.94	3.8	8.7	13.4
South basin	0.37	0.45	0.05	0.87	0.78	5.9	4.4	11.1

1993), but I did not find them on sections of *Phragmites* stems that had no sediment contact. Enchytraeids were generally found associated with leaf sheath material. They seem to penetrate the decomposing leaf sheath tissue of *Phragmites* as shown for *Spartina* (Healy & Walters, 1994). These findings suggest that in Lake Belau neither tubificids nor enchytraeids are true members of the aufwuchs community. Aeolosomatids are known to live in sediments as well, but like all naidids found in this study (maybe except *Pristinella* sp.) they inhabited the algal layer forming the aufwuchs matrix. Consequently, the Aeolosomatidae and Naididae reported in the present study can be counted as members of the aufwuchs community.

As in the littoral zone of Lake Belau, those aufwuchs communities of oligochaetes and aphanoneurans which have been studied in detail were often dominated by a small number (2–5) of mostly naidid species (Botts & Cowell, 1993; McElhone, 1978; Schönborn, 1987), with the genera *Chaetogaster*, *Nais* and *Stylaria* often predominating (McElhone, 1978; Meuche, 1939; Müller-Liebenau, 1956; Schönborn, 1987).

Seasonality

Due to the general 'lack of studies on seasonal development of macrophyte-naidid associations which include identification beyond the family level' (Beckett *et al.* 1992), comparison of the present results with literature data is limited. According to McElhone (1979, 1980) and Schönborn (1987), *Chaetogaster diastrophus* and *Nais* species occupy different food niches when occurring in the same habitat. Therefore, the succession of maxima of the dominant naidids *C. diastrophus*, *Nais* spp. and *S. lacustris* observed in Lake Belau may be the result of changes in the composition of available food. Characteristic seasonal changes in both the quality and quantity of potential food have been reported from Lake Belau (U. Müller, 1994): In spring diatoms develop abundantly on the reed stems and dominate the periphytic algal community, causing a biomass peak usually observed in April. Subsequently total algal biomass declines rapidly, and community composition changes with filamentous green algae becoming dominant. From about August on, diatoms dominate again, but total algal biomass remains low for the rest of the year.

Literature on the nutritional ecology of naidid species is to some extent contradictory, probably reflecting their plasticity rather than errors of researchers. Nevertheless, it appears that the dominant naidid taxa of Lake Belau use different food sources. *Chaetogaster diastrophus* seems to select bigger food items with a preference for large diatoms (McElhone, 1979, 1980; Smith & Kaster, 1986), rotifers (Streit, 1977), or ciliated protozoa (Schönborn, 1984). Although food selection has also been reported in *Nais variabilis* when discrete food qualities were offered (Bowker *et al.*, 1985), *Nais* are probably less selective than *C. diastrophus*. *Nais* species have been found to ingest diatoms (Bowker *et al.*, 1983; McElhone, 1979, 1980), unicellular green algae (Bowker *et al.*, 1983), detritus (McElhone, 1979, 1980; Schönborn, 1985), activated sludge (Ratsak *et al.*, 1993) and bacteria (Harper *et al.*, 1981; Lochhead & Learner, 1983), but seem to select against filamentous algae (Bowker *et al.*, 1983; McElhone, 1979, 1980). *Stylaria lacustris* was found to feed unselectively 'on the periphyton layer' (Streit, 1978), ingesting green algae (Wachs, 1967) or detritus (Finogenova, 1984). *Stylaria lacustris* assimilates diatoms more efficiently than green algae, but poorer assimilation may be compensated for by higher feeding rates (Streit, 1978).

Taking into account the reported food preferences and assuming an accumulation of detritus particles in the periphytic matrix, particularly when filamentous algae are abundant, the seasonal succession observed in Lake Belau could be interpreted as follows: in early spring an abundance of small prey and diatoms favours *C. diastrophus*. Subsequently, *Nais* species become dominant as a result of high diatom densities and detritus accumulation. Finally, in summer large amounts of filamentous green algae and detritus promote *S. lacustris*.

However, since control mechanisms in this food web are not known, other factors but food also have to be taken into consideration when looking for an explanation for the pattern observed in Lake Belau. The fact that over spring and summer successively larger naidid taxa reached their population peaks might also be caused by predation, reflecting changes in size selection by predators or in refuge availability for larger species in the three-dimensionsal structure of the aufwuchs matrix. Alternatively, abiotic factors such as temperature may be limiting at certain times. For example, *S. lacustris* did not survive in culture at 5 °C, but grew well at higher temperatures, an effect not observed in *C. diastrophus* or *Nais* (unpublished data).

Maxima of total naidid biomass were recorded in summer corresponding with relatively low algal biomass but highest algal productivity (U. Müller, 1994,

122

1995), indicating that algal biomass may have been limited by grazing at this time. Invertebrate grazing on algae is a central process in aufwuchs and has been reported not only to alter algal community composition (Botts, 1993), but also to limit algal biomass (e.g. Cattaneo, 1983; Hann, 1991; Kairesalo, 1984; Kairesalo & Koskimies, 1987). Whether naidids can indeed affect lentic aufwuchs algal communities and stimulate productivity to a significant extent requires further experimental testing. Given the similarity of patterns between sites (apart from the lack of an autumn maximum of *S. lacustris* at the south basin site) and in successive years (pers. obs.), Lake Belau offers a good opportunity to study both the influence of naidids on other members of the aufwuchs community and the mechanisms regulating population densities and seasonal succession of these oligochaetes.

Acknowledgements

I thank M. Dilge and A. Schmitt who helped processing the samples and S. Geisler for analysing water samples. R. Newzella collected field data with me. Data on *Phragmites* densities and distribution were kindly provided by B. Schieferstein. M. Gessner and R. Newzella commented on the manuscript. This work was supported by the 'Bundesminister für Forschung und Technologie' of the FRG as part of the project 'Ecosystem Research in the Bornhöved Lakes Region'.

References

Beckett, D. C., T. P. Aartila & A. C. Miller, 1992. Seasonal change in plant-dwelling Chironomidae and Naididae in a Wisconsin Lake. J. Freshwat. Ecol. 7: 45–57.

Botts, P. S., 1993. The impact of small chironomid grazers on epiphytic algal abundance and dispersion. Freshwat. Biol. 30: 25–33.

Botts, P. S. & B. C. Cowell, 1993. Temporal patterns of abundance of epiphytic invertebrates on *Typha* shoots in a subtropical lake. J. N. Am. Benthol. Soc. 12: 27–39.

Bowker, D. W., M. T. Wareham & M. A. Learner, 1983. The selection and ingestion of epilithic algae by *Nais elinguis* (Oligochaeta: Naididae). Hydrobiologia 98: 171–178.

Bowker, D. W., M. T. Wareham & M. A. Learner, 1985. A choice chamber experiment on the selection of algae as food and substrata by *Nais elinguis* (Oligochaeta: Naididae). Freshwat. Biol. 15: 547–557.

Brown, S.-D. & A. P. Austin, 1971. A method of collecting periphyton in lentic habitats with procedures for subsequent sample preparation and quantitative assessment. Int. Revue ges. Hydrobiol. 56: 557–580.

Cattaneo, A., 1983. Grazing on epiphytes. Limnol. Oceanogr. 28: 124–132.

Chauvet, E., N. Giani & M. O. Gessner, 1993. Breakdown and invertebrate colonization of leaf litter in two contrasting streams: significance of oligochaetes in a large river. Can. J. Fish. aquat. Sci. 50: 488–495.

Downing, J. A., 1986. A regression technique for the estimation of epiphytic invertebrate populations. Freshwat. Biol. 16: 161–173.

Finogenova, N. P. 1984. Growth of *Stylaria lacustris* (L.) (Oligochaeta, Naididae). Hydrobiologia 115: 105–107.

Hann, B. J., 1991. Invertebrate grazer-periphyton interactions in a eutrophic marsh pond. Freshwat. Biol. 26: 87–96.

Harper, R. M., J. C. Fry & M. A. Learner, 1981. A bacteriological investigation to elucidate the feeding biology of *Nais variabilis* (Oligochaeta: Naididae). Freshwat. Biol. 11: 227–236.

Healy, B. & K. Walters, 1994. Oligochaeta in *Spartina* stems: the miocrodistribution of Enchytraeidae and Tubificidae in a salt marsh, Sapelo Island, USA. Hydrobiologia 278: 111–123.

Kairesalo, T., 1984. The seasonal succession of epiphytic communities within an *Equisetum fluviatile* L. stand in Lake Pääjärvi, Southern Finland. Int. Revue ges. Hydrobiol. 69: 475–505.

Kairesalo, T. & I. Koskimies, 1987. Grazing of oligochaetes and snails on epiphtes. Freshwat. Biol. 17: 317–324.

Lochhaed, G. & M. A. Learner, 1983. The effect of temperature on asexual population growth of three species of Naididae (Oligochaeta). Hydrobiologia 98: 107–112.

McElhone, M. J., 1978. A population study of littoral dwelling Naididae (Oligochaeta) in a shallow mesotrophic lake in North Wales. J. anim. Ecol. 47: 615–626.

McElhone, M. J., 1979. A comparison of the gut contents of two co-existing lake-dwelling Naididae (Oligochaeta), *Nais pseudobtusa* and *Chaetogaster diastrophus*. Freshwat. Biol. 9: 199–204.

McElhone, M. J., 1980. Some factors influencing diet of coexisting, benthic, algal grazing Naididae (Oligochaeta). Can. J. Zool. 58: 481–487.

Meuche, A., 1939. Die Fauna im Algenbewuchs. Arch. Hydrobiol. 34: 349–520.

Müller, H. E., 1981. Vergleichende Untersuchungen zur hydrochemischen Dynamik von Seen im Schleswig-Holsteinischen Jungmoränengebiet. Kieler Geographische Schriften 53, 208 pp.

Müller, U., 1994. Seasonal development of epiphytic algae on *Phragmites australis* (Cav.) Trin. ex Sten. in a eutrophic lake. Arch. Hydrobiol. 129: 273–292.

Müller, U., 1995. Vertical zonation and production rate of epiphytic algae on *Phragmites australis* Freshwat. Biol. 34: 69–80.

Müller-Liebenau, I., 1956. Die Besiedlung der *Potamogeton*-Zone ostholsteinischer Seen. Arch. Hydrobiol. 52: 470–606.

Ratsak, C. H., S. A. L. M. Kooijman & B. W. Kooi, 1993. Modelling the growth of an oligochaete on activated sludge. Wat. Res. 27: 739–747.

Schönborn, W., 1984. The annual energy transfer from the communities of Ciliata to the population of *Chaetogaster diastrophus* (Gruithuisen) in the River Saale. Limnologica 16: 15–23.

Schönborn, W., 1985. Die ökologische Rolle der Gattung *Nais* (Oligochaeta) in der Saale. Zool. Anz. 215: 311–328.

Schönborn, W., 1987. Secondary production and energy transfer in the polluted River Saale (Thuringia, Southern GDR). Int. Revue ges. Hydrobiol. 72: 539–557.

Smith, M. E. & J. L. Kaster, 1986. Feeding habits and dietary overlap of Naididae (Oligochaeta) from a bog stream. Hydrobiologia 137: 193–201.

Streit, B., 1977. Morphometric relationships and feeding habits of two species of *Chaetogaster*, *Ch. limnaei* and *Ch. diastrophus* (Oligochaeta). Arch. Hydrobiol. Suppl. 48: 424–437.

Streit, B., 1978. A note on the nutrition of *Stylaria lacustris* (Oligochaeta: Naididae). Hydrobiologia 61: 273–276.

Wachs, B., 1967. Die Oligochaeten-Fauna der Fließgewässer unter besonderer Berücksichtigung der Beziehungen zwischen der Tubificiden-Besiedlungsdichte und dem Substrat. Arch. Hydrobiol. 63: 310–386.

Hydrobiologia **334**: 125–132, 1996.
K. A. Coates, Trefor B. Reynoldson & Thomas B. Reynoldson (eds), Aquatic Oligochaete Biology VI.
© 1996 *Kluwer Academic Publishers.*

Age, stage and size structure as population state variables for *Tubifex tubifex* (Oligochaeta, Tubificidae)

Andrea Pasteris[1], Giuliano Bonomi[1] & Carla Bonacina[2]
[1] *Dipartimento di Biologia Evoluzionistica Sperimentale, Università degli Studi di Bologna, Via Selmi 3, I-40126 Bologna, Italy*
[2] *CNR – Istituto Italiano di Idrobiologia, Largo Tonolli 50, I-28048 Pallanza NO, Italy*

Key words: population dynamics, population structure, age, size, stage, *Tubifex tubifex*

Abstract

Age based demography is not applicable to tubificid worms, because of the lack of any reliable method for assessing the age of individuals. 'Stage', namely the degree of reproductive development, and/or size could be used instead of age, if that knowledge improved prediction of an individual's fate and fecundity.

We measured, under laboratory conditions, the influence of age, size, and maturation stage on the fate and fecundity of individuals, by means of a modification of the standard cohort experiment. As expected all of the three state variables significantly affect the fecundity, though the residual variance is considerable.

As a first step in the evaluation of the effectiveness of a size and/or stage based demography for tubificids, the results are used to establish a size-stage structured matrix population model. The asymptotic behaviour of the model is studied and a basic sensitivity analysis is performed, to evaluate the effects of the observed variability of the matrix elements, on the eventual finite growth rate.

Introduction

The simplest population dynamics models only define population state in terms of total number of individuals or total biomass. However, state variables in mathematical models must satisfy certain requirements (Caswell, 1988). First, the state and the environment at time t must determine (at least stochastically), the response of the system at time t. Second, the state at time t_1 and the environment over the interval between t_1 and t_2 must determine (at least stochastically) the state at time t_2. Individuals making up a population differ greatly as far as growth rate, fecundity, survival probability and disposition for migration are concerned. Since these processes, ultimately, determine the variation of the density and biomass of a population, total abundance is often insufficient as a state variable, because it does not allow adequate prediction of the future conditions.

In classical demography, age structure (i.e., a frequency distribution of individuals over the values of age) is taken as a state variable, since demographic traits of individuals are considered functions of their age.

Tubificid worms have a potential life span of several years (Timm, 1984), however age based demography is not applicable to them. The most obvious reason is the lack of any reliable method for assessing the age of individuals; yet a second, less evident problem is even more fundamental: in fact, an organism's demographic traits are likely to depend on its size, physiological conditions and developmental stage rather than on age as such. In many organisms, age is highly correlated with size, condition and stage, and this leads to correlation between age and demographic traits (May, 1988; Ebenman & Persson, 1988). However, this is not true of tubificid oligochaetes; this is particularly evident if data from laboratory cohort cultures are considered. The coefficient of variation for the weight of individual worms of exactly the same age and reared in the same container can be as high as 65% for *Branchiura sowerbyi* (Bonacina *et al.*, 1994), 75% for *Tubifex tubifex* and 84% for *Limnodrilus hoffmeisteri* (Adreani & Bonomi,

1979). Differences among individuals of the same age, grown in different environments, are likely to be much larger.

Growth and development are continuous processes in oligochaetes; nonetheless, even if rather arbitrarily, we can sort the individuals into stages according to the maturation level of the reproductive organs. Again, laboratory cultures show that the time spent in each stage is highly variable among individuals of the same cohort.

Demographic traits of individuals of different size and reproductive status probably will not be the same. As a result, even if it were possible to determine the age structure of a population of tubificids, this would give little information on its fate.

Tubificids are not exceptions from this point of view: for similar reasons, many other organisms, both animals and plants, are not amenable to age based demography. To deal with such organisms, several authors have formulated demographic models that take into account size and/or stage structure, in addition to, or in place of, age structure; thus, a more general conceptual framework is emerging as opposed to traditional age based demography (Ebenman & Persson, 1988).

This approach seems to be applicable to tubificids as well; Bonomi & Di Cola (1980) proposed a 'bidimensional' model aimed at the assessment of mortality, fecundity and production of natural populations of *T. tubifex*, based on the classification of individuals both by weight and by maturation stage. However, the data they had available did not allow an estimation of the parameters of the model so they produced a simpler 'unidimensional' version of the model.

The aim of this work is to evaluate the effectiveness of weight and stage structure as population state variables for tubificids. For this purpose we compared the effect of age, weight, and maturation stage of the individual on the demographic traits of *T. tubifex*, in laboratory cultures.

The standard cohort experiment, applied to tubificids, does not allow collection of weight- or stage-classified data, as the worms cannot be reared separately if they have to breed. Neither can they be marked, so measures of survivorship and fecundity are averages over individuals of the same age, but of different weight and stage. Thus we established a modification of the standard cohort experiment: the cohort was periodically sorted into weight and stage classes and growth, survivorship and fecundity of these homogeneous groups were measured separately.

Materials and methods

Eggs and worms were reared in circular glass containers (internal diameter: 7.6 cm; height: 4.5 cm) half-filled with sand and then filled with aerated tap water. Frozen lettuce (*c.* 4 g w.w.) was put under the sand as food. Every week each container was checked, the sand washed, the residual food removed, and fresh lettuce added. Cultures were kept at 20 ± 0.1 °C.

The experimental design (Fig. 1) is rather complex; however, it allowed all the worms to be reared in similar containers and at a similar population density, both at the beginning, when the cohorts were not yet split in classes, and later on, when separated in classes of greatly different abundance. Meanwhile, through a weekly mixing, it was possible to reduce the segregation of individuals in small groups, and to keep homogeneity within each class.

A cohort (A) was started from cocoons produced in one week by a laboratory stock culture of *T. tubifex* which had been established using worms collected from Lake Orta, N. Italy, about one year before the beginning of the experiment. After one week a second cohort (B) was started. Cocoons were only used when the number of eggs was visible; this resulted in cocoons with more than 6 or 7 eggs being discarded. The total number of eggs was 607 in cohort A, 581 in cohort B. The cohorts were cultured separately from each other.

Cocoons were placed in several containers, 12 cocoons per container, and buried in the sand with a pipette. Once a week, cocoons were examined. Those that did not contain viable embryos were discarded; the others, from all containers, were mixed together and then re-distributed, 12 per container. The number of containers was gradually reduced as cocoons were discarded. Newly hatched worms were placed, in groups of 20, in containers identical to those used for cocoons. The number of containers was gradually increased as worms hatched.

Four weeks after 95% of the worms had hatched (i.e. two months after cohorts were started), worms were put in water, without sand and food for a few hours to allow the gut contents to clear. Then they were examined using a dissecting microscope and sorted in three stages: young (without a distinct clitellum), mature (with distinct clitellum), ovigerous (with visible oocytes). Afterward, each worm was dried with filter paper and weighted on an analytical balance at 0.1 mg. Thus, classes were obtained, defined by a specific combination of stage and weight. We adopted the

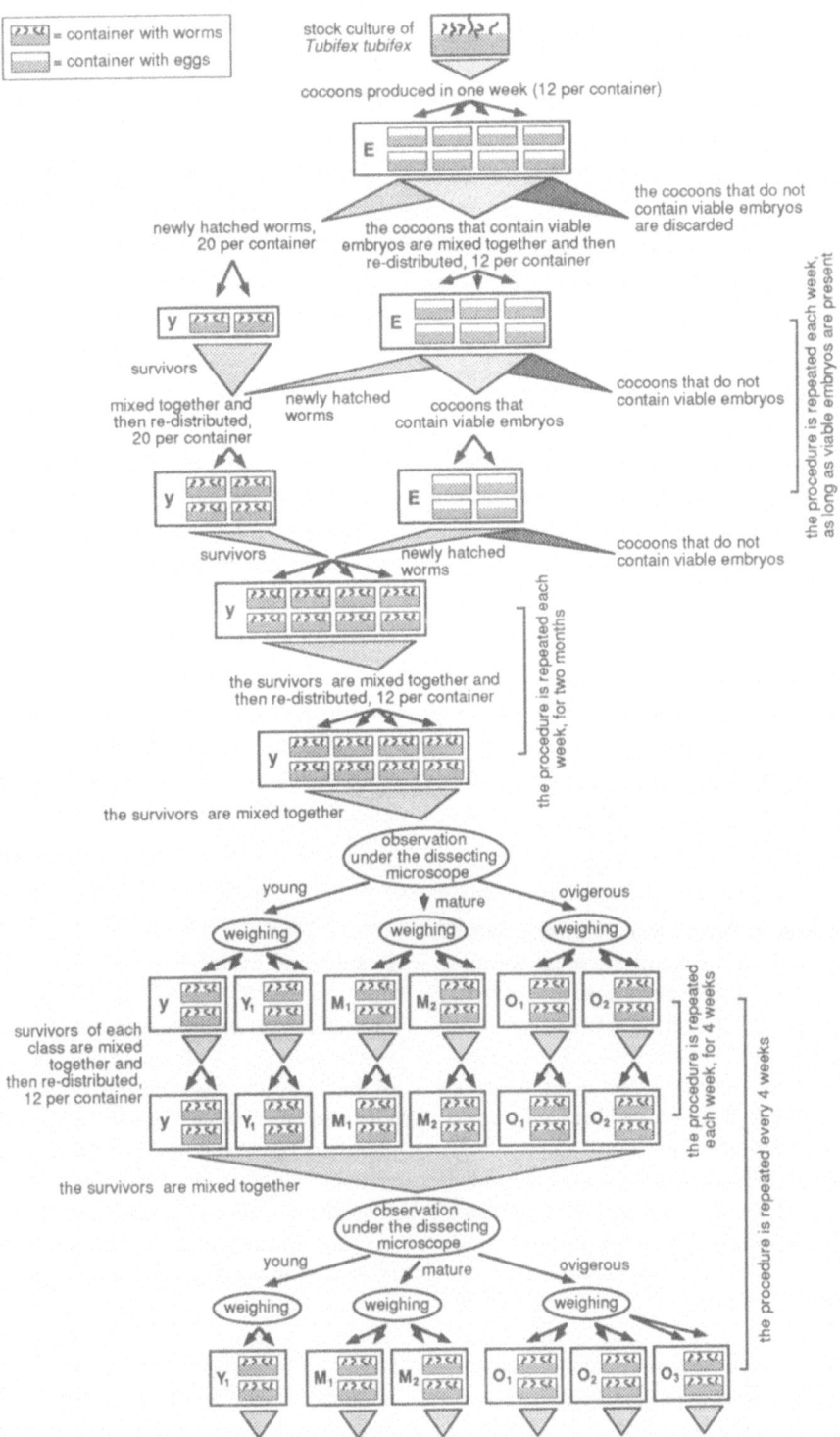

Fig. 1. Diagrammatic representation of the mixing and re-sorting procedure employed for each of the two experimental cohorts. Each symbol (y, Y_1, M_1, M_2, etc.) stands for a specific class of individuals (see Table 1). The figure gives only a qualitative description of the procedure; the number of containers and of classes which are represented do not correspond to the real experiment.

128

Table 1. Adopted stage-weight combinations for the classification of the individuals in the cohorts. Each symbol stands for a class that was actually observed during the experiment.

Stage	Weight class (mg)						
	≤ 1	1–3	3–6	6–10	10–15	15–25	> 25
Embryos	E						
Young	y	Y_1					
Mature		M_1	M_2	M_3	M_4		
Ovigerous		O_1	O_2	O_3	O_4	O_5	O_6

classification proposed by Bonomi & Di Cola (1980) with minor changes: all embryos were combined to one class and two further classes were included because of the presence of large individuals. Table 1 reports all the classes actually observed, their definition and the symbols by which they are referred to.

Worms, in sets of 20, from each class were reared in individual containers. Each week, all the individuals from a class were re-mixed together (but not with the individuals of the other classes) and sorted again into containers. During the weekly check the number of survivors and of laid eggs was counted.

Every four weeks, the worms were checked to determine their stage and weight. It was therefore possible to evaluate, for each class, the proportions of individuals that had (i) died, (ii) survived and stayed in the same class and (iii) survived and transferred to other classes. Finally (Fig. 1) all the worms were grouped according to their current class, regardless of their classification four weeks before. The new groups were reared separately for four weeks till the next weighing. Seven such weighings were performed at regular 4 week intervals. Shortening the period between two subsequent weighings would have significantly improved the description of the demographic processes, however 4 weeks were considered an acceptable compromise between precision and workload.

Since we worked with cohorts, the age of the individuals was known. Thus, for each specific combination of age, stage and weight, it was possible to assess fecundity, mortality and transition rate to other classes.

Each set of worms belonging to the same cohort and characterised by a specific age-stage-weight combination (i.e., all the worms that were mixed together weekly, during a specific 4-week period, between 2 subsequent weighings) will be called below a 'unit'. Indeed, such a set of individuals is the basic element on which data were collected in our experimental design; for instance we do know how many eggs each unit produced, but we do not know how many eggs each worm produced.

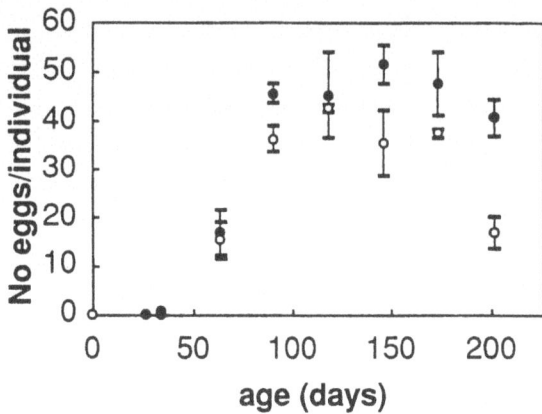

Fig. 2. Relationship between fecundity (average number of eggs produced per individual over four weeks) and age (solid circles: cohort A; open circles: cohort B). Bars represent ± standard error among all of the units with the same age.

Several units were actually formed by less than 20 individuals, even less in some cases. However, only units with at least 15 individuals are considered here; thus the density in the containers was kept between 15 and 20 individuals (corresponding to 3300–4400 ind. m^{-2}).

Results and discussion

The effects of age, weight and maturation stage on egg production are shown in Figs 2–4. As expected all of the three state variables affect the fecundity, though the residual variance is considerable. The effects are all significant ($p < 0.01$) if tested with the Kruskal-Wallis test, a non-parametric equivalent of one-way ANOVA (Siegel, 1956; Sokal & Rohlf, 1981).

As a first step in the evaluation of the effectiveness of a stage and/or size based demography for tubificids, the results of the experiment have been summarised by means of a stage-weight matrix model. Matrix models were first formulated to deal with age structured populations (Lewis, 1942; Leslie, 1945, 1948), but they were further developed to include populations structured by means of any other suitable descriptor like size or stage (Lefkovitch, 1965; Usher, 1972; Law, 1983; Hugues, 1984; Caswell, 1988, 1989).

Matrix models make two assumptions that fit our experimental design:

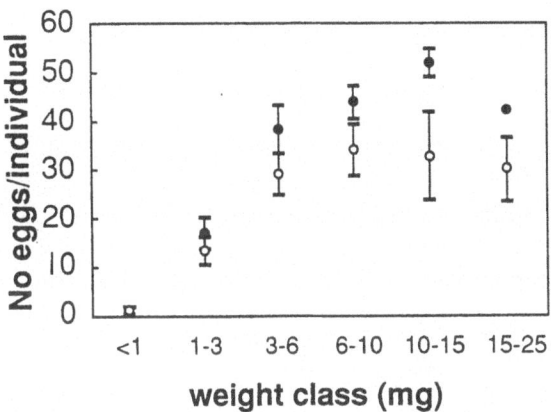

Table 2. Correspondence between stage-weight combination and number assigned to the class in the matrix model.

Stage	Weight class (mg)			
	≤ 1	1–3	3–10	> 10
Embryos	1			
Young	2	3		
Mature			4	5	6
Ovigerous			7	8	9

Fig. 3. Relationship between fecundity (average number of eggs produced per individual over four weeks) and weight (solid circles: cohort A; open circles: cohort B). Bars represent ± standard error among all of the units in the same weight class.

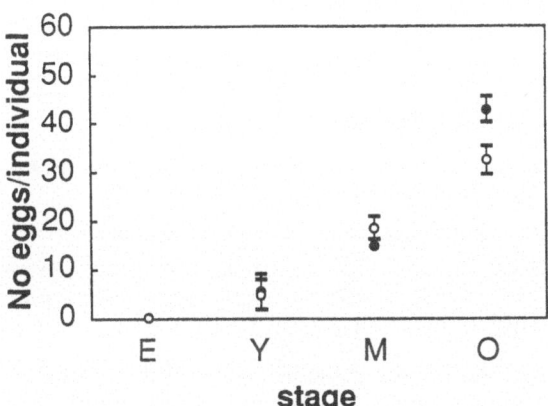

Fig. 4. Relationship between fecundity (average number of eggs produced per individual over four weeks) and stage (solid circles: cohort A; open circles: cohort B). Bars represent ± standard error among all of the units in the same maturation stage.

(1) the individuals in the population can be arranged in a number of discrete classes (e.g. age classes or size classes);

(2) time is a discrete variable, i.e. it assumes only values that are exact multiples of a basic unit (in other words: nothing is known about the population when time assumes fractional values).

In our model individuals are arranged in 9 classes; each class is a specific combination of stage and weight. In the model we did not adopt the classification of Table 1: in order to simplify the model we combined

some of the original weight classes. Each class in the model is referred to by a number (see Table 2). The time unit or *projection interval* is 4 weeks (28 days), the same interval occurring between two weighings in the experiment.

The state variable of the population is a column vector of 9 elements, the *population vector*. Each element is the number of individuals in a specific stage-weight class. Fecundity and transfer rates between classes are collected in a *projection matrix* (9 rows × 9 columns).

The model can be written as:

$$\mathbf{n}(t + 1) = \mathbf{A} \cdot \mathbf{n}(t)$$

where:

$\mathbf{n}(t)$ is the 9 element vector that describes the state of the population at a generic time t. Each element $n_j(t)$ in the vector is the number of individuals in the j-th class at time t;

$\mathbf{n}(t + 1)$ is the 9 element vector that describes the state of the population at time $t + 1$ (i.e. one time unit after t). Each element $n_i(t + 1)$ in the vector is the number of individuals in the i-th class present in the population at time $t + 1$;

$\mathbf{A}$ is the projection matrix.

Each generic element $a_{i,j}$ of the projection matrix (i.e. the element on the i-th row and on the j-th column) is a coefficient accounting for the average contribution of one individual in the class j at time t to the class i at time $t + 1$.

The coefficients estimated using our experimental data give the following projection matrix:

$$\mathbf{A} =$$

$$\begin{bmatrix} 0.06 & 0.92 & 11.45 & 15.36 & 20.56 & 0 & 23.07 & 40.62 & 41.21 \\ 0.59 & 0.21 & 0 & 0 & 0 & 0 & 0 & 0 & 0 \\ 0 & 0.23 & 0 & 0 & 0 & 0 & 0 & 0 & 0 \\ 0 & 0.24 & 0 & 0 & 0 & 0 & 0 & 1\cdot10^{-3} & 0 \\ 0 & 0.02 & 0 & 0 & 0 & 1.00 & 0 & 0.01 & 0.03 \\ 0 & 0 & 0 & 0 & 0 & 0 & 0 & 5\cdot10^{-4} & 2\cdot10^{-3} \\ 0 & 0.06 & 0 & 0 & 0 & 0 & 0 & 5\cdot10^{-4} & 0 \\ 0 & 0.17 & 1.00 & 1.00 & 1.00 & 0 & 1.00 & 0.69 & 0.13 \\ 0 & 0 & 0 & 0 & 0 & 0 & 0 & 0.29 & 0.83 \end{bmatrix}$$

All the elements in the matrix were estimated using *units* (see the materials and methods section for a definition) with at least 15 individuals. The only exception is column 6; only one individual belonging to class 6 (mature > 10 mg) was reared, so the coefficients on this column were estimated using this one individual.

All the elements on the first row, with the exception of $a_{1,1}$, are fecundities. They are the number of embryos (class 1) expected to be present at time $t + 1$ for one individual in the class j at time t. These elements were estimated as

$$\hat{a}_{1,j} = \frac{\text{number of eggs laid over 28 days by}}{\text{all the individuals belonging to the class } j} \over {\text{total number of individuals} \atop \text{belonging to the class } j}}.$$

The other elements on the matrix are all transfer rates between classes. Each element on the main diagonal of the matrix (i.e. if $i = j$) is the probability that one individual in the class j at time t survive and stay in the same class during the next time step. If $i \neq j$, $a_{i,j}$ is the probability that one individual in the class j at time t survive and move to the class i during the next time step. The transfer rates were estimated as

$$\hat{a}_{i,j} = \frac{\substack{\text{total number of individuals classified as } j \\ \text{at any weighing}}}{\substack{\text{number of individuals classified as } j \text{ at any weighing,} \\ \text{that were classified as } i \text{ at the next weighing}}}.$$

The model can also be written as a system of finite difference equations:

$$n_1(t+1) = 0.06n_1(t) + 0.92n_2(t) + 11.45n_3(t) +$$
$$+ 15.36n_4(t) + 20.56n_5(t) + 23.07n_7(t) +$$
$$+ 40.62n_8(t) + 41.21n_9(t)$$
$$n_2(t+1) = 0.59n_1 + 0.21n_2(t)$$
$$n_3(t+1) = 0.23n_2(t)$$
$$n_4(t+1) = 0.24n_2(t) + 1 \cdot 10^{-3}n_8(t)$$
$$n_5(t+1) = 0.02n_2(t) + 1.00n_6(t) + 0.01n_8(t) +$$
$$+ 0.03n_9(t)$$
$$n_6(t+1) = 5 \cdot 10^{-4}n_8(t) + 2 \cdot 10^{-3}n_9(t)$$

$$n_7(t+1) = 0.06n_2(t) + 5 \cdot 10^{-4}n_8(t)$$
$$n_8(t+1) = 0.17n_2(t) + 1.00n_3(t) + 1.00n_4(t) +$$
$$+ 1.00n_5(t) + 1.00n_7(t) + 0.69n_8(t) +$$
$$+ 0.13n_9(t)$$
$$n_9(t+1) = 0.29n_8(t) + 0.83n_9(t) .$$

Projection is the simplest form of analysis of such a model: beginning with an initial population vector $\mathbf{n}(0)$, the subsequent states $\mathbf{n}(1), \mathbf{n}(2), ...$ of the population are computed by repeated matrix multiplication:

$$\mathbf{n}(1) = \mathbf{A}\mathbf{n}(0)$$
$$\mathbf{n}(2) = \mathbf{A}\mathbf{n}(1) = \mathbf{A}\mathbf{A}\mathbf{n}(0) = \mathbf{A}^2\mathbf{n}(0)$$
$$\vdots$$
$$\mathbf{n}(t) = \mathbf{A}\mathbf{n}(t-1) = \mathbf{A}^t\mathbf{n}(0) .$$

Figure 5 shows the results of such a computation started from three different hypothetical $\mathbf{n}(0)$. The variation of $N(t)$ over the first ten time steps is reported in the graph, where $N(t) = \sum_j n_j(t)$ is the total number of individuals in the population at time t. In all of the three cases $N(0) = 9$, but the initial distribution of the individuals among the classes is different. The initial population vectors are:

- case a (all the individuals are in the embryo class):
  $\mathbf{n}(0) = [9\ 0\ 0\ 0\ 0\ 0\ 0\ 0\ 0]'$
- case b (there is one individual in each class):
  $\mathbf{n}(0) = [1\ 1\ 1\ 1\ 1\ 1\ 1\ 1\ 1]'$
- case c (all the individual are ovigerous > 10 mg):
  $\mathbf{n}(0) = [0\ 0\ 0\ 0\ 0\ 0\ 0\ 0\ 9]'$

At the beginning the behaviour of the three simulated populations is quite different from each other, depending on the different initial structure. However after a few time steps all of them grow linearly with the same slope. As the ordinate is in logarithmic scale the eventual growth of the populations is exponential and the *finite growth rate*, i.e. the ratio $N(t+1)/N(t)$, is a constant with the same value for all the considered cases. The same result is achieved if the projection is carried out from any initial population structure. The eventual finite growth rate is:

$$\lambda = 2.82 .$$

This value roughly means that the population triplicates in one month.

This kind of behaviour is not peculiar to this specific model: in fact it is common to a large class of models defined by a few, rather general, assumptions. Provided that the transition values don't vary through

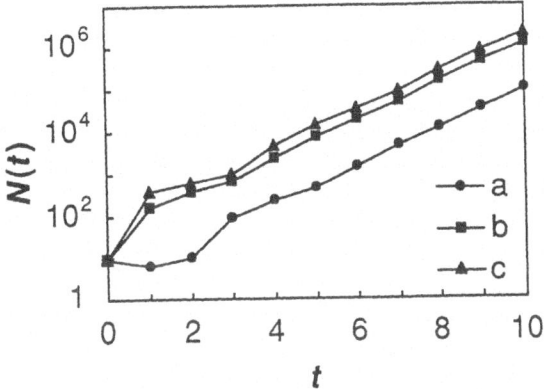

Fig. 5. Projection results using matrix **A** and three different hypothetical initial populations (cases a, b and c). The variation of the total number of individuals, $N(t)$, over the first ten time steps is reported.

time, the eventual finite growth rate, λ, is independent from the initial population structure and is determined only by the values of the demographic coefficients in the projection matrix (Caswell, 1989). Therefore λ is a useful integrative index of the performance of a population model; it is to be regarded as a potential long-term population growth rate, provided environmental conditions do not change and density effects do not take place.

Our estimate of λ cannot be regarded as the maximum finite growth rate, since better culture conditions are possible (for instance 20 °C is a suboptimal temperature); however finite growth rate of populations growing under natural conditions are usually lower.

An important part of the analysis of a matrix model is investigating how λ would vary in response to changes in the elements of the matrix. This *sensitivity analysis* may be of interest in several different contexts such as:

(1) measuring 'how important' a given vital rate is to population growth;
(2) quantifying the effects of environmental perturbations;
(3) evaluating alternative management strategies;
(4) predicting the intensity of natural selection;
(5) evaluating the effects of errors in estimation.

While points 1 to 4 assume that a reliable model is available, point 5 is related to the evaluation of this reliability and more relevant at this stage of the development of the model.

As explained above, each element $a_{i,j}$ in the matrix was estimated using all the units in the class j with more than 15 individuals. For some classes several units were observed, over the time span of the experiment. The rates estimated on single units of the same class can be rather different from each other. To assess the effect of the observed dispersion on λ, two further matrices were arranged. The first of them, $\mathbf{A}_{min}$, was set up as follows:

• As an estimate of fecundity of each class, the lowest value among all the units belonging to the class was used.
• As far as the transfer rates (elements from $a_{2,j}$ to $a_{9,j}$) are concerned, the following procedure was applied to every class: for each unit belonging to the class one projection matrix was arranged; in this matrix only the column corresponding to the considered class was different than in the original matrix **A**. In this column were put the transfer rates estimated on the single unit; then the asymptotic growth rate for each matrix was computed. The set of coefficients corresponding to the lowest growth rate was taken as an estimate of the transfer rates for the class.

The second matrix, $\mathbf{A}_{max}$, was set up in a similar way, using the highest fecundities and the sets of transfer rates corresponding to the highest growth rates. The asymptotic growth rates for $\mathbf{A}_{min}$ and $\mathbf{A}_{max}$ were computed, their values being, respectively:

$$\lambda_{min} = 2.24 , \qquad \lambda_{max} = 3.50 .$$

These two values are the lower and upper limit of the dispersion range for λ. The observed variability is due to two factors: random errors and age differences among units belonging to the same weight-stage class. Consequently the width of the variation range is also a measure of the effects of not taking age into account. Actually, the observed dispersion range is rather wide: predicting $N(t)$ using λ_{min} or λ_{max} would lead to completely different results after a few projection intervals. On the other hand one should bear in mind that range is a rough, maximal measure of dispersion and that it is heavily affected by outlying values. More sophisticated methods, such as *jacknife* and *bootstrap* (Caswell, 1989) could be used to estimate a confidence interval for λ, which is, of course, a more proper statistic of dispersion; however, at this stage of the work, we preferred the described method which is conceptually simpler and requires much less computing.

132

Conclusions

The analysis of the data provided by the experiment, is far from completed. Using the same data, several alternative classifications of the individuals could be formulated; a possible development of the work could be the comparison of models based on different classification criteria (i.e., age *vs* maturation stage *vs* weight *vs* combinations) and different number of classes. Moreover a thorough analysis of the effects of the estimation errors on λ is needed to supply a confidence interval and to separate clearly the effects of random variability from the effects of age differences.

More extensive work is required before size and/or stage structure can be adopted as a population state variable for *T. tubifex*; most of all it is necessary to verify if taking population structure into account sufficiently improves predictions about populations, in a range of environmental conditions, to justify the effort required for its assessment. However, before undertaking such laborious studies, some preliminary work is needed to test the consistency of the hypothesis and to refine it.

We think that this first analysis shows that using weight-stage structure as a population state variable for *T. tubifex* is a reasonable working hypothesis which is worth further investigation.

Acknowledgements

We would like to thank Mr Andrea Ferrari for his assistance in the collection of *Tubifex* from Lake Orta. We are much indebted to Miss Claudia Foschi for her help in laboratory operations.

References

Adreani, L. & G. Bonomi, 1979. Elementi per una dinamica di popolazione dei tubificidi. In 3° Congresso dell'Associazione Italiana di Oceanologia e Limnologia, Sorrento: 355–366.

Bonacina, C., A. Pasteris, G. Bonomi & D. Marzuoli, 1994. Quantitative observations on the population ecology of *Branchiura sowerbyi* (Oligochaeta, Tubificidae). Hydrobiologia 278: 267–274.

Bonomi, G. & G. Di Cola, 1980. Population dynamics of *Tubifex tubifex* studied by means of a new model. In R. O. Brinkhurst & D. G. Cook (eds), Aquatic Oligochaete Biology. Plenum Press, New York: 185–203.

Caswell, H., 1988. Approaching size and age in matrix population models. In B. Ebenman & L. Persson (eds), Size-Structured Populations. Springer-Verlag, Berlin: 85–105.

Caswell, H., 1989. Matrix Population Models: Construction, Analysis and Interpretation. Sinauer Associates, Sunderland, 328 pp.

Ebenman, B. & L. Persson, 1988. Introduction. Dynamics of size-structured populations: an overview. In B. Ebenman & L. Persson (eds), Size-Structured Populations. Springer-Verlag, Berlin: 3–9.

Hughes, T. P., 1984. Population dynamics based on individual size rather than age: a general model with a reef coral example. Am. Nat. 123: 778–795.

Law, R., 1983. A model for the dynamics of a plant population containing individuals classified by age and size. Ecology 64: 224–230.

Lefkovitch, L. P., 1965. The study of population growth in organisms grouped by stages. Biometrics 21: 1–18.

Leslie, P. H., 1945. On the use of matrices in certain population mathematics. Biometrika 33: 183–212.

Leslie, P. H., 1948. Some further notes on the use of matrices in certain population mathematics. Biometrika 35: 213–245.

Lewis, E. G., 1942. On the generation and growth of a population. Sankhya 6: 93–96.

May, R. M., 1988. Prolog. In B. Ebenman & L. Persson (eds), Size-Structured Populations. Springer-Verlag, Berlin: 1–2.

Siegel, S., 1956. Non Parametric Statistics for the Behavioral Sciences. McGraw-Hill, New York, 312 pp.

Sokal, R. R. & F. J. Rohlf, 1981. Biometry. Freeman & Co., New York, 859 pp.

Timm, T., 1984. Potential age of aquatic Oligochaeta. Hydrobiologia 115: 101–104.

Usher, M. B., 1972. Developments in the Leslie matrix model. In N. R. Jeffers (ed), Mathematical Models in Ecology. Blackwell, Oxford: 29–60.

Hydrobiologia **334**: 133–140, 1996.
K. A. Coates, Trefor B. Reynoldson & Thomas B. Reynoldson (eds), Aquatic Oligochaete Biology VI.
© 1996 *Kluwer Academic Publishers.*

Energy budget of Oligochaeta and its relationship with the primary production of Lake Sevan, Armenia

Karén Jenderedjian
Institute of Hydroecology and Ichthyology of the National Academy of Sciences of Armenia, Nairian 186, Sevan 378610, Armenia

Key words: Oligochaeta, energy budget, primary production, littoral, sublittoral, profundal

Abstract

From 1928 to 1991 the following oligochaete energy budget quantitative values: biomass (B), production (P), respiration (R), assimilation (A), ration (C) changed 8 to 12 times. With increasing depth the ratios of energy budget decreased: P/B ratio from 3.4 to 0.1, R/B ratio from 4.5 to 0.5, net production efficiency from 43 to 18%. A relationship was revealed between oligochaete biomass and the primary production (PP) of the lake. There is a delay in the response in oligochaete biomass to primary production. In Lake Sevan the 'delay' is 2 years in the littoral, 4 years in the sublittoral and 11 years in the profundal zone. The closest correlation was revealed between oligochaete energy budget quantitative values and the values of primary production of the preceding 10 years, which enables a prediction of the quantitative indices of the community of Oligochaeta. The values of energy budget ratios depend on temperature and oxygen regimes but not on the trophic status of the reservoir and were comparatively stable during the observed period.

Introduction

This paper describes oligochaete distribution and biomass changes over many years, with special reference to energy budget.

The material presented is based largely on previous studies. The research history of the benthos of Lake Sevan goes back to the 1928, when Arnoldi (1929) estimated the biomass of bottom animals for the first time. Later, investigations have been carried out by Fridman (1950), Markosian (1959, 1965, 1966, 1970), Meshkova (1976), Modern (1979), Nikolaev (1979, 1985), Ostrovsky (1980, 1985) and the author of this paper. Unfortunately, field studies were suspended in 1992 because of lack of funds.

Two thirds of a century of study show the most important benthic animal group to be Oligochaeta (21 to 77% of the zoobenthos biomass). Of this the family Tubificidae is found in the greatest number and biomass, the family Naididae is significant in number but not in biomass due to the small size of worms, the families Aeolosomatidae, Lumbriculidae and Enchy-traeidae are not abundant (Fridman, 1950; Churakova, 1972; Jenderedjian & Poddubnaya, 1987).

During the observed period the water level of Lake Sevan was artificially lowered by 19 m, which had an effect on the ecosystem of the lake, from primary production to fish community, including benthos and the Oligochaeta in particular. Because primary production is the most important characteristic of the reservoir (Odum, 1986), an attempt was made to find a relationship between the indices of oligochaete energy budget and the values of primary production of the lake.

The study area

Sevan (40°28′ N, 45°20′ E) is the largest lake of Transcaucasus Region and consists of the deeper Minor Sevan and comparatively shallow Major Sevan. Before the artificial outflow increase of 1933, the lake had been located at an altitude of 1916 m a.s.l., with maximum depth 99 m, surface 1416 km², volume 58.5 km³. By the 1980's the surface and volume were diminished by 12 and 42%, respectively (Gyosalian, 1984). Moreover, the living conditions had changed, of which characteristics important to the Oligochaeta are: substrata redistribution, anoxia near the bottom area, temperature range increase.

Eight main substrata are distinguished in Lake Sevan: river sediment, macrophyte zone, stones, sand, silty sand, yellow and black silt and crystalline profundal. Due to the water level drop the areas of macrophytes, stones and sand decreased, the areas of silt were enlarged.

According to temperature and oxygen conditions five depth zones are distinguished in the lake: (1) littoral, (2) sublittoral, (3) transitional zone in Minor Sevan, (4) profundal of Minor Sevan, and (5) profundal of Major Sevan (Jenderedjian, 1994a). In this paper the last three zones are considered together as a 'profundal'.

Materials and methods

A Petersen dredge enclosing 0.1 m² in 1928 to 1971 (one grab per sampling event) and 0.025 m² in 1976 to 1991 (two grabs) were used for sampling. The samples were taken along 12 to 27 transects, once in spring and along 4 to 8 transects monthly at 2 to 60 m depth (down to 70 m in 1955 to 1966 and to 80 m in 1928 to 1948). Samples were washed on a sieve and preserved in 4% formalin.

The benthic invertebrates were sorted to groups: Oligochaeta, Hirudinea, Gastropoda, Bivalvia, Amphipoda, Ephemeroptera, Trichoptera, Chironomidae, etc.

In 1984 to 1991 almost all specimens of Oligochaeta were identified to species including sexually immature worms. Only specimens of the families Lumbriculidae and Enchytraeidae were identified to families. The material includes approximately 107,000 specimens of oligochaetes from 2110 samples.

The following considerations apply to the oligochaete energy budget estimates.

Because animals were stored in formalin, weight loss was estimated at 20% (Borutsky, 1934). Dry matter content in the body of worms varied between 13 and 21% (Birger et al., 1967; Finogenova & Lobasheva, 1987). Energy content of oligochaete biomass varied between 17 and 25 J mg⁻¹ dry weight (Ivlev, 1938; Birger et al., 1967; Brinkhurst & Austin, 1979; Kosiorek, 1979; Calow & Riley, 1982; Gupta & Pant, 1983; Finogenova & Lobasheva, 1987). Using conver-

sion factors between wet (Wet weight), dry (DW) and formalin (FW) weight of 1.00:0.18:0.80 and energy content of 21 J mg⁻¹ DW, an energy equivalent of 4.7 J mg⁻¹ FW was established in estimating oligochaete energy budget.

Total annual biomass (B, J m⁻²) was calculated from the spring data with regard to seasonal biomass fluctuation and taking into account the surface of the areas between the depths for each observed year.

The total annual production (P, J m⁻²) was calculated by the same method using annual turnover (P/B) ratios of separate species. Hynes' method (Hynes & Coleman, 1968) was modified in estimating the productivity. This method is described in detail in a previous paper (Jenderedjian, 1994b). A P/B ratio of the dominant *Potamothrix alatus paravanicus* Poddubnaya & Pataridze for 1928 to 1983 was calculated (Jenderedjian, 1994b: 289, equation 13) showing a relationship between the annual P/B ratio and the sum of degree-days per year, the length of the anoxic period, depth and the size of clitellate animals. An annual P/B ratio of *Tubifex tubifex* (Müller) and *Limnodrilus hoffmeisteri* Claparéde was estimated at 1.6 to 2.2 and 1.5 to 2.2, respectively, in the mouth of heavily polluted River Gavaraget (Jenderedjian, 1989). An annual P/B ratio of 2.0 was used in estimating production of *T. tubifex* and *L. hoffmeisteri* for 1928 to 1983. Due to the low density there was no possibility to estimate production of the other species. An annual P/B of 2.0 (as found for allied *L. hoffmeisteri*) was used in estimating production of *Limnodrilus claparedeianus* Ratzel. Maximal biomass of *Rhyacodrilus coccineus* (Vejdovsky) occurred at depths of 4 to 15 m (Fridman, 1950; Churakova, 1972; Jenderedjian & Poddubnaya, 1987), where it seems to have a 1 or 2 years life span. That is why an annual P/B ratio of 2.2 was used in estimating production of *R. coccineus*. Other species were not considered because of their low biomass.

Respiration was calculated according to an equation for aquatic flatworms and annelids (Kamlyuk, 1974):

$$Q = 0.105W^{0.75}, \qquad (1)$$

where Q = oxygen consumption rate, ml O_2 h⁻¹ ind.⁻¹ at 20°.

Using conversion factors FW WW⁻¹ (1.25), month h⁻¹ (730) and J ml O_2^{-1} (20.3) equation (1) may be rewritten:

$$R = 1945W^{0.75} \qquad (2)$$

where R = respiration, J month⁻¹ ind.⁻¹, W = FW, mg. Respiration was calculated by equation (1) from the size frequency histograms using mean monthly temperatures and temperature factor $Q_{10} = 2.5$ for temperatures differing from 20 °C (Winberg, 1983). Yearly respiration was calculated as a sum of monthly values. Total annual metabolic expenditures (R, J m^{-2}) were estimated similar to biomass and production.

An assumption was made that only 60% of food is utilized:

$$C = A : 0.6, \qquad (3)$$

where C = ration, J m^{-2}, $A = P + R$ = assimilation, J m^{-2}.

Total primary production (PP, J m^{-2}) was estimated as a sum of plankton (PPP) and benthos (BPP) primary production. The PPP values after Gambarian (1968), Parparov (1990), Simonian (1991) were used in calculating a correlation between the oligochaete energy budget and the values of PP. For several years the values of PPP were calculated from the data of phytoplankton biomass seasonal dynamics (Meshkova, 1975) and P/B ratios (Parparov, 1979), and a temperature factor $Q_{10} = 2.5$. The values of BPP were calculated from the data of maximal biomass of macrophytes (Arnoldi, 1929; Markosian, 1951; Gambarian, 1979, 1984; archives material of the IHEI NASA). An energy content of 2.1 J mg^{-1} Wet weight was established for plants in estimating PP.

Results

Oligochaetes of 26 species (Jenderedjian & Poddubnaya, 1987) inhabit different biotopes and depths in Lake Sevan. Their distribution is unequal (Fig. 1). Naididae comprise almost 100% of the oligochaete fauna on stones covered by periphyton. *Limnodrilus hoffmeisteri* is the dominant species in river sediment (79%) and *R. coccineus* in the macrophyte zone (37%). *Trichodrilus* sp. is found exclusively in the crystalline profundal. On pure and silty sand, yellow and black silt *P. a. paravanicus* is the dominant species (40, 70, 85 and almost 100% of the oligochaete biomass, respectively).

Vertical distribution of the Oligochaeta is restricted by oxygen regime. When the minimal saturation of oxygen was more than half the maximum (Lyatti, 1932; Slobodchikov, 1955), five species (*Nais elinguis* Müller, *Amphichaeta leydigii* Tauber, *R. coccineus*, *P. a. paravanicus* and *Trichodrilus* sp.) occurred at depths over 40 m (Churakova, 1972). Since the 1970s, anoxic areas have been observed in the profundal of both Minor and Major Sevan (Gyosalian & Khorlashko, 1979; Parparova, 1985). Only *P. a. paravanicus* has been found from 1984 to 1991 at depths over 40 m in Minor Sevan and over 26 m in Major Sevan.

In 1984 to 1991 *P. a. paravanicus* made up on average 98% of oligochaete biomass in Lake Sevan. Only 0.8, 0.6, 0.3 and 0.2% of biomass fell to the share of the subdominant *L. claparedeianus*, *T. tubifex*, *R. coccineus* and *L. hoffmeisteri*, respectively. hierwasik

Changes in qualitative composition of the oligochaete fauna can be established from the frequency of occurrence of the most common species. *Potamothrix. a. paravanicus* was found in 1937 to 1939 (Fridman, 1950) in 85% and in 1984 to 1991 in 94% of samples, *L. claparedeanus* in 0 and 11%, *T. tubifex* in 1 and 20%, *R. coccineus* in 2 and 1%, *L. hoffmeisteri* in 2 and 19% of samples respectively.

Before and at the very beginning of water level drop (1928 to 1948) quantitative and qualitative indices of the benthos of Lake Sevan were stable. Total zoobenthos biomass has been estimated at 3.4 to 4.2 g m^2 (wet weight) of which Oligochaeta made up 48 to 55%. After 1948, a gradual increase of zoobenthos biomass occurred, up to 15.1 g m^2 (wet weight) in 1971. In this period oligochaete biomass remained comparatively stable (2.0 to 4.3 g m^{-2} wet weight from 1955 to 1971). In the second half of the 1970's the zoobenthos biomass increased sharply to 30.5–36.6 g m^{-2} (wet weight) and later dropped to 9.9–10.9 g m^{-2} (wet weight) in 1989 to 1991. The peak oligochaete biomass occurred in 1978 to 1985 (over 10 g m^{-2} wet weight). Later, oligochaete biomass decreased to 7.6 g m^2 (wet weight) in 1989 to 1991.

Over the study period the ratio between maximum and minimum biomass for Oligochaeta was 8:1, while for Hirudinea, mollusks, Amphipoda, Ephemeroptera, Trichoptera and Chironomidae the ratios were 100:1, 20:1, 80:1, 140:1, 150:1 and 55:1, respectively, and for overall zoobenthos community, it was 11:1.

From Figure 2b the biomass changes are seen to be more rapid in the littoral and sublittoral and slower in the profundal. Oligochaete biomass curves are similar to the PP curve (Fig. 2a). Correlation analysis shows a relationship between oligochaete biomass and the values of PP. A mean PPP or BPP value of the nearest year was used for those years when they were not estimated. Calculation revealed the closest correlation between biomass and PP of preceding years (i.e., PP$_{-10}$ means that the biomass of 1991 was compared with PP of 1981, of 1990 with 1980, etc.). In the profundal and in the sublittoral the greatest correlation between the

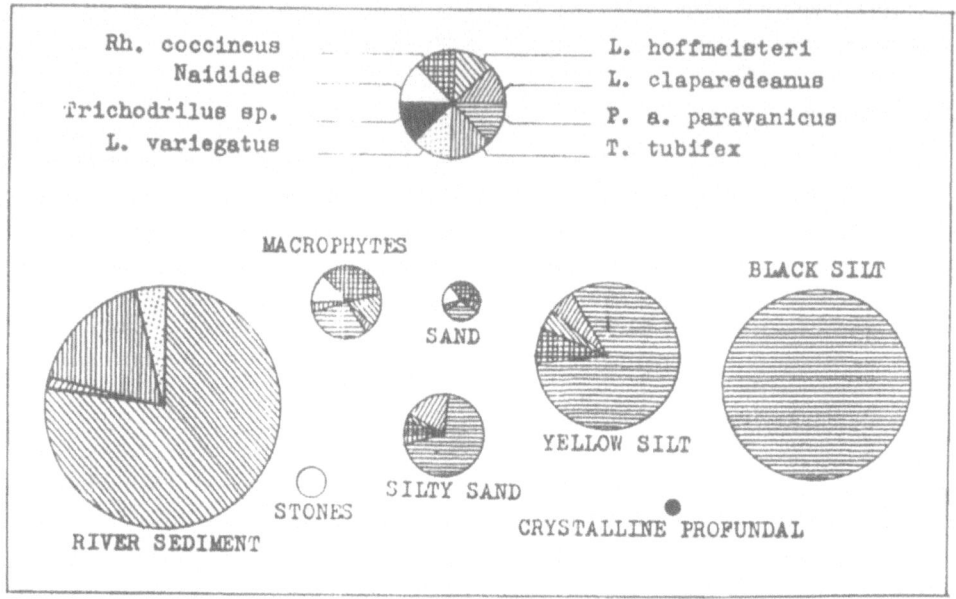

Fig. 1. Occurrence of *Oligochaeta* on different substrata of Lake Sevan expressed in percent of biomass (mean of 1984–1991). The surfaces of circles are proportional to the values of biomass.

biomass and PP values ($n = 22$) after 11 and 4 years, respectively, was established (the correlation between biomass and PP of the same year is given):

$$\ln B = -7.267 + 1.142 \ln PP_{-11} \qquad (4)$$

$$r^2 = 0.744 \ (r^2 = 0.144),$$

$$\ln R = -8.387 + 1.235 \ln PP_{-4} \qquad (5)$$

$$r^2 = 0.564 \ (r_0^2 = 0.329).$$

A high correlation was found between the oligochaete biomass in the littoral and PP of the preceding 2 years during 1928 to 1980 ($n = 11$):

$$\ln B = -9.231 + 1.245 \ln PP_{-2} \qquad (6)$$

$$r^2 = 0.889 \ (r_0^2 = 0.738),$$

but later in 1981 to 1991 the oligochaete biomass in the littoral was lower than expected. Perhaps, an occasional acclimatization of silver crucian (*Carassius auratus gibelio* Bloch) at the end of the 1970's was the cause of the reduced oligochaete (and the zoobenthos) biomass in the littoral zone (Oganesian & Smoley, 1985; Pivasian, 1990; Rubenian, 1993).

The highest correlation between total oligochaete biomass and PP values of the preceding 10 years was established ($n = 22$):

$$\ln B = -6.286 + 1.022 \ln PP_{-1} \qquad (7)$$

$$r^2 = 0.683 \ (r_0^2 = 0.230).$$

The 10-year 'delay' can be explained by the greater oligochaete biomass in the profundal and the morphometry of the lake (the profundal zone made up from 81 to 47% of the surface of the reservoir, depending on water level).

The energy budget data estimated for the oligochaete community of Lake Sevan are given in Table 1, together with data for BPP and PPP. The quantitative values of the energy budget as well as the values of BPP and PPP vary considerably. If the biomass depends on the PP, then production, respiration, assimilation and ration must also depend on the values of PP ($n = 22$):

$$\ln P = 6.981 + 1.078 \ln PP_{-10} \qquad (8)$$

$$r^2 = 0.629,$$

$$\ln R = -6.733 + 1.131 \ln PP_{-10} \qquad (9)$$

$$r^2 = 0.666,$$

$$\ln A = -6.186 + 1.114 \ln PP_{-10} \qquad (10)$$

$$r^2 = 0.655,$$

$$\ln C = -5.675 + 1.114 \ln PP_{-10} \qquad (11)$$

$$r^2 = 0.655.$$

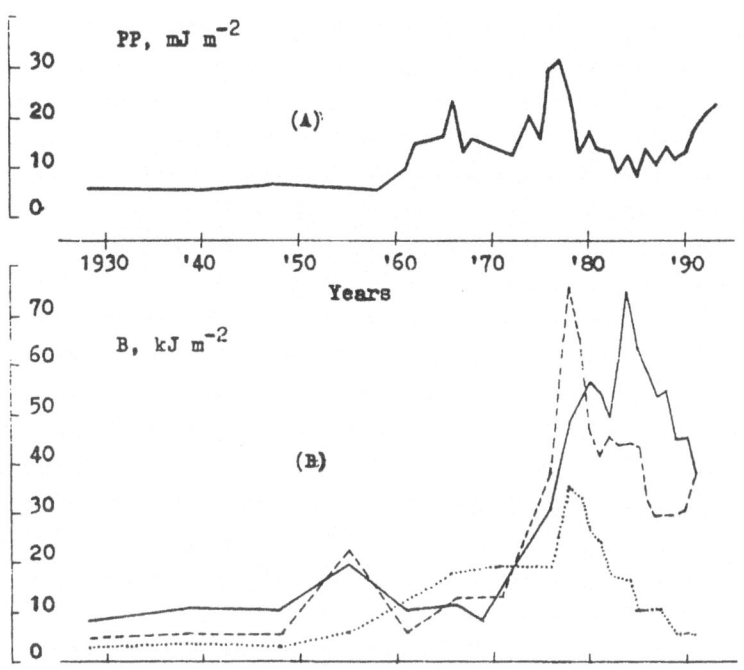

Fig. 2. Primary production (A) and interannual oligochaete biomass changes in the littoral (dashed line), in the sublittoral (dotted line) and in the profundal (continuous line) zones (B) of Lake Sevan.

From equation (11) it is presumed that the oligochaete community of Lake Sevan consumes about 1 to 2% of the PP of the lake which equals about 40% of the zoobenthos consumption without considering the proportion consumed by predators (Jenderedjian, 1993).

Although the quantitative values of the energy budget changed up to 8 to 12 times, the energy budget ratios retained comparatively stable (Table 1). The ratios of energy budget depend on temperature and oxygen conditions and decreased from shallow to deep water: P/B ratio from 3.4 to 0.1, R/B ratio from 4.5 to 0.5, net production efficiency (P/A) from 43 to 18%. The mean values of P/B, R/B and P/A ratios were 2.5, 3.5 and 0.41 in the littoral, 1.5, 2.5 and 0.37 in the sublittoral and 0.5, 1.3 and 0.26 in the profundal. The product of the mean P/B ratio by years of 'delay' gives a similar result of 5 to 6 in all depth zones.

Discussion

It is evident that the distribution of benthic animals depends on substratum. Fomenko (1972) has shown that it is possible to distinguish species pref-

erences for different types of organic matter in sediment. *Potamothrix a. paravanicus* prefers planktonogenous organic matter and its greatest biomass occurs in the profundal zone where sedimentation rate of autochthonal particulate matter is most significant (Glushchenko, 1989). *Rhyacodrilus coccineus* prefers organic matter of macrophyte origin, and a reduction in bottom algae (*Chara* spp.) and moss (*Fontinalis antipiretica* L.) over the period in Lake Sevan has resulted in its reduced abundance. *Limnodrilus hoffmeisteri* and *T. tubifex* prefer autochthonal, anthropogenic, organic matter and their distribution has expanded due to pollution at the beginning of the 1960s (Oganesian & Parparov, 1989). It seems that if the quantitative development of the tubificids at least depends on the quantity of organic matter in sediments the species composition depends mainly on the source of origin of organic matter.

A unique feature of the oligochaete fauna of Lake Sevan is the dominance of *P.a. paravanicus* and the comparative insignificance of other species. This simplified the oligochaete energy budget estimation.

As described above, the artificial drawdown of the lake has had the least effect on the oligochaete biomass compared to all other benthic animal groups. In

Table 1. Oligochaete energy budget and benthos and plankton primary production values in Lake Sevan. The designations are given as in the text. The values of B, P, R, A, C, BPP and PPP are in kJ m^{-2} yr^{-1}.

Year	B	P	R	A	C	P/B	R/B	P/A	BPP	PPP
1928	8	6	11	17	28	0.7	1.4	0.33	890	
1938	10	7	14	21	34	0.7	1.4	0.33	–	4650
1948	9	7	13	20	33	0.7	1.5	0.33	150	6420
1955	19	14	31	45	74	0.7	1.6	0.32	–	–
1958	–	–	–	–	–	–	–	–	–	5430
1961	10	8	16	24	40	0.8	1.7	0.33	90	9680
1962	–	–	–	–	–	–	–	–	–	14080
1965	–	–	–	–	–	–	–	–	–	15760
1966	12	13	24	37	61	1.0	1.9	0.35	70	23170
1967	–	–	–	–	–	–	–	–	–	12850
1968	–	–	–	–	–	–	–	–	–	14750
1969	–	–	–	–	–	–	–	–	–	14280
1971	15	13	28	41	69	0.9	1.9	0.32	–	–
1972	–	–	–	–	–	–	–	–	–	12310
1974	–	–	–	–	–	–	–	–	20	20070
1975	–	–	–	–	–	–	–	–	30	16260
1976	33	34	68	102	170	1.0	2.1	0.34	20	29590
1977	–	–	–	–	–	–	–	–	50	31090
1978	59	65	129	194	325	1.1	2.2	0.34	50	23870
1979	57	59	116	175	291	1.0	2.1	0.34	–	12790
1980	49	47	96	143	237	0.9	1.9	0.33	30	16640
1981	46	41	87	128	213	0.9	1.9	0.32	–	13350
1982	45	41	86	127	212	0.9	1.9	0.32	–	12480
1983	50	42	91	133	221	0.8	1.8	0.31	50	8680
1984	56	44	99	143	238	0.8	1.8	0.31	50	12120
1985	50	40	90	130	217	0.8	1.8	0.31	50	8220
1986	43	33	75	108	180	0.8	1.8	0.30	50	13260
1987	39	30	69	99	166	0.8	1.8	0.30	50	10660
1988	42	34	76	110	182	0.8	1.8	0.31	50	13310
1989	35	28	62	90	150	0.8	1.8	0.31	60	11390
1990	35	28	64	92	153	0.8	1.8	0.31	70	12570
1991	35	32	68	100	167	0.9	1.9	0.32	80	16760

part this can be explained by greater resistance of the worms to changes in the environment. However, the much longer life span of the dominant worm species, compared to other benthic invertebrates may also be a factor. It seems that only *Gammarus lacustris* Sars has a life span of up to 2 years in Lake Sevan (Markosian, 1947).

Phytoplankton and macrophytes are the most important source of organic matter in large limnic water bodies. There are few data from lakes where long-term studies of both PP and zoobenthos had been carried out. Similar PPP and zoobenthos biomass curves have been described in the lakes of Belorussia (Gavrilov,

1985; Kovalevskaya, 1985), and Alimov (1991) noted a disparity in time between these curves. Ostrovsky (1984) supposed that in Lake Sevan the 'delay' of zoobenthos biomass is dependent on the life span of the dominant species and suggested that four to six turnovers of organic matter were necessary to bring the biomass values in conformity with the values of PPP. The life span of the dominant *P.a. paravanicus* is estimated at one and a half year in the littoral (2-year 'delay'), between 2 and 4 years in the sublittoral (4-year 'delay') and up to 20 years in the profundal (11-year 'delay') (Jenderedjian, 1994a) and supports Alimov's suggestion. As stated above, about five or six turnovers of organic matter occurred in the community of Oligochaeta during the time of 'delay' and was irrespective of environmental conditions and the duration of 'delay'.

Half-century-long studies of the zooplankton of Lake Sevan have been carried out by Meshkova (1975) and Simonian (1991). The latter author showed a relationship between the quantitative values of the zooplankton community and the values of PPP of the lake. In zooplankton only the 'delay' in seasonal dynamics was appreciable, which can be explained by the multivoltine life cycles and high P/B ratios (from 14 to 32 per annum on average) of zooplankton. A four- year 'delay' against PPP and zooplankton has been found for the zoobenthos of Lake Sevan (Jenderedjian, 1993).

The patterns described above allow a prediction of the abundance of the Oligochaeta in future years. Judging from the PP values of the lake during the 1980s it is expected that oligochaete production will be lower in the next decade than in the 1989 to 1991 period (Table 1).

Acknowledgments

I am especially grateful to Dr Ashot Simonian for long discussions and helpful criticism of the work.

References

Alimov, A. F., 1991. Seasonal and many-year changes of zoobenthos biomass in the continental water bodies. Hydrobiol. J. 27: 3–9. [In Russian.]

Arnoldi, L. V., 1929. Materials on the bottom productivity studies of Lake Sevan. Transact. Sevan Hydrobiol. Station 2: 5–96. [In Russian.]

Birger, T. I., A. Y. Malyarevskaya & G. A. Olivari, 1967. Food value of the benthos of Dnieper, Kakhovskoe and Kremenchugskoe Reservoirs and its changes over influence of flow regulation. In Hydrobiological regime of Dnieper in condition of flow regulation, Kiev: 331–350. [In Russian.]

Borutsky, E. V., 1934. To question of technique of quantitative estimation of bottom fauna. To methods of processing of limnic benthos. Comparison of living and formalin weight. Transact. Limnol. Station Kosino, 22: 156–165. [In Russian.]

Brinkhurst, R. O. & M. J. Austin, 1979. Assimilation by aquatic oligochaetes. Int. Rev. ges. Hydrobiol. 64: 245–250.

Calow, P. & H. Riley, 1982. Observations on reproductive effort in British erpobellid and glossiphonid leeches with different life cycles. J. anim. Ecol. 51: 697–712.

Churakova, K. P., 1972. Objectives of taxonomy and distribution of Oligochaeta of Lake Sevan. In Aquatic Oligochaetes, Moscow: 75–82. [In Russian.]

Finogenova, N. P. & T. M. Lobasheva, 1987. Growth of Tubifex tubifex (Müller) (Oligochaeta, Tubificidae) under various trophic conditions. Int. Revue ges. Hydrobiol. 72: 709–726.

Fomenko, N. V., 1972. On ecological groups of Oligochaeta in the Dnieper. In Aquatic Oligochaetes, Moscow: 94–106. [In Russian.]

Fridman, G. M., 1950. Bottom fauna of Lake Sevan. Transact. Sevan Hydrobiol. Station 11: 7–92. [In Russian.]

Gambarian, M. E., 1968. Microbiological studies in Lake Sevan, Yerevan: 166 pp. [In Russian.]

Gambarian, P. P., 1979. Distribution of macrophytes in Lake Sevan. Transact. Sevan Hydrobiol. Station 17: 123–129. [In Russian.]

Gambarian, P. P., 1984. Macrophytes of Lake Sevan. In Limnology of Mountain Water Bodies, Yerevan: 55–56. [In Russian.]

Gavrilov, S. I., 1985. Macrozoobenthos. In Ecological System of Naroch Lakes, Minsk: 182–194. [In Russian.]

Glushchenko, L. O., 1989. Particulate organic matter and its role in the formation of water quality in Lake Sevan (Armenia). Arch. Hydrobiol. Beih. 33: 265–271.

Gupta, P. K. & M. C. Pant, 1983. Seasonal variation of the energy content of benthic macroinvertebrates of Lake Nainital, U.P. India. Hydrobiologia 99: 19–22.

Gyosalian, M. G., 1984. About some modern morphometrical and hydrobiological characteristics of Minor and Major Sevan. In Limnology of Mountain Water Bodies, Yerevan: 57–58. [In Russian.]

Gyosalian, M. G. & L. I. Khorlashko, 1979. About oxygen regime of Lake Sevan (from data of 1974–1976). In Transact. Sevan Hydrobiol. Station 17: 24–37. [In Russian.]

Hynes, H. B. N. & M. J. Coleman, 1968. A simple method of assessing the annual production of stream benthos. Limnol. Oceanogr. 13: 569–573.

Ivlev, V. S., 1938. Conversion of energy and growth of invertebrates. Bull. Nature Soc. Moscow, Biol. Dept. 47: 267–277. [In Russian.]

Jenderedjian, K., 1989. Peculiarities of biology and the production of Oligochaeta of Lake Sevan. Cand. Biol. Sci. Thesis, Leningrad: 24 pp. [In Russian.]

Jenderedjian, K., 1993. The changes in indices of zoobenthos biotic balance of Lake Sevan during changes in morphometrics and trophication of the reservoir. In Ecological Problems of Lake Sevan, Yerevan: 104–105.

Jenderedjian, K., 1994a. Population dynamics of Potamothrix alatus paravanicus Poddubnaya & Pataridze (Tubificidae) in different areas of Lake Sevan. Hydrobiologia 278: 281–286.

Jenderedjian, K., 1994b. Influence of environmental factors on the production of Potamothrix alatus paravanicus Poddubnaya & Pataridze (Tubificidae) in different areas of Lake Sevan. Hydrobiologia 278: 287–290.

Jenderedjian, K., T. L. Poddubnaya, 1987. Species composition and distribution of Oligochaeta in Lake Sevan. Biol. J. Armenia 40: 36–42. [In Russian.]

Kamlyuk, L. V., 1974. Energetic metabolism in free-living flatworms and annelid worms. J. General Biol. 35: 874–885. [In Russian.]

Kosiorek, D., 1979. Changes in chemical composition and energy content of Tubifex tubifex (Müll.), Oligochaeta, in life cycle. Pol. Arch. Hydrobiol. 26: 73–89.

Kovalevskaya, H. Z., 1985. Primary production of phytoplankton, its seasonal and interannual dynamics. In Ecological System of Naroch Lakes, Minsk: 93–99. [In Russian.]

Markosian, A. K., 1948. The biology of Gammarus of Lake Sevan. Transact. Sevan Hydrobiol. Station 10: 40–74. [In Russian.]

Markosian, A. K., 1951. Distribution and biomass of Chara algae and moss in Lake Sevan. Transact. Sevan Hydrobiol. Station 12: 29–34. [In Russian.]

Markosian, A. K., 1959. Benthos productivity in Lake Sevan. In Transact. 6th Conf. Probl. Inland Waters Biol., Leningrad: 139–145. [In Russian.]

Markosian, A. K., 1965. About some consequences of level lowering of Lake Sevan. Proc. Acad. Sci. Armenian SSR 18: 3–8. [In Russian.]

Markosian, A. K., 1966. Some results of influence of lowering of Lake Sevan on its regime. In Ecology of Aquatic Organisms, Moscow: 119–123. [In Russian.]

Markosian, A. K., 1970. Bioproductional development of Lake Sevan in connection with eutrophication provoked by lowering of the lake. Transact. 2nd All-Union Hydrobiol. Soc. Conf., Kishinev: 249. [In Russian.]

Meshkova, T. M., 1975. Regularities of zooplankton development in Lake Sevan, Yerevan: 276 pp. [In Russian.]

Meshkova, T. M., 1976. Eutrophication of Lake Sevan. Biol. J. Armenia 29: 14–22. [In Russian.]

Modern status of the zoobenthos of Lake Sevan, 1979. Transact. Sevan Hydrobiol. Station 17: 130–133. [In Russian.]

Nikolaev, S. G., 1979. Division of the benthal in Lake Sevan. J. Gen. Biol. 40: 143–151. [In Russian.]

Nikolaev, S. G., 1985. Structure of zoobenthos communities in Lake Sevan in years of sharp increase of trophic status and its changes in comparison to oligotrophic period. Cand. Biol. Sci. Thesis, Moscow: 24 pp. [In Russian.]

Odum, E. P., 1986. Basic ecology, Moscow: 706 pp.

Oganesian, R. O. & A. S. Parparov, 1989. The problems of Lake Sevan and ways of solution. In Conservation and Management of Lakes, Budapest: 367–378.

Oganesian, R. O. & A. I. Smoley, 1985. The crusian of Lake Sevan. Biol. J. Armenia 38: 725–726. [In Russian.]

Ostrovsky, I. S., 1980. Zoobenthos distribution in Lake Sevan in spring 1978. Biol. J. Armenia 33: 307–311. [In Russian.]

Ostrovsky, I. S., 1984. Zoobenthos biomass dependence on the primary production of Lake Sevan. In Limnology of Mountain Water Bodies, Yerevan: 216–218. [In Russian.]

Ostrovsky, I. S., 1985. Zoobenthos of Lake Sevan and its dynamics. Transact. Sevan Hydrobiol. Station 20: 132–186. [In Russian.]

Parparov, A. S., 1979. Primary production and chlorophyll *a* content in the phytoplankton of Lake Sevan. Transact. Sevan Hydrobiol. Station 17: 88–99. [In Russian.]

Parparov, A. S., 1990. Some characteristics of the community of autrotrophs of Lake Sevan in connection with its eutrophication. Hydrobiologia 191: 15–21.

Parparova, R. M., 1985. Peculiarities of the cycle of phosphorus in Lake Sevan on a background of the changes of its hydrochemical regime in relation with anthropogenic influence. Cand. Chem. Sci. Thesis, Rostovna-Donu: 24 pp. [In Russian.]

Pivasian, S. H., 1990. Nourishment of the silver crusian (*Carassius auratus gibelio* Bloch) in Lake Sevan. Biol. J. Armenia 43: 419–421. [In Russian.]

Rubenian, A., 1993. The problem of the silver crusian of Lake Sevan. In Ecological Problems of Lake Sevan, Yerevan: 108–109.

Simonian, A., 1991. Zooplankton of Lake Sevan. Yerevan: 299 pp. [In Russian.]

Slobodchikov, B. Y., 1955. Oxygen regime of Lake Sevan after data of 1947–1948. Transact. Sevan Hydrobiol. Station 14: 165–181. [In Russian.]

Winberg, G. G., 1983. Vant-Hoff's temperature factor Arrenius' equation in biology. J. Gen. Biol. 44: 31–42. [In Russian.]

Hydrobiologia **334**: 141–146, 1996.
K. A. Coates, Trefor B. Reynoldson & Thomas B. Reynoldson (eds), Aquatic Oligochaete Biology VI.
© 1996 *Kluwer Academic Publishers.*

Production and population dynamics of *Tubifex tubifex* in the profundal zone of a freshwater reservoir in N. Italy

Carla Bonacina[1], Andrea Pasteris[2], Giulio Di Cola[3], Giuliano Bonomi[2]
[1]*CNR-Istituto Italiano di Idrobiologia, Largo V. Tonolli 50, I-28048 Pallanza NO, Italy*
[2]*Dipartimento di Biologia evoluzionistica sperimentale, Università di Bologna, Via S. Giacomo 9, I-40126 Bologna, Italy*
[3]*Dipartimento di Matematica, Università di Parma, Via D'Azeglio 85, I-43100 Parma, Italy*

Key words: *Tubifex tubifex*, aquatic Oligochaeta, population dynamics, secondary production, pumped-storage reservoirs

Abstract

During a study of a pumped storage system from May 1979–June 1980 the profundal macrobenthos of the upper reservoir (Lago di S. Maria Valvestino) was sampled at a fixed station and the population of the tubificid *Tubifex tubifex* studied in detail. Eggs, embryos and the individuals living in an extra-cocoon stage were counted and individually weighed from monthly samples, according to the methods described in Bonomi & Di Cola (1980). Numerical recruitment during the study period was estimated as $257\,000$ ind m^{-2} yr^{-1}; of which $110\,000$ died either as eggs or as embryos, i.e. inside the cocoons, and a further $128\,000$ died before they attained sexual maturation. The data seem to confirm the typical demographic strategy of *T. tubifex* i.e. high fecundity and high mortality in the early life stages. The total annual production of the species was estimated at 91.7 g (w.w.) m^{-2}. The low P/B ratio (2.0 yr^{-1}) is considered to be mainly due to high population densities.

Introduction

The ever increasing number of freshwater reservoirs and the hydrological manipulations they undergo, provide limnologists with many new opportunities for studying such aquatic ecosystems, particularly the possible effects of some existing or projected water use. These opportunities often permit part of the research programmes to conduct detailed studies of selected reservoir communities. These communities typically display low structural complexity, thus allowing an evaluation of the population dynamics of species that is far more difficult to investigate in more complex biological communities. Over the last fifteen years we have been engaged in several research programmes, funded by Italian companies or local administrations (Acquedotto Pugliese, ENEA, ENEL, Regione Toscana), describing the limnology of freshwater reservoirs. This provided the opportunity to conduct supplementary research on the profundal benthos, particularly the oligochaete Tubificidae. A series of experiments were conducted, mainly using *Tubifex tubifex* and *Limnodrilus hoffmeisteri*, that have established specific biological traits and have been used in the estimation of population parameters for field populations of these species (Bonomi & Di Cola, 1980; Adreani *et al.*, 1981; Bonacina *et al.*, 1989a, 1989b, 1991a; Pasteris *et al.*, 1994). The Lago di Santa Maria Valvestino (N.E. Italy), provided a special case for studying its profundal benthos community as the hydrological balance, the thermal and chemical condition of its water, together with its phyto- and zooplankton were also studied. The present paper deals with the results of the study of the secondary production and the population dynamics of *T. tubifex* in the profundal zone of this reservoir.

142

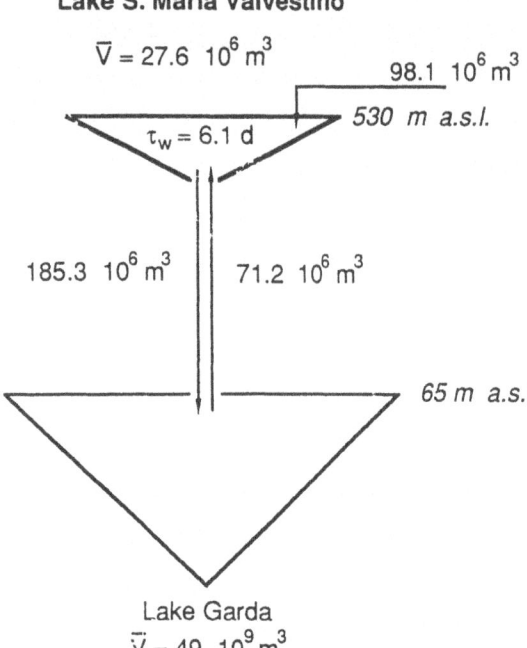

Fig. 1. Simplified scheme of the pumped-storage system Lake Garda-Lake Santa Maria Valvestino. Arrows relate to the annual water discharges for the period of our study.

The reservoir

Lake S. Maria Valvestino is the upper reservoir (530 m a.s.l.) of a pumped storage system with Lake Garda, N. Italy (Fig. 1). Its maximum volume is 52.5 10^6 m^3, the maximum depth 95 m. In addition to the supply from the natural upstream catchment (Torrente Toscolano) and from an artificial uptake on a stream (Torrente S. Michele) situated in a parallel valley, a considerable hydraulic load is pumped daily (week days) from Lake Garda. The discharge down to Lake Garda is used for electricity production to cover the peak demand from industry. The ratio: total annual water supply/mean reservoir volume, gives a theoretical water renewal rate of 6.1 yr^{-1}, i.e. an average renewal time of about 2 months.

Methods

From June 1979 to May 1980 routine monthly limnological measurements were taken at selected depths, including the maximum depth, at which profundal benthos was sampled. Measurements were: temperature

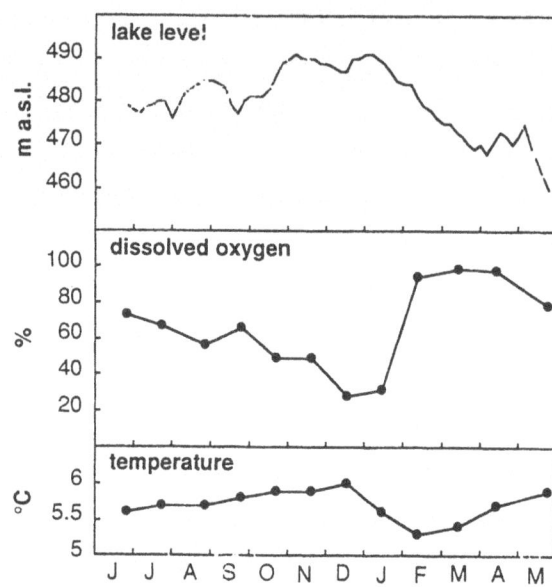

Fig. 2. Lake level variations, dissolved oxygen and temperature at the benthos sampling station.

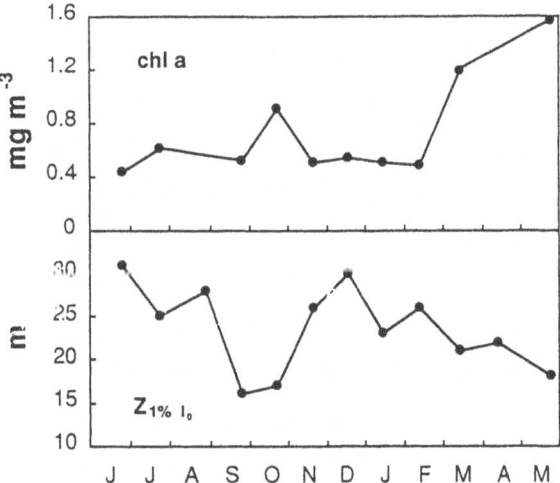

Fig. 3. Average water column chlorophyll concentration and depth of 1% of surface water irradiance in Lake Valvestino.

(reversing thermometer and thermistor), transparency (Secchi disk), light attenuation (photocell), dissolved oxygen, pH, conductivity, total alkalinity, nutrients, sestonic chlorophyll a. Phytoplankton and zooplankton were also studied, but the results are not yet published. A buoy was placed in a quasi-central position (confluence of the two main valleys) in the reservoir, to which the boat was moored while sampling the profundal benthos. On each monthly occasion 12 replicate Ekman (15 × 15 cm) samples were taken; the mud was sieved through 150 μm mesh sieve and the residue preserved in formalin. Particular care was given to the sieving procedure: a very delicate sieving increases dramatically the proportion of intact (not broken) worms in the sample. Sorted animals were exclusively represented by Tubificidae (*T. tubifex* and *L. hoffmeisteri*) and Chironomidae (*Chironomus* gr. *anthracinus*, *C.* gr. *plumosus*, *Polypedilum* sp., *Paratendipes* sp., *Paracladopelma* sp., *Paralauterborniella* sp., *Limnochironomus* sp., *Stempellina* sp., *Tanytarsus* sp., *Micropsectra* sp., and *Procladius* sp.).

Cocoons and free-living specimens of *T. tubifex* and of *L. hoffmeisteri* can be distinguished under the stereomicroscope. Cocoons of *T. tubifex* have an ellipsoidal shape and a transparent wall that allows an easy count of their egg or embryo content, while those of *L. hoffmeisteri* are more elongated and their sticky wall is covered by fine sediment particles so that only a dissection permits the above counts. Distinguishing between the two species of worm is easy as *Limnodrilus* lacks hair setae. Each individual *T. tubifex* was classified after binocular observation as ovigerous (O, with oocytes in the ovisac), mature (M) or immature (Y); while egg wet weight was considered to be 0.02 mg ind^{-1}, all hatched individuals were weighed individually and the immature ones divided into small immature, (Y$_1$, with individual weight ≤ 1.0 mg) and large immature (Y$_2$, with individual weight >1.0 mg) according to their weight (formalin wet weight). For the sake of simplicity, however, the total abundance of the young individuals (Y$_1$ + Y$_2$), is used in the model adopted here. Similarly, eggs (E$_1$) and embryos (E$_2$), originally counted separately are considered here as a single cumulative category (E). The procedure follows the one described in Bonomi & Di Cola (1980). The data for abundance, biomass and residence times were smoothed. The production was calculated following Petrowich *et al.* (1964) and the population parameters estimated according to a linearized version of the model of Bonomi & Di Cola (1980). Birth rates (b) were calculated assuming that eggs and embryos undergo a 40% mortality during their development (cf. Bonacina *et al.*, 1989a, b).

Results

Lake level variations were considerable during the period of our study (Fig. 2) and a remarkable decreasing trend occurred from January to May (ΔH = 31 m, corresponding to a ΔV of 25 10^6 m^3 = 38–13 10^6 m^3). In

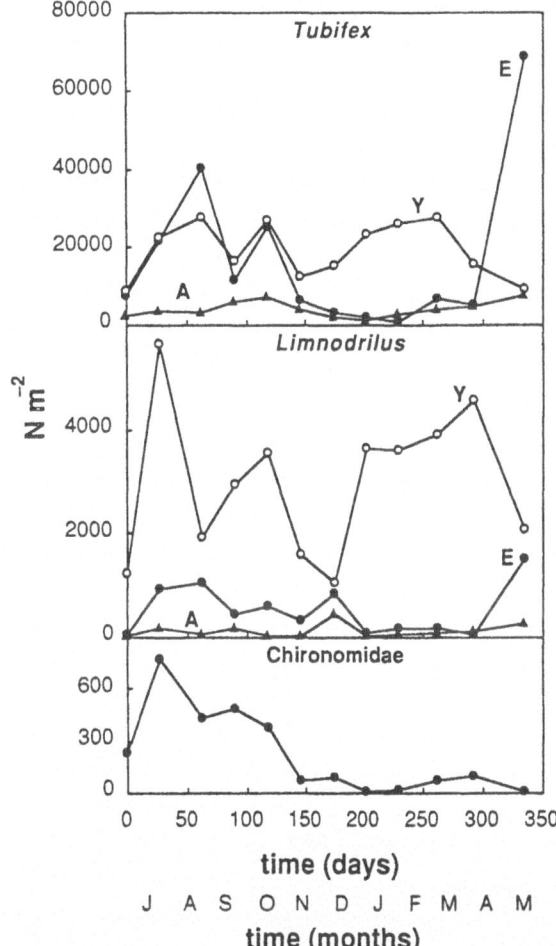

Fig. 4. Abundance of *Tubifex tubifex*, *Limnodrilus hoffmeisteri* (E = eggs + embryos, Y = immatures, A = mature + ovigerous) and the Chironomidae.

spite of the high mean flushing rate of the reservoir, the bottom temperature was very low during the period of our observations and the annual range was small (5.2–6.1 °C). Complete overturn began between December and January and in February the bottom temperature attained its minimum value (5.2 °C), while dissolved oxygen from the December minimum (28% saturation) peaked to complete saturation in February. The waters were rather transparent, the depth of 1% surface solar irradiance being in the range: 16–31 m and the average concentration of chlorophyll *a* in the water column being in the range 0.4–1.6 mg m^{-3} (Fig. 3).

The population of *T. tubifex* was abundant (Fig. 4) and largely dominant over that of *L. hoffmeisteri* and the Chironomidae (*Tanytarsus* sp. and *Procladius* sp. being dominant). *Tubifex* was represented by individ-

144

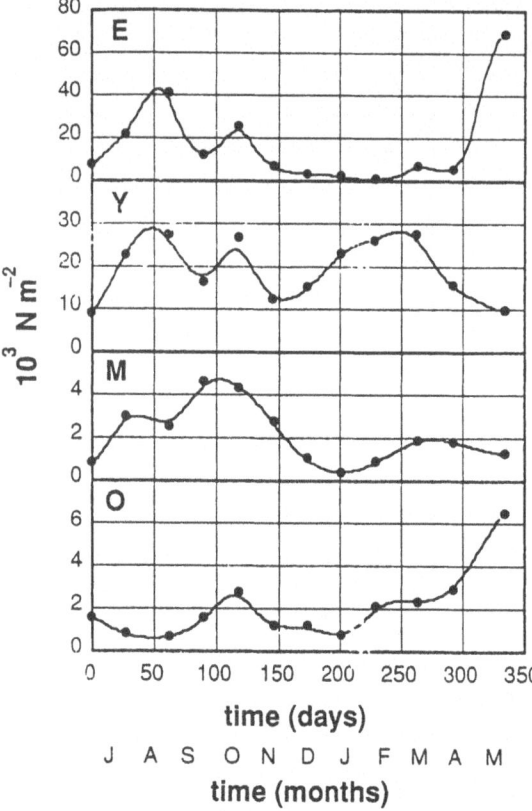

Fig. 5. Abundance of eggs + embryos (E), immatures (Y), matures (M) and ovigerous (O) of *T. tubifex*. Dots: experimental values; continuous lines: smoothed data.

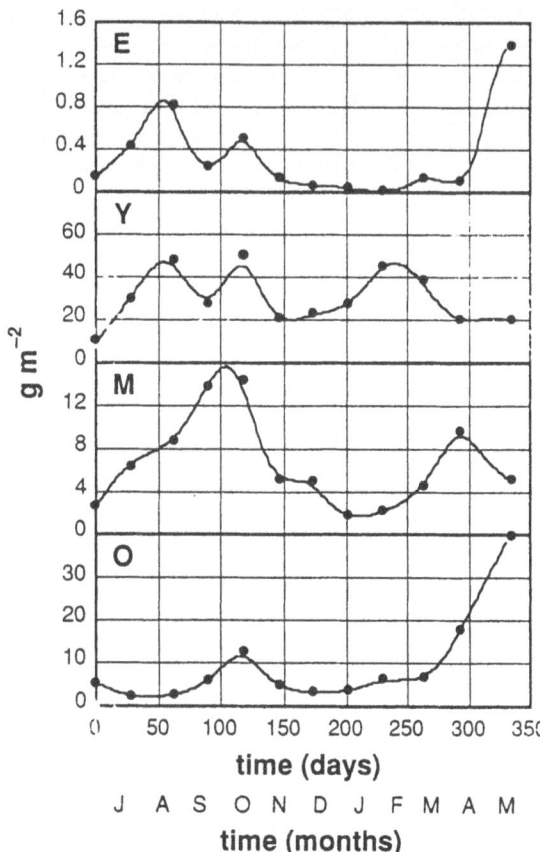

Fig. 6. Biomass of eggs + embryos (E), immatures (Y), matures (M) and ovigerous (O) of *T. tubifex*. Dots: experimental values; continuous lines: smoothed data.

uals of all the life stages through the year. Egg and embryo (E) numbers were lowest in January–February, but the continuous presence of eggs and ovigerous individuals suggests a continuous reproductive period. This may well be an effect of the high oxygen levels but also of a not excessively high population abundance at our profundal benthos station (Poddubnaya, 1980; Bonacina *et al.*, 1989b).

The abundance and biomass values for the E, Y, M and O categories are shown in Figs 5 and 6, respectively. The analysis of the population of *T. tubifex* shows that birth rate appears to be higher in periods when bottom temperatures tend to increase and dissolved oxygen is very high (Figs 2 & 7). The spring period, during which a dramatic lake level decrease takes place, and a corresponding increase in phytoplankton biomass occurs (Figs 2 & 3), is a very important season for the population recruitment, apparently continuing all through August (Figs 5, 7 & 8). The increase of N_Y in December–March is clearly the result of the high

birth rates during the first 120 days of our observations (July–October). Indeed an egg laid at the lowest temperature of this period of the year requires 45 days for its embryonic development.

The annual degree-day cumulation in our benthos station may be estimated as 2000–2100 degree-days. As an egg of *T. tubifex* requires (depending on the growth rate it will undergo) 1750–3000 degree-days to develop and grow up to enter the ovigerous stage O, we may assume that the E peak of May is partly caused by ovigerous individuals that were born at the very beginning of our observations. So in this, as in other reservoirs (Archipova, 1976; Poddubnaya, 1980; Bonomi & Di Cola, 1980) part of the population of *T. tubifex* may have one generation per year. Of course at shallower bottom layers the share of individuals of this voltinism may be greater. On the other hand, one must remember that temperature is only one of the factors controlling growth and development of this species; for instance,

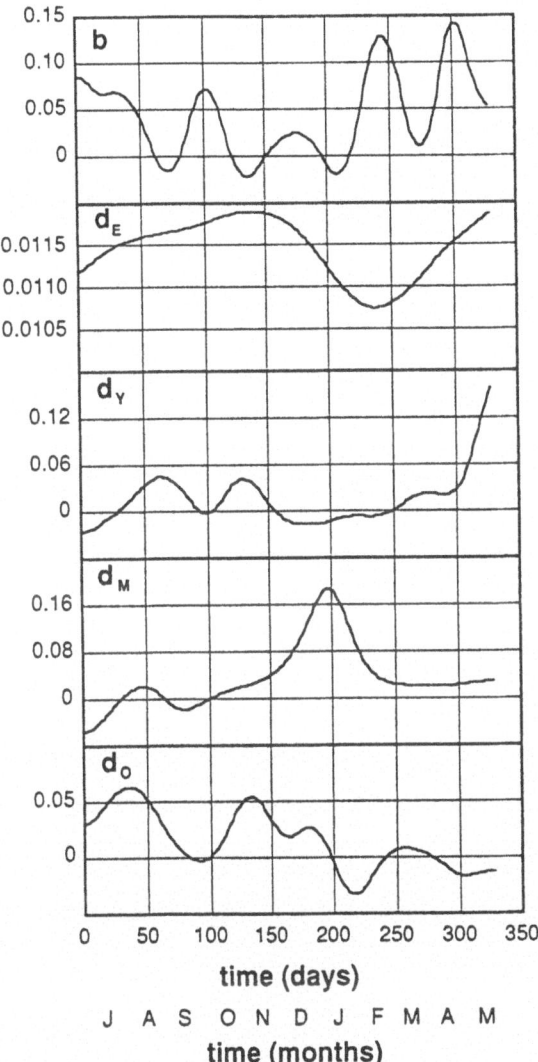

Fig. 7. Birth rates (b) and death rates of embryos (d_E) immatures (d_Y), matures (d_M) and ovigerous (d_O) of *T. tubifex*.

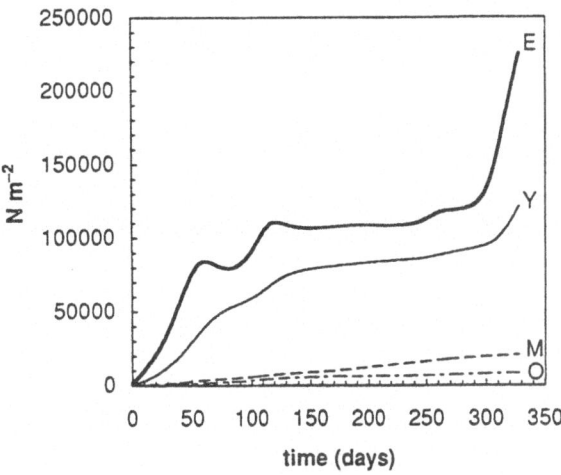

Fig. 8. Cumulative recruitment curves for the E, Y, M and O compartments.

Table 1. Mean biomass (B), annual production (P), and P/B of immature (Y), mature (M), ovigerous (O) and of the whole population (*: 9.3 yr^{-1} for the < 1.0 mg ind.$^{-1}$ compartment).

	B (g m^{-2})	P (g m^{-2} yr^{-1})	P/B (yr^{-1})
Y	30.7	57.0	1.9*
M	6.9	27.1	3.9
O	8.9	7.5	0.8
Total	46.5	91.6	2.0

several experiments have shown that in *T. tubifex* the generation time, as well as the specific fecundity R_0, is density dependent (Bonacina *et al.*, 1989a, b).

Cumulative egg production is estimated at 235 000 ind m^{-2} (333 d); in the same period of time about 100 000 individuals die as eggs or embryos and about 117 000 before attaining the egg laying stage (Fig. 8). Of course this is not a numerical budget: in fact, as already pointed out, a large part of the population requires at least 1.5 years for the entire life cycle. Table 1 shows that total production, in terms of formalin wet weight, is estimated at 91.6 g m^{-2} yr^{-1}, 57.0 g of which, i.e. 62%, are produced in the young

stage (Y). The overall P/B ratio is 2 yr^{-1}. Data from other Italian reservoirs, Pietra del Pertusillo Reservoir (Bonomi, 1979; Bonomi & Di Cola, 1980) and Lake Suviana (Comini, 1981; Bonacina *et al.*, 1991b) indicate that in *T. tubifex* the P/B ratio tends to remain low in populations with high or very high population densities, but seems to be higher in converse situations. Indeed in the first reservoir, characterised by low population densities, a P/B = 5.2 yr^{-1} (48.92 g m^{-2} yr^{-1}/9.32 g m^{-2}) was calculated, but in the second one , with high population numbers, the P/B ratio was 2.3 yr^{-1} (243.67 g m^{-2} yr^{-1}/128.42 g m^{-2}). These conclusions are in accordance with those of Poddubnaya (1980).

Analogous calculations for *L. hoffmeisteri* in Lake S. Maria Valvestino (Colombo, 1982) indicate a lower value for production (17.7 g m^{-2} yr^{-1}), but a very similar value for the P/B ratio (2.3 yr^{-1}). Young mortality was found to be lower than in *T. tubifex*, again confirming that the share of annual production resulting from mortality in the pre-maturation stages is comparatively very high in *T. tubifex*.

Acknowledgments

We would like to thank Mrs L. Adreani for her technical help, Mr A. Ferrari and Mr A. Pranzo for collaboration in the field. Dr I. Maestrini was very helpful in the solution of problems of data analysis and set the numerical algorithm for the smoothing of experimental data. The National Electricity Company of Italy (ENEL, Brescia) kindly supplied many precious facilities and aids, and the Forestry Service of the Ministry for Agriculture put at our disposal a boat for the field work.

References

Adreani, L., C. Bonacina & G. Bonomi, 1981. Production and population dynamics in profundal lacustrine Oligochaeta. Verh. int. Ver. Limnol. 21: 967–974.

Arkhipova, N. R., 1976. Peculiarities of the biology and production in *Limnodrilus hoffmeisteri* Clap. (Oligochaeta, Tubificidae) on grey muds in Rybinskoe Reservoir. Proc. Reservoir Biol. Inst.34: 5–15. [In Russian.]

Bonacina, C., G. Bonomi & C. Monti, 1989a. Density-dependent processes in cohorts of *Tubifex tubifex*, with special emphasis on the control of fecundity. Hydrobiologia 180: 135–141.

Bonacina, C., G. Bonomi & C. Monti, 1989b. Population analysis in mass cultures of *Tubifex tubifex*. Hydrobiologia 180: 127–134.

Bonacina, C., G. Bonomi, G. Di Cola & C. Monti, 1991a. An improved model for the study of population dynamics in *Tubifex tubifex* (Oligochaeta, Tubificidae). Verh. int. Ver. Limnol. 24: 2764–2767.

Bonacina, C., G. Bonomi, S. Gazzera & A. Pasteris, 1991b. Struttura, dinamica e distribuzione dei popolamenti macrobentonici in un sistema lago artificiale – lago serbatoio. Riv. Idrobiol. 30: 103–135.

Bonomi, G., 1979. Ponderal production of *Tubifex tubifex* Müller and *Limnodrilus hoffmeisteri* Claparède (Oligochaeta, Tubificidae), benthic cohabitants of an artificial lake. Boll. Zool. 46: 153–161.

Bonomi, G. & G. Di Cola, 1980. Population dynamics of *Tubifex tubifex* by means of a new model. In R. O. Brinkhurst & D. G. Cook (eds), Aquatic oligochaete biology. Plenum Press, New York: 185–203.

Colombo, E., 1982. Produzione e dinamica di popolazione di *Tubifex tubifex* e *Limnodrilus hoffmeisteri* (Oligochaeta, Tubificidae) nel lago di Santa Maria Valvestino. M.S. Thesis, University of Milano, 124 pp.

Comini, P., 1981. Dinamica di poplazione di *Tubifex tubifex* (Oligochaeta, Tubificidae) nel lago di Suviana. M.S. Thesis, University of Bologna, 99 pp.

Pasteris, A., C. Bonacina & G. Bonomi, 1994. Observations on cohorts of *Tubifex tubifex* cultured at different food levels, using cellulose substrate. Hydrobiologia 278: 315–320.

Petrowich, P. G., E. A. Shushkina & G. A. Pechen, 1964. Computation of zooplankton production. Dokl. Akad. Nauk SSSR, 139: 1235–1238. [In Russian.]

Poddubnaya, T. L., 1980. Life cycles of mass species of Tubificidae (Oligochaeta). In R. O. Brinkhurst & D. G. Cook (eds), Aquatic oligochaete biology. Plenum Press, New York: 175–184.

Hydrobiologia **334**: 147–155, 1996.
K. A. Coates, Trefor B. Reynoldson & Thomas B. Reynoldson (eds), Aquatic Oligochaete Biology VI.

Superficial and hyporheic oligochaete communities as indicators of pollution and water exchange in the River Moselle, France

M. Lafont, J.C. Camus & A. Rosso
Cemagref, Biology of Aquatic Ecosystems Division, 3 bis quai Chauveau, CP 220, F-69336 Lyon Cedex 09, France

Key words: Oligochaeta, river, pollution, benthos, hyporheos

Abstract

Benthic oligochaetes were sampled on three occasions (June, August and October 1992) in the upper (0–10 cm) and hyporheic (35–45 cm depths) sediments at five sites of the River Moselle, from upstream of the town of Epinal to Velle-sur-Moselle. The first site (upstream from Epinal) is considered as unpolluted and the four remaining sites are polluted by industrial effluents. The most polluted stations were generally dominated by the pollution tolerant taxon *Limnodrilus*. Numbers of individuals of this taxon decreased at the less polluted last site in recovery zone, and were also scarce in the first unpolluted site. It is noteworthy that these tendencies were observed in both superficial and hyporheic substrates and to the greatest degree in hyporheic ones. At the unpolluted site, the hyporheic habitat is dominated by the groundwater species *Propappus volki*, *Pristina* spp., *Pristinella* spp. At the less polluted site (last site), the deep sediments are dominated by groundwater species and the Tubificidae without hair setae decrease from June to October. As a result of water exchange between superficial and subterranean waters, superficial substrates of the first and the last stations tend to be colonised by a high proportion of hyporheic species that suggests that flow is primarily from subterranean to superficial waters. The contrary is the case at other polluted stations which are characterised by the invasion of hyporheic substrates by the pollution tolerant superficial taxa *Limnodrilus*. This suggests that water flows from the river to the deeper groundwater. These two stations are located near drinking water plants which utilise groundwater, thus increasing the vulnerability of groundwater to surface contaminants.

Introduction

The pollution of groundwater has to become a major concern of managers, as groundwater is a major drinking water supply. Consequently management agencies, such as French Water Agencies, require tools for the assessment of the vulnerability of subterranean waters to surface water contamination.

It is well established that water exchange between surface and groundwater can result in aquifer contamination by surface pollution. Such areas are usually detected by hydraulic methods, and are sometimes detectable from field observation such as visual observation of water percolation at the bottom of pits, or from the examination of aquatic macrophyte communities (Carbiener *et al.*, 1990).

However, these active exchange zones cannot be identified from direct field observation in homogeneous areas of coarse sediment, which frequently constitute the common substrates in the main river channel (Lafont *et al.*, 1992). In such cases, the examination of coarse sediment dwelling organisms may provide an indirect measurement, even if the focus is hydraulic research (Durbec, 1986).

When rivers are contaminated by industrial or urban effluent, the detection of active exchange zones is urgently needed. Therefore, the concept of using invertebrates as indicators of both exchange zones and pollution impact has been developed (Durbec & Lafont, 1991) to complement the use of indicator macrophyte communities (Carbiener *et al.*, 1990).

The ecology of oligochaetes suggests that they would be a good study tool. Several authors have point-

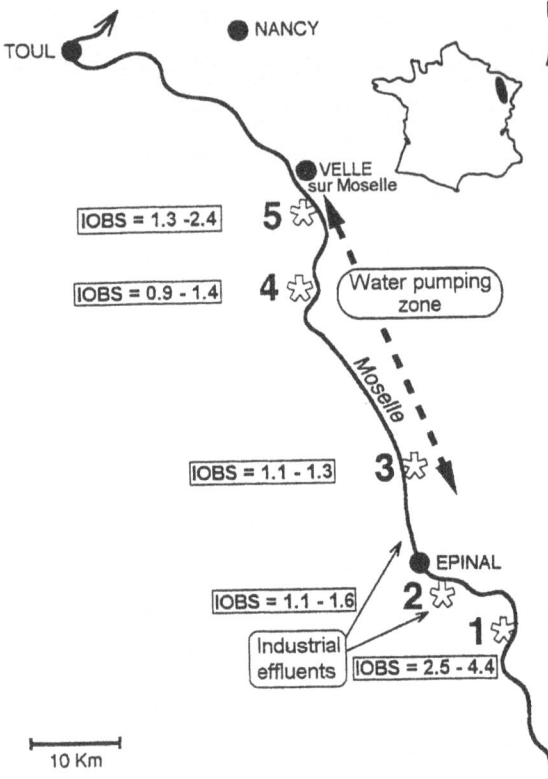

Figure 1. Map of study sites in the River Moselle; values of the IOBS index: the first value (left) is June and the second (right) is October.

ed out the interaction of these worms with subterranean waters (Ferrarese & Sambugar, 1976; Gaschignard, 1984; Gaschignard-Fossati, 1986; Giere, 1993; Juget, 1984, 1987; Juget & Dumnicka, 1986; Lafont, 1989; Lafont & Durbec, 1990; Lafont *et al.*, 1992; Strayer & Bannon O'Donnell, 1988). In fact subterranean waters are considered as major dispersion route for species and the primary reservoir of many species (Lafont, 1989; Juget & Lafont, 1994). The presence and abundance of species such as *P. volki* in surface sediments are good indicators of exchange zones between surface and subterranean waters (Lafont *et al.*, 1992) and the value of oligochaetes as pollution indicators has been demonstrated.

The investigation of the vulnerability of subterranean waters of a reach of the River Moselle was initiated. The area receives industrial effluents (paper industry) and the subterranean waters of the alluvial plain are utilized for drinking water supply. However, no detailed hydraulic measurements were available and the assessment of the vulnerability of this reach was to be based on this biological study. Therefore, a new

sampling strategy was developed, based on sampling of both surface and hyporheic sediments.

Description of sites studied

The study area is a section of the French part of the River Moselle, from upstream of the town of Epinal to Velle-sur-Moselle (Fig. 1). Five sites were chosen for pollution monitoring (sites 1 to 5) and four for the assessment of the vulnerability of groundwater (sites 1, 3, 4, 5), as coarse sediments were absent at the site 2, which is impounded but receives the first major inflows from paper industry.

At site 1 the river is approximately 30 metres wide and 50 metres wide at sites 3, 4 and 5. Sites 1, 3, 4, 5 are characterized by coarse sediment (stones and gravels) and a moderate current velocity. Permanent layers of sand were found on the banks, except at site 3 where sand layers were absent in August. Site 2 (60 metres wide) contains only sand and mud.

For the entire study area the river flows on sandy and silty alluvium, where aquifers are usually not very deep, have high productivity, and may be fed with water by the river (Anonymous, 1992).

The pattern of water discharges of the River Moselle during 1992 is such that June was at the end of high flow period, and the lowest water flow periods occurred from July to October (Lafont & Camus, 1993; Agence de l'Eau Rhin-Meuse, pers. comm.).

The river waters are fresh and well-oxygenated. From site 1 to site 4, the calcium and magnesium content varies from 7.4 mg l^{-1} to 20.5 mg l^{-1}, pH varies from 6.7 to 7.5, conductivity from 82 to 180 μS cm^{-1}, total phosphorus from 0.14 to 0.19 mg l^{-1} and ammonium salts from 0.09 to 0.07 mg l^{-1}, with the highest value at site 3 (0.46 mg l^{-1}) (Agence de l'Eau Rhin-Meuse, pers. comm.; Lafont & Camus, 1993).

Industrial effluents are discharged to the river (sites 2 and 3) and water withdrawal from the aquifer occurs in the vicinity of sites 3, 4 and 5 (Fig. 1).

Materials and methods

The five sites were sampled three times (June, August, October 1992). Surface coarse sediments (S) were sampled using a Surber type net (400 cm^2 aperture; mesh size of the net: 0.160 mm) and coarse sediments were removed with a small shovel (about 10 cm depth). Each sample comprised three replicates (3 × 400 cm^2), taken randomly in the coarse substrata and pooled in a glass jar.

The hyporheic sediments (L) were sampled at the same place as the surface sediments. Ten litres of material (sediment and interstitial water) were pumped at −35 to −45 cm depths in the coarse substrata by using a Bou-Rouch pump (Bou & Rouch, 1967). Each sample consisted of three replicates (3 × 10 litres).

The sandy sediments at sites 1, 3, 4 and 5 were also sampled with a Surber type net (100 cm^2 aperture, mesh size of the net: 0.500 mm). Each sample consisted of three replicates taken randomly in the 10 cm surficial layers of sand (3 × 100 cm^2) and combined in a glass jar. The shore sediments at the dam (site 2) were sampled using a Rofes & Savary type corer (Rofes & Savary, 1981) of 25 cm^2 aperture. Each sample consisted of 6 cores (6 × 25 cm^2).

A total of 24 samples from coarse sediments and 14 samples from sandy sediments (sand layers were absent at site 3 in August) were taken and preserved in the field with 6% formaldehyde.

In the laboratory, after previously separating by decanting the mineral particles (stones, gravel, sand) from the organic fraction (organic fragments and invertebrates), the 24 coarse sediment samples were washed through 0.160 mm sieves and the 14 sandy sediment samples through 0.500 mm sieves.

The residue from the sieving of each sample was then poured into sub-sampling boxes. A total of 100 oligochaete specimens were sorted by hand under a binocular microscope, from sub-samples taken randomly with a dropping tube from the squares in sampling box. Worm specimens were mounted on slides with a mixture of lactic acid and glycerin and identified to species level where possible.

As different sampling devices were used in the superficial and deep sediments, the abundance data for the oligochaetes were converted to percentages. The analysis of the 24 samples from coarse sediments was done using a Factorial Correspondence Analysis (FCA).

The oligochaete index IOBS (IOBS = 10S.T^{-1}, S = species richness, T = percentage of predominant group of Tubificidae, either with or without hair setae), based on the examination of sandy sediments washed through a sieve with 0.500 mm aperture (Rosso et al., 1994), was used to assess the occurrence of pollution. Index values less than 1 are considered as indicative of high toxicity, values between 1 and 2 are indicative of high pollution, values between 2 and 3 indicate moderate pollution and values greater than 3 are typical where biological quality is good.

Results

Species occurrence

Of the 36 taxa identified, 21 are very frequent and occurred more than six times at the 24 samples from coarse sediments (see Fig. 2). The other 15 taxa present are *Tubifex ignotus*, *Psammoryctides barbatus*, *Quistadrilus multisetosus*, *Haber* sp., *Potamothrix moldaviensis*, *Rhyacodrilus subterraneus*, *Branchiura sowerbyi*, *Aulodrilus pluriseta*, *Chaetogaster langi*, *Uncinais uncinata*, *Dero digitata*, *Vejdovskyella comata*, *Cognettia* sp., *Marionina argentea* and *Marionina riparia*.

The presence of *Q. multisetosus* and *R. subterraneus* is noteworthy in the River Moselle, as the former species had only been recorded in France from the Rhône basin (Lafont, 1989), the latter from caves or springs in karstic systems (Juget & Dumnicka, 1986; Lafont, 1989).

Assessment of pollution in sandy sediment (IOBS index)

Three main biological quality zones can be identified (Fig. 1):

Zone 1, represented by site 1, is of moderate to good quality and receives intermittent effluents (hydrocarbons and PCB in sediments, Lafont & Camus, 1993).

Zone 2, represented by sites 2–4, is heavily polluted, mainly by organic matter, mercury (sites 3 and 4), lead (site 3), hydrocarbons and PCB (sites 2, 3, 4).

Zone 3 (site 5) is a recovery zone but sediments are contaminated with PCB and chlorinated pesticides.

The chemical quality of surface waters supports the biological diagnosis, but chemical recovery of the river, both in waters and sediments, is documented from site 4 rather than from site 5 as shown by the biological index. The crustacean communities show the same pattern as the oligochaete response (Lafont & Camus, 1993), as well as the diatom communities (M. Coste, pers. comm.). It has been suggested that site 4 is impacted by diffuse sources of agricultural pollution and/or toxic material from site 3, which have not been recorded by our chemical analyses.

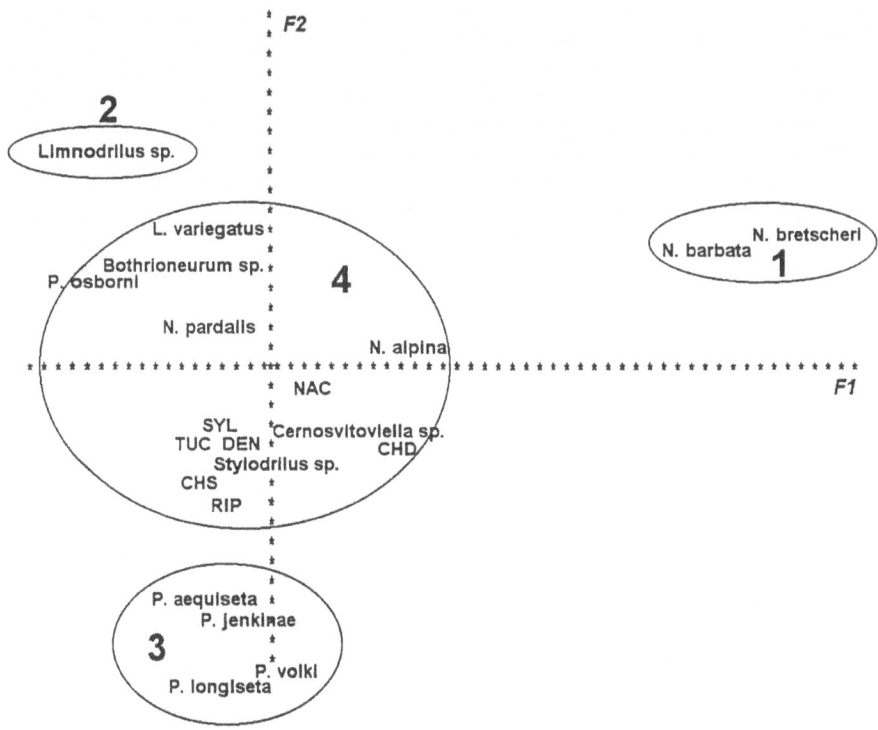

Figure 2. Plots of the four groups of species on the first plane of the Factorial Correspondence Analysis (FCA); NAC: *N. communis*; SYL: *S. lacustris*; TUC: immatures of Tubificidae with hair setae; DEN: *D. nivea*; CHD: *C. diaphanus*; CHS: *C. diastrophus*; RIP: *R. parasita*.

Organisation of the oligochaete species and sediment samples in the Factorial Correspondence Analysis (FCA)

The use of FCA on the relative abundance of the oligochaete species provides information on the distribution of the species in relation to substrate (S, surface, L, hyporheic) and location (1, 3, 4, 5). Four species groups can be identified on the first plane of the FCA (axes 1 and 2, 37% of the total inertia, Fig. 2).

Group 1

This group includes two species contributive to the axis 1, *Nais barbata* and *Nais bretscheri* (Fig. 2). These species characterize site 1 in June (Fig. 3).

Group 2

This group includes the genus *Limnodrilus* (*L. hoffmeisteri* and immature worms). This highly pollution-tolerant taxon is a strong contributor to axes 1 and 2 (Fig. 2). It characterizes surficial and deep coarse sediments at site 4, and deep sediments at site

3 in August and October and site 5 in June (Fig. 3). Group 2 is most distant from the other species groups on the first plane of the FCA.

Group 3

This includes four species which have a relative contribution to the second factor of more than 5%: *Propappus volki*, *Pristina longiseta*, *Pristina aequiseta foreli*, and *Pristinella jenkinae*. These species are good describers of active water exchange between surficial and subterranean waters (Lafont *et al.*, 1992).

Group 4

This group is represented by 14 species or taxa which contribute little to factors 1 and 2 (less than 5%). These species are frequently present at all the sites but in low numbers (*Chaetogaster diastrophus*, immature worms of Tubificidae with hair setae, *Stylaria lacustris*). Only *Bothrioneurum* sp. predominates at one site for one season (3L3). Several species are relatively scarce and not abundant (*Ripistes parasita, Dero nivea, Chaeto-*

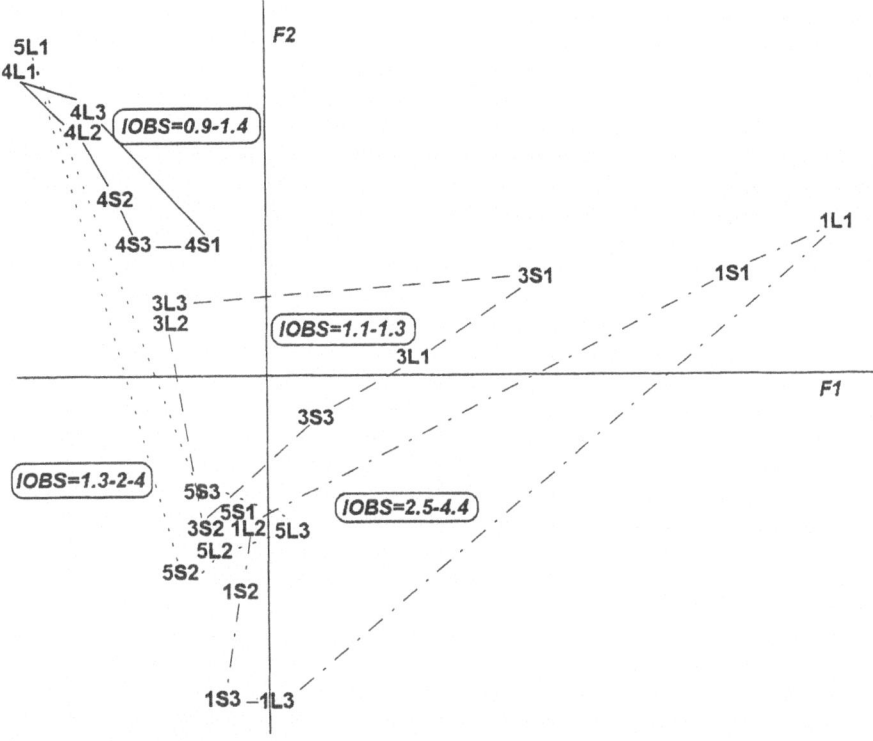

Figure 3. Distribution of samples on the first plane of the FCA; the values of the IOBS index (June, October) are plotted for each of the four sites studied; the first numbers indicate the site; the letters indicate the type of sediment (S= superficial, L= hyporheic); the last numbers indicate the season: June (1), August (2) and October (3); example: 1 S1 indicates the first site in superficial sediments in June.

gaster diaphanus, *Nais communis, Nais alpina, Nais pardalis, Lumbriculus variegatus...*).

It is noteworthy to find that *L. variegatus* and *Bothrioneurum* sp. are associated with the pollution-tolerant *Limnodrilus* (Fig. 2). This supports the pollution tolerance of these two species (Lafont, 1989; Rosso *et al.*, 1994).

The distribution of the four sites on the first factorial plane (Fig. 3) indicates some interesting trends.

First, the structure of the F1-F2 plane does not differentiate between the surficial (S) and hyporheic coarse sediments (L).

Second, the effect of pollution is represented on the F1-F2 plane by the differentiation of species in Group 1 and Group 2 (factor F1, spring) and Group 3 (subterranean species, factor F2, Fig. 2). This trend was confirmed by the examination of the IOBS index values (Figs 1 and 3). A gradient of increasing pollution is seen from the upstream site (1) to the downstream sites (3 and 4), with the last site (5) being of intermediate biological quality.

Third, the locations of site 1 samples are scattered and distinctly separate from those of samples from sites 3 and 4 (Fig. 3). The distribution of samples from site 3 is less scattered than site 1 and the site 4 samples are proximal. Site 5 shows similarities to the three other sites.

Seasonal evolution of sites and water discharges (Fig. 4)

Site 1

The oligochaete communities change from being a superficial naidid population (group 1, June) to a hyporheic population (group 3, August and October, Fig. 4) both in surface (S) and deep sediments (L). At the end of the high water period (June), the deep sediments are invaded by typical superficial species, whereas the superficial sediments are invaded by stygophilous species when surface waters are low (August, October, Fig. 4).

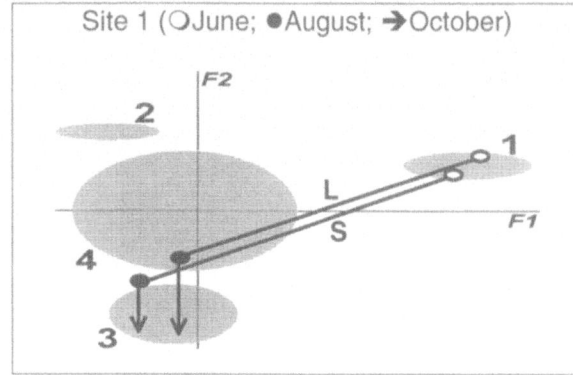
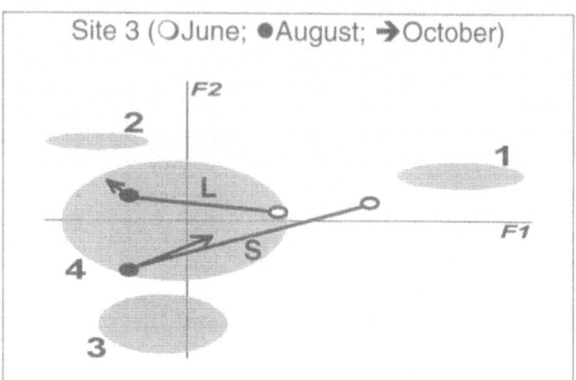
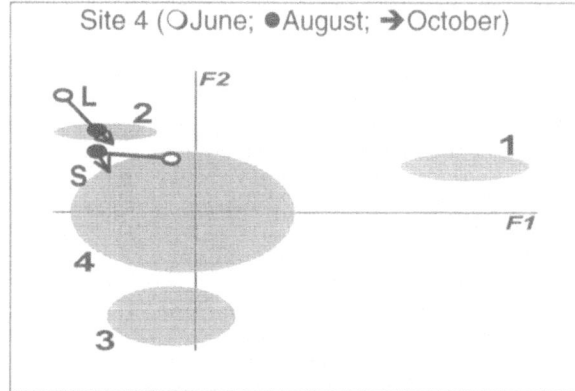
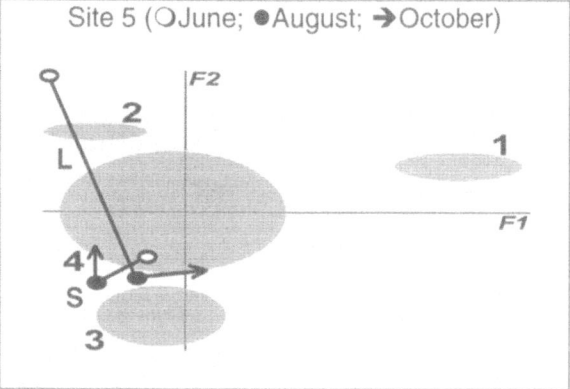

Figure 4. Seasonal evolution of the four study sites. Numbers 1 to 4: groups of species (see Fig. 2). S: superficial sediments; L: hyporheic, sediments.

Table 1. Biological diagnosis of the water exchanges between surface sediments (S) and hyporheic sediments (L) and of the vulnerability of subterranean water to surface water contamination.

Sites	Water pumping plants	End of high waters (June)	Low waters (August)	End of low waters (October)	Pollution	Vulnerability
1	No	S → L	L → S	L → S	Low	Low (spring)
3	Yes	S → L	S → L	S → L	High	High (summer, fall)
4	Yes	S → L	S → L	S → L	Very high	High (all seasons)
5	Yes	S ↔ L	L → S	L → S	Moderate to high	High (spring)

Site 3

The hyporheic sediments are dominated by superficial pollution tolerant taxa (including *Limnodrilus*) during the low water period (August and October).

Site 4

Surface and bottom sediments are always grouped together and are dominated by the pollution tolerant genus *Limnodrilus* (Fig. 4). The site is strongly polluted and the groundwater is contaminated by polluted surface waters.

Site 5

The hyporheic sediments are dominated by the genus *Limnodrilus* in June (Fig. 4) and the contamination by polluted surface water seems evident. On the contrary, stygophilous species, such as *P. aequiseta foreli*, *P. jenkinae* and *P. volki*, together with several species from group 4 (Fig. 4), predominate in surface (June, August, October) and deep sediments (August, October). A biological situation similar to that at site 1 is found again.

The general diagnosis of the vulnerability of undergroundwater to a surface pollution (Table 1) is only provided as an overview of the response of the oligochaete communities, because available hydrologic data are presented as monthly discharge measurements.

Discussion

The main difficulty with this type of study is the amount of time required when managers need a rapid response. It is therefore necessary to provide managers with summary such as Table 1. However, risk of such a summary is the reduction of the information and consequent incomplete description of the true situation. Nevertheless, some trends shown in this data set are noteworthy. The ecological validity of oligochaete species as water exchange describers is difficult to prove without more detailed hydraulic studies than those reported here (Table 1), and also probably because absence or scarcity of stygobiont taxa such as *Trichodrilus* or *Rhyacodrilus*. These last taxa generally show higher occurrence in aquifers on a larger spatial scale (Juget, 1984, 1987) than in our study area, restricted to the river bed.

The water exchange describer status has previously been proved for some of the species contributing most to the Factorial Correspondence Analysis (FCA), such as *P. volki*, *P. aequiseta foreli*, *P. longiseta* or *P. jenkinae* (Lafont *et al.*, 1992). The last species has recently been found at 25–35 metres depth in a karstic site (Malard *et al.*, 1994). The status has been demonstrated for taxa with little contribution to the FCA, such as *Stylodrilus* sp. (Gaschignard, 1984), and only empirically demonstrated for taxa like *Cernosvitoviella* sp. and *P. osborni* (Lafont *et al.*, 1992). There is no further indication

allowing us to verify or contradict the ecological status of these last two taxa in the investigated sites.

To examine these exchange dynamics and characteristics of the ecological function, both superficial and hyporheic sampling is necessary. For example, the predominance of two species of the superficial genus *Nais* in the deep sediments of site 1 in June (Figs 2–4) demonstrates the importance of water exchange from the river down to the underflow. At the end of the spring, there had been an extended period of high water levels in the River Moselle, and it was not surprising to find that the river was supplying the underflow with water. On the contrary, the superficial sediments are invaded by stygophilous species when surface waters are low (August, October), that demonstrates a supply of surface water with groundwater. All these observations are consistent with the hydrogeological patterns at site 1.

At sites 3 (August, October) and 4 (June, August, October), the superficial pollution-tolerant genus *Limnodrilus* had invaded the underflow, while it was still predominant in the superficial substrates, even when superficial water discharge was low (Fig. 4). These observations indicate that water pumping plants in the groundwater have disturbed the dynamics of water exchanges. The predominant direction of water exchange is from the river down to the underflow during periods of high or low superficial water levels.

At site 5, the situation is rather complex in June as water exchange is facilitated in both vertical directions (river – groundwater, groundwater – river). It is also possible that the degree of pollution drives up hyporheic species to the surface in spring. When surface water flow is low (August, October), a biological situation similar to that at site 1 is found.

The surface and deep coarse sediments are closely connected in the study area, which is consistent with literature data (Dole-Olivier & Marmonier, 1992; Giere, 1993; Ward & Palmer, 1994). It is evident that hyporheic sediments do not provide here a refuge zone for superficial invertebrates, as recorded by Rouch & Danielopol (1987) and Bretschko (1991). For example, the invasion of hyporheic waters by the superficial pollution-tolerant taxon *Limnodrilus* sp. can be considered as an extension of its habitat as it follows the movement of pollutants in the sediment. This is enhanced by the supply of groundwater with surface water in the area of water pumping plant operations.

On the other hand, the predominance of *Limnodrilus* in the deep sediments of station 5 at spring (June) suggests that subterranean species have migrated to superficial sediments, which does not support the concept of a refuge zone in hyporheic waters (Griffith & Perry, 1993). We would suggest that here the surface acts as a refuge for hyporheic species which escape the inflow of polluted waters within the alluvial habitat (Giere, 1993). This also means that, either for pollution or vulnerability assessments, it is necessary to consider both surface and deep oligochaete communities when coarse sediments predominate.

These observations also fit the results from studies of microcrustacean communities in the River Moselle (Lafont & Camus, 1993) and of a heavily polluted section of the River Rhône, the groundwater of which is operated for drinking water supply (Durbec *et al.*, 1992). It has been suggested that the underflow records present and past pollution events, whereas the superficial sediments have a greater instability (Giere, 1993). For example, the substrate disturbance due to spates is less in deeper levels (McElravy & Resch, 1991). It is clear that the artificial activation of water exchange from the river down to the groundwater, by water abstraction from aquifers, magnifies the temporal record of a pollution event in hyporheic sediments.

As has already been stated by several authors (Ferrarese & Sambugar, 1976; Hynes, 1983; Rouch & Danielopol, 1987; Danielopol, 1989; Ward & Palmer, 1994), it is important to consider subterranean sediments, even when studying the ecology of superficial invertebrates in rivers. This is also true of oligochaete communities if one considers their ecology. The sampling protocol proposed here is now being operated on several sites of the Rhône River where hydraulic measurements are available.

Acknowledgements

Financial support for this work was provided by the Agence de l'Eau Rhin-Meuse. We thank Mrs C. Riou, Dr M. Babut, Mr C. Breuzin and Mr G. Demortier (Agence de l'Eau Rhin-Meuse) for their help on the field and the communication of original physicochemical data; we thank Dr T. B. Reynoldson for having nicely revised the English text. The authors also thank the referees for comments that improved the paper.

References

Anonymous, 1992. Carte hydrogéologique du Bassin Rhin-Meuse et données sur les échanges Moselle-aquifère. Document de l'Agence de l'Eau Rhin-Meuse, France, 4 pp.

Bou, C. & R. Rouch, 1967. Un nouveau champ de recherches sur la faune aquatique souterraine. C. R. Acad. Sci., Paris 265: 369–370.

Bretschko, G., 1991. Bed sediments, groundwater and stream limnology. Verh. int. Ver. Limnol. 24: 1957–1960.

Carbiener, R., M. Trémolières, J. L. Mercier & A. Ortsheit, 1990. Aquatic macrophyte communities as bioindicators of eutrophication in calcareous oligosaprobe stream waters (Upper Rhine plain, Alsace). Vegetatio 86: 71–98.

Danielopol D. L., 1989. Groundwater fauna associated with riverine aquifers. J. N. Am. Benthol. Soc. 8: 18–35.

Dole-Olivier, M. J. & P. Marmonier, 1992. Patch distribution of interstitial communities: prevailing factors. Freshwat. Biol. 27: 177–191.

Durbec, A., 1986. Sectorisation des berges des ballastières en eaux. Application à l'étude des échanges hydrodynamiques avec la nappe phréatique d'Alsace, au nord de Strasbourg. Thèse de Doctorat, Université Louis Pasteur, Strasbourg, 202 pp.

Durbec, A. & M. Lafont, 1991. Sensibilité à la pollution des zones de captage en site alluvial. Complémentarité de l'Hydraulique et de la Biologie. Infotech. CEMAGREF 83, note 4: 1–8.

Durbec A., M. Lafont & J.C. Camus, 1992. Etude de la vulnérabilité à la pollution du champ captant de Chasse-sur-Rhône (38). Rapport BUGEAP-CEMAGREF, 16 pp.

Ferrarese U. & B. Sambugar, 1976. Ricerche sulla fauna interstiziale iporreica dell'Adige in relazione allo stato di inquinamento del Fiume. Riv. Idrobiol. 15: 47–127.

Gaschignard, O., 1984. Impact d'une crue sur les invertébrés benthiques d'un bras mort du Rhône. Verh. int. Ver. Limnol. 22: 1997–2001.

Gaschignard-Fossati, O., 1986. Répartition spatiale des macroinvertébrés benthiques d'un bras vif du Rhône; Rôle des crues et dynamique saisonnière. Thèse de Doctorat, Université Lyon I, 197 pp.

Giere, O., 1993. Meiobenthology. The Microscopic Fauna in Aquatic Sediments. Springer-Verlag Berlin, Heidelberg, 328 pp.

Griffith, M. B. & S. A. Perry, 1993. The distribution of macroinvertebrates in the hyporheic zone of two small Appalachian headwater streams. Arch. Hydrobiol. 126: 373–384.

Hynes, H. B. N., 1983. Groundwater and stream ecology. Hydrobiologia 100: 93–99.

Juget, J., 1984. Oligochaeta of the epigean and underground fauna of the alluvial plain of the French upper Rhône (biotypological try). Hydrobiologia 115: 175–182.

Juget, J., 1987. Contribution to the study of Rhyacodrilinae (Tubificidae, Oligochaeta), with the description of two new stygobiont species from the alluvial plain of the French Upper Rhône, *Rhyacodrilus amphigenus*, sp. n. and *Rhizodriloides phreaticola*, g. n., sp. n. Hydrobiologia 155: 107–118.

Juget, J. & E. Dumnicka, 1986. Oligochaeta (incl. Aphanoneura) des eaux souterraines continentales. In L. Botosaneanu (ed.), Stygofauna mundi. E. J. Brill, Leiden: 234–244.

Juget, J. & M. Lafont, 1994. Theoretical habitat templets, species traits, and species richness: aquatic Oligochaetes in the Upper Rhône River and its floodplain. Freshwat. Biol. 31: 327–340.

Lafont, M., 1989. Contribution à la gestion des eaux continentales: utilisation des oligochètes comme descripteurs de l'état biologique et du degré de pollution des eaux et des sédiments. Thèse de Doctorat ès Sciences, Université Lyon I, 311 pp.

Lafont, M. & A. Durbec, 1990. Essai de description biologique des interactions entre eau de surface et eau souterraine: application à l'évaluation de la vulnérabilité d'un aquifère à la pollution d'un fleuve. Ann. Limnol. 26: 119–129.

Lafont, M., A. Durbec & C. Ille, 1992. Oligochaete worms as biological describers of the interactions between surface and groundwater: a first synthesis. Reg. Rivers Res. Manag. 7: 65–73.

Lafont, M. & J. C. Camus, 1993. Utilisation des oligochètes et des microcrustacés dans l'évaluation des échanges nappe-rivière et dans l'appréciation de la contamination du milieu. Application à la Moselle à l'amont de Nancy. Rapport Agence de l'Eau Rhin-Meuse, 34 pp.

McElravy E. P. & V. Resch, 1991. Distribution and seasonal occurrence of the hyporheic fauna in a northern California stream. Hydrobiologia 220: 233–246.

Malard F., J. L. Reygrobellet, J. Mathieu & M. Lafont, 1994. The use of invertebrate communities to describe groundwater flow and contaminant transport in a fractured rock aquifer. Arch. Hydrobiol. 131: 93–110.

Rofes, G. & M. Savary, 1981. Description d'un nouveau modèle de carottier pour sédiments fins. Bull. fr. Piscic. 283: 102–113.

Rouch, R. & D. L. Danielopol, 1987. L'origine de la faune aquatique souterraine, entre le paradigme du refuge et le modèle de la colonisation active. Stygologia 3: 345–372.

Rosso, A., M. Lafont & A. Exinger, 1994. Impact of heavy metals on benthic oligochaete communities in the River Ill and its tributaries. Wat. Sci. Tech. 29: 241–248.

Strayer, D. & E. Bannon O'Donnell, 1988. Aquatic Microannelids (Oligochaeta and Aphanoneura) of undergroundwaters of Southeastern New-York. Am. Midl. Nat. 119: 327–335.

Ward, J. V. & M. A. Palmer, 1994. Distribution patterns of interstitial freshwater meiofauna over a range of spatial scales, with emphasis on alluvial river-aquifer systems. Hydrobiologia 287: 147–156.

Hydrobiologia **334**: 157–161, 1996.
K. A. Coates, Trefor B. Reynoldson & Thomas B. Reynoldson (eds), Aquatic Oligochaete Biology VI.
© 1996 *Kluwer Academic Publishers.*

Reversal of eutrophication in four Swiss lakes: evidence from oligochaete communities

Claude Lang & Olivier Reymond
Conservation de la faune, Marquisat 1, CH 1025 St-Sulpice, Switzerland

Key words: biomonitoring, eutrophication, indicator species, lake, oligochaetes, recovery, zoobenthos

Abstract

Following the limitation of phosphorus inputs, total phosphorus concentrations decreased substantially between 1980 and 1990 in four lakes of western Switzerland. Tubificid and lumbriculid communities of Lakes Geneva and Neuchâtel responded clearly to this decrease. Indeed, mean relative abundance of species typical of oligotrophic conditions (mostly *Stylodrilus heringianus* and *Spirosperma velutinus*) doubled in oligochaete communities of both these deep lakes (> 40 m). These changes indicated that both lakes were meso-eutrophic around 1980, but mesotrophic since 1990. In contrast, oligochaete communities of Lakes Morat and Joux, which consisted mostly of tolerant species (*Tubifex tubifex*, *Potamothrix hammoniensis*, and *Limnodrilus hoffmeisteri*), did not indicate an improvement of environmental conditions between 1980 and 1990. In Lake Joux the ratio of chironomid to oligochaete biomass was a more simple indicator of change than the species present in oligochaete communities.

Introduction

Until now the biological responses of lakes to the increase of phosphorus concentrations in the water are better documented than those following its decrease (Levine & Schindler, 1989). But, as more and more lakes of the northern hemisphere begin to recover from man-made eutrophication, this aspect attracts more attention. However, studies are in general concentrated on the responses of phytoplankton (Saas, 1989), whereas the zoobenthos is often neglected.

The recovery of a lake is not complete as long as its sediments, especially those of the profundal, are not recolonized by the species (or at least by their ecological equivalents) which prevailed therein, before the onset of eutrophication. Indeed, the sediment is located on the receiving end of all processes going on within a lake (Levine & Schindler, 1989). Therefore, the restoration of its pristine state, as far as this goal is realistic, is the landmark of a successful recovery of the whole lake.

Following the removal of phosphorus by sewage treatment plants and the ban of phosphorus in detergents, total phosphorus concentrations decreased between 1980 and 1990 in the water of Lakes Geneva,

Neuchâtel, Morat, and Joux (Table 1). In the present study, we compare the responses of zoobenthic communities (especially those of oligochaetes) to the abatement of eutrophication in these four lakes of western Switzerland (Lang & Reymond, 1992, 1993a, b, c, 1995). Our goal is to answer the following question: how to monitor the recovery of the studied lakes from eutrophication with zoobenthos?

Stations and methods

Studied lakes can be divided into two groups according to their morphometry (Table 1): two large and deep lakes (Lakes Geneva and Neuchâtel), two small and shallow lakes (Lakes Morat and Joux). In each lake, sampling sites for zoobenthos were located (Fig. 1) as far as possible from the main external inputs of organic matter (large rivers and sewage treatment plants). In that way, they were mostly exposed to organic sedimentation derived from phytoplankton whose intensity varies according to the trophic state (Baines & Pace, 1994). Sampling sites, located in the profundal, have been visited before and after the abatement of eutrophication (Tables 2, 3). Detailed results of these

Table 1. Characteristics of the studied lakes. Sources: Blanc *et al.*, 1991; Liechti, 1989; Pokorni, 1991; Vioget, 1991.

Characteristic		Lake			
		Geneva	Neuchâtel	Morat	Joux[1]
Depth	mean (m)	153	64	25	21
	max. (m)	310	153	45	33
Lake surface (km^2)		582	215	23	9
Volume (km^3)		89.0	13.8	0.55	0.16
Altitude of lake surface (m)		372	429	429	1004
Total phosphorus (mg m^{-3}) in 1980		82 (37.5)[2]	67 (37.1)[2]	100 (71.5)[2]	45 (34.0)[2]
in 1990		55 (25.2)	29 (16.0)	50 (35.6)	25 (18.9)
More than 4 mg O_2 l^{-1}					
up to a depth of (m)		260	153	10–15	10–15
No. of total circulation per year		0.1[3]	2	2	2
Theoretical residence time (year)		11.9	8.2	1.2	0.85
Area of drainage basin (km^2)		7975	2672	693	211
Average altitude of the basin (m)		1670	780	–	–
Number of inhabitants (thousand)		760	245	66	7

[1] Covered by ice two or three months every winter.
[2] Phosphorus concentration divided by the mean depth (log transformed).
[3] Total circulation only after a cold winter (i.e. every ten years); total circulation down to 150 m every winter.

Table 2. Changes in tubificid and lumbriculid communities of Lakes Geneva and Neuchâtel before (1982–84) and after (1990–1992) abatement of eutrophication. Sources: Lang & Reymond (1992, 1993 b, 1995). FR = frequency (percentage of cores with species); RA = relative abundance (%); A = species absent, + = present; NS = not significant; * = significant P = 0.05 or less.

Code	Species	Unit	Geneva 40 m 1982	Geneva 40 m 1991	Neuchâtel 40 m 1984	Neuchâtel 40 m 1992	Geneva 150 m 1990
1	*Bichaeta sanguinea* Bretscher	FR	+	+	+	+	+
2	*Bythonomus lemani* Grube		10	24	A	A	3
3	*Stylodrilus heringianus* Claparède		39	44	37	47	33
4	*Spirosperma velutinus* (Grube)		46	65	17	19	8
	Oligotrophic species 1–4		66	78*	44	59*	45
5	*Potamothrix vejdovskyi* (Hrabe)		93	62	0	23	33
6	*Spirosperma ferox* (Eisen)		49	22	2	13	0
7	*Psammoryctides barbatus* (Grube)		23	28	0	0	0
8	*Potamothrix moldaviensis* (Vejdovsky, Mrazek)		A	A	0	9	A
	Mesotrophic species 5–8		95	78*	2	33*	33
9	*Limnodrilus hoffmeisteri* (Claparède)		+	+	+	+	+
10	*Limnodrilus profundicola* (Verrill)		+	+	+	+	+
11	*Potamothrix heuscheri* (Bretscher)		+	+	A	A	+
12	*Potamothrix hammoniensis* (Michaelsen)		+	+	+	+	+
13	*Tubifex tubifex* (Müller)		+	+	+	+	+
	Eutrophic species 9–13		87	67*	100	77*	84
	Oligotrophic species 1–4	RA	17	41*	16	33*	24
	Mesotrophic species 5–8		51	29*	0	15*	16
	Eutrophic species 9–13		32	30 NS	84	52*	60
	No. of 16 cm^2 cores		61	192	64	171	189

Table 3. Mean relative abundance (%) of tubificid species and mean biomass (mg/16 cm²) of tubificids, *Chaoborus* and chironomid larvae in Lakes Morat and Joux before (1984, 85) and after (1991, 92) the abatement of eutrophication. Source: Lang & Reymond (1993 a, c). + = present; NS = not significant. * = significant P = 0.05 or less.

Zoobenthic taxa	Unit	Morat 40 m		Joux 25 m	
		1984	1991	1985	1992
Tubifex tubifex	Abundance	50.0	50.0 NS	23.8	63.6*
Potamothrix hammoniensis		50.0	50.0	0	0
Limnodrilus hoffmeisteri		0	0	76.2	36.4
Tubificids	Biomass	40.8	31.9 NS	43.2	42.7 NS
Chaoborus		+	+	7.1	18.4 NS
Chironomids		+	+	39.9	15.1*
All		40.8	31.9 NS	90.3	76.2*
No. of 16 cm² cores		16	16	120	120

surveys are presented elsewhere (Lang & Reymond, 1992, 1993a, b, c, 1995).

Zoobenthos samples (16 cm²/sample) were obtained with a corer, operated from the water surface, in Lakes Geneva and Neuchâtel, Morat, and by a diver in Lake Joux. Sediments were sieved (mesh size aperture: 0.2 mm) and the retained material preserved in 5% formalin. Oligochaetes, chironomids, and *Chaoborus* were picked, counted, and weighed after removing excess water with blotting paper. Oligochaetes were mounted (Reymond, 1994) and identified to species, whereas chironomids and *Chaoborus* were not identified.

Three categories of species were identified following the guidelines of Lang (1991): species whose numerical dominance in tubificid and lumbriculid communities indicates oligotrophic, mesotrophic and eutrophic conditions, are designated in Table 2. Relative abundance of oligotrophic species was calculated as a percentage of the total number of tubificid and lumbriculid worms present in each core. Juvenile worms (diameter less than 0.3 mm) were excluded from calculations to reduce the effects of seasonal variability on species abundance (Lang, 1991).

Relative abundance of oligotrophic worm species was used to monitor the recovery of Lakes Geneva and Neuchâtel from eutrophication (Lang & Reymond, 1992, 1993b, 1995). As these species were absent from the profundal of Lakes Morat and Joux, other benthic indicators were used: total composition of worm communities in Lake Morat (Lang & Reymond, 1993a); in addition to that indicator, biomass of chironomids,

tubificids, and *Chaoborus* was used in Lake Joux (Lang & Reymond, 1993c).

Results

Macrofauna consisted mainly of tubificid and lumbriculid worms in the samples collected in Lakes Geneva and Neuchâtel, and mostly of tubificid worms in Lake Morat (Tables 2, 3). In contrast, chironomids and *Chaoborus* larvae were abundant in samples collected in Lake Joux.

Increase of oligotrophic worm species in Lakes Geneva and Neuchâtel after the abatement of eutrophication indicated an improvement of conditions within the sediments (Table 2). According to the recorded abundances, these lakes were deemed meso-eutrophic around 1980, but mesotrophic since 1990 (Lang & Reymond, 1992, 1993b). In Lake Geneva however, the response of oligotrophic species was less clear at a depth of 150 m than at a depth of 40 m (Lang & Reymond, 1995). Indeed, the recovery of the deepest areas takes more time because distances to travel to recolonize patches without oligotrophic species are greater in the profundal than in the sublittoral.

In Lake Morat, relative abundance and biomass of the eutrophic worm species were the same after and before the abatement of eutrophication (Table 3). In contrast, relative abundance of *Tubifex tubifex* increased whereas biomass of chironomid larvae and total zoobenthos decreased in Lake Joux. These trends, which indicated deteriorating conditions in the profun-

160

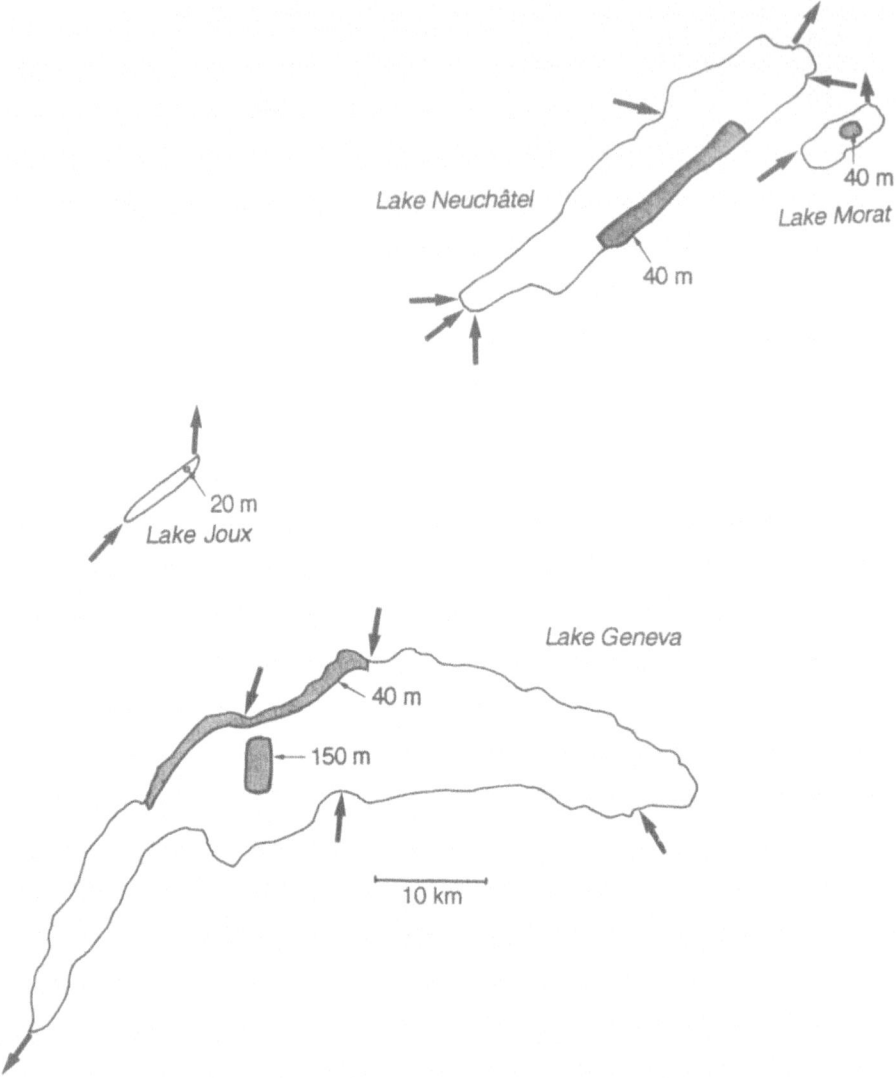

Fig. 1. Location (shaded areas) and depth of sampling sites in four studied lakes. Arrows indicate main organic inputs and outflows from the lakes.

dal, were attributed to the persistent dominance of the cyanobacteria *Oscillatoria rubescens* (Lang & Reymond, 1993c). This interpretation was confirmed by the increase of chironomid larvae in 1994 following the disappearance of *Oscillatoria* in 1993 (Lang & Reymond, unpublished data).

Discussion

The above results suggest that the response of zoobenthos to the abatement of eutrophication depends,

among other factors, on the morphometry of lakes. In the two large and deep lakes, the oligochaete communities responded clearly and positively whereas, in the two small and shallow lakes, zoobenthic communities were unchanged (Lake Morat) or even degraded (Lake Joux). However, as morphometry affects the extent of eutrophication (Saas, 1989), this result reflects also the previous history of these lakes. For instance, Lake Morat was eutrophic in 1936 and *T. tubifex* and *P. hammoniensis* were already the only species present in the profundal (Rivier in Lang & Reymond, 1993a). In Lakes Morat and Joux, the composition of zoobenthos was affected by the shortage of oxygen which persisted in the profundal after the abatement of eutrophication (Table 1). But the increase of zoobenthic biomass with depth indicated good environmental conditions for tolerant species (Lang & Reymond, 1993a, c).

Depth of sampling sites, relative to the maximum depth of each lake, could also affect the outcome of comparison between the four lakes. For instance, a sampling depth of 40 m in Lake Morat corresponds, expressed as a percentage of the maximum depth (Table 1), to a sampling depth of 275 m in Lake Geneva. And, as discussed previously (Lang & Reymond, 1995), zoobenthos responded more slowly to the abatement of eutrophication at a depth of 150 m than a depth of 40 m. Therefore, to detect the first signs of recovery, it could be necessary to sample zoobenthos in the littoral (5–10 m) of Lakes Morat and Joux. For instance, the mesotrophic species *Spirosperma ferox* was observed in 1994 at a depth of 5 m in Lake Joux (Lang & Reymond, unpublished data). The gradual recolonization of the profundal by this species can be used to track the recovery of Lake Joux.

The above results suggest some empirical rules to monitor the recovery of the four studied lakes from eutrophication with zoobenthos. Firstly, if oligotrophic worm species are present, use them: they are easy to identify and they give clear indication on the trophic state. For instance, the relative abundance of the oligotrophic species was significantly different between the three troughs present in the western basin of Lake Geneva (Petit Lac) whereas total biomass of zoobenthos was the same (Lang, 1986). In addition, their relative abundance can be predicted from total phosphorus concentration recorded in the water (Lang, 1990). Therefore, observed and predicted values can be compared and the analysis of discrepancies provided useful insights on the speed of recovery. Secondly, if oligotrophic worm species are absent from the profundal, the percentage of *Tubifex tubifex*, *Potamothrix hammoniensis* and *P. heuscheri* in tubificid communities can be used as an indicator of environmental conditions. Since these species can withstand the most extreme situations (Milbrink, 1980), the decrease of their relative abundance in tubificid communities indicates an improvement of conditions. However, in some cases as in Lake Joux (Table 3), total biomass of zoobenthos, divided into biomass of oligochaetes, chironomid and *Chaoborus* larvae, can be a more simple indicator of change than the species present in oligochaete communities.

Acknowledgments

Comments by Reinmar Grimm, Trefor Reynoldson, and one anonymous referee have improved this text.

References

Baines, S. B. & M. L. Pace, 1994. Relationships between suspended particulate matter and sinking flux along a trophic gradient and implications for the fate of planktonic primary production. Can. J. Fish. aquat. Sci. 51: 25–36.

Blanc, P., C. Corvi & F. Rapin, 1991. Evolution physico-chimique des eaux du Léman. Commission internationale pour la protection des eaux du Léman. 23–44. CH-1000 Lausanne 12.

Lang, C., 1986. Eutrophisation du Léman indiquée par les communautés d'oligochètes: campagne 1982–1985. Schweiz. Z. Hydrol. 48: 230–239.

Lang, C., 1990. Quantitative relationships between oligochaete communities and phosphorus concentrations in lakes. Freshwat. Biol. 24: 327–334.

Lang, C., 1991. Decreasing phosphorus concentrations and unchanged oligochaete communities in Lake Geneva: how to monitor recovery? Arch. Hydrobiol. 122: 305–312.

Lang, C. & O. Reymond, 1992. Reversal of eutrophication in Lake Geneva: evidence from the oligochaete communities. Freshwat. Biol. 28: 145–148.

Lang, C. & O. Reymond, 1993a. Eutrophisation du lac de Morat indiquée par les communautés d'oligochètes: tendance 1980–1991. Revue suisse Zool. 100: 11–18.

Lang, C. & O. Reymond, 1993b. Recovery of Lake Neuchâtel (Switzerland) from eutrophication indicated by the oligochaete communities. Arch. Hydrol. 128: 65–71.

Lang, C. & O. Reymond, 1993c. Trends in phytoplanktonic and zoobenthic communities after the decrease of phosphorus concentrations in Lake Joux. Revue suisse Zool. 100: 907–912.

Lang, C. & O. Reymond, 1995. Contrasting responses of oligochaete communities to the abatement of eutrophication in Lake Geneva. Hydrobiologia 308: 77–82.

Levine, S. N. & D. W. Schindler, 1989. Phosphorus, nitrogen, and carbon dynamics of experimental lake 303 during recovery from eutrophication. Can. J. Fish. aquat. Sci. 46: 2–10.

Liechti, P., 1989. L'état du lac de Morat. Bulletin de l'OFEFP, Berne 2/89: 32–36.

Milbrink, G., 1980. Oligochaete communities in pollution biology: the European situation with special reference to lakes in Scandinavia. In: Aquatic Oligochaete Biology. R. O. Brinkhurst & D. G. Cook (eds), Plenum Press, New York.

Pokorni, B., 1991. Surveillance des eaux de surface du lac de Neuchâtel. Rapport du service cantonal de la protection de l'environnement, CH-2034 Peseux, Neuchâtel.

Reymond, O., 1994. Préparations microscopiques permanentes d'oligochètes: une méthode simple. Bull. Soc. Vaud. Sc. Nat. 83: 1–3.

Saas, H., 1989. Lake restoration by reduction of nutrient loading. Academia Verlag, H. Richarz. Sankt Augustin, 497 pp.

Vioget, Ph., 1991. Qualité du lac de Joux. Rapport du service des eaux et de la protection de l'environment. CH-1006 Epalinges, Suisse.

Hydrobiologia **334**: 163–168, 1996.
K. A. Coates, Trefor B. Reynoldson & Thomas B. Reynoldson (eds), Aquatic Oligochaete Biology VI.
© 1996 *Kluwer Academic Publishers.*

163

A comparison of two tubificid oligochaete species as candidates for sublethal bioassay tests relevant to subtropical and tropical regions

Mercedes R. Marchese[1] & Ralph O. Brinkhurst[2]
[1] *Instituto Nacional de Limnología, José Maciá 1933, 3016 Santo Tomé, Santa Fe, Argentina*
[2] *Aquatic Resources Center, Box 680818, Franklin, TN 37068-0818, USA*

Key words: Sediment bioassay, Oligochaeta, *Tubifex tubifex*, *Branchiura sowerbyi*, growth, reproduction

Abstract

Two tubificid oligochaetes, *Branchiura sowerbyi* and *Tubifex tubifex*, were evaluated as potential test organisms for sediment bioassays. We attempt to reproduce the sediment bioassay proposed by Reynoldson *et al.* (1991) using his strain of *Tubifex tubifex* and his reference sediment and to compare this technique using *Branchiura sowerbyi*. This species was chosen because it is more common and dominant in tropical and subtropical environments than *Tubifex tubifex*. Data on survival and reproduction were obtained for both species, and growth estimates were obtained for *B. sowerbyi*. The sublethal bioassay with *T. tubifex* confirmed earlier estimates of a test duration of 4 weeks at 22.5 °C. *B. sowerbyi* cultures can produce usable estimates in 21 days at 30 °C.

Introduction

Contaminated sediments are a potential source of toxic elements that may be released into the water column or transmitted to higher trophic levels. Measuring the chemical concentrations of contaminants in sediments does not reveal bioavailability and negative effects on the biology of organisms, and so it is necessary to test for this directly. Many acute aquatic toxicity tests have been described, using various benthic organisms such as chironomids, amphipods, mayflies and oligochaetes. Sediment bioassays with oligochaetes have been reported by Chapman *et al.* (1982a, 1982b), Wiederholm *et al.* (1987), Milbrink (1987), Casellato & Negrisolo (1989), Reynoldson *et al.* (1991), Ankley *et al.* (1991) and Dermott & Munawar (1992).

The acute toxicity test endpoints most frequently used are mortality or immobilisation (Buikema *et al.*, 1982). On the other hand, the chronic toxicity test endpoints include mortality, growth, reproduction and behaviour (Buikema & Voshell, 1993). Reproduction and growth are useful endpoints for predicting ecological conditions. Phipps *et al.*, 1993 have used *Lumbriculus variegatus* (Oligochaeta, Lumbriculidae) for bioaccumulation studies. In this paper we attempt

Table 1. Physical and chemical characteristics of field sediments.

SiO_2	%	36.23
Al_2O_3	%	6.99
Fe_2O_3	%	2.83
MgO	%	1.85
CaO	%	17.25
Na_2O	%	1.15
K_2O	%	1.76
TiO_2	%	0.32
MnO	%	0.12
P_2O_5	%	0.21
Pb	ppm	25.8
As	ppm	1.9
Zn	ppm	96.6
Cu	ppm	12.1
Cr	ppm	41.5
V	ppm	25.2
S	ppm	9100.0
Cd	ppm	0.4
sand	%	14.5
silt	%	76.8
clay	%	8.7
LOI	%	17.3

Table 2. Mean values (S.D.) for reproduction parameters for *T. tubifex* 22.5 °C without sieving the sediment. The ratio young/cocoon is based on empty cocoons.

	Week 1	Week 2	Week 3	Week 4
Cocoons	9.2 (1.1)	15.4 (1.6)	32.4 (1.3)	36.4 (2.6)
% Hatch	0	0	30.2 (4.5)	39.8 (5.3)
Young	0	1.4 (0.5)	48.2 (6.0)	93.2 (15.6)
Cocoons/adult	2.3 (0.3)	3.9 (0.4)	8.2 (0.4)	9.1 (0.6)
Young/cocoon	0	0	5.0 (0.9)	6.4 (0.5)
Young/adult	0	0.3 (0.1)	12.0 (1.5)	23.3 (3.9)
Per day				
Cocoon/adult	0.32 (0.0)	0.28 (0.0)	0.39 (0.0)	0.33 (0.0)
Young/adult	0.0 (0.0)	0.02 (0.0)	0.57 (0.1)	0.83 (0.1)

Table 3. Mean values (S.D.) for reproduction parameters for *Branchiura sowerbyi* at 25 °C and 30 °C.

	Week 1	Week 2	Week 3	Week 4	Week 5
25 °C					
Cocoons	0.8 (0.7)	2.6 (1.3)	16.4 (2.5)	23.4 (2.9)	19.4 (5.4)
% Hatch	0	20.6 (19.4)	7.4 (2.4)	27.2 (9.0)	34.4 (12.6)
Young	0	0.6 (0.8)	1.8 (0.7)	9.4 (4.2)	14.0 (3.6)
Cocoons/adult	0.2 (0.1)	0.5 (0.2)	3.3 (0.5)	4.7 (0.6)	3.8 (1.1)
Young/cocoons	0	1.5 (0.5)	1.3 (0.4)	1.5 (0.3)	2.6 (1.0)
Young/adult	0	0.1 (0.2)	0.3 (0.1)	1.9 (0.8)	2.8 (0.7)
Per day					
Cocoons/adult	0.03 (0.0)	0.03 (0.0)	0.16 (0.0)	0.17 (0.0)	0.11 (0.0)
Young/adult	0.0 (0.0)	0.007 (0.0)	0.01 (0.0)	0.07 (0.0)	0.11 (0.0)
30 °C					
Cocoons	5.0 (2.2)	11.4 (1.6)	11.6 (1.5)	13.0 (3.6)	13.0 (4.0)
% Hatch	0	18.2 (10.6)	78.0 (4.9)	86.6 (14.9)	94.4 (7.5)
Young	0	3.8 (1.9)	9.2 (1.9)	10.4 (0.9)	11.8 (2.8)
Cocoons/adult	1.0 (0.4)	2.3 (0.3)	2.3 (0.3)	2.6 (0.7)	2.6 (0.8)
Young/cocoons	0	2.0 (0.6)	1.0 (0.2)	1.0 (0.2)	1.0 (0.1)
Young/adult	0	0.8 (0.4)	1.8 (0.4)	2.1 (0.2)	2.4 (0.6)
Per day					
Cocoons/adult	0.14 (0.1)	0.16 (0.0)	0.11 (0.0)	0.09 (0.0)	0.07 (0.02)
Young/adult	0.0 (0.0)	0.06 (0.0)	0.09 (0.0)	0.07 (0.0)	0.07 (0.02)

to reproduce the work of Reynoldson *et al.* (1991) using his strain of *Tubifex tubifex* and his reference sediment. We then tested *Branchiura sowerbyi* as a bioassay organism because it is a common species in tropical environments enriched with organic matter. It is also easier to handle and identify than other tubificids because of the prominent gills.

The present paper reports results on the survival and reproduction of these two species and also growth of *B. sowerbyi* tested in an unpolluted reference sediment from Big Creek Marsh, Lake Erie using the methods of Reynoldson *et al.* (1991).

Materials and methods

The sediment was collected in Big Creek Marsh, Lake Erie (provided by T. B. Reynoldson) and it was immediately shipped to the Aquatic Resources Center, where it was stored in the dark at 4 °C until used in bioassays. Physical and chemical characteristics for this sediment are displayed in Table 1.

The sediment was used within 2 days of receipt for the first experiment with *B. sowerbyi* and the *T. tubifex* test was run 18 days later. Subsequent tests were used after the sediment had been stored longer, but Reynoldson *et al.* (1991) indicated that storage time did not affect their results. Twenty four hours before testing, the sediment was filtered through a 250 μm sieve. The tests were carried out in 250 ml beakers with 100 ml of sieve sediment and 100 ml of dechlorinated tap water. Twenty beakers were used in tests with *T. tubifex*, and 25 with *B. sowerbyi*. The beakers were placed in the incubator at 22.5 °C (± 1 °C) for *T. tubifex* and at 25 °C (± 1 °C) in one test series and at 30 °C (± 1 °C) in another for *B. sowerbyi*.

Mature worm specimens were maintained at the appropriate temperature in the incubator 24 h before beginning the bioassay. Worms were added to each of the bioassay beakers which were then returned to the incubator. In work with *T. tubifex*, we used 4 worms per beaker, as in Reynoldson *et al.* (1991). In tests with *B. sowerbyi*, 5 worms were used. At the end of each week, 5 beakers were removed from the test incubator, and the adults, young, full and empty cocoons and eggs per cocoon were counted. The ratio cocoons/adult was calculated using the total number of cocoons. The ratio young/cocoon was calculated using the number of empty cocoons.

Eighty-five cocoons were cultured at 25 °C and 30 °C to determine the hatch time for *B. sowerbyi* and ninety-four cocoons at 22.5 °C for *T. tubifex*.

The entire sediment was examined for worms and cocoons without sieving. In a preliminary test using *T. tubifex* the sediment was sieved (250 μm), and both the material retained on the sieve and the sieved residues were examined for worms and cocoons. Only one or two empty cocoons were ever found in the sieved residues and so the oligochaete material was all retained by the sieve.

Experiments with *T. tubifex* lasted 4 weeks, those with *B. sowerbyi* were extended an extra week. The beakers were not aerated and food was not added throughout the experiments. Water level in the beakers was maintained by adding dechlorinated tap water as required every 2–3 days.

Other experiments conducted to analyse growth in *Branchiura sowerbyi* were run for 56 days at 25 °C (± 1 °C) in two series of five replicates each. Ten small immature worms were added to each beaker. The worms were counted and weighed every week in one test and every two weeks in another one in order to compare the stress of handling. In contrast to the reproduction experiments, the worms were returned to the beakers in the same sediments and examined again the following week. The worms, with gut empty (they were kept in water for 15–30 min), were put on a piece of aluminium foil and blotted dry with paper tissue. They were then weighed on an OHAUS 110 microbalance to the nearest 0.01 mg and then returned to the sediment.

The mean specific daily growth rate ($G_w\%$) was calculated according to Reynoldson (1987):

$$G_w\% = LnW_2 - LnW_1 \times 100 \ t, \qquad (1)$$

where, W_1 = biomass in mg at start of experiment; W_2 = biomass in mg at end of experiment; and t = time in days.

Results

Reproduction

In experiments with *T. tubifex*, survival was 100%. The worms started to lay cocoons during the first week and young appeared after 2 weeks (Table 2). The average number of eggs per cocoon was 4.48 (+/− 0.5). Development from eggs to newly hatched worms took 7–21 days at 22.5 °C. Very few cocoons were found with non-viable decomposing eggs. The application of ANOVA to the cocoon/adult showed that only the differences between third and fourth week were not significant, with $F = 5.63$ ($p > 0.05$). The difference between means of young/adult were significant between all the weeks without stabilising the production of young. Both cocoon and offspring production were converted to daily production rates. Cocoon production stabilised at 0.39 (± 0.02) cocoons·adult^{-1}·d^{-1} after 3 weeks. The results for % hatch and young/cocoon for

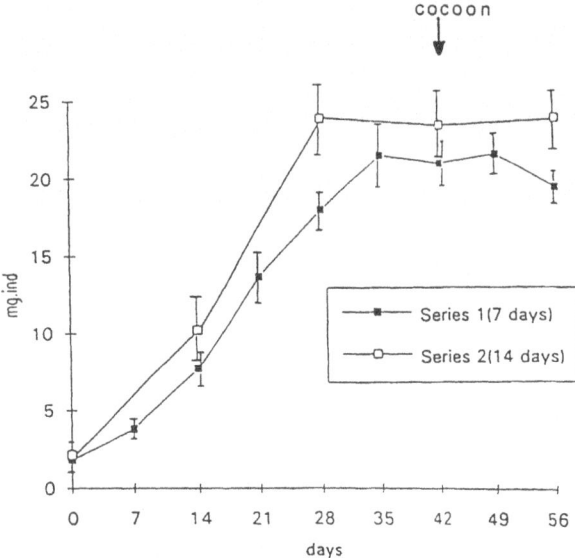

Fig. 1. Average growth of *B. sowerbyi* immature worms as determined every 7 days (Series 1) and every 14 days (Series 2). Reproduction began between 28 and 35 days.

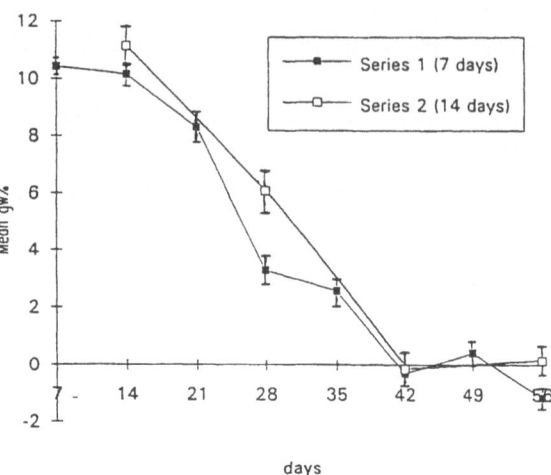

Fig. 2. Mean specific daily growth rate (G_w %) of immature *B. sowerbyi* at 25 °C determined weekly (Series 1) and every 14 days (Series 2).

Week 2 in Table 2 were affected by the hatching of some, but not all, young from the cocoons.

In a test period of five weeks with *B. sowerbyi*, sexually mature individuals were able to produce both cocoons and offspring in two or three weeks (Table 3). ANOVA showed that there were not significant differences between the production of cocoons and young/adult only between the fourth and fifth weeks ($F = 1.69$, $p > 0.05$ and $F = 2.72$, $p > 0.05$, respectively) at 25 °C. The average daily cocoon and offspring production rate stabilised at 0.17 ($\pm$ 0.02) cocoons·adult^{-1}·d^{-1}, and at 0.07 ($\pm$ 0.03) young·adult^{-1}·d^{-1} after four weeks.

The differences between the means of cocoon and young production from the second and third week were not significant, respectively, with $F = 0.42$ ($p > 0.01$) and $F = 1.61$ ($p > 0.05$) at 30 °C. The average daily cocoon and production rate stabilised at 0.16 ($\pm$ 0.02) cocoons adult^{-1} d^{-1} after two weeks and three weeks were required for the average daily rate of offspring production to stabilise at 0.09 ($\pm$ 0.02) young·adult^{-1}·d^{-1}.

In cultures of separated cocoons, full embryo development from an egg being laid until the young leaves the cocoon took approximately three weeks at 25 °C but only 1 week at 30 °C.

The fecundity (number eggs per cocoon) of *Branchiura sowerbyi* during the whole period was 1.94

($\pm$ 0.13), with variation between cocoons being from 1 to 3 eggs at 25 °C. At 30 °C the value was 1.82 ($\pm$ 0.15). Usually, in cocoons with 2 eggs, only one was viable.

The egg production per surviving worm was 6.36 ($\pm$ 0.83).

Growth

Immature *Branchiura sowerbyi* grew rapidly until sexual maturity and then its weight remained constant with a slight decrease when the first cocoons were recorded (Fig. 1). From an initial weight of 1.82 mg ind^{-1} the worms grew about 22 mg in 35 days (when cocoon production begins), 0.58 mg day^{-1} on the average.

The specific growth rate of *Branchiura sowerbyi* from age 0–30 days (before reproduction starts) was 0.081. Daily mean growth rate (G_w%) was high during the first weeks, but became negative when the worms are breeding (Fig. 2). Growth rates obtained for animals weighed every two weeks were slightly higher than those for worms weighed every week. Survival at the end of the experiment was 100%.

Discussion

We attempted to duplicate the experiments reported by Reynoldson *et al.* (1991) for *T. tubifex*. In our study, average fecundity (number of eggs per cocoon) was 4.48, similar to the value reported by Kosiorek (1974) and lower than that reported by Finogenova & Loba-sheva (1987), which was 9.7 at 20.5 °C. Reynoldson *et al.* (1991) did not report egg numbers. Our experiments produced higher numbers of cocoons and of young estimated per adult per day after three weeks than those reported by Reynoldson *et al.* (1991) which were 0.17 (0.04) cocoons/adult and 0.21 (0.03) young/adult, probably because we have removed them every week.

Embryonic development took 7–21 days in our experiments, Reynoldson *et al.* (1991) reported 7–14 days, Paoletti (1989), 15 days at 20 °C and higher, and Kosiorek (1974) quoted 10–12 days.

The specific growth rate of *Branchiura sowerbyi* obtained was similar to growth rates calculated by Finogenova & Lobasheva (1987) for *T. tubifex* of 0.078 (0–30 days) and 0.044 (0–70 days) in natural silt at 20 °C. Reynoldson (1987) reported 0.011 for *T. tubifex*, 0.013 for *L. hoffmeisteri* and 0.014 for *I. templetoni* at 12 °C.

The egg production per surviving worm in *B. sowerbyi* was lower than the value of 11.2 (± 2.6) reported by Aston & Milner (1981) in mud at 25 °C. The number of eggs per cocoon was 1.94 (± 0.13), in our experiments but Aston *et al.* (1982) reported 2.82 (± 0.47). Aston (1968) concluded that 25 °C was the optimum temperature in his experiments but in our results, cocoon production and number of young worms because of the higher hatching rate were better at 30 °C than at 25 °C. However, the number of young per cocoon is higher at 25 °C.

The differences between means of cocoons and young/adult from week 2 are not significant at 30 °C, but at 25 °C the differences are significant before week 4.

We would suggest that the bioassay protocol using *B. sowerbyi* could be reduced to 21 days at 30 °C, by which time the number of hatched worms should be high enough to be a useful endpoint, but at 25 °C tests this could be extended one week.

Growth may be a good alternate endpoint for a test when young worms are used.

We consider according to the results obtained with *Branchiura sowerbyi* in comparison with *Tubifex tubifex*, that this is a good alternative test species for monitoring tropical and subtropical environments. It is more common indigenous species than *T. tubifex* in enriched organic environments in South America, where *blanchardi* form of *T. tubifex* is dominant in environments with high conductivity (Varela *et al.*, 1986; Marchese, 1988; Wiedenbrug, 1993; Rodriguez, pers. comm.)

Acknowledgements

We are indebted to Dr T.B. Reynoldson for financial assistance, and much help and encouragement. Professor Marchese worked at Aquatic Resources Center under a fellowship from the National Council of Scientific and Technological Research (CONICET) of Argentina.

References

Ankley, G. T., G. L. Phipps, E. N. Leonard, D. A. Benoit, V. R. Mattson, P. A. Kosian, A. M. Cotter, J. R., Dierkes, D. J. Hansen & J. D. Mahony, 1991. Acid-volatile sulfide as a factor mediating cadmium and nickel bioavailability in contaminated sediments. Envir. Toxicol. Chem. 10: 1299–1307.

Aston, R. J., 1968. The effect of temperature on the life cycle, growth and fecundity of *Branchiura sowerbyi* (Oligochaeta: Tubificidae). J. Zool. Lond. 154: 29–40.

Aston, R. J. & A. G. P. Milner, 1981. Conditions required for the culture of *Branchiura sowerbyi* (Oligochaeta: Tubificidae) in activated sludge. Aquaculture 26: 155–160.

Aston, R. J., K. Sadler & A. G. P. Milner, 1982. The effect of temperature on the culture of *Branchiura sowerbyi* (Oligochaeta:Tubificidae) on activated sludge. Aquaculture 29: 137–145.

Buikema, A. L., Jr., B. R. Niederlehner & J. Cairns, Jr., 1982. Biological monitoring. Part IV- Toxicity testing. Wat. Res. 16: 239–262.

Buikema, A. L., Jr. & J. R. Voshell, Jr., 1993. Toxicity studies using freshwater benthic macroinvertebrates. In D. M. Rosenberg & V.H. Resh (eds), Freshwater Biomonitoring and Benthic Macroinvertebrates, 344–398. Chapman & Hall, N.Y.

Casellato, S. & P. Negrisolo, 1989. Acute and chronic effects of an anionic surfactant on some freshwater tubificid species. Hydrobiologia 180: 243–252.

Chapman, P. M., M. O. Farrell & R. O. Brinkhurst, 1982a. Relative tolerances of selected aquatic oligochaetes to combinations of pollutants and environmental factors. Aquat. Toxicol. 2: 47–67.

Chapman, P. M., M. O. Farrell & R. O. Brinkhurst, 1982b. Relative tolerances of selected aquatic oligochaetes to individual pollutants and environmental factors. Aquat. Toxicol. 2: 69–78.

Dermott, R. & M. Munawar, 1992. A simple and sensitive assay for evaluation of sediment toxicity using *Lumbriculus variegatus*. Hydrobiologia 235/236: 407–414.

Finogenova, N. P. & T. M. Lobasheva, 1987. Growth of *Tubifex tubifex* Muller (Oligochaeta, Tubificidae) under various trophic conditions. Int. Revue ges. Hydrobiol. 72: 700–726.

Kosiorek, D., 1974. Development cycle of *Tubifex tubifex* Muller in experimental culture. Pol. Arch. Hydrobiol. 21: 411–422.

Marchese, M. R., 1988. New record of the *blanchardi* form of *Tubifex tubifex* (Muller, 1774) (Oligochaeta, Tubificidae) in Argentina and its relationship to suggested synonymous species. Physis 46: 55–58.

Milbrink, G., 1987. Biological characterization of sediments by standardized tubificid bioassays. Hydrobiologia 155: 267–275.

Paoletti, A., 1989. Cohort cultures of *Tubifex tubifex* forms. Hydrobiologia 180: 143–150.

Phipps, G. L., G. T. Ankley, D. A. Benoit & V. R. Mattson, 1993. Use of the aquatic oligochaete *Lumbriculus variegatus* for assessing

168

the toxicity and bioaccumulation of sediment-associated contaminants. Envir. Toxicol. Chem. 12:269–279.

Reynoldson, T. B., 1987. The role of environmental factors in the
ecology of tubificid oligochaetes. An experimental study. Holarct.
Ecol. 10: 241–248.

Reynoldson, T. B., S. P. Thompson & J. L. Bamsey, 1991. A sediment
bioassay using the tubificid oligochaete worm *Tubifex tubifex*.
Envir. Toxicol. Chem. 10: 1061–1072.

Varela, M. E., J. A. Bechara & N. Andreani, 1986. El macrobentos y
su relación con las fluctuaciones de salinidad en ríos y esteros del
Chaco Oriental (Argentina). Ambiente subtropical 1: 134–147.

Wiedenbrug, S., 1993. Aspectos da estrutura espacial da macrofauna
béntica da Lagoa Emboaba, R.S. Tesis. Universidade Federal do
Río Grande do Sul, 157 pp.

Wiederholm, T., A. Wiederholm & G. Milbrink, 1987. Bulk sediment
bioassays with five species of fresh-water oligochaetes. Wat. Air
Soil Pollut. 36: 131–154.

Hydrobiologia **334**: 169–183, 1996.
K. A. Coates, Trefor B. Reynoldson & Thomas B. Reynoldson (eds), Aquatic Oligochaete Biology VI.
© 1996 *Kluwer Academic Publishers.*

Oligochaetes and eutrophication; an experiment over four years in outdoor mesocosms

Piet F.M. Verdonschot

Institute for Forestry and Nature Research, P.O. Box 23, 6700 AA Wageningen, The Netherlands

Key words: Oligochaete, eutrophication, large-scale experiment, ditch, phosphorus, nitrogen, oxygen

Abstract

Eight experimental ditch mesocosms were used to study the effect of eutrophication over four years. The experimental ditches had a sand or clay bottom. The ditches were treated with additions of phosphorus, phosphorus and nitrogen, or without additions (controls). Oligochaetes were sampled by deploying trays with substratum for colonization over twenty weeks. Both the important variables phosphorus, nitrogen and oxygen as well as the oligochaete species and numbers are presented. The effects of nutrient additions on phosphorus, nitrogen and oxygen concentrations were described together with changes in oligochaete species composition and numbers. The results were further analyzed by redundancy analysis (RDA). In the clay-lined ditches nutrient addition coincided with fluctuation in oxygen concentration. The higher the nutrient addition levels the longer the period of oxygen depletion became. During oxygen depletion the number of oligochaetes was strongly reduced or even became zero. The low nutrient status of the sandy bed in the sand-lined ditches slowed down the rate of colonization. Only a few tubificids were collected. Eutrophication effects were only observed at the highest nutrient addition level. Considerable variation is attributed to stochastic factors in the sand-lined ditches. Whether oligochaete species were present was related to the length of the colonization period. The substratum composition and food together with oxygen regime decided whether they become more or less abundant in ditches. Large-scale mesocosm experiments require time to develop. Only after the first colonization period variables of species presences and abundances can be employed to detect changes associated with eutrophication. Oligochaetes can be used to measure colonization as well as eutrophication processes.

Introduction

In The Netherlands, ditches with a total estimated length of 300 000 km are a major water type. Ditches are up to 10 m wide and less than a metre deep. They are dug to improve water drainage mainly for agricultural activities. Therefore, the influence of current agricultural techniques on water quality, for example the use of pesticides and fertilizers, is large. This article focuses on the effect of adding fertilizers on oligochaetes in ditches. This knowledge should support ditch water quality assessment and management techniques.

Eutrophication is the process resulting from the addition of nutrients to an ecosystem. Oligochaete presence and abundance is often used as an indicator of trophic state (e.g. Wiederholm, 1980) and

depends on substratum type (Wachs, 1967; Lafont, 1977), food quality (Pasteris *et al.*, 1994) and availability (Brinkhurst & Cook, 1974; Brinkhurst & Austin, 1979), oxygen regime (Palmer, 1968), other physical and chemical variables and, last but not least, biotic interactions (Chua & Brinkhurst, 1973).

Oligochaetes do not directly respond to nitrogen and phosphorus concentrations. They respond to the effect of production rate of organic material. They directly feed on the organic matter available when plants die off. When nutrient concentrations are high, a high amount of organic matter is produced which offers food for oligochaetes. On the other hand, oxygen conditions will become less favourable which in its turn limits oligochaete numbers. When nutrient concentra-

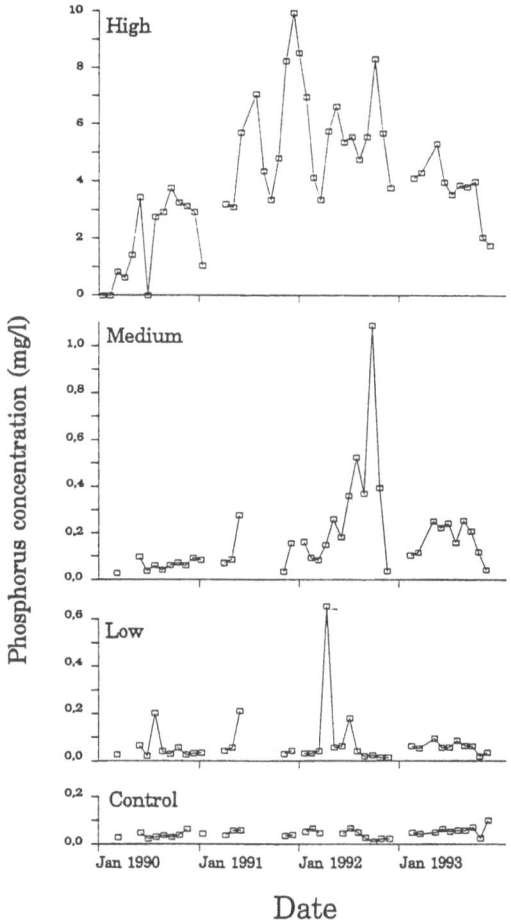

Fig. 1. Phosphorus concentrations (mgP l^{-1}) in clay-lined ditches in time based on monthly observations during 1990–1993.

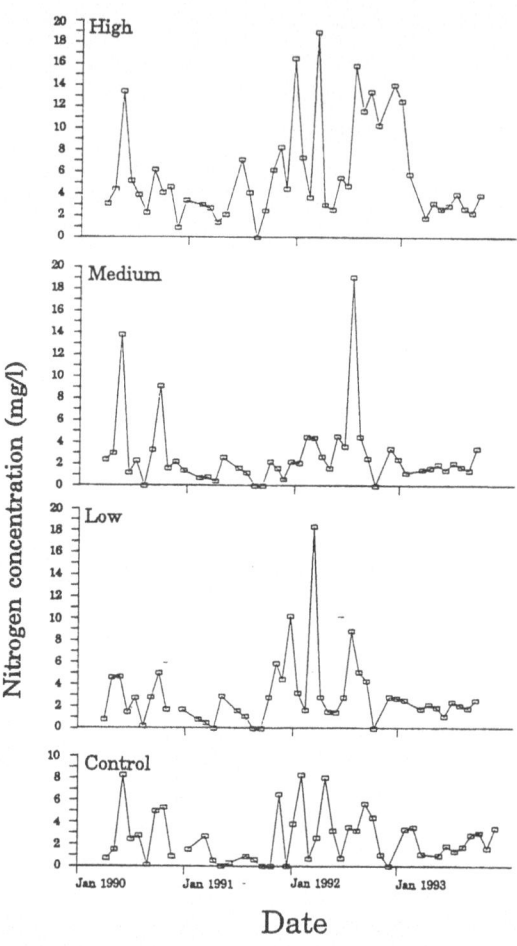

Fig. 2. Nitrogen concentrations (mgN l^{-1}) in clay-lined ditches in time based on monthly observations during 1990–1993.

tions are low, less organic matter is produced and less food will be available for oligochaetes.

The response of aquatic oligochaetes to the process of eutrophication is, among others described (Brinkhurst & Jamieson, 1971), mostly inferred from field studies. This field information is useful to interpret species distributions but does not inform about the correlation between observed distributions and specific environmental variables. The alternative is based on laboratory experiments. But they often ignore the complex biotic (e.g. competition, predation) and abiotic (e.g. sublethal stressors, combined effects) interactions. A third possibility are field experiments whereby the complexity of the ecosystem is maintained and studied under controlled circumstances in so-called mesocosms.

In a series of experimental ditches, a large-scale outdoor experiment was performed to study the effect of eutrophication on the ditch ecosystem. We studied the effect of adding nitrogen and phosphorus to ditches on oligochaete assemblages (Verdonschot & Ter Braak, 1994).

Materials and methods

The experimental ditches are fully described by Drent & Kersting (1992). Eight experimental ditches were used for this experiment. Each ditch is 40 m long, 3.5 m wide at the water surface and 2 m wide at the bottom. The water depth is 0.5 m and the depth of the sediment bed is 0.25 m. The total water volume is 55 m^3 per ditch. Four ditches had a clay bottom (the

171

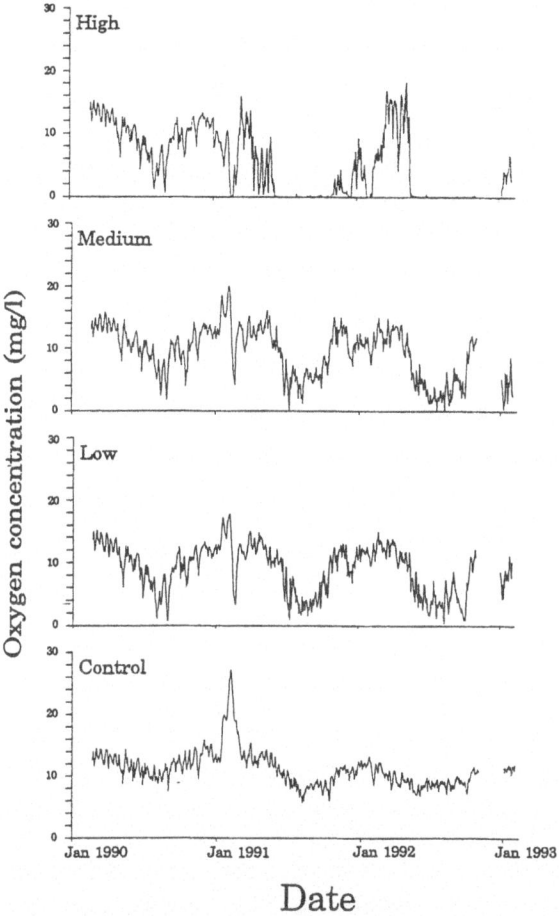

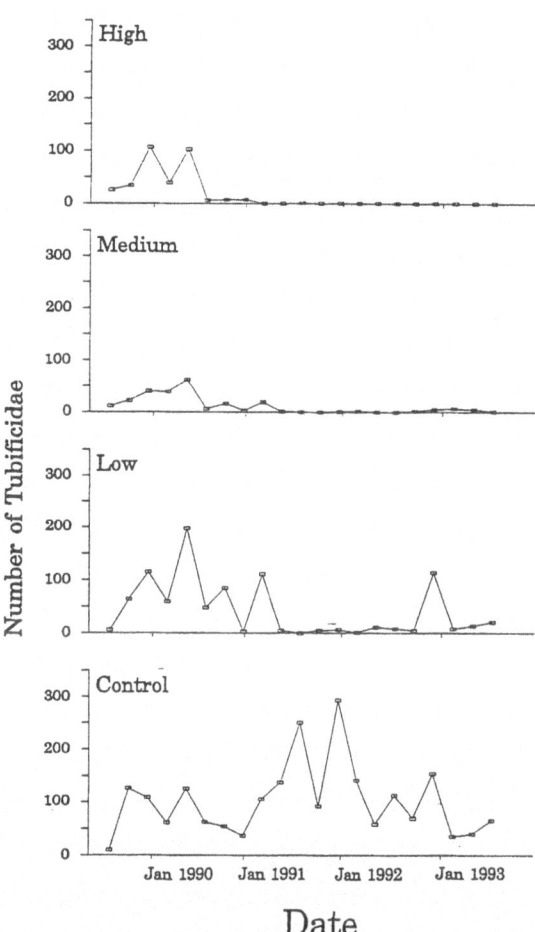

Fig. 3. Oxygen concentrations (mg l^{-1}) in clay-lined ditches in time based on continuous observations during 1990–1993.

Fig. 4. Numbers of Tubificidae in clay-lined ditches in time based on ten weeks observations during 1989–1993.

clay was extracted from an uncontaminated lake) and four had a sand bottom (the sand was taken from a deep, sterile sand layer). All ditches were filled with water in February 1988 and the experiment started on 23 May 1989. The original water source was deep groundwater with a total-P concentration of 0.052 mg l^{-1} and total–N of 0.304 mg l^{-1}. The sediment of the clay-lined ditches contained 0.48 mgN/g dry weight and 0.23 mgP/g dry weight and the sand-lined ditches 0.043 mgN/kg dry weight and 0.10 mgP/g dry weight. Through precipitation and groundwater supply in dry periods about 0.401 kgN y^{-1} and 0.004 kgP y^{-1} is added a year in each experimental ditch.

The study design (Verdonschot & Ter Braak, 1994) included a gradient of four treatment levels in both sand-lined and clay-lined ditches:

Ditch treatment	Nutrient addition
Control ditch	No nutrient additions
Low loaded ditch	0.049 kgP y^{-1}, supplied each half year
Medium loaded ditch	0.155 kgP y^{-1} and 0.559 kgN y^{-1}, supplied each half year
High loaded ditch	1.488 kgP y^{-1} and 8.928 kgN y^{-1}, supplied each month

The experimental design followed a simple and straight set-up of increasing levels of nutrient addition. No replicates were used.

172

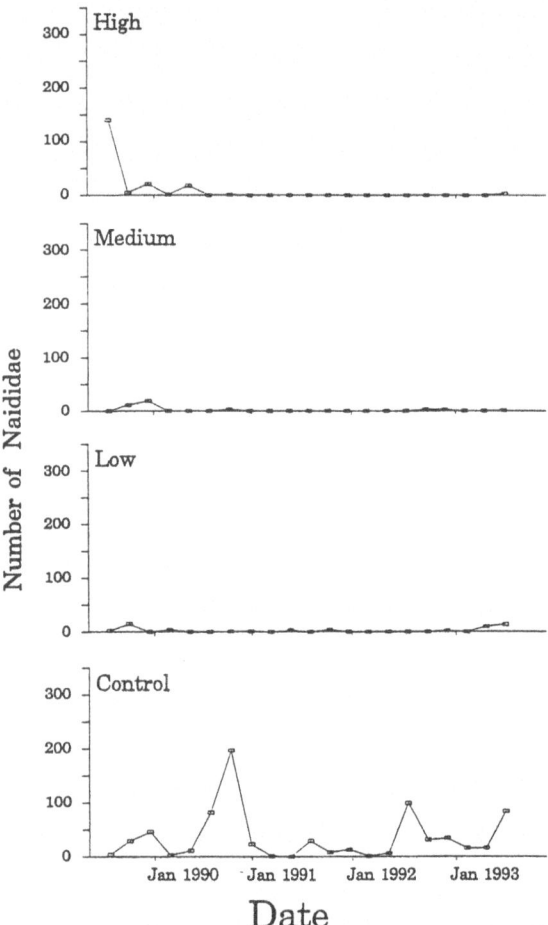

Fig. 5. Numbers of Naididae in clay-lined ditches in time based on ten weeks observations during 1989–1993.

Oligochaetes were collected by deploying trays with substratum and retrieving the trays after 20 weeks (Verdonschot & Ter Braak, 1994). Each tray consisted of an inner cylinder (height 0.06 m, surface area of 0.0085 m^2) closed at the bottom and placed in an outer cylinder which fitted just around the first. The outer cylinder was set into the ditch bottom. The inner cylinder was filled with the same material as was used for the ditch bed. During sampling only the inner cylinder was removed without disturbing the outer one nor the surrounding sediment. Before sampling, the tray was enclosed by a gauze cover (mesh size 300 μm) to collect the oligochaetes living upon and above the tray and to prevent individuals from escaping during sampling. After sampling the tray was newly filled and replaced. Thus, every second sampling period the same tray was sampled.

Sampling was undertaken every ten weeks starting in July 1989 up to July 1993. In total 18 samples were taken from each ditch. The samples were sieved and sorted by eye. The oligochaetes were preserved in formalin (4%), identified and counted.

Nitrogen and phosphate concentrations were analyzed monthly from the water surface; oxygen, temperature and pH were measured continuously at a depth of 10 cm below the water surface as well as above the water bottom; daily averages are represented. Regular samples of ammonium, chloride and sulphate were also used in the data analysis. Season, year of sampling and ditch were defined as nominal variables. Nutrient supply was defined as an ordinal variable. The number of individuals as well as the number of species per sample were also used as environmental variables.

Oligochaete abundances were transformed to a logarithmic scale and ordinated by redundancy analyses (RDA) with the program CANOCO 3.1 (Ter Braak, 1988, 1990). The data analyses are fully described by Verdonschot & Ter Braak (1994). RDA assumes a linear model for the relationship between the response of each taxon and the ordination axes and is used if the gradient length in the data is short (< 4 units of standard deviation [s.d.]; Ter Braak, 1988). In our case the gradient length was smaller then 3 s.d. which implies that the data are quite homogeneous. RDA is the constraint form of PCA of taxon data, in which the components (axes) are constrained by linear combinations of environmental variables. The ordination results are presented as correlation biplots of taxa, sites and environmental variables (Verdonschot & Ter Braak, 1994). The eigenvalue of an ordination axis in RDA is the proportion of the total variance explained by that axis and indicates its relative importance. An unrestricted permutation test is used to test the validity of the total ordination. This technique is fully explained by Ter Braak (1990) and Verdonschot & Ter Braak (1994).

Results

Clay-lined ditches

The substratum of the clay-lined ditches consists of clay taken from an uncontaminated lake, mixed and put in the ditches. By this a number of seeds, cocoons, and other biological material was introduced to the ditch. Before the experiment started the clay-lined ditches were inter-connected to permit the exchange of water and organisms. All four clay-lined ditches were equal

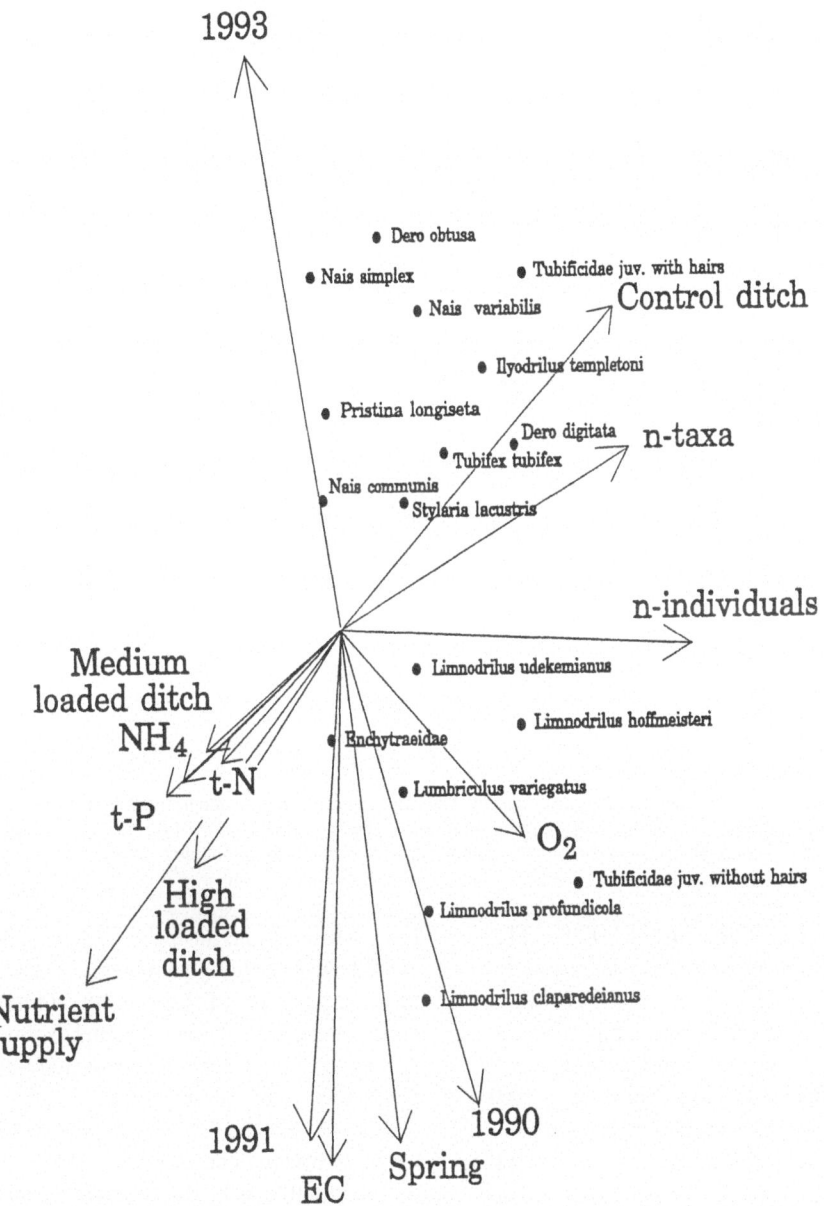

Fig. 6. Ordination (RDA) diagram of the axes 1 and 2 showing the variation in the transformed abundances of taxa (dots) among environmental variables (arrows) in clay-lined ditches.

at the beginning of the experiment. First the development of the variables phosphorus, nitrogen, oxygen as well as tubificids and naidids during the experiment will be described. Afterwards, the similarities and dissimilarities between samples indepent from time will be illustrated by ordination.

Phosphorus concentrations in the control clay-lined ditch hardly changed over the four years and fluctuated around 0.02 to 0.05 mgP l^{-1} (Fig. 1) with an annual average in 1993 of 0.05 mgP l^{-1} (s.d. 0.02). The series from control ditch to high loaded ditch showed an increase in phosphorus concentrations between ditch as well as in time. In 1993 the annual averages were 0.06 (s.d. 0.02), 0.17 (s.d. 0.07) and 3.69 (s.d. 0.99) mgP l^{-1}, respectively. Both low and medium loaded ditch showed peaks in phosphorus concentration

174

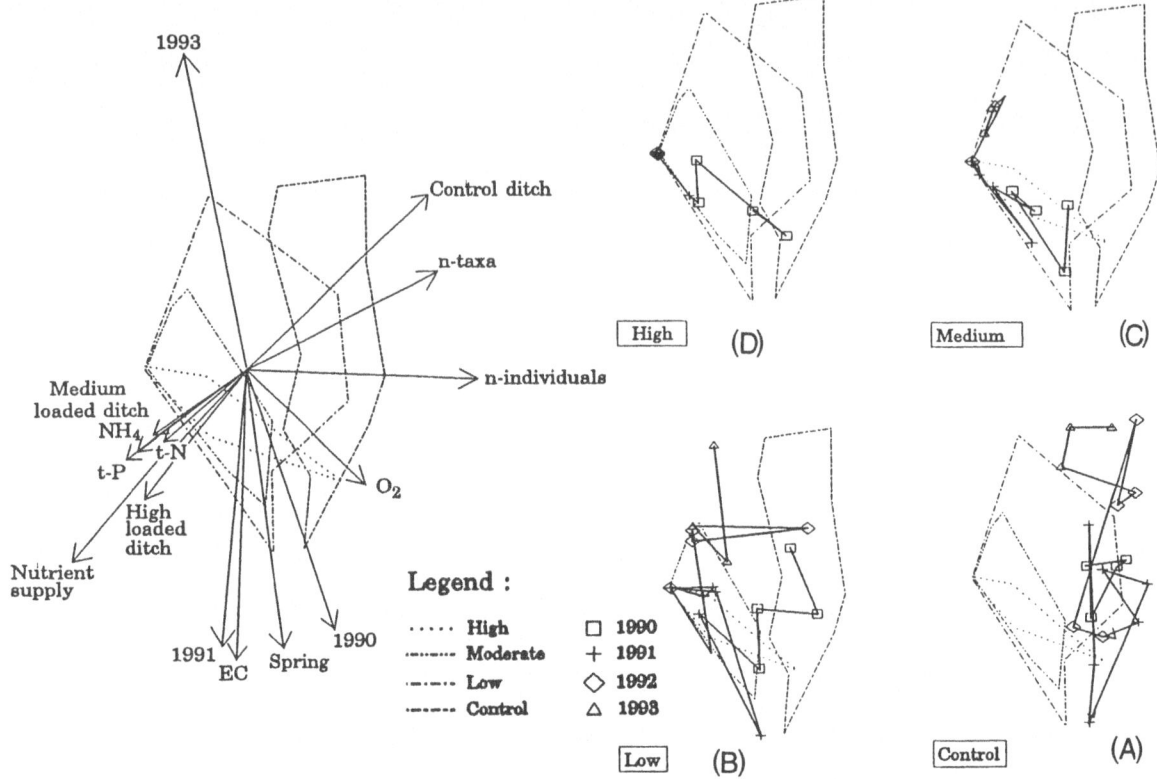

Fig. 7. Ordination (RDA) diagram of the axes 1 and 2 showing the variation in sampling sites (contour lines) among environmental variables (arrows) in clay-lined ditches. The contour lines indicate groups of samples per year per ditch. The bold line indicates the succession pattern in sites over the years for A. control ditch, B. low loaded ditch, C. medium loaded ditch, D. high loaded ditch. Years are indicated.

around April–May and October–November. The high loaded ditch showed a strong fluctuation in phosphorus concentration. This is mainly because of the addition of phosphorus in April and October in both, the low and medium loaded ditches, and the monthly nutrient addition in the high loaded one. Also season, with algae blooms and macrophyte development in summer taking up nutrients affected these patterns. In general, the clay-lined ditches followed linearly the intended levels of phosphorus in time.

Nitrogen concentrations showed a less clear pattern (Fig. 2). No nitrogen was added to the control and low loaded ditch. Nitrogen concentrations fluctuated between 0–8 mgN l^{-1} in the control ditch and 0–10 mgN l^{-1} in the low loaded one; the annual averages in 1993 were 2.23 (s.d. 0.94) and 2.23 (s.d. 0.49) mgN l^{-1}, respectively. The medium loaded ditch showed on average the same pattern as the former ones with an annual average in 1993 of 2.02 (s.d. 0.74) mgN l^{-1}. Only some peaks were higher because measurements coincided with times of nitrogen addition. The nitro-

gen concentrations in the high loaded ditch were on average higher, though a number of peaks again could be a consequence of the monthly addition. The lower concentrations in 1993 (an annual average of 5.10 [s.d. 4.00] mgN l^{-1}) indicate an incorporation of part of the supplied nitrogen in the system.

The oxygen concentrations of the control ditch fluctuated between 8 and 13 mg l^{-1} (Fig. 3). As expected, this fluctuation results from season, with lower concentrations in summer and higher ones in winter. This seasonal pattern is also visible in the loaded ditches. In the low loaded ditch oxygen concentration lowered to about 2–4 mg l^{-1} in the summer periods and went up to about 13 mg l^{-1} in winter. The peak in oxygen in the control ditch and the corresponding troughs in the loaded ditches relate to ice-formation. Algae under the ice produced oxygen in the control ditch. The higher organic load in the loaded ditches caused the decomposition process to dominate and lower the oxygen level. During the successive years the period with low concentrations repeated and extended each

Table 1. Yearly total numbers of oligochaetes in the clay-lined ditches.

| | Ditch | | | | | | | | | | | | | | | |
| | High loaded | | | | Medium loaded | | | | Low loaded | | | | Control | | | |
Period: –	I	II	III	IV	I	II	III	IV	I	II	III	IV	I	II	III	IV
Limnodrilus claparedeianus	53	–	–	–	18	–	–	–	6	1	–	–	2	5	–	–
Limnodrilus hoffmeisteri	41	6	–	–	11	9	1	–	37	6	–	10	21	69	31	31
Limnodrilus profundicola	–	–	–	–	–	1	–	–	–	2	–	–	–	2	4	–
Limnodrilus udekemianus	–	–	–	–	–	1	1	–	2	–	4	1	7	2	2	1
Tubifex tubifex	–	–	–	–	1	–	–	–	–	–	–	8	16	4	6	8
Ilyodrilus templetoni	4	–	–	–	13	–	–	1	36	–	–	31	36	5	2	43
Tubificidae juv. with hs	15	2	–	–	15	2	–	19	106	18	3	83	118	31	123	202
Tubificidae juv. without hs	194	12	1	–	117	32	1	–	256	225	16	37	233	279	666	191
Dero digitata	185	1	–	–	28	–	–	7	21	4	4	2	93	295	51	91
Dero obtusa	–	–	–	–	–	–	–	–	–	–	–	2	–	–	–	103
Stylaria lacustris	–	–	–	–	2	2	–	–	–	–	–	4	–	8	–	3
Nais simplex	–	–	–	–	–	1	–	–	–	–	–	17	–	–	–	10
Nais variabilis	–	–	–	–	–	–	–	–	–	1	–	–	–	–	6	73
Nais communis	–	–	–	1	–	–	–	–	–	–	–	–	–	–	–	–
Pristina longiseta	–	–	–	2	–	–	–	–	–	–	–	1	–	–	–	–
Enchytraeidae	–	1	–	1	–	2	–	–	–	–	–	–	–	–	–	–
Lumbriculus variegatus	–	–	–	–	–	–	–	–	–	–	–	–	–	2	5	–
Tubificidae total	307	20	1	–	175	45	3	20	443	252	23	170	433	397	834	476
Naididae total	185	1	–	3	30	3	–	7	21	5	4	26	93	303	57	280
Oligochaeta total	492	22	1	4	205	50	3	27	464	257	27	196	526	702	896	756

Legend: Period: I = July 1989 – July 1990; II = July 1990 – July 1991; III = July 1991 – July 1992; IV = July 1992 – July 1993; hs = hair setae.

year. This pattern is also visible in the medium loaded ditch. In 1990 the oxygen pattern of the high loaded ditch followed that of both low and medium loaded ones. In later years anoxic periods occurred and this anoxic period extended each successive year.

The main vegetation patterns were qualitatively noted in the field. They showed a clear response to nutrient loading. The control ditch was covered, for a longer period, with the submersed macroalgae Chara sp. The low and medium loaded ditches were covered with the macrophyte Elodea sp. and filamentous algae. In the first year the filamentous algae also occurred in the high loaded ditch but was afterwards replaced by the floating macrophyte Lemna sp.

Numbers and abundances of oligochaetes were highest in the control clay-lined ditch (Table 1). Tubificids were abundant throughout the research period (Fig. 4), though numbers fluctuated. Most numerous were Limnodrilus hoffmeisteri, Ilyodrilus templetoni and juvenile tubificids. During the first one and a half year, tubificids were also abundant in the low loaded ditch. Again numbers were dominated by Limnodrilus hoffmeisteri and Ilyodrilus templetoni. Numbers col-

lapsed in the following years but somewhat recovered temporarily in the last winter. The tubificids were present in the first year of the experiment in the medium and high loaded ditches but their numbers decreased to few or zero in the following years.

Naidids were abundant in the control ditch with peaks in summer to autumn (Fig. 5). Numbers were dominated by Dero digitata (Table 1). Especially in the last year more species were collected. Naidids were scarce in the low loaded ditch with somewhat higher number of Dero digitata at the beginning and of Nais simplex at the end of the experiment. After the first year with a reasonable abundance of Dero digitata, naidids were scarcely collected in both medium and high loaded ditch.

Multivariate analysis techniques (RDA-ordination) were used to study the similarities and dissimilarities between samples based on oligochaetes and environmental variables in the clay-lined ditches. The eigenvalues of the first four axes were 0.53, 0.08, 0.03 and 0.02, respectively. Most of the variation (54%) in the data is explained in the first axis, and 62% in the first two axes (Figs 6 and 7). All variables with a correlation

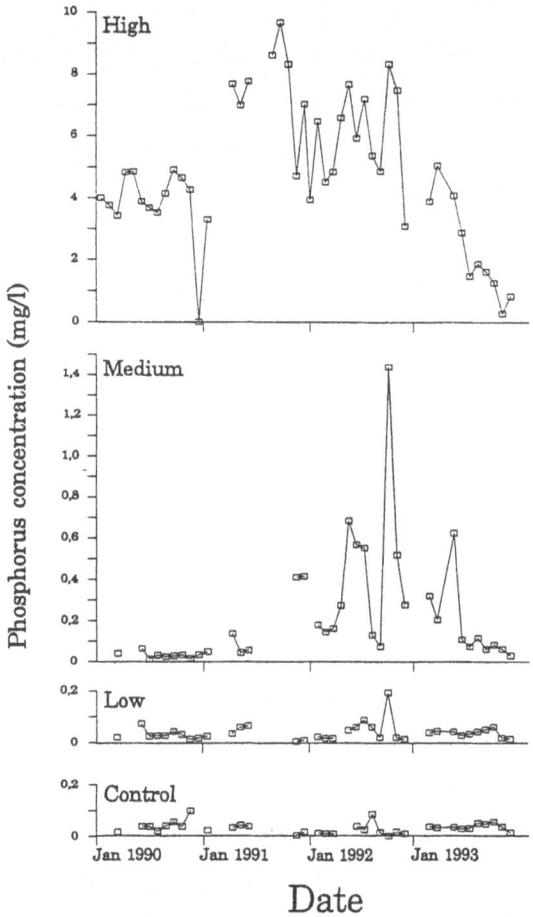

Fig. 8. Phosphorus concentrations (mgP l⁻¹) in sand-lined ditches in time based on monthly observations during 1990–1993.

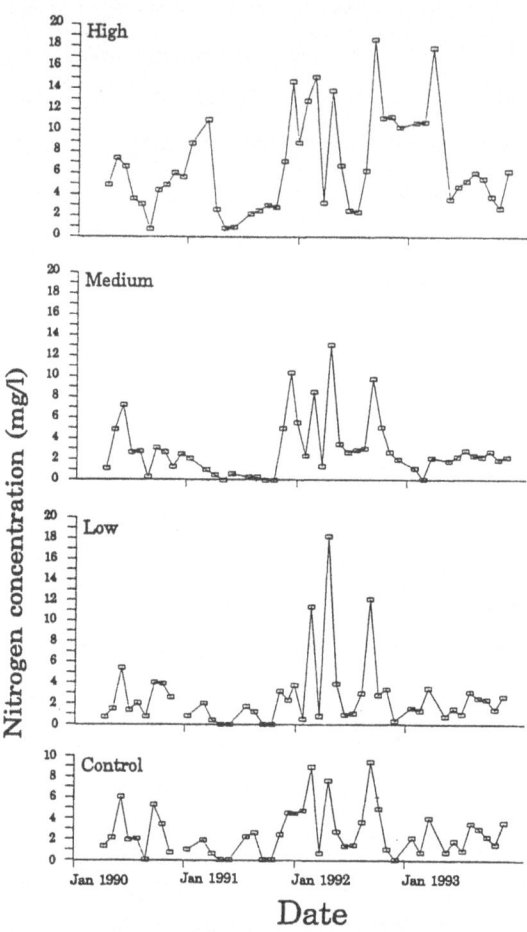

Fig. 9. Nitrogen concentrations (mgN l⁻¹) in sand lined ditches in time based on monthly observations during 1990–1993.

(Ter Braak, 1988) higher than 0.3 with one of the ordination axes are shown. A permutation test showed that the result of the ordination of the clay-lined ditches was significant ($p < 0.01$). Along the first axis (Figs 6 & 7) the control ditch with high numbers of individuals and taxa, and high oxygen concentrations are projected on the right versus the high and medium loaded ditches with high phosphorus and nitrogen concentrations on the left. The year 1990 is also pointing to the right. At the beginning of the experiment all ditches were equal, thus all samples of this period are projected close to those of the control ditch. The second axis, though of minor importance, shows a difference between the years 1990 and 1991 (together with spring samples and a lower conductivity) at the bottom and the year 1993 at the top.

A closer look at the samples (Fig. 7) indicates that this pattern occurs for all ditches except the high loaded one. In this ditch almost no oligochaetes were found anymore in the last two years of the experiment. Samples of the control ditch show a repeated pattern (Fig. 7A) of a one year cycle and are projected per year as a circle in the ordination diagram. In the years 1990 and 1991 below and from summer 1992 on until the end above axis one. This circle corresponds to seasonal differences. The change over the years is related to a late colonization of several taxa (Table 1). The first two samples of all three loaded ditches overlap the 'circles' of the control ditch samples below axis one. The conditions of all ditches at the beginning of the experiment were comparable. Thereafter, all three loaded ditches tended to move to the left whereby a circling or seasonal pattern remains more and less visible in

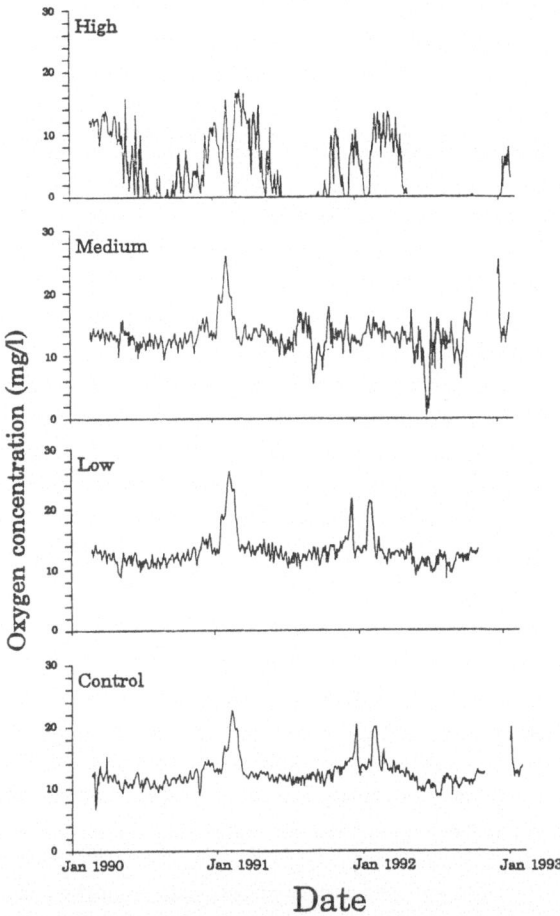

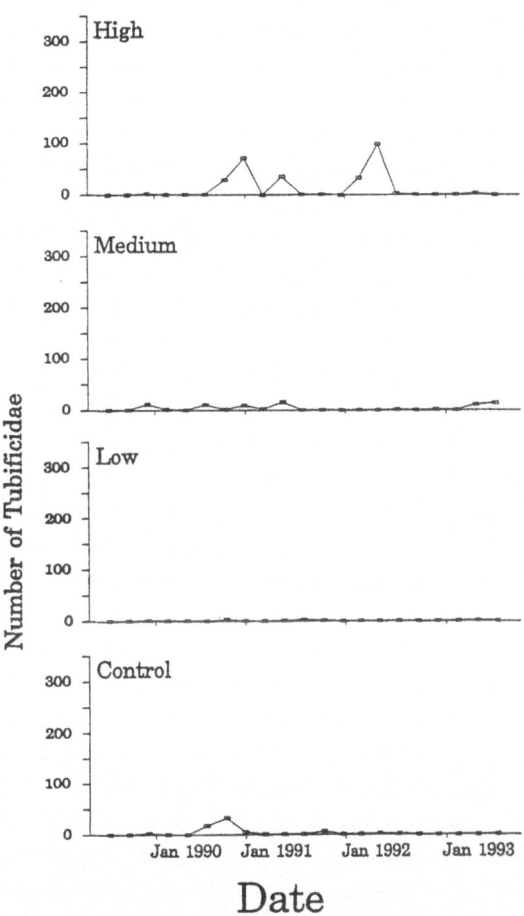

Fig. 10. Oxygen concentrations (mg l^{-1}) in sand-lined ditches in time based on continuous observations during 1990–1993.

Fig. 11. Numbers of Tubificidae in sand-lined ditches in time based on ten weeks observations during 1989–1993.

the low and medium loaded ditch, respectively. Also the change between the first and the latter two years is more and less clear in the low and medium loaded ditch, respectively. The high loaded ditch develops to one point where few or no oligochaetes were found.

The distribution of taxa in the ordination diagram (Fig. 6) is also related to the above described pattern. The more common taxa of the control ditch occur in the right of the diagram with *Limnodrilus hoffmeisteri* and the juvenile tubificids without hair seta below axis one. This position is partly because of the occurrence of both in all ditches at the beginning of the experiment. *Tubifex tubifex, Ilyodrilus templetoni, Dero digitata* and the juvenile tubificids with hair seta characterize the control ditch. On the left in the diagram, the family of Enchytraeidae, together with the 1993-occurring species of *Pristina longiseta* and *Nais communis*, char-

acterize the high loaded ditch. This ditch is, as already stated, further mainly characterized by an absence of oligochaetes. At the top of the diagram *Nais simplex, Dero obtusa* and *Nais variabilis* can be related to the year 1993. At the bottom *Lumbriculus variegatus, Limnodrilus profundicola* and *L. claparedeianus* represent the earlier years. With their more scattered distribution *Stylaria lacustris* and *Limnodrilus udekemianus* take an intermediate position.

Sand-lined ditches

The substratum of the sand-lined ditches was taken from a 40 meters deep, sterile sand layer. At the beginning of the experiment the sand-lined ditches did not contain any live biological material.

The phosphorus concentrations of the control and low loaded sand-lined ditch were constantly low, about

178

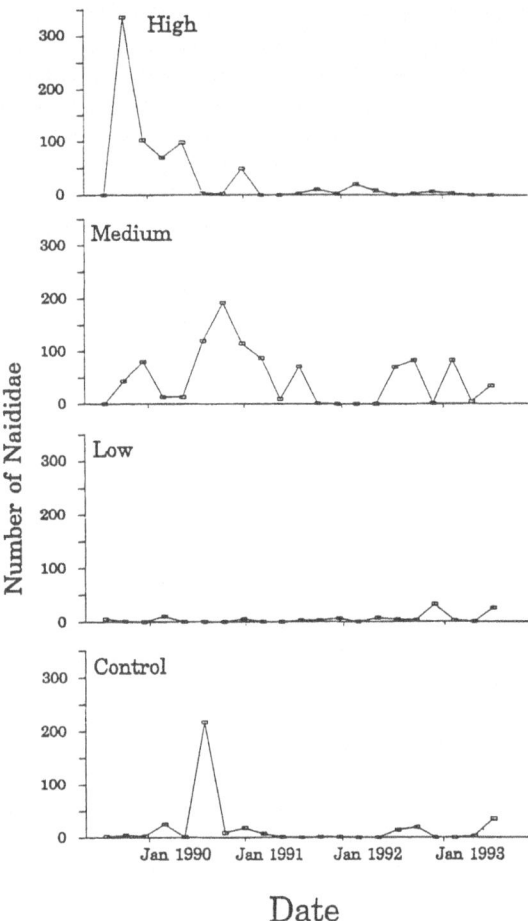

Number of Naididae

Fig. 12. Numbers of Naididae in sand-lined ditches in time based on ten weeks observations during 1989–1993.

0.02 mgP l^{-1} on average, despite the addition of phosphorus in the latter (Fig. 8). The annual averages in 1993 were 0.04 (s.d. 0.01) mgP l^{-1}, in both. The response of the medium loaded ditch on phosphorus addition was more clear in 1992 and the total concentration was higher in 1993, on annual average 0.17 (s.d. 0.17) mgP l^{-1}. Phosphorus concentrations in the high loaded ditch raised quickly and fluctuated strongly, partly because of monthly additions. The lowering in 1993 (the annual average of 2.32 (s.d. 1.49) mgP l^{-1}) could be a response to plant development.

No nitrogen was added to the control and low loaded ditches. Nitrogen concentrations fluctuated between 0–6 mgN l^{-1} in both the control and low loaded ditch (Fig. 9). The annual averages in 1993 were 2.18 (s.d. 1.13) and 1.95 (s.d. 0.86) mgN l^{-1}, respectively. In 1992 peaks occurred up to 8 mgN l^{-1} in the control and 12–18 mgN l^{-1} in the low loaded ditch. The medi-

um loaded ditch shows on average the same pattern with an annual average in 1993 of 2.00 (s.d. 0.75) mgN l^{-1}. Only some peaks, especially those of 1992, were higher compared to the low loaded one. This is partly because of times of nitrogen addition. Nitrogen in the high loaded ditch was on average high and fluctuated strongly. The annual average in 1993 was 7.02 (s.d. 4.23) mgN l^{-1}.

Oxygen concentrations in the control and low loaded sand-lined ditch fluctuated around 10–14 mg l^{-1} (Fig. 10). There is a minor seasonal pattern visible. The peaks up to more than 20 mg l^{-1} each January–February are remarkable. These peaks are probably related to an algae bloom on the ditch bottom. Oxygen strongly fluctuated in the high loaded ditch and over the four years the period with low or no oxygen extends successively. Oxygen levels decreased quickly in the high loaded ditch after the beginning of the experiment. The periods of anoxic conditions extended through the following years.

Except for a thin algae layer at the end of the experiment, the bottom of the control ditch was bare. The low loaded ditch also lacked vegetation except for a thin algae layer covering the bottom. The medium loaded ditch also lacked macrophytes, algae blooms occurred in the latter years of the experiment. The high loaded ditch was characterized by a floating layer of the macrophyte *Lemna* sp.

Tubificids were scarcely present in the control, low and medium loaded ditches (Table 2, Fig. 11). Mainly juveniles and an incidently occurring mature specimen of especially *Limnodrilus hoffmeisteri* occurred. In the high loaded ditch some tubificid peaks occurred, again mainly juveniles and specimens of *Limnodrilus hoffmeisteri* and *L. claparedeianus*.

Several naidid species were collected in the control ditch during the study (Table 2). The abundances were mostly low, except for a peak in 1990 for *Nais variabilis* and in 1993 for *Pristina longiseta* (Fig. 12). Several naidids were present in the low loaded ditch though all in low abundances. The medium loaded ditch was abundantly inhabited by naidids. Most dominant was *Dero digitata* with incidental peaks in abundance of *Nais simplex*, *N. variabilis* and *Slavina appendiculata*. This is possibly related to the occurring algae blooms in this ditch. After a high abundance peak of *Uncinais uncinata* in the high loaded ditch the naidids there tended to decrease. Interesting is the colonization and population development in numbers of *Lumbriculus variegatus*, it inhabits the floating duckweed layer.

Table 2. Yearly total numbers of oligochaetes in the sand-lined ditches.

| | Ditch | | | | | | | | | | | | | | | |
| | Control | | | | Low loaded | | | | Medium loaded | | | | High loaded | | | |
Period	I	II	III	IV	I	II	III	IV	I	II	III	IV	I	II	III	IV
Limnodrilus claparedeianus	–	–	–	–	–	–	–	–	–	–	–	–	–	3	40	–
Limnodrilus hoffmeisteri	–	8	1	–	–	2	–	1	6	9	–	–	–	34	55	1
Limnodrilus udekemianus	–	–	–	–	–	–	–	–	–	1	–	–	–	–	4	–
Tubifex tubifex	1	–	–	–	–	–	–	–	–	–	–	–	–	–	1	–
Aulodrilus pluriseta	–	–	–	–	–	–	–	–	–	–	–	–	–	2	–	–
Tubificidae juv. with hs	–	2	–	3	1	1	1	–	–	2	–	24	–	1	–	–
Tubificidae juv. without hs	2	48	11	1	1	1	2	–	6	25	–	1	3	96	33	8
Dero digitata	–	–	1	1	–	–	–	–	43	252	–	262	3	14	39	5
Dero obtusa	–	–	–	–	–	–	–	–	–	–	–	13	–	–	–	–
Stylaria lacustris	1	1	–	–	–	–	–	–	–	–	–	–	28	4	–	–
Uncinais uncinata	6	5	–	–	14	–	–	–	3	–	–	–	568	–	–	–
Nais simplex	1	6	–	2	–	5	9	36	103	6	1	–	1	–	–	–
Nais variabilis	27	238	1	6	2	–	2	–	–	160	–	–	7	36	2	6
Nais communis	–	2	–	–	–	–	1	–	–	1	–	–	–	–	–	–
Slavina appendiculata	–	–	–	–	–	–	1	4	–	104	71	–	–	–	1	–
Pristina longiseta	–	–	–	57	–	–	6	26	–	–	–	–	–	–	–	–
Pristina foreli	–	–	–	5	–	–	–	–	–	–	–	–	–	–	–	–
Lumbriculus variegatus	–	–	–	–	–	–	–	1	–	–	–	–	–	6	29	30
Tubificidae total	3	58	12	4	2	4	3	1	12	37	–	25	3	136	133	9
Naididae total	35	252	2	71	16	5	19	66	149	523	72	275	607	54	42	11
Oligochaeta total	38	310	14	75	18	9	22	68	161	560	72	300	610	196	204	50

Legend: Period: I = July 1989 – July 1990, II = July 1990 – July 1991, III = July 1991 – July 1992, IV = July 1992 – July 1993, hs = hair setae

RDA-ordination is also used to further analyse the results of the experiment in the sand-lined ditches. The eigenvalues of the first four axes were 0.25, 0.11, 0.08 and 0.4, respectively. The first two axes together explained 36% of the variance (Fig. 13), with the third axis cumulatively 44% (Fig. 14). Again the permutation test showed that the results were significant ($p < 0.01$). The first two axes were related to the four ditches, whereby the low loaded ditch was poorest in number of individuals and taxa. The control and high loaded ditch both point to the bottom, and the medium loaded one to the top. As the third axis is almost as important as the second, it is also shown together with the first one. Here, the eutrophication gradient between the high loaded ditch together with phosphorus and nitrogen load is projected versus the control ditch together with high oxygen concentrations and a higher pH.

In contrast with the clay-lined ditches, the sand-lined ditches showed no seasonal pattern in the ordination of samples. The samples were split in four groups, the four ditches, but within each group showed a random distribution.

The poor development of numbers of oligochaetes in the control and low loaded sand-lined ditches is illustrated in the diagrams (Figs 13 & 14). Except for both species of the genus Pristina, no taxa were projected close to the ditch variables. Taxa explicitly related to the high loaded ditch are Limnodrilus claparedeianus, L. udekemianus, L. hoffmeisteri, Lumbriculus variegatus, Tubifex tubifex, Aulodrilus pluriseta and the juvenile tubificids without hair seta. In the sand-lined ditches tubificids dominated the high loaded ditch. In the medium loaded ditch Dero obtusa, D. digitata, Slavina appendiculata and juvenile tubificids with hair seta were more characteristic. Here the naidids dominated. All other taxa (all naidids) took a more intermediate position.

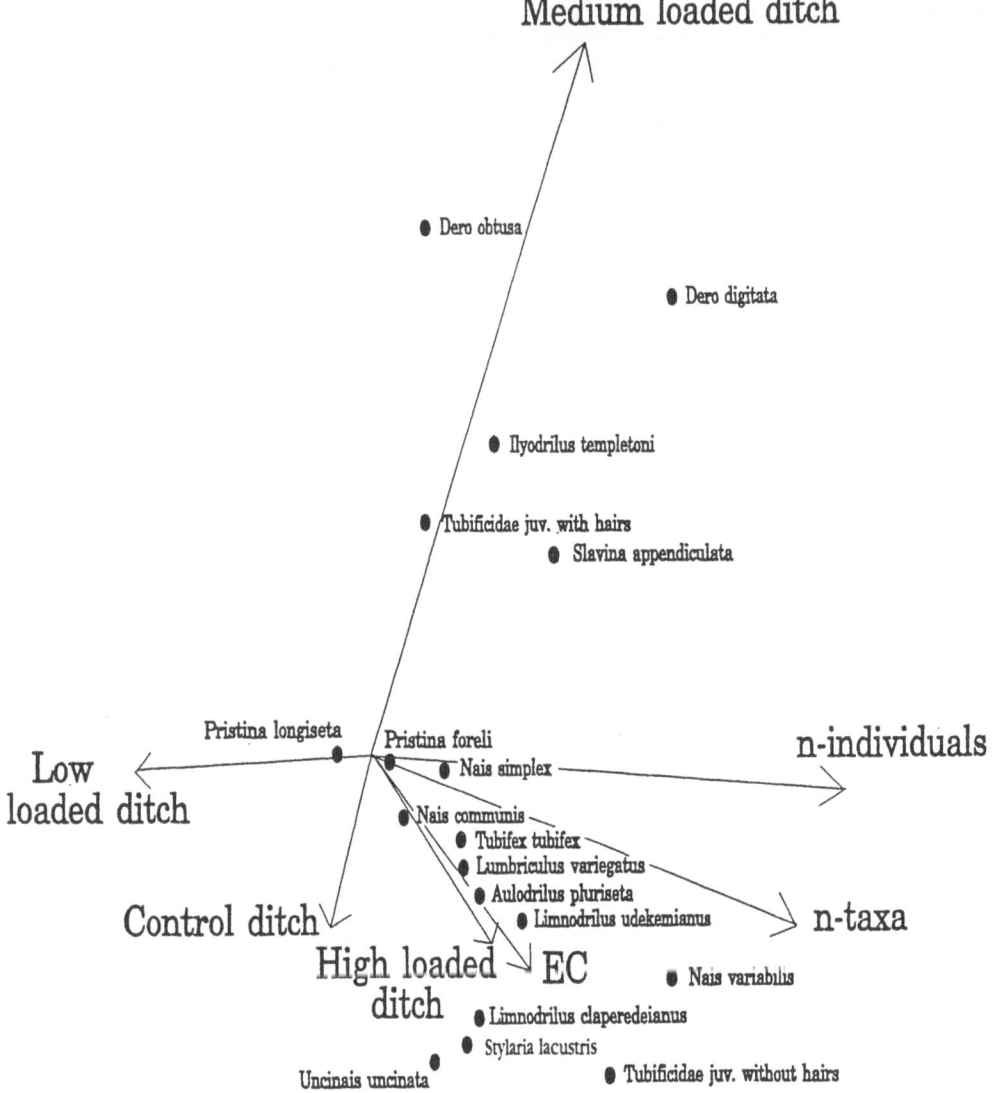

Fig. 13. Ordination (RDA) diagram of the axes 1 and 2 showing the variation in the transformed abundances of taxa (dots) among environmental variables (arrows) in sand-lined ditches.

Discussion

Most of the collected oligochaetes have quite wide amplitudes on the eutrophy gradient (Särkkä, 1987; Verdonschot, 1989). Still, differences in tolerance to the effect of nutrient loading were visible in this experiment. Most striking was the disappearance of species through nutrient loading in the clay-lined ditches. In comparing the four clay-lined ditches the tubificids *Limnodrilus profundicola*, *L. udekemianus* and *Tubifex tubifex* appeared somewhat more sensitive in comparison to the very tolerant *Limnodrilus hoffmeis-*

teri and *Ilyodrilus templetoni*. The high tolerance of *Limnodrilus hoffmeisteri* confirms earlier observations (e.g. Chapman *et al.*, 1982). Concerning *Tubifex tubifex*, records in literature of tolerance limits differ a lot. This is probably because of the presence of different ecological varieties. *Limnodrilus claparedeianus* colonized the ditches but disappeared after the first two years. The distribution and abundance of oligochaetes in clay-lined ditches was strongly related to the oxygen regime. Incidental low oxygen concentrations did not affect the oligochaete assemblages. But the longer the period of oxygen depletion, the

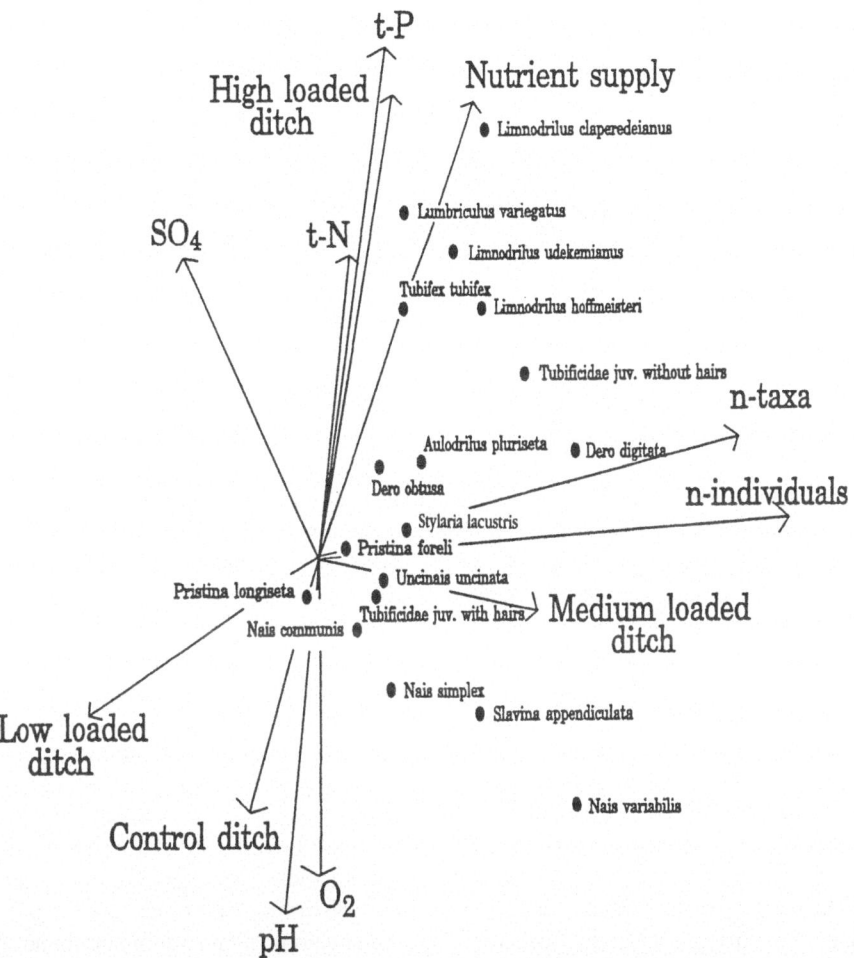

Fig. 14. Ordination (RDA) diagram of the axes 1 and 3 showing the variation in the transformed abundances of taxa (dots) among environmental variables (arrows) in sand-lined ditches.

stronger the oligochaetes were affected. In general, tubificids were more abundant and always present in the control clay-lined ditch. The higher the nutrient load, the longer the period of oxygen depletion and the poorer the tubificid population was developed. The reduced numbers of tubificids were directly related to the increased levels of nutrient loading. This is in contrast with the common opinion that oligochaetes tend to increase when waters become eutrophied (Milbrink, 1973; Wiederholm, 1980). Naidids were scarce in the clay-lined ditches except for the bottom inhabiting *Dero digitata*. This tube building species seemed to behave like a tubificid with respect to nutrient loading. It was frequently collected and its sensitivity to the increase of nutrient load is comparable with that of *Limnodrilus hoffmeisteri*. Only in the control clay-lined ditch, with a well developed vegetation and a

suitable oxygen regime, other naidid species established. At the end of the experiment, this control ditch probably reached a more mature stage in succession which is reflected in an increase in number of naidid species (Verdonschot & Ter Braak, 1994). In the high loaded clay-lined ditch, *Nais communis* and *Pristina longiseta* occurred in very low numbers in the last year. This is probably because of their trophic preference but is just coincidental colonization.

Sand is very poor in minerals compared to clay. From the beginning of the experiment, this mineral poor state was reflected in the minimal amount of vegetation development and low animal colonization. Sand-lined ditches were scarcely inhabited by tubificids because a suitable bottom substratum (because of the low nutrient status of the sandy bed) and thus food was lacking. The effect of nutrient loading was

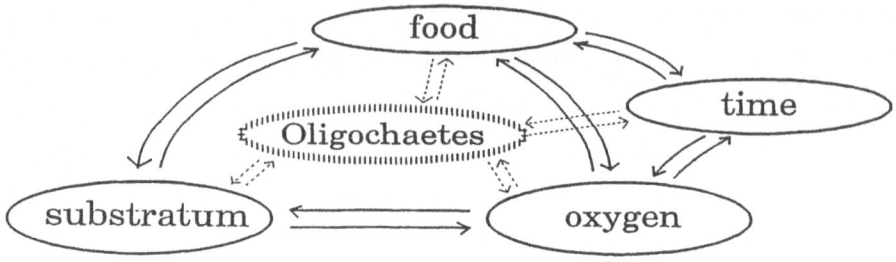

Fig. 15. The relations between distribution and abundance of oligochaetes and major key-factors in fresh, permanent, neutral, stagnant ditches.

only shown in the increase of tubificids in the high loaded ditch. In this ditch enough organic matter was produced to sustain a tubificid population. Apparently the ditches were not colonized by oligotrophic species. Probably their colonization speed is low. Naidids more frequently occurred in the sand-lined ditches, especially in the medium loaded one, though their numbers as well as the species composition varied strongly. *Uncinais uncinata* and *Stylaria lacustris* colonized the sand-lined ditches but disappeared again within the first two years. Late colonizers were *Dero obtusa, Slavina appendiculata, Pristina longiseta* and *P. foreli.* Still, the occurrence of naidids in the sand-lined ditches was not obvious. Striking was the colonization and development in numbers of *Lumbriculus variegatus* in the high loaded ditch. In general, each ditch showed an individual characteristic succession. As replicates were not studied, it is not known if this individual characteristic succession was real or coincidental. The low nutrient status of the sandy bed and of the nutrient poor groundwater which filled the ditches at the beginning of the experiment, most probably caused a stochastical pattern of colonization as was already concluded by Verdonschot & Ter Braak (1994).

At a macro-geographic scale, the distribution of oligochaetes is controlled by factors like temperature, chlorinity, current velocity, duration of drought, acidity and bottom type (Brinkhurst & Jamieson, 1971; Wachs, 1967). Within permanent, neutral, fresh, stagnant ditches other factors determine presence and abundance of oligochaetes (Fig. 15). The difference between sand and clay as initial substratum is of great importance on the effect of eutrophication. The initial substratum is directly related to the initial eutrophy level and thus to the production of organic material. The latter is an important food source for the oligochaetes. The poorer sand with a low production rate offers little food and is thus scarcely inhabited by oligochaetes. While in the more productive clay-lined

ditches oligochaetes can flourish. Directly related to the amount of organic material is the oxygen concentration (Fig. 15). Oxygen decides on the distribution and abundance of oligochaetes, as their ability to withstand complete oxygen depletion for a longer time is different (Chapman *et al.,* 1982). The composition of the substratum and thus food together with oxygen regime decide whether oligochaetes become more or less abundant. Time is the fourth important factor (Fig. 15). The presence or absence of oligochaete species depends on time. Species need time to colonize waters and assemblages need time to establish. Furthermore, ecological succession is an important factor to take into account in mesocosm experiments. Dependent on the conditions in the beginning of the experiment, a colonization or stabilization time is necessary.

The ordination proved to be a useful tool to detect the developmental patterns in ditches and stresses the importance of eutrophication and succession time in this experiment.

Acknowledgements

The author thanks Joke Schot, Tina Webber and Tjeerd-Harmen van den Hoek for assistance in the field and the laboratory. Tjeerd-Harmen drew all the figures. Further thanks go to several colleagues of the Winand Staring Centre for Integrated Land, Soil and Water Research, the Agricultural University Wageningen and the Institute for Forestry and Nature Research for their contribution, especially for providing the chemical data.

References

Brinkhurst, R.O. & B.G.M. Jamieson, 1971. Aquatic Oligochaeta of the World. Oliver & Boyd, Edinburgh, 860 pp.
Brinkhurst, R.O. & D.G. Cook, 1974. Aquatic Earthworms (Annelida: Oligochaeta). In C.W. Hart, Jr. & S.L.H. Fuller. Pollution

Ecology of Freshwater Invertebrates. Academic Press, New York 143–156.

Brinkhurst, R.O. & M.J. Austin, 1979. Assimilation by aquatic Oligochaeta. Int. Rev. ges. Hydrobiol. 64: 245–250.

Chapman, P.M., Farrell M.A. & Brinkhurst R.O., 1982. Relative tolerances of selected aquatic oligochaetes to individual pollutants and environmental factors. Aquatic Toxicol. 2: 47–67.

Chua, K.E. & R.O. Brinkhurst, 1973. Evidence of interspecific interactions in the respiration of tubificid oligochaetes. J. Fish. Res. Board Can. 30: 617–622.

Drent, J. & K. Kersting, 1992. Experimental ditches for ecotoxicological experiments and eutrophication research under natural conditions. Wat. Res. 27: 1497–1500.

Lafont, M., 1977. Les oligochètes et la détection des pollutions dans les cours d'eau. Eaux Ind. 17: 84–85.

Milbrink, G., 1973. On the use of indicator communities of Tubificidae and some Lumbriculidae in the assessment of water pollution in Swedish lakes. ZOON 1: 125–139.

Palmer, M.F., 1968. Aspects of the respiratory physiology of *Tubifex tubifex* in relation to its ecology. J. zool., Lond. 154: 463.

Pasteris, A., C. Bonacina & G. Bonomi, 1994. Observations on cohorts of *Tubifex tubifex* cultured at different food levels, using cellulose substrate. Hydrobiologia 278: 315–320.

Särkkä, J., 1987. The occurrence of oligochaetes in lake chains receiving pulp mill waste and their relation to eutrophication on the trophic scale. Hydrobiologia 155: 259–266.

Ter Braak, C.J.F., 1988. CANOCO – A FORTRAN program for canonical community ordination by [partial] [detrended] [canonical] correspondence analysis, principal component analysis and redundancy analysis (version 2.1). Report LWA-88–02. Agricultural Mathematics Group, Wageningen.

Ter Braak, C.J.F., 1990. Update notes: Canoco version 3.10. Agricultural Mathematics Group, Wageningen, The Netherlands.

Verdonschot, P.F.M., 1989. The role of oligochaetes in the management of waters. Hydrobiologia 180: 213–227.

Verdonschot, P.F.M. & C.J.F. Ter Braak, 1994. An experimental manipulation of oligochaete communities in mesocosms treated with chlorpyrifos or nutrient additions: multivariate analyses with Monte Carlo permutation tests. Hydrobiologia 278: 251–266.

Wachs, B., 1967. Die Oligochaeten Fauna der Fliessgewässer unter besonderer Berücksichtigung der Beziehungen zwischen der Tubificiden Besiedlung und dem Substrat. Arch. Hydrobiol. 63, 310.

Wiederholm, T., 1980. Use of benthos in lake monitoring. J. Wat. Pollut. Cont. Fed. 62, 537 –547.

Hydrobiologia **334**: 185–191, 1996.
K. A. Coates, Trefor B. Reynoldson & Thomas B. Reynoldson (eds), Aquatic Oligochaete Biology VI.

Oligochaete communities at the mouth of the Neva and their relationship to anthropogenic impact

N. P. Finogenova

Zoological Institute, Russian Academy of Sciences St. Petersburg, 199034 Russia

Key words: anthropogenic impact, changes in oligochaete community, Neva Mouth, Russia

Abstract

Changes in the oligochaete community of the Neva Mouth for the period 1982–1993 are discussed in relation to anthropogenic impacts. A number of community parameters such as species composition, biomass and diversity were generally stable, although local changes were noted. Predominance of eutrophic species, i.e. *L. hoffmeisteri, T. tubifex, P. hammoniensis* was enhanced, whereas mesotrophic and oligotrophic species – *L. isoporus, S. heringianus, T. newaensis* disappeared or became less numerous at some stations. The reverse trend was observed at other stations. These phenomena were accompanied by a sharp increase in the variability of biomass and/or diversity, which is regarded as exceeding normal annual fluctuations. The oligochaete community is recognized as a promising indicator of the ecological situation in the Neva Mouth.

Introduction

The Neva Mouth is the eastern part of the Gulf of Finland in the Baltic Sea. It is bounded by the Neva River delta to the east and to the west it is connected to the Gulf of Finland by northern and southern straits off Kotlin Island. Situated on the Neva Mouth is St. Petersburg with its suburbs and health resorts, numerous human settlements, industrial and agricultural enterprises. The Neva Mouth has been strongly effected by anthropogenic impacts which continue to increase. More than 735 000 tons of organic carbon (including 10% of labile organic matter), 79 000 tons of nitrogen (phosphorus and other chemical components are 1–2 orders of magnitude lower), 3500–4400 tons of synthetic compounds, heavy metals and other toxic substances enter the Neva Mouth with effluents annually (Shiklomanov *et al.*, 1989). In the 1980s construction of a dam for protection of St. Petersburg from flooding was started in the western part of the estuary, off Kotlin Island. The project involves the construction of a series of dams with a total length of 22.2 km with 2 locks for ships and 6 outflows. Construction of the dam in the area of the northern strait was completed by 1986. Dams are currently being built in the area of the south-

ern strait. Regular observations on the state of the Neva Mouth were started during the period of construction of the dam. Detailed hydrobiological studies were conducted in 1982–1984 (Winberg & Gutelmacher, 1987) and monitoring of some parts of the estuary was continued. In this paper I have examined changes in the oligochaete community of the Neva Mouth, which is the most abundant group in the benthos.

Area, material and methods

The Neva Mouth covers an area of 329 km^2, is 21 km long and has a maximum width of 15 km and together with the eastern part of the Gulf of Finland forms the estuary of River Neva. The average annual flow of the Neva River is equal to 79 km^2. Water in the Neva Mouth is renewed every 2.5–5.5 days. The transition zone of the Neva River waters has a current velocity of 8–30 cm s^{-1} in the central part of the bay. Two thirds of the Neva River outflow is directed through the northern strait. Water in the Neva Mouth is fresh. Sometimes a weak flow of brackish water from the Gulf of Finland occurs. The commonest depths are 3–5 m; in some areas depths of 7–8 m occur. The bottom of the bay consists mostly of fine muddy sand, and with a low mud content in the eastern area of the Neva Mouth; in the central and western parts the bottom changes to a muddy sand and mud while in the south-eastern corner blue clay with mud occurs. About 60 −70% of the sediments in the bay contain less than 5% organic matter. Oxygen content of the water in May-September is 7.3–10.5 mg l^{-1} (75–113% of saturation). The waters of the Neva Mouth are characterized by low concentrations of polluting substances even during periods of high loading as a result of dilution with Neva River water. Because of the high loads of allochthonous organic matter the trophic structure of the bay's ecosystem is altered. The contribution of primary production to the first trophic level here constitutes less than 10% of the total. Allochthonous organic matter enters into the biotic cycling mostly through bacterial action, partly through filtrators and sedimentors (Finogenova & Lobasheva, 1987; Shishkin, 1991).

In this paper material from the collections of 1982–1993 is used. For the characterization of oligochaete communities I chose 10 stations in different parts of the Neva Mouth for which data are available for several years (Fig. 1):

St. 1 – clayey mud, depth 5.5–7 m, collections of 1982–1984, 1992;

St. 2 – slightly muddy sand, depth 3.5–5 m, collections of 1982–84, 1990–93;

St. 3 – muddy sand, depth 2.5–3.5 m, collections of 1982–83, 1992–93;

St. 4 – grey mud, depth 3.5–5 m, collections of 1982–84, 1992–93;

St. 5 – grey mud, depth 4–5 m, collections of 1982–84, 1992–93;

St. 6 – mud, depth 5 m, collections of 1992–93;

St. 7 – grey mud with sand, depth 3.5–5 m, collections of 1982–84, 1992–93;

St. 8 – grey mud, clay, depth 4–5.5 m, collections 1982–84, 1992–93;

St. 9 – mud, clay with sand and gravel, depth 5.5–6 m, collections of 1982–84, 1992–93;

St. 10 – grey-black mud, sand, pebble, depth 4–5 m, collections of 1982–84, 1992–93.

Stations were sampled for four months over the open water period (July–October) using a Petersen grab 0.025 m^{-2}, with 2–5 replicates per station. Samples were preserved with 4% formaldehyde. Benthic animals were separated from the sediment with a 200 μm mesh sieve. All the sample or a sub-sample (1/2–1/64) was used for calculating the absolute density, biomass (fresh weight) and species composition of animals.

For characterization of the oligochaete community the Shannon-Weaver diversity index (d) as modified by Wihlm & Dorris (1968) was used.

Results and discussion

As reported previously (Finogenova *et al.*, 1987) the current structure of the benthic fauna of the Neva Mouth was established in the 1920s when numbers of the relict crustaceans *Pallasea quadrispinosa* Sars, and *Pontoporeia affinis* Lindstr. declined abruptly and the abundance of oligochaetes increased. The crustaceans had completely disappeared by the 1990s. The macrobenthos of the Neva Mouth as a whole was still rich and diverse at the beginning of the 1980s. Along with oligochaetes small bivalve molluscs of the family Sphaeriidae were predominant in the benthos, primarily on the muddy sand in the eastern and central part of the transition zone.

The oligochaetes occurring at stations 1–10 are shown in Table 1. The anthropogenic addition of large amounts of allochthonous organic matter has favoured the development and distribution of the eutrophic species *Limnodrilus hoffmeisteri, L. udekemianus, Tubifex tubifex, Potamothrix hammoniensis, Tubifex ignotus, Aulodrilus japonicus* in the Neva Mouth. Large populations of mesotrophic and oligotrophic species also occurred, such as *Lamprodrilus isoporus, Spirosperma ferox, Tubifex newaensis*. Many species responded positively to the increase of organic matter, as shown by the greater abundance at the eastern end of the Mouth, where most of organic matter is discharged (Table 2). However, *Stylodrilus heringianus*, apparently a strictly oligotrophic species, is suppressed in the Neva Mouth. Its abundance is very low (4–1600 m^{-2}) comprising only 0.05–4.3% of the total abundance of oligochaetes. In the 1990s the distributional range of *T. tubifex* declined. At the present time this species is concentrated primarily in the south-eastern corner of the Neva Mouth, although in the 1980s it was wide spread over the entire Neva Mouth; and in the area of Station 2 and the outlet zone of the Neva River was among the most abundant species. *Potamothrix hammoniensis* has become more common and abundant in the eastern part of the Neva Mouth in recent years. Previously it was concentrated in the central and western areas. Despite quantitative and qualitative annual variation the oligochaete community, as a whole, was relatively stable in the Neva Mouth over the study period. The level of biomass was maintained at many sta-

Table 1. Mean relative density (%) of worm species at stations 1–10 through full period of observations: + up to 1%, 1 < 1–2, 2 < 2–5, 3 < 5–10, 4 < 10–20, 5 < 20–30, 6 < 30–40, 7 < 40–50, 8 < 50–70, 9 > 70.

Taxa Stations	1	2	3	4	5	6	7	8	9	10
Stylaria lacustris (L.)	+	+		1	+		2	+	+	1
Arcteonais lomondi (Martin)			+	+	+		+		1	+
Ripistes parasita (Schmidt)			1	2	+					
Vejdovskyella intermedia (Bret.)	+	+	2	+	2	+	+	+	3	+
Slavina appendiculata (Udekem)				+	+					
Dero sp.	+									+
Nais barbata Mull.		+								
Nais simplex Piguet		+								
Nais communis Piguet		+							+	
Nais elinguis Mull.		3	+							
Nais pardalis Piguet		+	+							
Specaria josinae (Vejd.)					+		+		+	
Piguetiella blanci (Piguet)		+	1	+	+		+			+
Ophidonais serpentina (Mull.)	+									
Uncinais uncinata (Oersted)	+	+	1	1			2	+		+
Paranais litoralis (Mull.)			+							
Amphichaeta leydigi Tauber		+	+							
Chaetogaster diaphanus (Gruith.)	+	+				+		+	+	+
Aulodrilus japonicus Yamaguchi	+	+		1	2	2	3	2	3	4
Aulodrilus limnobius Bret.				1				2	+	+
Aulodrilus pigueti Kov.				+	+				+	+
Rhyacodrilus coccineus (Vejd.)	+									
Limnodrilus hoffmeisteri Clap.	7	6	6	8	7	9	8	9	7	9
Limnodrilus udekemianus Clap.	1	+	+	+	2	+	+			
Limnodrilus claparedeanus Ratz.			2		+					
Limnodrilus profundicola (Verril)	+			+		+	+			
Potamothrix hammoniensis (Mich.)	1	2	2	+	3	3	3	3	3	3
Potamothrix moldaviensis Vejd.et Mr.		2		+						2
Potamothrix heuscheri (Bret.)					+		+			
Tubifex tubifex (Mull.)	7	2	1	2	4	+		+	3	+
Tubifex newaensis (Mich.)	+	3	4	3	3	+	3	+	+	2
Tubifex ignotus (Stolc)	+	+		+	+		+		+	
Isochaetides michaelseni (Last.)		1	+	+	+					
Psammoryctides albicola (Mich.)	+	+					+			
Psammoryctides barbatus (Grube)		1			+		+			
Spirosperma ferox Eisen	3	1	3	4	4	+	3	4	4	1
Spirosperma velutinus (Grube)		+								
Ilyodrilus templetoni (South.)										+
Lumbriculus variegatus (Mull.)	+	3	1	+	+		+		1	
Stylodrilus heringianus Clap.		+	+	+	+		1			
Lamprodrilus isoporus Mich.		5	5	2	2		2		1	
Enchytraeidae				+	+					

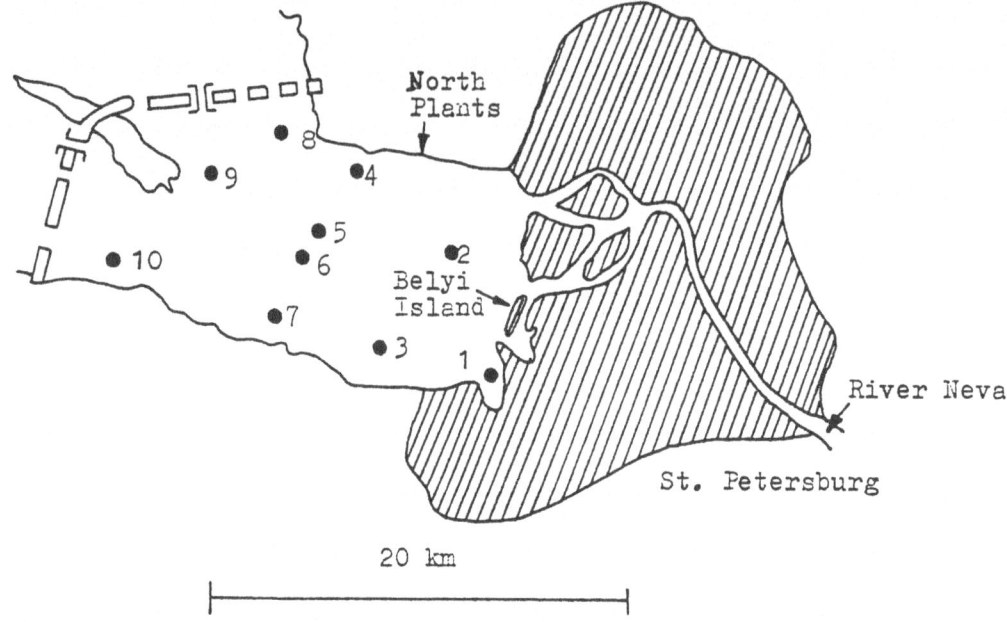

Fig. 1. Map of the Neva Mouth and the sampling stations (1–10).

Table 2. Abundance of worm species (No.m^{-2}) in different areas of the Neva Mouth.

Species	East Region St 1–3	Central Region St 4–7	West Region St 8–10
Limnodrilus hoffmeisteri	46,600	5,200	4,700
Tubifex tubifex	12,400	1,000	200
Tubifex newaensis	2,300	570	200
Tubifex ignotus	1,000	15	10
Spirosperma ferox	3,900	1,340	400
Lamprodrilus isoporus	8,900	230	15
Vejdovskyella intermedia	2,000	140	240

tions (Table 3, St. 1, 3, 5, 7, 9). In fact, comparison with historic data from the transition zone (stations 6 and 7) using data from Gurjanova (1949) and materials of 1975–1976 (Salazkin, 1982) shows consistency in biomass over a long period (Fig. 2).

The slight variations in the diversity index at St. 1, 3, 5, 9 (Table 4) also suggest long term stability in the overall oligochaete community. However, the oligochaete community of this body of water is very mobile and is capable of responding to local changes in the environmental conditions. Most frequently this is shown by increased dominance of *Limnodrilus hoffmeisteri* and *L. udekemianus*, which led to an increase of oligochaete biomass and a decline in the diversity index as at St. 10 in 1992 or to a decline of the diversity index as at St. 7 in 1992–1993 (Tables 3, 4). Sometimes the reverse was observed, with a biomass decline the number of species increased, the dominance of *L. hoffmeisteri* declined, and diversity increased. This was observed at St. 4 (Tables 3, 4) where the maximum number of oligochaetes was recorded in 1993, i.e. 17 species as compared to 6–11 species in previous years; at the same time the biomass of oligochaetes was lower and the contribution of *L. hoffmeisteri* to the total abundance declined from 70% to 45%. In the same year *Stylodrilus heringianus, Lamprodrilus isoporus, Ripistes parasita* appeared for the first time (it should be noted that *R. parasita* in the Neva River and the Neva Mouth appears to behave as a mesotrophic species). At this time the diversity of the entire macrobenthos increased and *Unio* was first observed at this station. This all supports a local improvement of conditions. The most dramatic changes in the oligochaete community were observed in the centre of the transition zone (St. 2). Here the communities changed twice over the past 10 years (Table 5). At the beginning of the 1980s the eutrophic species *L. hoffmeisteri, Tubifex tubifex*, together with *Lumbriculus variegatus* were dominant and the biomass of oligochaetes was large, this suggested sedimentation of large amounts of labile organic matter (Table 3). In

Table 3. Biomass of Oligochaeta (g wet weight m^{-2}) at stations 1–10 from 1982–93.

Station	1	2	3	4	5	6	7	8	9	10
Year										
1982	187.6	160.9	4.4	8.3	23.6		4.0	3.8	5.9	3.4
1983	147.3	238.0	4.3	7.0	11.0		3.2	5.5	4.5	8.7
1984	105.8	376.0		11.1	12.6		2.1	6.5		
1990		63.2								
1991		60.7				48.1				
1992	92.1	64.9	2.4	32.2	11.9	18.4	3.0	11.7	10.1	20.2
1993		18.6	2.3	14.2	16.7		5.0	6.2	9.2	17.8
Mean	133.0	140.3	3.4	14.6	15.1		3.4	6.8	7.4	12.5
S.D.	43.2	128	1.2	10.3	5.2		1.1	3.0	2.6	7.8
C.V.	32	91	35	70	35		32	44	36	62

Table 4. Diversity indices (d) at stations 1–10 from 1982–93.

Station	1	2	3	4	5	6	7	8	9	10
Year										
1982	1.1	2.8	2.1	1.3	2.1		2.8	1.7	2.4	1.9
1983	1.2	2.1	2.4	1.4	2.6		2.9	1.1	2.3	1.8
1984	1.1	2.9						1.6		
1990		2.3								
1991	1.4	2.2								
1992	1.5	1.5	2.5	1.3	2.5	0.8	1.6	1.0	2.1	0.8
1993		2.6	2.6	2.8	2.0	0.5	1.7	1.7	2.6	1.5
Mean	1.3	2.4	2.4	1.7	2.3		2.2	1.4	2.4	1.5
S.D.	0.2	0.5	0.2	0.7	0.3		0.7	0.3	0.2	0.5
C.V.	14	20	9	43	12		31	23	9	33

1990–92 the oligochaete biomass declined and *L. isoporus* became the dominant species; the oligotrophic species *Stylodrilus heringianus* also appeared in the community and the oligotrophic species *Spirosperma velutinus* was found close to this station. Also, the abundance of the mesotrophic species *Tubifex newaensis* increased. In 1993 the importance in the community of the oligochaetes *L. hoffmeisteri*, *P. hammoniensis* increased, *S. heringianus* disappeared and the abundance of *L. isoporus* declined. It is suspected that some oligochaete species inhabiting the muddy sand of St. 2, such as *P. barbatus* and *S. ferox* cannot compete with *L. isoporus* and that their abundance decreased less from environmental change than competition from this species.

Structural changes in other groups of benthos also occurred at the same time as changes in the oligochaete community. The biomass of sphaeriids declined abruptly in the 1990s, particularly in the vicinity of St. 2 where numbers became two orders of magnitude lower. At the same time *Unio* was spreading throughout the Neva Mouth. The development of the oligochaete community as a whole and the entire benthic community is evidence of the dynamics of processes occurring in this body of water. Of major importance to the Neva Mouth ecosystem is the input of allochthonous organic matter settling to the bottom and the hydrological regime responsible for its distribution. In 1978 treatment facilities were put into operation on the Belyi Island in the river outfall area and subsequently on the northern coast in the area of St. 4. These plants began discharging waste water into the Neva Mouth. Undoubtedly their discharge regime affected the transition zone of the bay, particularly in the region of Sts 2, 4, 5. Benthic communities may also have been affected by local changes in the hydrology from the construction of the dam. However, up to the present time the high water flow and high dissolved oxygen levels in the water have allowed species with different ecological requirements to co-exist in

Table 5. Abundance in thousands of individuals m^{-2} (left) and percent of total numbers (right) of dominant and indicator species at Station 2.

Species	Years					
	1982	1984	1990	1991	1992	1993
Limnodrilus hoffmeisteri	84.5 / 62.6	61.7 / 31.8	3.0 / 8.2	4.5 / 8.3	5.0 / 10.6	10.5 / 47.6
Tubifex tubifex	15.0 / 11.1	4.3 / 2.2	0.3 / 0.9	0.2 / 0.3		0.2 / 1.0
Potamothrix hammoniensis	4.2 / 3.4					2.3 / 10.1
Tubifex newaensis	3.3 / 2.4	2.8 / 1.4	9.8 / 26.1	4.5 / 8.2	5.0 / 10.5	3.0 / 13.9
Spirosperma ferox	2.7 / 2.0	4.3 / 2.2	0.2 / 0.4		0.001 / 0.02	0.5 / 2.2
Psammoryctides barbatus	1.9 / 1.4	8.3 / 4.3	0.4 / 1.2	0.2 / 0.4	0.05 / 0.1	0.1 / 0.6
Lamprodrilus isoporus	0.03 /0.02	5.3 / 2.8	16.6 / 44.2	39.2 / 72.8	33.7 / 70.4	1.0 / 4.7
Stylodrilus heringianus		0.4 / 0.2	1.6 / 4.3	0.3 / 0.5	0.2 / 0.5	
Lumbriculus variegatus	16.5 / 12.2	44.0 / 22.7	3.5 / 9.3	0.6 / 1.0	0.08 / 0.2	0.2 / 0.8

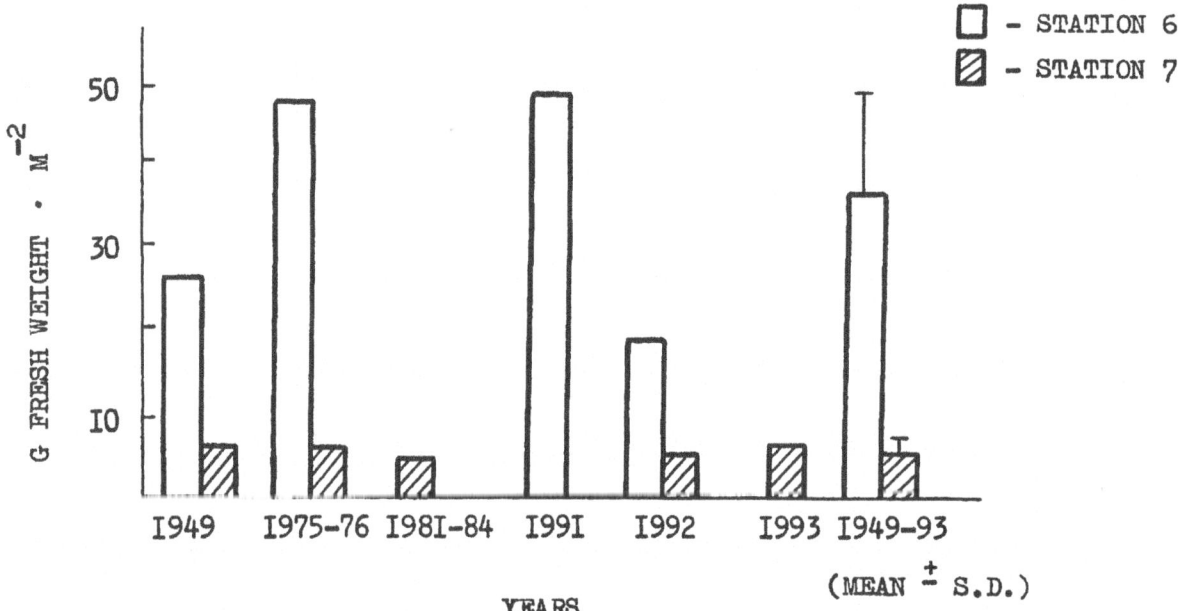

Fig. 2. Biomass of Oligochaeta at stations 6 and 7 through time.

the river mouth. It is suspected that the large amounts of organic material have neutralized the effects of toxic substances accumulating in the sediment.

Examination of oligochaete species associations using correlation analysis of the percentage of species from the collections of 1982–1993 revealed the strongest negative relations between *L. hoffmeisteri* and *L. isoporus, T. newaensis* (r = −0.56 and r = −0.65) and a positive relationship between *T. newaensis, Propappus volki* and *S. heringianus* (r = 0.57), *P. barbatus* and *S. heringianus* (r = 0.55), *L. variegatus* and *P. barbatus* (r = 0.75). These associations agree with the generally accepted concept of ecological groups

of oligochaetes. These data were kindly provided by Dr V. V. Menshutkin.

Monitoring large European and American lakes has previously shown that oligochaete populations respond to changes in trophic status even when oxygen stress is not apparent (Probst, 1987; Lang, 1989, 1990). However, a direct comparison of the Neva Mouth to such large lakes is inappropriate. The Neva Mouth is no longer a natural system and it probably represents an ideal habitat for the development of a rich oligochaete community. The appearance of a community dominated by the mesotrophic species *L. isoporus* and *T. newaensis* and with increased numbers of *S.*

heringianus would be a good indicator of improved conditions. In the Neva Mouth community there is evidence that *S. ferox* is behaving as a eutrophic species. It may, like *T. tubifex*, have a variable ecological significance, i.e. when abundance is low it is an indicator of oligotrophic conditions and when abundance is high it is an indicator of eutrophic conditions. Increased dominance of *L. hoffmeisteri* and *T. tubifex,* lower numbers of species and increased dominance of oligochaete biomass are good indicators of reduced water quality, these are all related to an increase in sedimentation of organic matter on the bottom which, dependent on local water movement may lead to secondary pollution and deterioration of water quality. Oligochaete communities are sufficiently sensitive to environmental change and could constitute the basis of monitoring the ecological situation in the Neva Mouth.

Acknowledgments

The author is grateful to Dr V. V. Menshutkin for data on correlational analysis of worm species relative density. I wish to thank also V. G. Vlasova and O. G. Parshkova for technical assistance.

References

Finogenova, N. P., S. M. Golubkov, V. E. Panov, E. V. Balushkina, V. Y. Pankratova, T. M. Lobasheva & A. M. Pavlov, 1987. Macrobenthos. In G. G. Winberg & B. L. Gutelmacher (eds), The Neva Mouth. Hydrobiological investigations. Trudy Zool. Inst. AN SSSR 151, Nauka Leningrad: 111–120. [In Russian.]

Finogenova, N. P. & T. M. Lobasheva, 1987. Transformation of organic matter by Oligochaeta and their growth rate. In: G. G. Winberg & B. L. Gutelmacher (eds) The Neva Mouth. Hydrobiological Investigations. Trudy Zool. Inst. AN SSSR 151, Nauka, Leningrad: 135–145. [In Russian.]

Gurjanova, E. F. (ed.), 1949. Materials to the study of the Neva Mouth's benthos. Uchen. zap. Leningr. univ., ser biol. nauk 21: 107–141. [In Russian.]

Lang, C., 1989. Eutrophication of Lake Neuchatel indicated by the oligochaete communities. Hydrobiologia 174: 57–65.

Lang, C., 1990. Quantitative relationships between oligochaete communities and phosphorus concentrations in lakes. Freshwat. Biol. 24: 327–334.

Probst, L., 1987. Sublittoral and profundal Oligochaeta fauna of the Lake Constance (Bodensee-Obersee). Hydrobiologia 155: 277–282.

Salazkin, A. A., 1982. Bottom fauna of the Neva Mouth and some peculiarities of its distribution. Sbornik nauch. tr. GosNIORKh 192: 70–77. [In Russian.]

Shiklomanov, I. A., L. Y. Preobrazhenskii, B. G. Skakalskii & B. A. Shishkin, 1989. Hydrobiological and ecological Investigations of water system the Ladoga Lake – the Neva – the Neva Mouth. Izvest. VNIIG im. Vedeneeva 213: 14–29. [In Russian.]

Shishkin, B. A., 1991. Integral estimation of biota's role in formation of quality of water under different types of anthropogenic impact. Trudy 5 Vsesoyuznogo Gidrolog. syezda 5: 435–442. [In Russian.]

Winberg, G. G. & B. L. Gutelmacher (eds), 1987. The Neva Mouth. Hydrobiological Investigations. Trudy Zool. Inst. AN SSSR, 151, Nauka, Leningrad: 3–205. [In Russian.]

Wilhm, J. L. & J. C. Dorris, 1968. Biological parameters of water quality. Bioscience 18: 477–481.

Hydrobiologia **334**: 193–198, 1996.
K. A. Coates, Trefor B. Reynoldson & Thomas B. Reynoldson (eds), Aquatic Oligochaete Biology VI.
© 1996 *Kluwer Academic Publishers.*

Upstream-downstream movement of macrofauna (with special reference to oligochaetes) in the River Raba below a reservoir

Elżbieta Dumnicka
Karol Starmach Institute of Freshwater Biology, Polish Academy of Sciences, Sławkowska 17, 31–016 Kraków, Poland

Key words: oligochaetes, macrofauna, effect of the dam, upstream–downstream movements

Abstract

Oligochaetes are the dominant group of macrofauna in a river reach (1–450 m) below the Dobczyce dam on the River Raba. The Oligochaeta made up from 60.2% to 78.7% of the community. The next most abundant group were the Chironomidae comprising 20.5%–38.8% of the fauna. Upstream-downstream movements of the macrofauna were studied at one station, with an oligochaete density of 50 000 ind. m^{-2}, and a chironomid density about 20 000 ind. m^{-2}. In the drift the proportions of these two groups were reversed – Chironomidae represented 59% of the drift fauna, and the oligochaetes – about 40%. On the river bottom and in the drift a similar number of species (12–13) of Naididae were identified.

In order to compare the intensity of the movement of the various taxonomic groups, the percentage of animals moving upstream and downstream over 24 h was calculated. The highest percentage of the population migrating were dipterans (7.3%). The Chironomidae and Ephemeroptera had similar proportions migrating (3.2% and 2.6% respectively). The Oligochaeta had the lowest value – only 0.6% of animals were displaced in a 24 h period.

The Naididae appear to be strongly attached to the substratum and, under normal environmental conditions, their reproduction sufficiently compensated for the decrease in their number associated with the drift or predation.

Introduction

Reservoirs have a major effect on the composition and structure of the bottom fauna in rivers downstream by reducing the density of certain groups of the fauna and increasing the density of other groups (see review made by Ward & Stanford, 1979). The density of oligochaetes, and especially that of Naididae, usually increases, but little is known about species preferring this environment and of the adaptations to it. Studies of bottom fauna drift, (especially of insects), have a long tradition and large literature (see reviews by Waters, 1972; and Statzner *et al.*, 1984).

Investigations of upstream movement is less frequent – see review by Soderstrom (1987). At present little is known of the upstream – downstream movement of oligochaetes which in most streams and rivers represent only a small percentage of the fauna (Waringer, 1992; Bergey & Ward, 1989; Bish-

op & Hynes, 1969; Cellot, 1989). In these studies oligochaetes are often ignored (Elliott, 1971; Hobbs & Butler, 1981), or not determined (Williams & Williams, 1993).

Authors (Armitage, 1977; Dumnicka, 1987) have shown rivers downstream of reservoirs to be useful sites to examine migration of oligochaetes. Furthermore, the composition of the reservoir communities are completely different (Dumnicka, 1987, 1993) to the river community.

The hydrochemistry, hydrobiology and ichthyobiology of the River Raba and the Dobczyce Reservoir have been investigated by The Karol Starmach Institute of Freshwater Biology (Bednarz & Starzecka, 1993; Dumnicka, 1993; Fleituch, 1992; Gwiazda, 1989; Starzecka & Bednarz, 1994).

This paper describes the distance downstream of the dam over which the composition of zoocoenosis of the river bottom is effected and provides a ten-

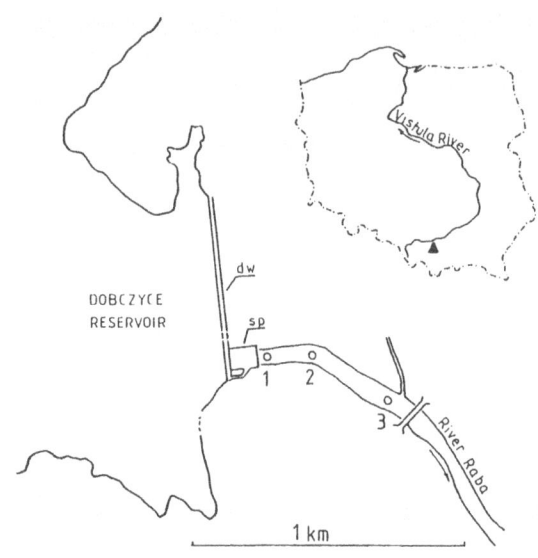

Fig. 1. Map of the River Raba below Dobczyce Reservoir showing the sampling stations.

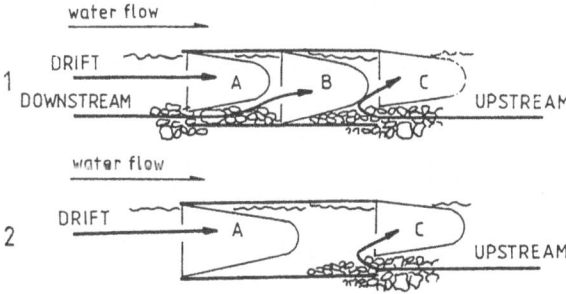

Fig. 2. Design of the original and modified samplers used for capturing the invertebrates moving upstream and downstream.

tative estimation of the drift extent and intensity of upstream movements of the macrofauna – especially of the oligochaetes.

Study area

The investigations were carried out on the River Raba, the right bank tributary of the River Vistula, below the dam reservoir in Dobczyce. A constant volume (1.8 m³ sec⁻¹) of water from the reservoir is discharged through bottom sluices to a concrete stilling pool (of

2–3 m depth). The stations were located at a distance: 1 m (station 1), 100 m (station 2) and 450 m (station 3) from the stilling pool (Fig. 1). Investigations of the upstream–downstream movements of macrofauna were conducted in the vicinity of station 2. Below station 3, a small polluted stream enters the River Raba, which affects the composition and the structure of the bottom fauna, and for this reason no samples were collected below this point. At all stations the river substrate consisted of stones of various size (5–20 cm) overgrown with filamentous algae (*Cladophora sp.*). Nearer to the dam the stones are closely packed (resembling a paved road); the bottom at station 3 is of natural character.

Materials and methods

The drift and the upstream movements of macrofauna were measured twice: on May 26 and June 8, 1993, using a modified box sampler (Bergey & Ward, 1989). The modification consisted of removing the middle net and lowering the front net (Fig. 2). The nets were rinsed every 2 h, during 24 h, but only four samples collected during daylight and four samples collected during the night were taken into consideration. Samples of the bottom fauna were collected (each time five samples) using the bottom scraper. The mesh size of the net used in the box sampler and the scraper was 0.3 mm. The samples were preserved in 4% formalin, sorted in the laboratory using a stereoscopic microscope, and transferred to 40% ethanol. Whole specimens of oligochaetes were mounted in Canada balsam.

Results

Bottom fauna

The composition and the density of the fauna in the study reach of the River Raba were similar on both dates (differences in the density and domination were statistically insignificant at $P > 0.4$ and $P > 0.3$ respectively), and for this reason the results obtained from both dates were combined.

The oligochaetes were the dominant group at each station, although their proportion declined downstream from 78.7% at station 1 to 60.2% of the fauna at station 3. However, the abundance remained similar at all three stations: at station 1 it was equal to about

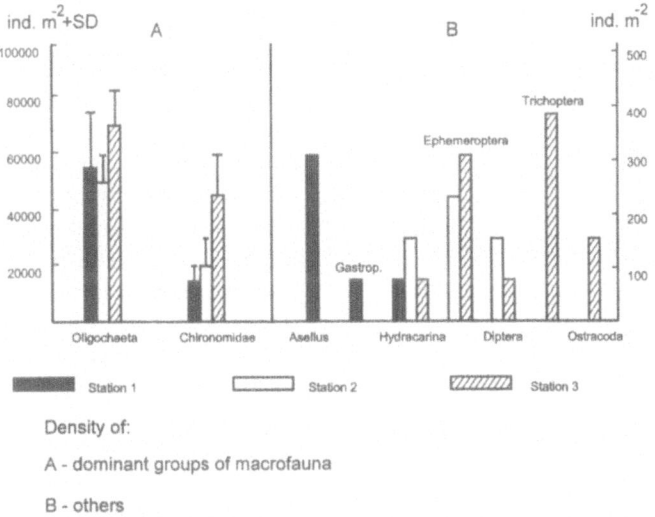

Fig. 3. Density of macrofauna in the River Raba below Dobczyce Reservoir.

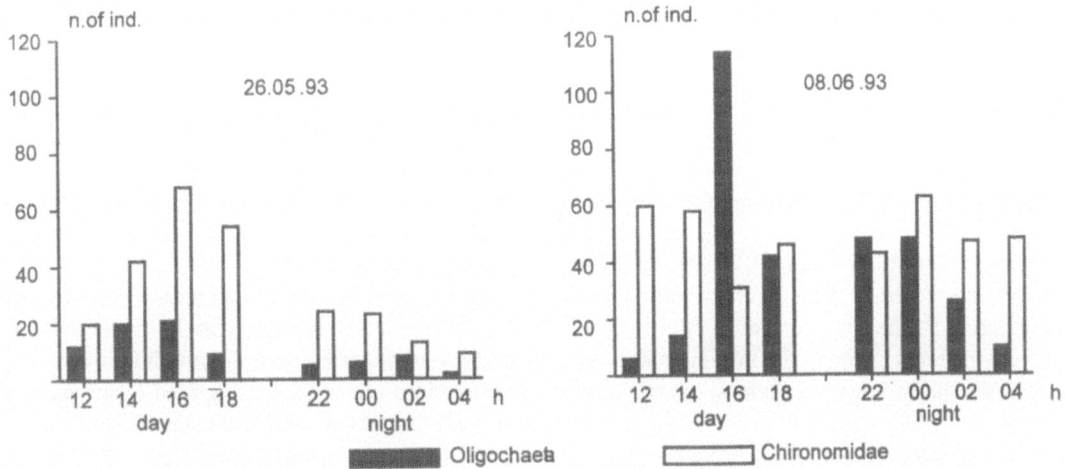

Fig. 4. Numbers of invertebrates drifting during 2 h. Numbers of invertebrates moving upstream during 2 h.

54 000m⁻², at station 2 it decreased slightly to about
50 000m⁻², while at station 3 it increased to 69 000 ind.
m⁻² (Fig. 3); these differences were not statistically
significant.

The second most numerous group were the Chironomidae, which represented 20.5% of the fauna at
station 1 to 38.8% at station 3. Their density increased
downstream. At stations 1 and 2 there were fewer
than 20 000 ind. m⁻², while at station 3 the number
increased to 44 500 ind. m⁻². The difference in the
density between stations 2 and 3 is significant at the
level P < 0.05, and between stations 1 and 3 – at the
level P < 0.01.

The remaining taxa occurred in small numbers
comprising at most a total 1% of the fauna. At station 1 they were represented by species found in the
stilling pool (*Asellus aquaticus* and Gastropoda: *Physa
fontinalis* and *Lymnaea (Galba) sp.*). At station 2
Ephemeroptera and Diptera from the families Ceratopogonidae and Limoniidae occurred. In addition, Trichoptera and Ostracoda were observed at station 3. At
all stations *Hydra sp.* and Nematoda were found in low
numbers.

Among the oligochaetes living on the bottom of
the river 13 species of Naididae were identified and
juvenile specimens of the family Enchytraeidae were
also found. The number of species changed a little –

196

Table 1. Percent species composition of the oligochaete fauna of the Raba River.

Species	Station 1	Station 2	Station 3	Drift
Chaetogaster diastrophus	22.2	42.0	68.0	75.4
Nais bretscheri	28.1	19.0	4.0	4.9
N. elinguis	16.6	15.2	15.8	1.2
N. barbata	15.6	4.6	3.2	2.8
N. pardalis	4.6	5.7	0.4	1.9
N. pseudobtusa	3.9	9.6	3.3	6.2
N. alpina	3.1	0.7	1.5	1.2
Chaetogaster diaphanus	2.9	0.2	0.9	0.3
Nais variabilis	2.6	0.7	0.4	0.6
N. communis	0.4	1.2	0.5	1.2
N. christinae		0.2	0.2	1.6
Amphichaeta leydigii		0.9		
Nais simplex			1.6	0.9
Enchytraeidae juv.			0.2	0.9
Tubificidae juv.				0.6
Stylaria lacustris				0.9

at station 1: 10 species, and at station 3: 12 species were identified, but the proportions changed considerably (Table 1). This is best shown by *Chaetogaster diastrophus* which represented 22.2% of the community at station 1 to 68% at station 3. The proportion of *Nais bretscheri* declined from 28.1% at station 1 to only 4% at station 3. The proportion of *N. barbata* also decreased downstream. Among the dominant species only *N. elinguis* showed no changes between stations.

Drift and upstream movements of macrofauna

In the drift net the crustacean zooplankton (primarily Copepoda) washed out from the reservoir were most abundant. They were particularly numerous in the first sampling event (1.43 g dry weight, on the average, during 2 h). On the second sampling occasion they were much less numerous – on the average 0.17 g of dry weight during the same period of time.

The box samples were only taken at station 2, where the density of oligochaetes was approximately 50 000 ind. m^{-2}, and Chironomidae about 20 000 ind. m^{-2}. In the drift the proportion of these two taxonomic groups was reversed; Chironomidae represented 59% of the drift fauna, and the oligochaetes, 41%. Although the densities of oligochaetes and chironomids collected from bottom samples were similar in both events, the amount of drift was lower on the first occasion (Fig. 4). However, only the Chironomidae showed a statistical-

ly significant reduction ($P < 0.05$) in numbers present in the drift on the second sampling event.

Besides Chironomidae and Oligochaeta other taxa were also observed in the drift: Diptera, Ephemeroptera, Nematoda, Hydrozoa, Coleoptera larvae and even the larvae of *Chaoborus sp.*, but only the first two taxa occurred regularly.

The amount of drift changed irregularly over the 24 h. There were large oscillations in the number of animals captured in the successive periods (Fig. 4). The mean drift of the Naididae and Chironomidae was greater on both occassions in the day, however, the differences in the amount of the drift between day and night were not statistically significant.

Thirteen Naididae species were caught with a drift net (Table 1). In addition to the taxa occurring in the substratum were *Stylaria lacustris* and Tubificidae juv. – both found in the stilling pool and the reservoir (Tubificidae). The most frequently encountered species in the drift was *Chaetogaster diastrophus*, which comprised as much as 75.4% of the oligochaete drift fauna. *Nais pseudobtusa* was also numerous, and represented 6.2% of the captured animals. The proportion of some species: *Nais bretscheri*, *N. elinguis*, *N. barbata* and *N. pardalis* in the drift was much lower than in the bottom fauna. One of the species observed in the drift, *N. simplex* only occurred at station 3.

A very small number of animals were found migrating upstream. An average of eight individual chironomids were captured each 2 h, with no diurnal difference. More oligochaetes migrated upstream at night (1.4 ind. 2 h^{-1}) than during the day (0.7 ind. 2 h^{-1}), but these differences were not statistically significant. The species migrating upstream were primarily *Chaetogaster diastrophus* and *Nais bretscheri*; migration was not observed in *N. elinguis*.

To compare the migration intensity in the various taxonomic groups the number of animals migrating upstream and drifting downstream over 24 h was calculated as the percentage of the fauna living on the bottom. The greatest percentage of migrating population were members of the two dipteran families, the Ceratopogonidae and Limoniidae (7.3%); while not numerous in the benthos, they were often found both in the drift and in the process of migrating upstream. Similar proportions of the Chironomidae and Ephemeroptera were found migrating, 3.2% and 2.6% respectively. In the case of the oligochaetes only 0.6% of animals occurring in a bottom area of 1 m^2 had migrated over a 24-hour period.

Discussion

The macrofauna of the Raba River was studied above and below the planned reservoir, before flooding (Fleituch, 1992). The Oligochaeta and Chironomidae were the dominant components of the bottom fauna; in addition the Ephemeroptera, Trichoptera and Plecoptera were very numerous. Over a period of a few years after the reservoir construction these latter three groups of insects almost completely disappeared from the river fauna below the reservoir for a distance of some hundreds of meters. Similar changes have been observed in other rivers below reservoirs (Armitage, 1978). Among the oligochaetes the first to be eliminated were the Tubificidae, which were more numerous in the earlier investigations of Fleituch (1992). This could be due to the loss of habitats suitable for this taxon, silted up river sections with slow current.

The physical and chemical characteristic of River Raba below the reservoir are more stable than in unregulated rivers (Mazurkiewicz, in press), and for this reason they have little influence on changes in the intensity of the drift. In many insect taxa drift was greater at night; in the cases of Naididae and Chironomidae no changes in the 24 h intensity of the drift were observed (Bergey & Ward, 1989).

Among the Naididae occuring in the Raba River below the dam swimming species (*Nais elinguis, N. barbata, N. pardalis, N. pseudobtusa* and *N. variabilis*) were dominant, but the non swimming *N. bretscheri* was also present. It would appear that this ability did not influence occurrence in drift. The high abundance of *Chaetogaster diastrophus* indicates that very small species were able to live just below the reservoir. The Naididae, which often comprise a considerable part of the bottom fauna in rivers below reservoirs (Armitage, 1978; Dumnicka, 1987), rarely undertake upstream migration and represent a small percentage of drift in Polish rivers (Fleituch, 1985). These attributes allow them to occur in large numbers on the river bottom. On the other hand, in the Yr Ogof stream (North Wales) drift compensation in oligochaetes was high – more than 40% at two stations and 22% at one station (Williams & Williams, 1993). It appears that in the River Raba the reproduction is a factor sufficiently compensating the losses in population, and that individuals occurring in the drift may be treated as migrating. Only after ecological catastrophes such as flood and drought are migrating individuals able to settle in unoccupied habitats. These migrations may occur also in hyporheic waters or into back waters (O'Leary *et al.*, 1992), but given the small size of these animals this will be a slow process.

References

Armitage, P. D., 1977. Invertebrate drift in the regulated River Tees, and an unregulated tributary Maize Beck, below Cow Green dam. Freshwat. Biol. 7: 167–183.

Armitage, P. D., 1978. Downstream changes in the composition, numbers and biomass of bottom fauna in the Tees below Cow Green Reservoir and in an unregulated tributary Maize Beck, in the first five years after impoundment. Hydrobiologia 58: 145–156.

Bednarz, T. & A. Starzecka, 1993. The production and destruction of organic matter in the water and surface layer of bottom sediments on the stream-estuary – Dobczyce Dam Reservoir line (southern Poland). Acta Hydrobiol. 35: 109–119.

Bergey, E. A. & J. V. Ward, 1989. Upstream-downstream movements of aquatic invertebrates in a Rocky Mountain stream. Hydrobiologia 185: 71 - 82.

Bishop, J. E. & H. B. N. Hynes, 1969. Upstream movements of the benthic invertebrates in the Speed River, Ontario. J. Fish. Res. Bd. Can. 26: 279–298.

Cellot, B., 1989. Macroinvertebrate movements in a large European river. Freshwat. Biol. 22: 45–55.

Dumnicka, E., 1987. The effect of dam reservoirs on oligochaete communities in the River Dunajec (Southern Poland). Acta Hydrobiol. 29: 25–34.

Dumnicka, E., 1993. Profundal macrofauna of the Dobczyce Reservoir (southern Poland) in the fifth year after its filling. Acta Hydrobiol. 35: 329 - 340.

Elliott, J. M., 1971. Upstream movements of benthic invertebrates in a lake district stream. J. Anim. Ecol. 40: 235–252.

Fleituch, T. Jr., 1985. Macroinvertebrate drift in the middle course of the River Dunajec (Southern Poland). Acta Hydrobiol. 27: 49–61.

Fleituch, T. M., 1992. Evaluation of the water quality of future tributaries to the planned Dobczyce reservoir (Poland) using macroinvertebrates. Hydrobiologia 237: 103–116.

Gwiazda, R., 1989. Initial stage of bird settlement on the Dobczyce dam reservoir (Vistula basin, southern Poland). Acta Hydrobiol. 31: 373–384.

Hobbs, H. H. & M. J. Butler, 1981. A sampler for simultaneously measuring drift and upstream movements of aquatic macroinvertebrates. J. Crust. Biol. 1: 63–69.

Mazurkiewicz, G., unpublished data. Hydrochemical characteristics of the water of the River Raba.

O'Leary, P., P. S. Lake, R. Marchant & T. J. Doeg, 1992. Macroinvertebrate activity in the water column of backwaters in an upland stream in Victoria. Aust. J. mar. Freshwat. Res. 43: 1403–1407.

Soderstrom, O., 1987. Upstream movements of invertebrates in running waters – a review. Arch. Hydrobiol. 111: 197–208.

Starzecka, A. & T. Bednarz, 1994. Decomposition of organic matter in bottom sediments of a stream and the Dobczyce dam-reservoir in the area of the Wolnica creek (southern Poland). Arch. Hydrobiol. 129: 327–337.

Statzner, B., C. Dejoux & J. M. Elouard, 1984. Field experiments on the relationship between drift and benthic densities of aquatic insects in tropical streams (Ivory Coast). I. Introduction: a review of drift literature, methods, and experimental conditions. Revue Hydrobiol. trop. 17: 319–334.

Ward, J. V. & J. A. Stanford, 1979. Ecological factors controlling stream zoobenthos with emphasis on thermal modification of

198

regulated streams. In J. V. Ward & J. A. Stanford (eds), The Ecology of Regulated Streams. Plenum Press, New York: 35–55.

Waringer, J. A., 1992. The drifting of invertebrates and particulate organic matter in an Austrian mountain brook. Freshwat. Biol. 27: 367–378.

Waters, T. F., 1972. The drift of stream insects. Ann. Rev. Ent. 17: 253–272.

Williams, D. D. & N. E. Williams, 1993. The upstream/downstream movement paradox of lotic invertebrates: quantitative evidence from a Welsh mountain stream. Freshwat. Biol. 30: 199–218.

Hydrobiologia **334**: 199–206, 1996.
K. A. Coates, Trefor B. Reynoldson & Thomas B. Reynoldson (eds), Aquatic Oligochaete Biology VI.
©1996 *Kluwer Academic Publishers.*

A comparison of reproduction, growth and acute toxicity in two populations of *Tubifex tubifex* (Müller, 1774) from the North American Great Lakes and Northern Spain

Trefor B. Reynoldson[1], Pilar Rodriguez[2] & Maite Martinez Madrid[2]
[1]*National Water Research Institute, Environment Canada, CCIW, 867 Lakeshore Rd, Burlington, Ontario L7R 4A6, Canada*
[2]*Dpto. Biologia Animal y Genetica, Faculdad de Ciencias, Universidad del Pais Vasco, Bilbao, 48080, Spain*

Key words: toxicity, cultures, oligochaetes, *T. tubifex*, growth, reproduction

Abstract

Reproduction in *Tubifex tubifex* is being used as part of a suite of indicators of sediment toxicity in Canada and Spain, and reproduction of *T. tubifex* is being considered as a component of sediment objectives for environmental regulation and clean-up in the Canadian Great Lakes. The data being used to set these reproductive targets have been developed from a single culture of *T. tubifex* from Lake Erie. The plasticity of this particular species is well known and before it can be adopted widely as a test organism it is necessary to determine whether a single culture source should be used or if cultures derived from different populations respond similarly. A series of experiments with two cultures, one from Lake Erie the second from a small mountain stream in Northern Spain have shown that the Spanish worms appear to produce fewer cocoons per adult (mean 8.6 S.D. 1.0) than those from Lake Erie (mean 10.4 S.D. 0.3) at 22.5 °C, a standard test temperature. The number of young produced per adult by the Spanish culture is also lower (mean 19.0 S.D. 4.6) than the L. Erie population (mean 30.6 S.D. 2.3), however, the Spanish population has higher reproductions rates at a lower temperature. The Spanish worms also have lower and more variable growth rates than the Canadian population. There also appear to be slight differences in the sensitivities to toxicants, with the Canadian worms having higher LC50s for copper, chromium and cadmium. While there are differences in the responses in the two cultures these are not considered to be sufficient to invalidate the use of either population in a standard bioassay protocol as long as appropriate calibration and validation are undertaken.

Introduction

Benthic macroinvertebrates are frequently used for the classification and monitoring of natural environments and in laboratory toxicity tests. However, the aquatic Oligochaeta often are not considered as useful indicators or as test species because of perceived difficulties in their identification and the perception that they are tolerant to pollution. These perceptions are inaccurate. Identification keys are available, their sensitivity has been demonstrated in laboratory toxicity tests and detailed standard test protocols are available (Naididae: Learner & Edwards, 1963; Chapman & Mitchell, 1986; Tubificidae: Chapman & Mitchell, 1986; Milbrink, 1987; Wiederholm et al., 1987; Casellato & Negrisolo,

1989; Reynoldson et al., 1991; Enchytraeidae: Roembke and Knacker, 1989; Lumbriculidae: Bailey & Liu, 1980; Keilty et al., 1988; Dermott & Munawar, 1992).

The oligochaete *Tubifex tubifex* (Müller, 1774) can adapt to a wide range of environmental conditions and occurs in both oligotrophic and eutrophic conditions and is found in both non-polluted and very polluted waters (Milbrink, 1983; Lauritsen et al., 1985). A 28 day reproductive toxicity test with *T. tubifex* is being used in Canada as part of a remediation program on the Great Lakes (Bailey et al., 1995; Reynoldson et al., 1995). The test also is being used in Spain to determine the toxicity of river sediments in the industrial area of Bilbao. The objective of this study was to determine whether a single culture should be used for future work to avoid differences in test responses resulting from the source of the culture populations. To address this, separate and paired experiments were conducted examining survival, reproduction and growth in cultures derived from Spanish and Canadian populations of *T. tubifex*.

Methods

Characteristics of culture animals and sediment

The Spanish population of worms and culture sediment were obtained from Barazar in Gorbea Natural Park, Bizkaia, Spain. This is a mountain stream at an altitude of 570 m, the stream is approximately 2 m wide and less than 50 cm deep (43°03′ N, 2°43′ W). The Canadian culture sediment came from Big Creek Marsh, Long Point, Lake Erie (42°36′ N, 80°27′ W), a United Nations biosphere preserve, which is a shallow embayment of 850 ha with a water depth of 0.75–1 m. Canadian cultures of *T. tubifex* were derived from populations from western Lake Erie and Hamilton Harbour, Lake Ontario. Sediments from Barazar used for cultures and experiments were collected in December 1993 (BAR) and two separate batches of sediment were collected from Long Point in Lake Erie in the spring (LP7) and the fall (LP8) of 1992. Sediment from both sources (Table 1) are predominantly fine grained silts, the Long Point sediment is higher in organic material and generally higher in metal concentrations (Table 1).

Laboratory culture

Canadian and Spanish populations of *T. tubifex* were maintained at the National Water Research Institute (NWRI), Burlington, Ontario, on sieved (250 μm pore size), previously frozen sediment from Long Point (LP) (Table 1). The culture vessels were covered $20 \times 20 \times 20$ cm aquaria with a 5 cm layer of sediment, and 10 cm of overlying dechlorinated, City of Burlington, Lake Ontario, tap water (pH 7.8–8.3, conductivity 439–578 μS, hardness 119–137 mg l^{-1}). The aquaria were gently aerated, maintained on an 8:16 h light dark cycle and at ambient room temperature (20 ± 2 °C). The cultures were checked regularly, and when the proportion of sexually mature individuals declined the sediment was replaced, approximately every three months.

Spanish cultures were maintained at the Universidad del Pais Vasco (UdVP), in Bilbao, Spain. The sediment from Barazar (BAR) (Table 1) was sieved through a 250 μm pore-size sieve and kept in the dark at 4 °C prior to use. The cultures were maintained in

Table 1. Characteristics of culture sediments from Spain and Canada.

	Barazar - Spain	Long Point - Canada
Sediment Characteristics (% composition)		
Sand	5.6	14.5
Silt	73.4	76.8
Clay	20.9	8.7
LOI	8.8	17.0
Metals & Major Elements (g g^{-1} dry weight)		
Cr	41.5	41.5
Zn	213	96.6
Cd	9.5	0.4
Pb	89.6	25.8
Ni	45.9	16.9
Mn	217	1200
Fe	30992	28300
Cu	20.7	12.1
S	646.2	9100
P	367.5	2100
Ca	2269	172500

aquaria (h 25 cm, w 40 cm, d 20 cm) with a 7 cm layer of sediment and 10 cm of overlying dechlorinated tap water (pH 7.1, conductivity 180.5 μS, CaCO$_3$ alkalinity 50 mg l^{-1}). The aquaria were kept in an incubator at 22.5 ± 1 °C with gentle aeration in the dark.

Reproduction tests

Differences in reproduction in the two populations of *T. tubifex* were investigated in two series of experiments. First, Spanish and Canadian populations were tested separately at UdVP and at the NWRI with the different, respective, culture sediments (BAR and LP, Table 1) and overlying tap water. Second, paired experiments with Spanish and Canadian worms, both in culture at NWRI, were conducted with LP sediment.

In both experiments tests were performed using the protocol of Reynoldson et al. (1991). For each test five replicates were established. Each replicate contained 100 ml of sediment and 100 ml of overlying tap water in a 250 ml glass beaker. The sediment was sieved through a 250 μ m pore-size sieve to eliminate the indigenous invertebrates (Reynoldson et al., 1994). Four sexually mature worms were added to each replicate container. The tests were maintained in the dark at the test temperature (20 or 22.5 ± 0.5 °C) and gently aerated; pH, dissolved oxygen and temperature were checked weekly in each beaker. At the start of each test, each replicate beaker received 80 mg of Tetramin® as a food supplement. Reproduction was measured by counting the number of cocoons and young produced in each replicate beaker and expressed as the number produced per adult.

Growth experiments

Differences in growth rates of the two populations were examined. First, the effect of individual size on growth was examined at a single temperature (25 ± 0.5 °C) and second, the effect of temperature on growth rate was examined. In the first experiment growth of the two populations was measured using individuals ranging from 0.3–6.0 mg (wet weight), this range represents individuals that are a few days old (hatched from the cocoon) to those approaching their first reproduction. In the second experiment differences in growth response to temperature (5, 10, 15, 20, 25 and 30 ± 0.5 °C) were examined in eight week old animals that were sexually immature and had not reproduced. Animals in the 1.0–4.0 mg (wet weight) size range were used and all experiments lasted for 10 days. Animals were maintained separately in 100 ml glass containers containing 50 ml of LP sediment and 50 ml of Lake Ontario tap water.

All worms were sexually immature at the start of the experiment. Wet weights were measured for each individual at the beginning and the end of the experiment. It was necessary to use wet weight rather than dry weight as growth rates were calculated for individual worms. There is a high correlation between wet and dry weight ($r^2 = 0.997$) in *T. tubifex*, and therefore wet weight measurements were considered acceptable. All data are expressed as wet weights. Growth was calculated as % growth per day ($G_w\%$):

$$G_w\% = (\ln FW - \ln SW) \times 100 d^{-1},$$

where FW = final weight (mg), SW = start weight (mg) and d = number of days.

Sensitivity to reference toxicants

Paired experiments were conducted with the two populations to study their relative sensitivity to selected toxicants. Acute toxicity was examined in dechlorinated, Lake Ontario tap water over a 96 h period using methods similar to those reported by Chapman et al. (1982). The animals were examined every 24 hours

and were considered dead if they did not respond to tactile stimulation. At each dose, five replicate 250 ml beakers containing 100 ml of water were established to which five individual worms were added (wet weight = 3.1–10.0 mg). The experiments were conducted at 22 ± 0.5 °C and maintained in the dark, the beakers were not aerated. Oxygen, pH and conductivity were measured daily. Dead animals were counted and removed daily.

Three metals, cadmium ($CdCl_2$), chromium ($K_2Cr_2O_7$) and copper ($CuSO_4$) and an organic compound, (hexachlorocyclohexane – lindane) were tested in acute toxicity experiments using a series of concentrations (Table 2) with dechlorinated Lake Ontario tap water ($CaCO_3$ alkalinity 90–120 mg l^{-1}, hardness 119–137 mg l^{-1}). Nominal concentrations (Table 2) of trace metals were confirmed by inductively coupled plasma-atomic emission spectroscopy (ICP-AES) on a multichannel Jarrell-Ash (Franklin, Ma) Atom Comp® 1100. Lindane analysis was performed by dual capillary column gas/liquid chromatography with dual and single electron capture detection using a Hewlett Packard GC/ECD (model 5890). Confirmation analyses were done by GC/MS. The margin of error is estimated to be 10% from extraction, isolation and instrumentation.

Calculation of LC50s was conducted by the least squares probit method using a programme developed by the University of Guelph (T. James).

Results

Reproduction

Reproduction was first compared from the results of a number of repeated bioassays using the Spanish (BAR) and Canadian culture sediments (LP7 and LP8). These data (Table 3) provide an indication of the range in reproductive performance of *T. tubifex* from several separate runs on each culture sediment. A comparison of reproduction in the two cultures, in their respective sediment using the Mann Whitney test with mean values for each run showed that the numbers of cocoons produced by the Spanish worms on BAR sediment was significantly less ($P<0.05$) than the Canadian worms in both LP7 and LP8 sediments. There was no significant difference ($P>0.05$) between the number of young produced at BAR and LP7 but the difference was significant ($P<0.05$) between the Spanish worms and the Canadian worms in LP8. Because these tests

Table 2. Dose ranges for four toxicants in 96 h acute toxicity tests with *T. tubifex*, showing nominal and measured concentrations (mg l^{-1}).

Copper		Cadmium		Chromium		Lindane	
nominal	actual	nominal	actual	nominal	actual	nominal	actual
0	<0.002*	0	<0.003*	0	<0.005*	0	0.008
0.025	<0.004*	0.1	0.128	1	0.724	0.5	0.346
0.050	0.056	0.2	0.248	2	2.456	1.0	0.593
0.100	0.114	0.4	0.564	4	5.360	1.5	1.246
0.200	0.198	0.8	1.028	8	9.960	2.0	3.258
0.400	0.368	1.6	2.004	16	17.560	3.0	1.373
0.800	0.772	3.2	4.040	32	35.960	4.0	3.113
1.600	1.882	6.4	7.840	64	74.400	8.0	4.426
3.200	3.660	12.8	15.000	128	142.000	16.0	5.972
6.400	8.160			256	370.400		

* indicates below detection.

Table 3. Results of repeated bioassays with Spanish and Canadian cultures on different culture sediments, showing numbers of cocoons (CCAD) and young (YGAD) produced per adult *T. tubifex* (standard deviation in parentheses).

BARAZAR		LONG POINT 7		LONG POINT 8	
CCAD	YGAD	CCAD	YGAD	CCAD	YGAD
6.1 (0.6)	5.1 (7.0)	8.3 (0.9)	15.8 (2.2)	6.6 (0.6)	9.6 (1.0)
8.4 (0.4)	21.0 (4.0)	8.7 (0.8)	9.1 (0.6)	10.9 (0.5)	28.8 (3.3)
6.3 (1.9)	12.5 (11.1)	7.3 (1.4)	15.1 (1.6)	10.8 (0.5)	38.3 (3.5)
6.9 (0.3)	13.9 (4.3)	9.9 (0.8)	18.2 (3.2)	9.4 (1.3)	22.8 (2.0)
8.7 (0.5)	27.4 (4.9)	8.8 (0.4)	15.5 (0.9)	9.5 (1.2)	25.7 (4.7)
9.0 (0.1)	26.6 (8.6)	10.1 (1.1)	20.5 (3.0)	6.6 (0.6)	9.6 (1.0)
6.9 (0.8)	8.0 (3.9)	8.3 (0.9)	18.4 (2.1)	10.7 (0.5)	32.8 (3.6)
		7.9 (0.4)	14.0 (2.1)	10.3 (0.8)	19.1 (3.3)
		10.8 (0.7)	31.1 (3.8)	9.0 (0.7)	26.7 (2.0)
		10.5 (0.5)	18.4 (3.2)	10.4 (1.6)	45.4 (1.7)
				10.6 (1.3)	36.2 (9.1)
				9.6 (1.4)	29.1 (5.5)

were performed at different laboratories using different culture sediment the variability in response could be attributed to differences between the populations, the laboratories or the sediment.

To exclude methodological differences or effects due to differences in sediment quality, reproduction was compared in four paired tests on LP sediment using individuals from Canadian and Spanish populations. The Spanish worms were maintained in culture at NWRI for several months prior to the experiment, using LP sediment. These experiments were conducted at two temperatures: 20 ° and 22.5 ± 0.5 °C.

The results from two experiments at each temperature are presented in Table 4. At the standard test temperature (22.5 °C) the results are in the range observed in the different culture sediments (BAR, LP7 and LP8 in Table 3), with the exception of Spanish cocoon production in the first experiment. The two populations were compared using the Mann Whitney test. In both experiments at 22.5 °C cocoon production was significantly ($P<0.05$) lower for the Spanish worms, and in the second experiment the number of young produced was significantly ($P<0.05$) lower. At 20 °C the opposite occurred; cocoon production was significantly ($P<0.05$) lower in the Canadian population in both experiments and young production lower ($P<0.05$) in the first experiment.

Spanish *T. tubifex* are smaller than those from the Canadian culture population. For 11 sexually mature animals randomly selected from the two populations

203

Table 4. Comparison in number of cocoons (CCAD) and young (YGAD) produced by Spanish and Canadian cultures of *T. tubifex* in sediment from Long Point at 20 °C and 22.5 °C (standard deviation in parentheses).

	20 °C CCAD	20 °C YGAD	22.5 °C CCAD	22.5 °C YGAD
Exp 1				
Spain	7.0 (0.6)	22.3 (3.4)	4.3 (2.2)	18.8 (9.4)
Canada	5.0 (0.4)	4.6 (2.2)	10.3 (1.3)	27.4 (5.6)
Exp 2				
Spain	6.7 (0.8)	17.2 (8.6)	8.6 (1.0)	19.0 (4.6)
Canada	4.9 (0.2)	7.0 (1.2)	10.4 (0.3)	30.6 (2.3)

Table 5. Summary of reproduction in Spanish and Great Lakes reference sites, with some sediment characteristics.

	Mean	Range
Great Lakes ($n = 163$)		
Cocoons per adult	9.1	3.3–11.8
Young per adult	24.6	1.1–48.9
LOI %	11.4	1.0–38.7
Sand%	32.8	0.0–99.8
Silt%	32.0	0.0–86.3
Spain ($n = 4$)		
Cocoons per adult	8.6	6.9–10.5
Young per adult	12.9	1.2–35.5
LOI %	5.2	2.4–8.8
Sand%	46.1	5.8–91.6
Silt%	46.5	5.8–73.4

the mean wet weights were respectively, 5.6 mg (S.D. 2.6) and 8.1 mg (S.D. 1.4) for Spanish and Canadian worms. However, the biomass of mature worms does not appear to influence number of cocoons ($r^2 = 0.03$, $n = 56$) or young ($r^2 = 0.01$, $n = 56$) produced per adult in Canadian worms. In Spanish worms a similar relationship between initial biomass and number of cocoons ($r^2 = 0.01$, $n = 35$) and young ($r^2 = 0.001$, $n = 35$) was observed.

As part of a programme to develop biological sediment objectives, a large number of reference sites have been sampled in the North American Great Lakes (Reynoldson et al., 1995). From these data we have provided information on the range of cocoon and young production in *T. tubifex* in a variety of sediments from 163 Great Lakes sites as well as four Spanish sites (Table 5). The results from the few unpolluted Spanish sites so far examined show the average number of cocoons produced per adult again to be lower than

the average for the Great Lakes but well within the range observed in the much larger Great Lakes data set. Average young production at Spanish reference sites is about half that observed in the Great Lakes. The organic content (percent LOI) at the Spanish reference sites is also generally lower, the average value being half that of the Great Lakes sites.

Growth rates

We have also investigated the growth behaviour of the two populations of *T. tubifex*. Figure 1 shows the general pattern in somatic growth of worms at 25 °C. There is a strong relationship between initial size and somatic growth and smaller individuals have much higher rates of growth. In the size range of individuals tested (Spain 0.30–5.91 mg; Canada 0.37–6.24 mg), the Spanish population appears to show a logarithmic relationship with initial size ($r^2 = 0.67$), while a linear relationship ($r^2 = 0.46$) provided the best fit for the Canadian culture, although we would expect the logarithmic growth phase to occur if smaller Canadian worms had been included. The wet weights of newly hatched (1 day old) individuals are, for Spanish and Canadian worms respectively, 0.07 (S.D. 0.01) mg and 0.08 (S.D. 0.00) mg.

To compare growth differences between the Spanish and Canadian populations over a range of temperatures we used animals that were larger than 1.0 mg initial weight and excluded those individuals that became sexually mature over the period of the experiment. The Spanish population has a consistently lower growth than the Canadian, and at lower temperatures (10–15 °C) the Canadian worms have growth rates that are 2–4 times greater (Figure 2), however, these differences are not statistically significant (*t*-test; $P > 0.05$) except at 30 °C.

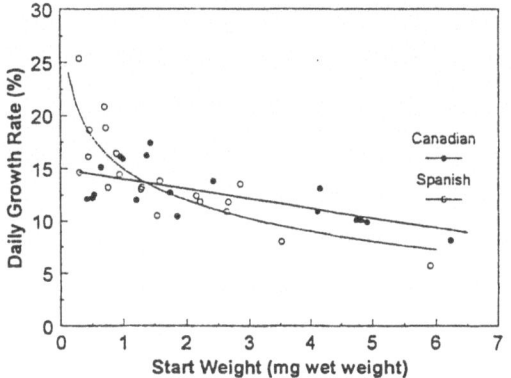

Figure 1. Effect of size on growth rates of two populations of *T. tubifex*.

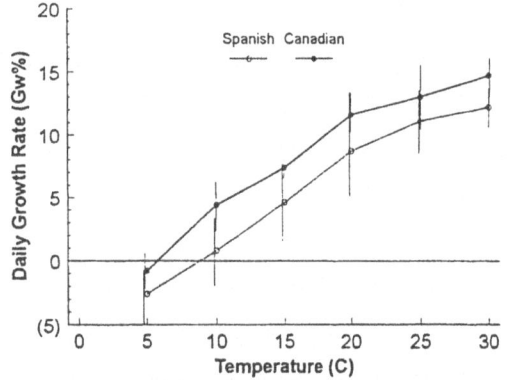

Figure 2. Relationship between temperature and growth rate (mean with S.D.) in two populations of *T. tubifex*.

Sensitivity to reference toxicants

To compare relative sensitivities to toxicants we examined the acute toxicity of the two populations to four selected chemicals. LC50 values were calculated for 24, 48, 72 and 96 h and are presented in Table 6 for each time period and toxicant. In addition, calculated LC50 values for cadmium and copper for worms from the Spanish population estimated at the Universidad del Pais Vasco are also presented together with selected relevant literature values (Table 6).

Based on LC50 data, the Canadian *Tubifex* are slightly more tolerant to all four contaminants. The response of the two populations in lindane shows both to be relatively insensitive up to 72 h.

Discussion

There are only a few published examples (Maltby, 1984; Mearns et al., 1986; Munzinger & Monicelli,

1991) of intercalibration tests for bioassays performed by different laboratories using different cultures of the same species. The performance of *Daphnia magna* cultures is examined periodically in a number of laboratories of the European Community (EC) in 'ring tests', using a reference chemical and standard protocols (Cabricenc, 1986).

Differences in growth rates and reproduction in lentic and lotic populations of *T. tubifex* have been described by Poddubnaya (1980). Lotic forms (Spanish worms) would be expected to be more eurybiotic, with more rapid growth and higher reproduction than lentic forms (Canadian worms) which should be adapted to the relatively stable conditions in the deep and colder waters of lakes. These experiments have shown that there are differences in growth and reproduction in the two populations. The growth rates showed the same general pattern (Figure 1). However, the Spanish worms had lower growth rates at all temperatures (Figure 2). This difference may be important if an endpoint based on growth is used in bioassay studies. If such an endpoint is considered then standard size individuals should be used as small animals (<1.0 mg) have higher growth rates than large animals. Two points are also of ecological interest. First, the different growth rates (Figure 2) suggest that these two populations are adapted to different environmental conditions. Second, the greater variability of growth rates of the Spanish worms (Figure 2) may indicate an adaptation of the Spanish population to the fluctuating conditions that occur in stream environments.

Reproduction is lower in the Spanish worms at 22.5 °C but higher at 20 °C. The ability to reproduce and grow over a broader range of temperature is an adaptation one would expect in a population exposed to the greater temporal variability in temperature observed in stream compared to lake environments. Despite the lower number of young produced by the Spanish worms it is within the range of values obtained from extensive work on reference sites in the Laurentian Great Lakes. Whether the differences between the two populations of *T. tubifex* have been affected by several generations in standardized laboratory conditions is unknown.

There are practical concerns arising from variability in sensitivity to toxicants in different populations. First, the reliability of chemical criteria derived from standard LC50 data in dose response tests; second, accurately establishing a difference between a test response and literature values for the same response. These two populations appear to be more tolerant to

Table 6. 24–96 h LC50 (mg l^{-1}) values calculated for cultures of *T. tubifex* from Spain and Canada with four chemicals.

Chemical	24 h LC50	48 h LC50	72 h LC50	96 h LC50
Copper				
Spain (NWRI)	0.49	0.18	0.11	0.07
Canada (NWRI)	0.86	0.26	0.16	0.09
Brkovic-Popovic 1977	1.38	0.89		
Cadmium				
Spain (UdPV)	1.2	0.9	0.6	0.4
Spain (NWRI)	6.5	3.6	2.4	1.7
Canada (NWRI)	9.8	6.5	5.4	3.2
Brkovic-Popovic 1977	1.2	0.7		
Chapman et al. 1982				0.32
Chromium				
Spain (UdPV)	57.4	33.2	16.5	9.8
Spain (NWRI)	122.1	54.5	28.7	15.5
Canada (NWRI)	137.8	95.5	49.5	38.1
Brkovic-Popovic 1977	86.0	4.6		
Lindane				
Spain (NWRI)	none	none	5.0	3.5
Canada (NWRI)	none	none	4.4	3.9

UdPV - tests conducted in Bilbao, Spain.
NWRI - tests conducted in Burlington, Canada.

cadmium and chromium than reported by Brkovic-Popovic & Popovic (1977) and Chapman et al. (1982) but more sensitive to copper (Table 6). Unpublished data of similar acute tests performed in Spain with immature worms of several sizes show lower but very similar LC50 values for cadmium and chromium to the values obtained at NWRI with Spanish worms (Table 6). These data suggest that the implications for developing criteria are small given the safety margins incorporated into most procedures for setting criteria from toxicity data. Similarly the implications for other comparisons may be relatively minor given the amount of variability which presently exists in such tests. Literature data show considerable variation usually because tests differ from each other in the temperature used, the alkalinity of water, the age of organisms or the conditions of the test.

These data show differences in the responses in the measured endpoints in the two culture populations. Applying the responses from data derived from one population to results from a second population should be done cautiously. For example, the use of reference data on *T. tubifex* for the Great Lakes would be inappropriate for determining whether or not reproduction has been impaired in sediments from Spanish rivers. By rearing separate cultures from two different populations in the same laboratory we have eliminated differences in handling and maintenance. Therefore, the differences observed are attributes of the different populations. One reason given for the variability in responses observed in different laboratories is the genetic variability of the cultures derived from different natural populations (Baird et al., 1989; Soares et al., 1992). Anlauf (1994) has described genetic differences in *T. tubifex* populations from lacustrine habitats in Germany. These variants showed differences in growth and reproduction, indicating that physiological differences should be established in a test population before it is used in toxicity testing. The Canadian and Spanish populations do show differences in the genome that could be responsible for the difference in relative variability in the two populations. Starch gel electrophoresis data (Coates, pers. comm.) showed the Canadian population to be homozygotic at 19 loci for 13 allozymes, and the Spanish population to be homozygotic at 13 loci for 10 allozymes.

In order to avoid confounding effects due to the source population of test organisms, quality criteria for survival, reproduction and reference toxicants should be established permitting comparison of results from different laboratories. In aquatic oligochaete toxicity tests, as far as we know, only Roembke (1989) and ASTM (1994) have proposed quality criteria for validating such tests. In order to produce consistent data in toxicity testing with *T. tubifex* at a minimum: animals of a standard age or size range should be used; reproductive tests should measure their first reproduction; a standard temperature should be used in all laboratories; and, a standard sediment should be used for culturing and as a positive control. Finally, we suggest that limits be established for selected reference toxicants to allow intercalibration of different cultures of *T. tubifex* being used in toxicity tests, as is done for other indicator species. This would provide a basis for establishing the condition and sensitivity of the worms used in sediment toxicity testing.

Acknowledgements

The authors wish to thank Ms Sherri Thompson for her assistance in the laboratory, and the Basque Government for financial support (project AMB 93 0460) to the junior authors.

References

Anlauf, A., 1994. Some characteristics of genetic variants of *Tubifex tubifex* (Müller, 1774) Oligochaeta: Tubificidae) in laboratory cultures. Hydrobiologia 278: 1–6.

ASTM, 1994. Standard Guide for Conducting Sediment Toxicity Tests with Freshwater Invertebrates. Designation E 1383-94. American Society for Testing and Materials, Philadelphia.

Baird, D. J., I. Barber, M. Bradley, P. Calow & A. Soares, 1989. The *Daphnia* bioassay: a critique. Hydrobiologia 188/189: 403–406.

Bailey, H. C. & D. H. W. Liu, 1980. *Lumbriculus variegatus*, a benthic oligochaete as a bioassay organism. In J. C. Eaton, P. R. Parrish & A. C. Hendricks (eds), Aquatic Toxicology, ASTM STP 707, American Society for Testing and Materials: 205–215.

Bailey, R. C., K. E. Day, R. H. Norris & T. B. Reynoldson, 1995. Macroinvertebrate community structure and sediment bioassay results from nearshore areas of North American Great Lakes. J. Great Lakes Res. 21: 42–52.

Brkovic-Popovic, I. & M. Popovic, 1977. Effects of heavy metals on survival and respiration rate of tubificid worms: Part I: Effects on survival. Envir. Pollut. 13: 65–72.

Cabricenc, R., 1986. Exercise d'intercalibration concernant une methode de determination de l'ecotoxicite a moyen terme des substances chimiques vis-a-vis des daphnies. Unpublished EC Report. Contract W/63/476 (214). Ref. I.R.C.H.A.D. 8523. Vert-le-Petit, 20 pp.

Casellato, S. & P. Negrisolo, 1989. Acute and chronic effects of an anionic surfactant on some freshwater tubificid species. Hydrobiologia 180: 243–252.

Chapman, P. M. & D. G. Mitchell, 1986. Acute tolerance tests with the oligochaetes *Nais communis* (Naididae) and *Ilyodrilus frantzi* (Tubificidae). Hydrobiologia 137: 61–64.

Chapman, P. M., M. O. Farrel & R. O. Brinkhurst, 1982. Relative tolerances of selected aquatic oligochaetes to individual pollutants and environmental factors. Aquat. Toxicol. 2: 47–67.

Dermott, R. & M. Munawar, 1992. A simple and sensitive assay for evaluation of sediment toxicity using *Lumbriculus variegatus*. Hydrobiologia 235/236: 407–414.

Keilty, T. J., D. S. White & P. F. Landrum, 1988. Short-term lethality and sediment avoidance assays with endrin contaminated sediment and two oligochaetes from Lake Michigan: *Stylodrilus heringianus* (Lumbriculidae) and *Limnodrilus hoffmeisteri* (Tubificidae). Arch. Envir. Contam. Toxicol. 17: 95–101.

Lauritsen, D. D., S. C. Mozley & D. S. White, 1985. Distribution of oligochaetes in Lake Michigan and comments of their use as indices of pollution. J. Great Lakes Res. 11: 67–76.

Learner, M. A. & R. W. Edwards, 1963. The toxicity of some substances to *Nais* (Oligochaeta). Proc. Soc. Wat. treat. Exam. 12: 161–168.

Maltby, L., 1984. Responses of *Gammarus pulex* (Amphipoda, Crustacea) to metalliferous effluents-identification of toxic components and the importance of interpopulation variation. Envir. Pollut. 84: 45–52.

Mearns, A. J., R. C. Swartz, J. M. Cummins, P. A. Dinnel, P. Plesah & P. M. Chapman, 1986. Interlaboratory comparison of a sediment toxicity test using the marine amphipod, *Rhepoxynius abronius abronius*. Mar. Envir. Res. 25: 99–124.

Milbrink, G., 1983. An improved environmental index based on the relative abundance of oligochaete species. Hydrobiologia 102: 89–97.

Milbrink, G., 1987. Biological characterization of sediments by standardized tubificid bioassays. Hydrobiologia 155: 267–275.

Munzinger, A. & F. Monicelli, 1991. A comparison of the sensitivity of 3 *Daphnia magna* populations under chronic heavy metal stress. Ecotoxicol. Envir. Safety 22: 24–31.

Poddubnaya, T. L., 1980. Life cycles of mass species of Tubificidae. In R. O. Brinkhurst & D. G. Cook (eds), Aquatic Oligochaete Biology. Plenum Press, New York: 175–184.

Reynoldson, T. B., S. P. Thompson & J. L. Bamsey, 1991. A sediment bioassay using the tubificid oligochaete worm *Tubifex tubifex*. Environ. Toxicol. Chem. 10: 1061–1072.

Reynoldson, T. B., K. E. Day, C. Clarke & D. Milani, 1994. The effects of indigenous animals on chronic endpoints in freshwater sediment toxicity tests. Envir. Toxicol. Chem. 13: 973–977.

Reynoldson, T. B., K. E. Day, R. C. Bailey & R. H. Norris, 1995. Methods for establishing biologically based sediment guidelines for freshwater quality management using benthic assessment of sediment. Aust. J. Ecol. 20: 198–219.

Roembke, J., 1989. Study of the toxicity for *E. albidus* (Enchytraeidae). Generic Report, Battele Institute. V. Projekt Nr-R66 238.

Roembke, J. & Th. Knacker, 1989. Aquatic toxicity test for enchytraeids. Hydrobiologia 180: 235–242.

Soares, A. M. V. M., D. J. Baird & P. Calow, 1992. Interclonal variation in the performance of *Daphnia magna* Straus in chronic bioassays. Envir. Toxicol. Chem. 11: 1477–1483.

Wiederholm, T., A. M. Wiederholm & G. Milbrink, 1987. Bulk sediment bioassay with five species of fresh-water oligochaetes. Wat. Air Soil Pollut. 36: 131–154 .

Hydrobiologia **334**: 207–217, 1996.
K. A. Coates, Trefor B. Reynoldson & Thomas B. Reynoldson (eds), Aquatic Oligochaete Biology VI.
© 1996 *Kluwer Academic Publishers.*

Morphogenesis of helical fibers in haplotaxids

Magda de Eguileor[1], Roberto Valvassori, Giulio Lanzavecchia[1] & Annalisa Grimaldi[1]
Dipartimento di Biologia, Università di Milano, via Celoria 26, I-20133 Milano, Italy
[1]*present address: Facoltà di Scienze di Varese, via Ravasi 2, I-21100 Varese, Italy*

Key words: haplotaxids, helical muscles, ultrastructure, growth

Abstract

There are two different muscle fiber types in haplotaxids. The pseudo-circomyarian type is typical of *Haplotaxis gordioides* and the flattened circomyarian type of *Pelodrilus leruthi*. The mechanisms of growth in fiber size and in fiber number of the two fiber types in the hindmost region of adult specimens have been studied ultrastructurally. The increase in length and girth of the muscle fiber is always the result of the insertion of new myofilaments in the peripheral zones of the muscle cells. The increase in the number of fibers seems to be due to division of differentiated muscle cells.

Introduction

The ultrastructure of the helical fibers in oligochaetes is well known, and the different variations have been described (Lanzavecchia, 1977; Lanzavecchia *et al.*, 1985, 1987, 1989; Valvassori & de Eguileor, 1991). In the round circomyarian fibers the closed circle of contractile material is organized in sarcomeres disposed helically around a longitudinal cytoplasmic axis. In the flattened circomyarian fibers, the sarcomeres are similarly placed to form helical bands, but the contractile material is arranged on the two opposite flat sides of the fiber so that in each muscle cell the contractile layers are parallel.

The round circomyarian fibers form a system of closed helices that are continuous along the entire surface of the cell. In contrast, the flattened circomyarian fibers have only a partially closed system, with an opening in the nuclear region. These fibers grow lengthwise in the same way as cross-striated fibers and, therefore, do not pose problems of interpretation. It is easy to explain the growth in diameter of the open systems, but difficult to explain it for continuous closed systems. In the circomyarian fibers, structural restrictions are imposed by the specific geometry of the helices and it is not easy to imagine how the new contractile material is inserted and organized within this closed geo-metric system. An additional problem is the increase in number of fibers during the life of the worm.

The two problems were investigated in *Haplotaxis gordioides* Hartmann, 1821 and *Pelodrilus leruthi* Hrabe, 1953 by observing the changes in number and structure of the longitudinal fibers in the zone of growth, specifically in those subterminal metameres in which, even in the adult, almost all of the major processes of morphogenesis of different structures takes place. Of course, in the terminal metameres of the worm, typical images of myoblast differentiation are seen.

The muscle fibers of *H. gordioides* are pseudo-circomyarian (due to the connection between the central cytoplasmic axis and the cytoplasmic peripheral zone) and those of *P. leruthi* (Fig. 13) are flattened circomyarian as shown in a previous paper (de Eguileor *et al.*, 1990).

Materials and methods

Pieces of *H. gordioides* and *P. leruthi* were fixed in 2% glutaraldehyde in 0.1 M cacodylate buffer (pH 7.2) for two hours, washed overnight in 0.1 M cacodylate buffer (pH 7.2). The samples were post-fixed for two hours with 1% osmic acid in the same buffer (pH 7.2), washed in distilled water and stained *en bloc* for 2 h in the dark in 2% aqueous uranyl acetate. After standardized dehydration in an ethanol series, specimens were embedded in an Epon 812-Araldite mixture.

Sections were obtained with an LKB Ultrotome V. Semi-thin sections were stained with crystal violet and basic fuchsin and then observed under a Jenaval light microscope.

Thin sections stained with uranyl acetate and lead citrate were observed with a JEOL 1200 EX electron microscope. Glycogen was demonstrated according to the method of Thiery (1967).

Terminology: The helical fiber, in cross section, shows the same sequence of bands (I-A-I) as seen in longitudinal sections of cross-striated muscles. Accepting the terminology proposed in Rosenbluth (1965), we call these regions sarcomeres.

Results

Haplotaxis gordioides

The size of the fiber is increased by insertion of the new sarcomeres between those already present in the peripheral contractile ring. Increase of fiber size occurs in all the fibers of the body wall. It is massive (Figs 4, 5, 7, 8) in the fibers of subterminal metameres and in those fibers that are going to divide, while it is on a small scale in fibers located in the central metameres of the worm (Fig. 3; Fig. 1B). When the process is massive, newly synthesized filaments of myosin build up into large wedges, with as many as 350 elements. These do not interact with the actin filaments (Figs 4, 5, 7) which appear later. The actin filaments accumulate in adjacent zones (Fig. 4) and then these filaments are organized to form the typical sarcomeres. When the newly formed sarcomeres are organized, the I bands are divided by the Z-elements and the vesicles of the sarcoplasmic reticulum (Figs 7 & 8).

The number of fibers increases in all the metameres, although by slightly different modalities. In the subterminal metameres there are 3 to 4 layers of fibers (Fig. 2), with diameters increasing progressively from the epidermis to the body cavity.

The increase in fiber number appears to occur in the gross polygonal fibers of the innermost layer. In nearly all these fibers, there are wide zones composed of Z-elements, sarcotubules and filaments of actin and myosin, in no apparent order (Fig. 4). There are also masses of glycogen granules (Fig. 6). In individual fibers there are organized blocks of contractile material, with sarcomeres placed at angles to each other, as if there had been different planes of crystallization (Figs 9 & 10). Therefore, it seems that there may be some smaller sub-fibers within a single fiber. These hypothetical, adjacent sub-fibers are often separated by regular rows of continuous, closely-positioned vesicles (Figs 4, 9 & 10), which appear to originate in great wedges of vesicles near the sarcolemma (Fig. 9).

In the central metameres, there are 7 to 8 layers of fibers (Fig. 1) progressively larger from the circular muscle layer to the coelom. Images ascribable to an increase in number are seen above all in external fibers of the central metameres, close to the circular fibers (Figs 1 & 11). The small external fibers frequently form groups made up of a few, tightly packed elements. Each group of fibers is surrounded by the basal lamina (Fig. 11), which does not invade the spaces between one fiber and another (Fig. 12). In contrast, all of the larger fibers are surrounded by their own basal lamina and a connective tissue layer (Fig. 1).

Myoblasts cannot be seen in these two situations but do appear in the metameres of the pygidium.

Pelodrilus leruthi

The increase in fiber size is seen as an accumulation of myosin and actin filaments in the cytoplasmic zone containing the nucleus and mitochondria, turned toward the coelomic cavity (Figs 18 & 19), and in the appearance of new Y-shaped sarcomeres between those already existing on the faces of the fiber (Figs 14, 16 & 17; Fig. 24). The last condition produces a pattern of chevrons made up of several sarcomeres fitting within each other.

The increase in fiber number follows division of the fibers along planes longitudinal and radial with respect to the body wall. At the end of the process, each fiber has given rise to two adjacent parallel fibers (Figs 22 & 23). The division of the fiber starts on the coelomic oriented side in the form of two widely separated laminae with simple striation (Fig. 15). Each lamina elongates by a proliferation of the new sarcomeres that curve inward (Fig. 20). In time, this results in a system of double oblique striation, with each muscle fiber giving rise to two parallel fibers (Figs 21 & 22).

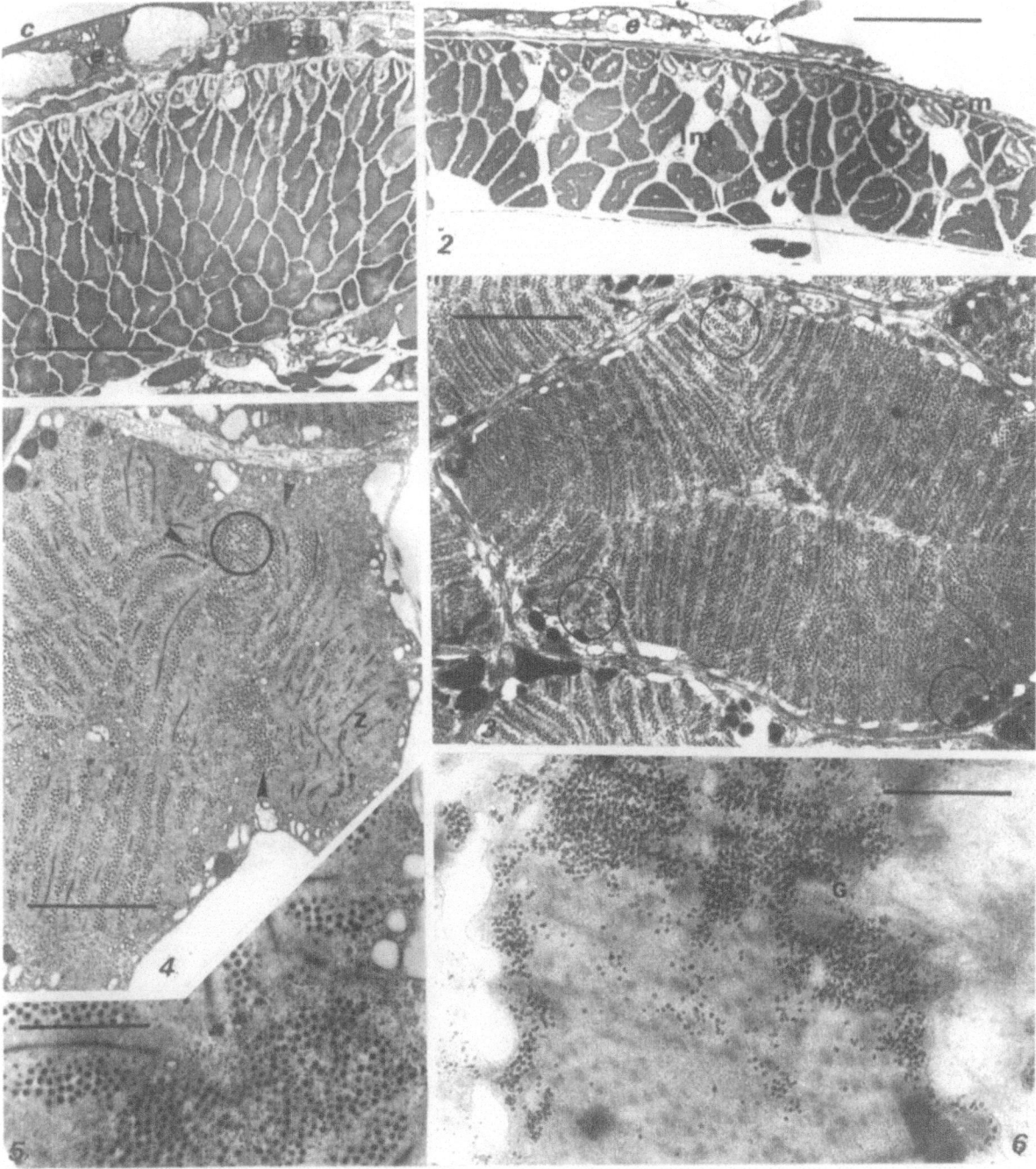

Figures 1–6. Haplotaxis gordioides.

Figs 1 & 2. Light micrographs of cross-sections of the body wall. Fig. 1, numerous series of longitudinal muscle fibers are present in the central region of the animal; Fig. 2, a smaller number of series is visible in the subterminal region. c, cuticle; e, epidermis; cm, circular muscle; lm, longitudinal muscle. Scale bar = 66 μm.

Fig. 3. Longitudinal fiber in cross-section. Growing areas (encircled) are visible. Scale bar = 3 μm.

Fig. 4. In a region of the fiber the spatial and numeric ratios of myofilament to Z-elements (z) and sarcotubules are greatly altered. The muscle fibers show, in this area, a disordered appearance (arrowheads). Scale bar = 2.8 μm.

Fig. 5. Detail of the encircled area in Fig. 4. Scale bar = 1.1 μm.

Fig. 6. Thiery reaction. A large amount of dark granules (glycogen, G) are visible. Scale bar = 1.3 μm.

210

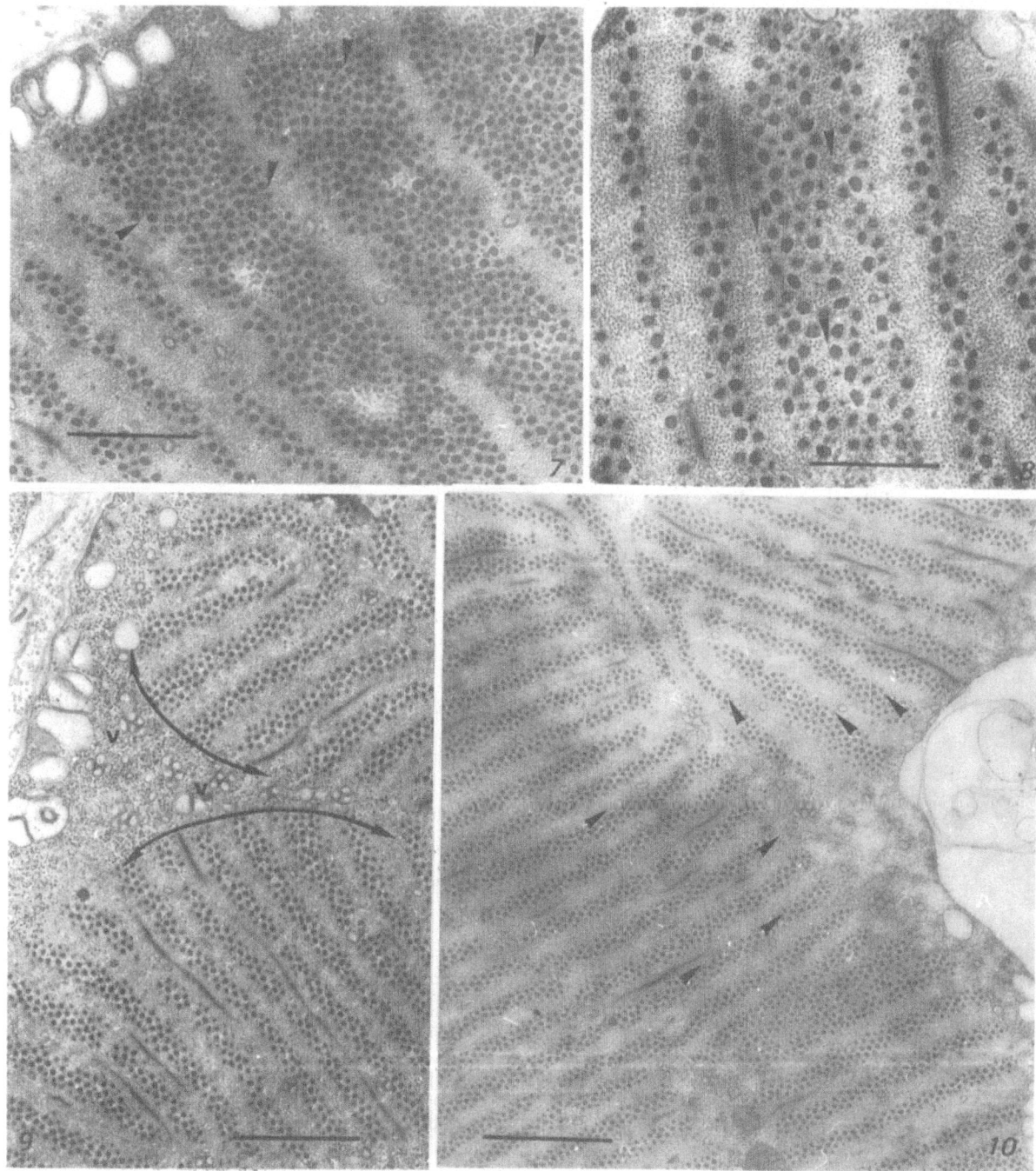

Figures 7–10. Haplotaxis gordioides.

Fig. 7. Few actin filaments are visible among myosin filaments crowded into large masses (arrowheads). Scale bar = 1 μm.

Fig. 8. The myosin filaments are subdivided into sarcomeres (arrowheads). Scale bar = 0.5 μm.

Fig. 9. Sarcoplasmic reticulum vesicles divide the contractile material into two structural blocks (curved arrows). Scale bar = 1.2 μm.

Fig. 10. In the fiber, two blocks of contractile material are easily recognizable because adjacent sarcomeres show different 'planes of crystallization' (arrowheads). Scale bar = 0.45 μm.

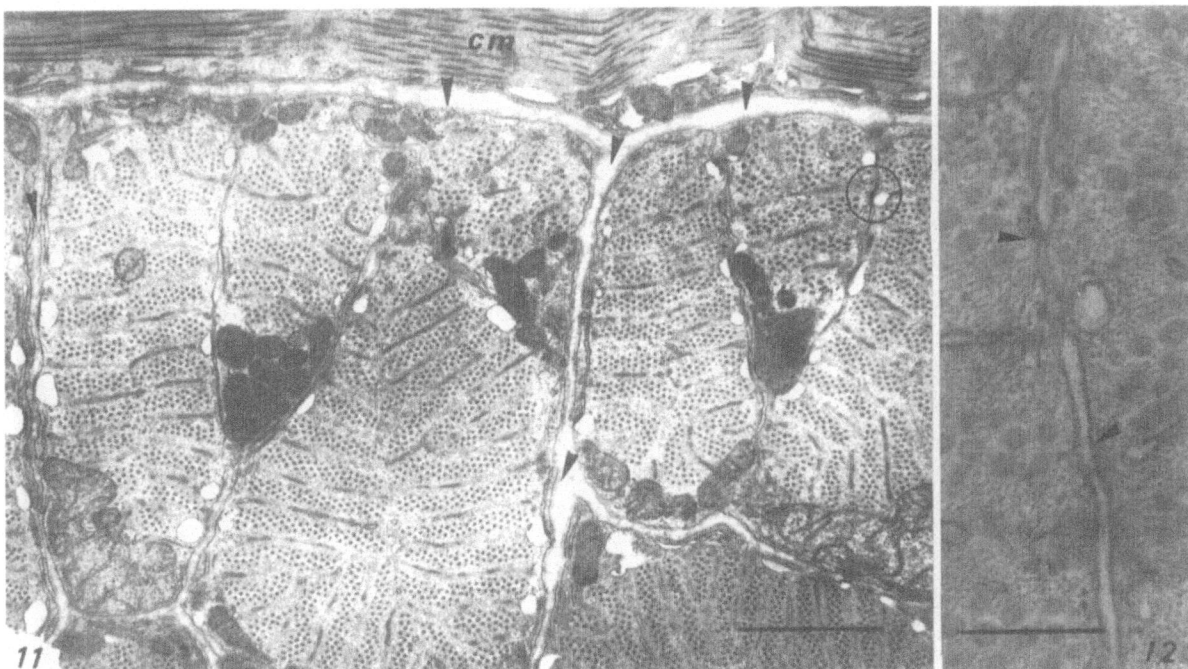

Figures 11–12. Haplotaxis gordioides.
Fig. 11. Groups of longitudinal fibers, close to the circular ones (cm), are bound by the basal lamina and connective tissues (arrowheads). Inside each group the fibers are in close contact. Scale bar = 2.8 μm.
Fig. 12. Detail of the encircled area of Fig. 11. The membrane of two fibers are in close contact (arrowheads). Scale bar = 0.35 μm.

Discussion

The growth of haplotaxid helical fibers involves both an increase of the mass of a single fiber and an increase of their number. The two phenomena are largely correlated, since an increased number of fibers is almost always preceded by a considerable increase in volume. As a consequence, even though the two processes are quite distinct, they occur more or less simultaneously, especially in the subterminal regions of the worms (Fig. 4).

Elongation of the fibers is always the result of insertion of new contractile material into the terminal regions and is similar to that reported in fibers with cross-striation (Fischman, 1972; Goldspink, 1979; Muntz, 1990; Naidoo, 1993). However, an increased diameter of the fiber is not the result of peripheral deposition of new material, since new sarcomeres must be inserted into a geometrically defined system. It is the result instead, of insertion of new contractile material from the periphery of the fiber toward the center (Fig. 3). This same type of centripetal growth is quite common for all embryonic helical fibers (de Eguileor *et al.*, 1987, 1993; Lanzavecchia *et al.*, 1989), and it

is also the system of growth for many organisms with radial symmetry.

In a cross-section of *H. gordioides,* the muscle cells which are simply growing in diameter show in the periphery of the fiber Y-shaped, i.e., chevron-shaped, structures. In *P. leruthi* there is an analogous process: on the lateral margins of the flattened circomyarian fibers there are more or less rounded protuberances in which the sarcomeres look like chevrons (Figs 14, 16 & 17).

The numbers of fibers increase in quite different ways in the two species, but in neither *H. gordioides* nor *P. leruthi* is there multiplication followed by differentiation of the myoblasts. Myoblasts are found only in the terminal metameres, and there are definitely none in the subterminal or central ones in which the increased number of fibers are seen. As a consequence, this increase occurs in perfectly differentiated cells, which grow and divide. In *H. gordioides* the multiplication process is characterized by an accumulation of myosin and of actin filaments as wedge-shaped or rectangular masses; a new formation of Z-elements and of sarcotubules; a gradual organization of the structures into sarcomeres; a notable accumulation of glycogen gran-

212

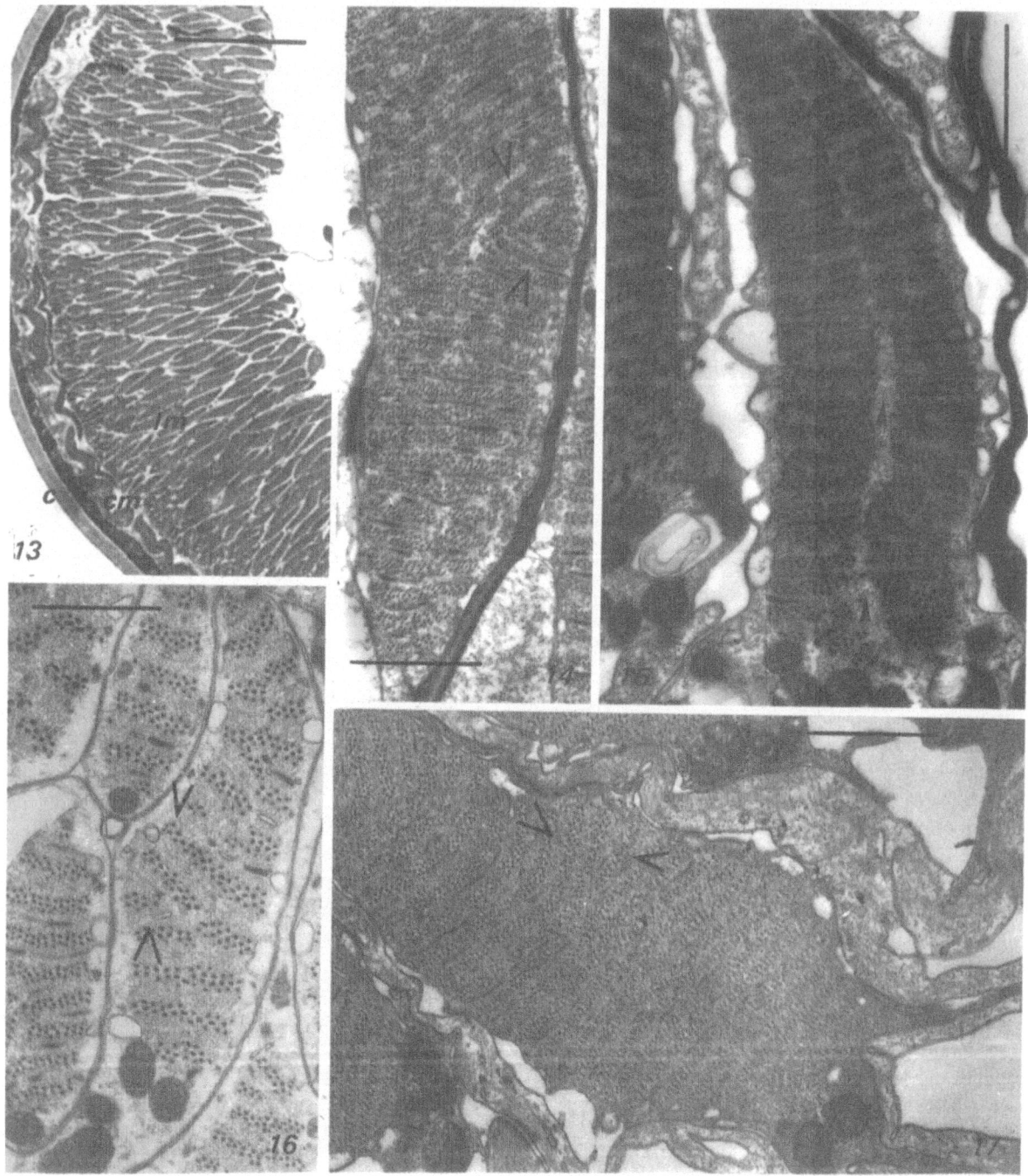

Figures 13–17. Pelodrilus leruthi

Fig. 13. Light micrograph of cross-sectioned body wall. c, cuticle; e, epidermis; cm, circular muscle; lm, longitudinal muscle. Scale bar = 100 μm.

Figs 14, 16 & 17. Sarcomeres forming chevron patterns (arrowheads) are frequently visible on the lateral margins of the fibers. Scale bars = 1.4 μm for Figs 14 & 17; = 1.2 μm for Fig. 16.

Fig. 15. During the division process, the two halves of the fiber become separated by a cytoplasmic layer (arrowheads). Scale bar = 1.3 μm.

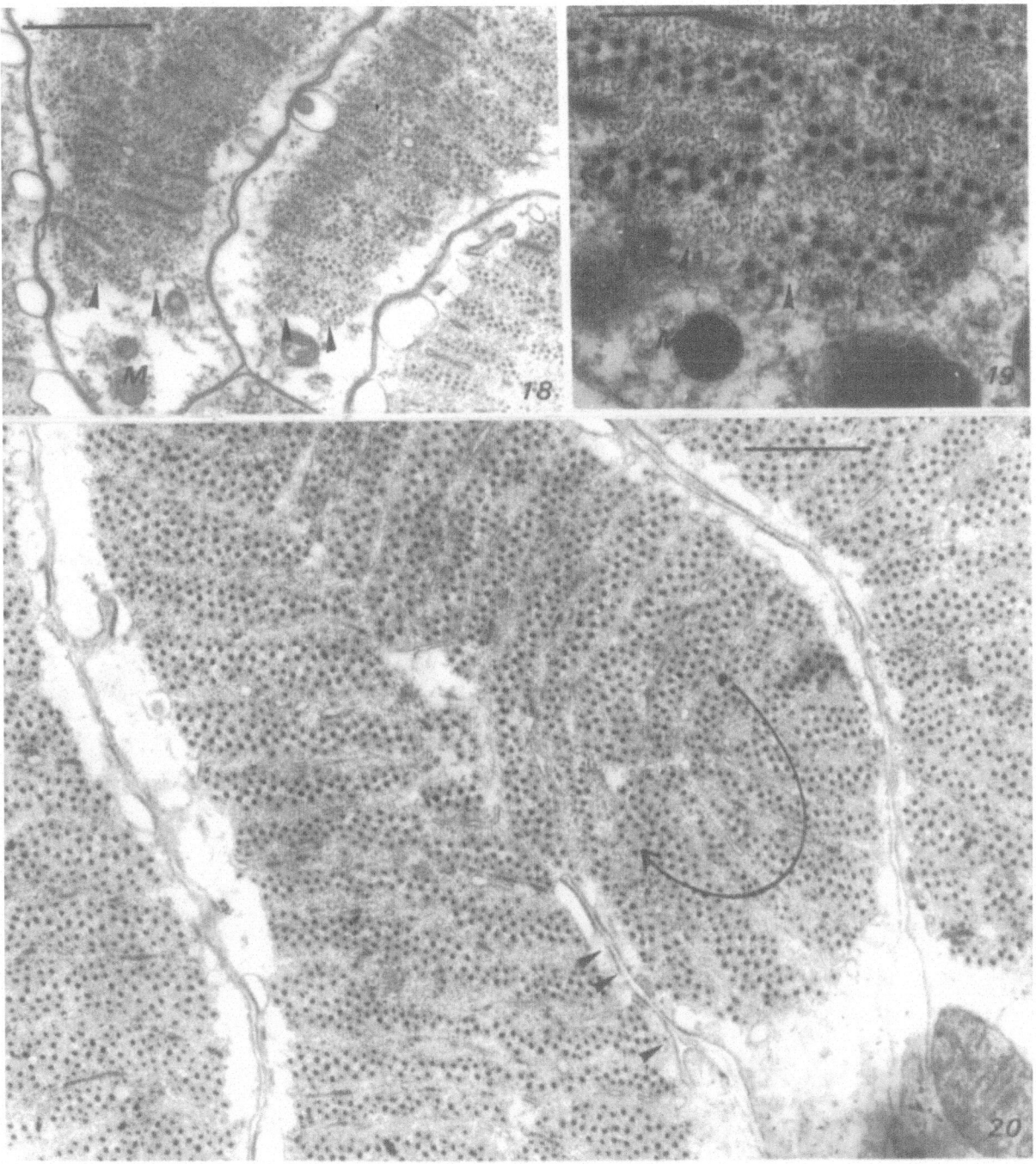

Figures 18–20. Pelodrilus leruthi
Figs 18 & 19. The new filaments (arrowheads) occupy the region of the fiber turned toward the coelom, easily recognizable due to the presence of mitochondria (M). Scale bars = 0.9 μm for Fig. 18; = 0.35 μm for Fig. 19.
Fig. 20. The newly formed sarcomeres fold inward (curved arrow). A fusion of vesicles is visible (arrowheads). Scale bar = 0.8 μm.

214

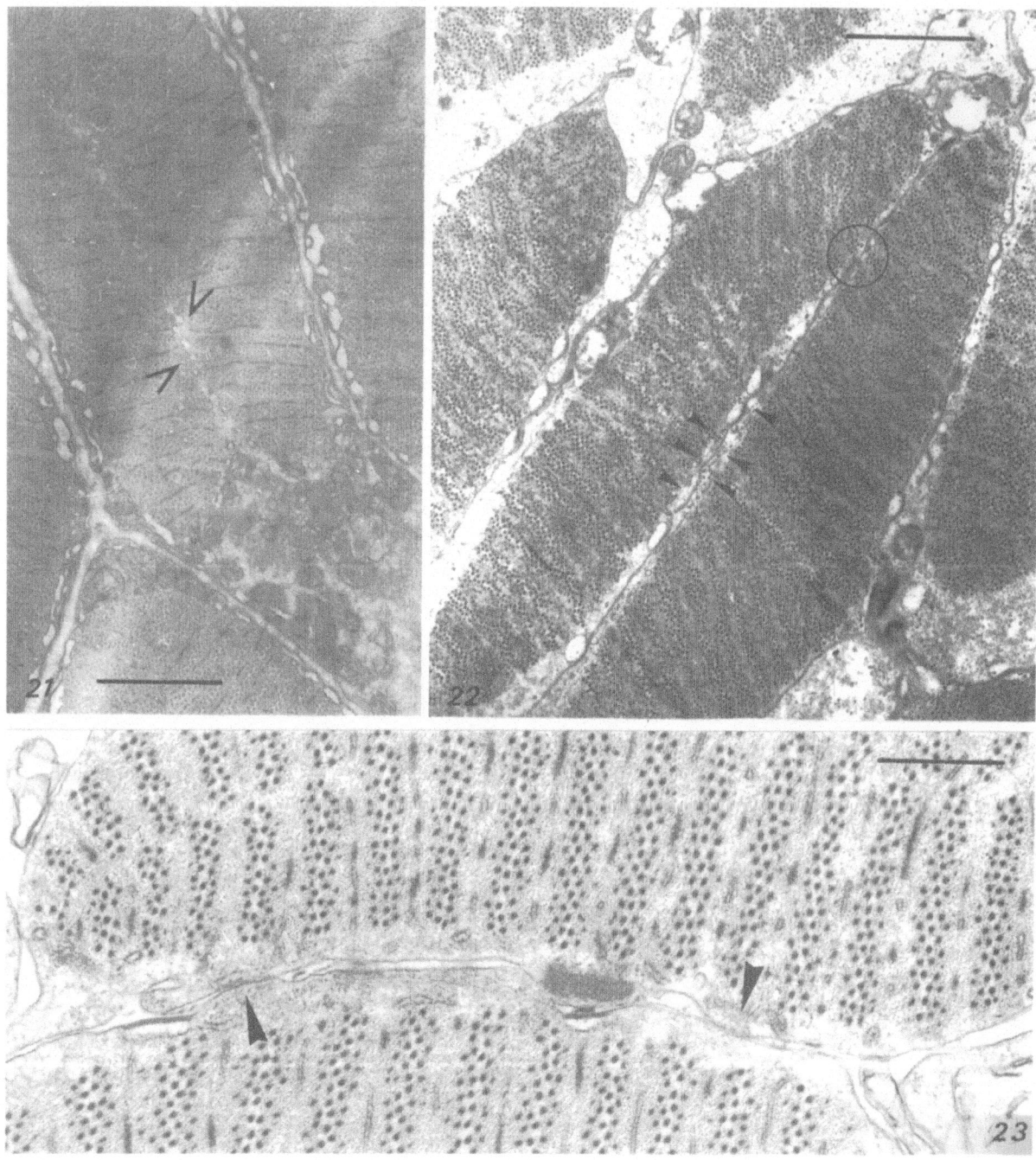

Figures 21–23. Pelodrilus leruthi
Fig. 21. The row of sarcoplasmic reticulum vesicles (arrowheads) underlies the two double obliquely striated systems inside a single fiber. Scale bar = 1.6 μm.
Fig. 22. Two 'daughter fibers', completely separated but in close contact (arrowheads). Scale bar = 1.6 μm.
Fig. 23. Detail of Fig. 22 (encircled area), with mechanical junctions (arrowheads). Scale bar = 0.6 μm.

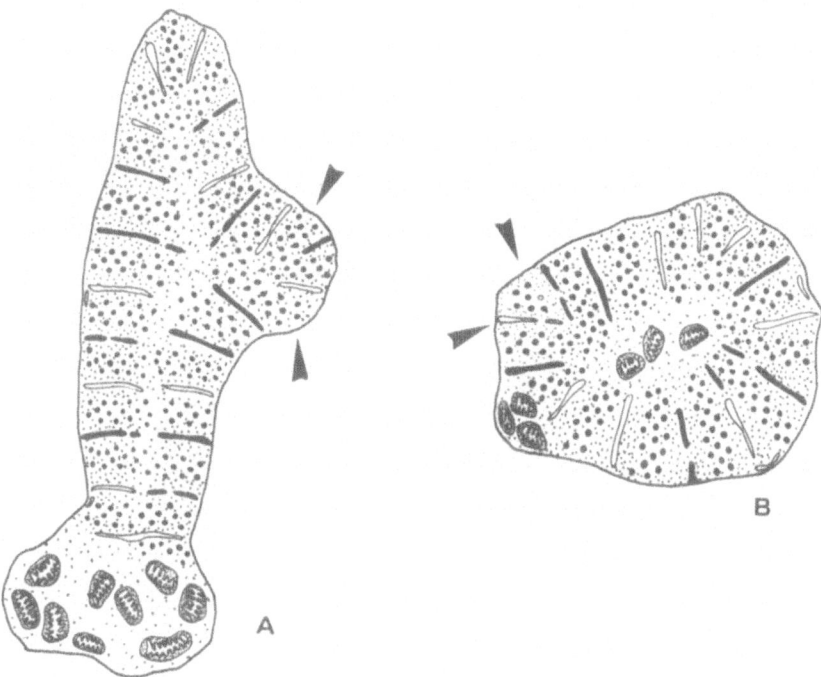

Fig. 24. Drawing representing the hypothetical ways of growth (arrowheads) of muscle fibers of *P. leruthi* (A) and *gordioides* (B).

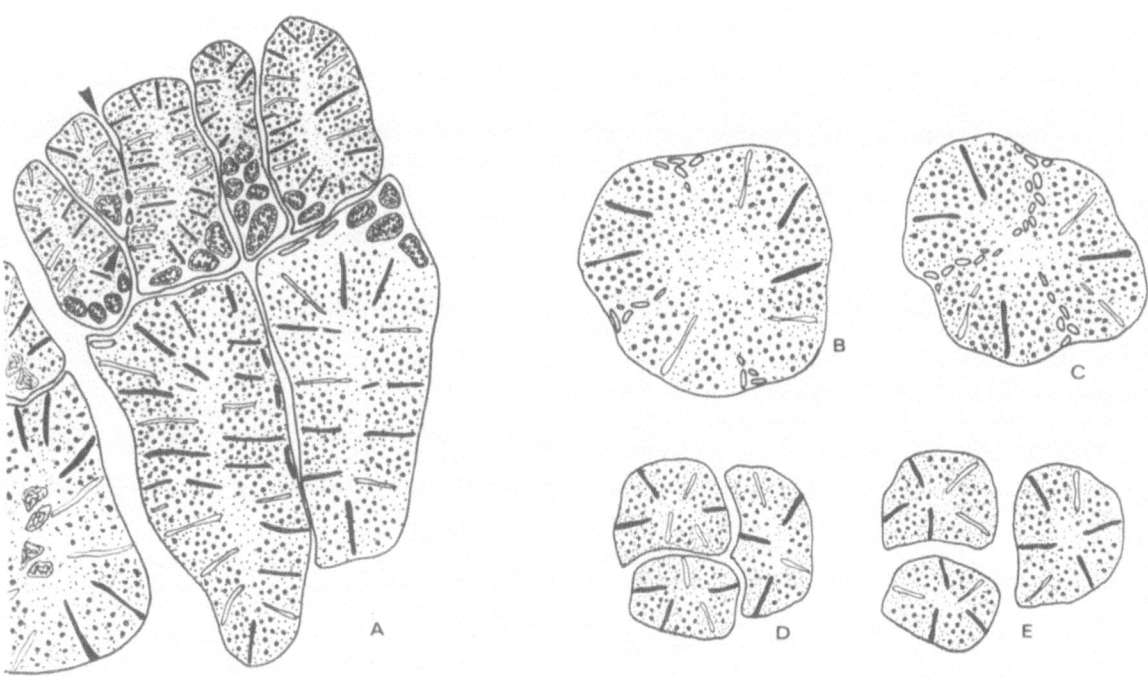

Fig. 25. H. gordioides. A: new formation of the outer longitudinal fibers underneath the circular muscles, in central segments (arrowheads). B, C, D, E: sequence of events leading to multiplication of fibers in subterminal segments. This sequence is typical of the large, inner, longitudinal fibers. After a great storage of new contractile material and glycogen, the fiber (B) divides itself into some daughter fibers (D–E).

216

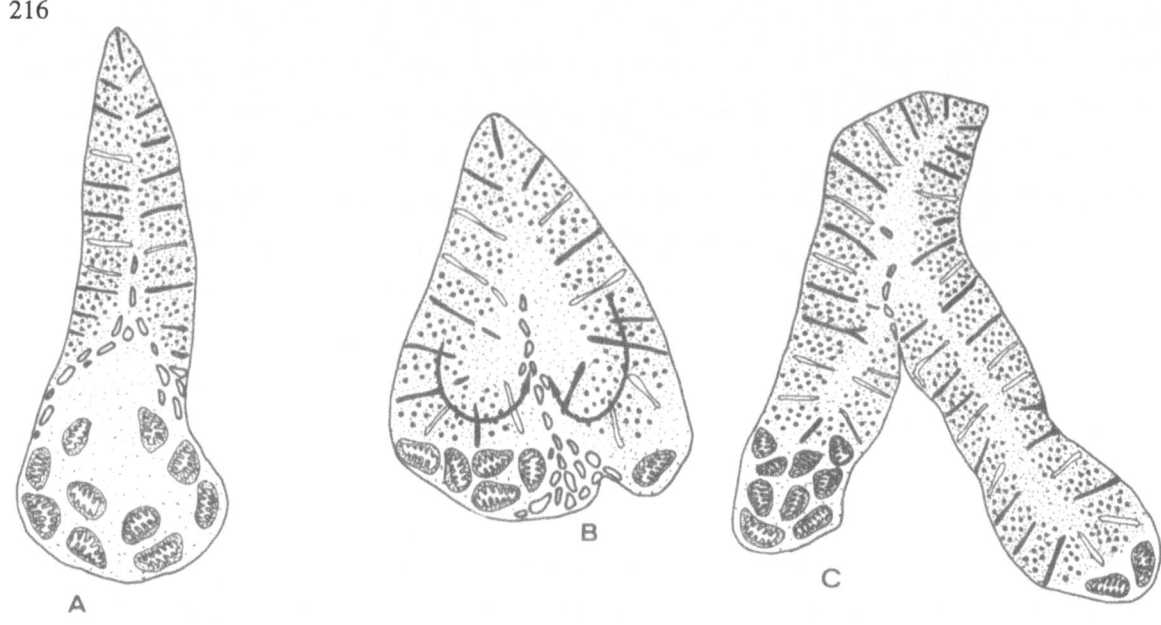

Fig. 26. P. leruthi. Process of division of the flattened circomyarian muscle fiber. The distancing of the two halves of the contractile material (A) is followed by a folding of newly formed sarcomeres into the inside (B). The double oblique striation is reassembled (C).

ules; a considerable production of membrane forming series of vesicles between the contractile and cytoplasmic material, defining the structural blocks (Figs 4, 9 & 10); a fusion of the membranes of the vesicles and formation of fibers smaller than the original ones; and a distancing of the newly formed fibers as a result of production of a basal lamina and insertion of collagen material. The series of events is summarized in Fig. 25.

In *P. leruthi*, the fibers begin to divide in the region toward the coelom (Fig. 15). The division appears to proceed along the longitudinal plane by a mechanism resembling the opening of a zipper (Fig. 20). In this case, glycogen granules (Fig. 6) and vesicles of sarcolemmal origin mark a line of division through the fibers. The following phenomena are then seen: a progressive distancing of the two halves of the fiber from each other; a synthesis of new contractile material along the two newly exposed halves; a folding of the newly formed sarcomeres into the inside of pre-existing ones; a reconstruction of the double oblique striation; and a production of vesicles that fuse and become the two new membranes of fibers (Figs 20 & 26).

In *H. gordioides*, the 'mother fiber' divides into several 'daughter fibers' of different dimensions while in *P. leruthi* the division results in two fibers of similar size and shape.

The construction of new fibers in *H. gordioides* and in *P. leruthi* can be related to the difference in shape of their respective fiber types.

References

de Eguileor, M., R. Valvassori & G. Lanzavecchia, 1987. Muscle development in enchytraeids Acta embriol. Morph. Exp. 8: 329–332.

de Eguileor, M., R. Valvassori, G. Lanzavecchia & G. Scarì, 1990. Body wall muscles in haplotaxids *Haplotaxis gordioides* and *Pelodrilus leruthi* (Annelida, Oligochaeta). Zoomorphology 110: 27–36.

de Eguileor, M., D. Barberis, G. Lanzavecchia, G. Scarì & R. Valvassori, 1993. Development of *Glossiphonia complanata* (Annelida, Hirudinea) and morphogenesis of helical muscle fibers. Rep. Develop. 24: 178–188.

Fischman, D. A., 1972. Development of striated muscle. In G. H. Bourne (ed), Structure and Function of Muscle. Vol. I Academic Press, New York: 75–148.

Goldspink, G., 1979. Development of muscle. In G. Goldspink (ed.), Differentiation and Growth of Cells in Vertebrate Tissue. Chapman & Hall, London: 69–99.

Lanzavecchia, G. 1977. Morphological modulations in helical muscles (Aschelminthes and Annelida). Int. Revue Cytol. 51: 133–186.

Lanzavecchia, G., M. de Eguileor & R. Valvassori, 1985. Superelongation in helical muscles of leeches. J. Muscle Res. Cell Motility 6: 569–584.

Lanzavecchia, G., M. de Eguileor, R. Valvassori & P. Lanzavecchia, 1987. Analysis and reconstruction of unusual obliquely striated fibers in lumbriculids (Annelida, Oligochaeta). J. Muscle Res. Cell Motility 8: 209–219.

Lanzavecchia, G., M. de Eguileor, R. Valvassori, L. Di Lernia & C. Cambiaso, 1989. Morphogenesis of body wall muscle fibers in *Enchytraeus minutus*. Hydrobiologia 180: 91–97.

Muntz, L., 1990. Cellular and biochemical aspects of muscle differentiation. Comp. Biochem. Physiol. 97B: 215–225.

Naidoo, P. R., 1993. EM evidence of myoblast origin in regenerating human skeletal muscle explants. Cell Biol. Int. 17: 825–831.

Rosenbluth, J., 1965. Ultrastructural organization of obliquely striated muscles in *Ascaris lumbricoides*. J. Cell. biol. 25: 495–515.

Thiery, J. P., 1967. Mise en évidence des polysaccharides sur coupes fines en microscopie electronique. J. Microscop. 6: 987–1018.

Valvassori, R. & M. de Eguileor, 1991. Helical muscles: models and adaptations. In G. Lanzavecchia & R. Valvassori (eds), Form and Function in Zoology. Mucchi, Modena. Selected Symposia and Monographs U.Z.I. 5: 141–162.

Hydrobiologia **334**: 219–227, 1996.
K. A. Coates, Trefor B. Reynoldson & Thomas B. Reynoldson (eds), Aquatic Oligochaete Biology VI.
© 1996 *Kluwer Academic Publishers.*

Histochemical characterization of secretions in the reproductive systems of two species of branchiobdellidans (Annelida: Clitellata): a new character for the phylogenetic matrix?

Stuart R. Gelder
Department of Biology, University of Maine at Presque Isle, Maine 04769, USA

Key words: Annelida, branchiobdellidans, histochemistry, reproductive system, phylogeny

Abstract

The species description of *Cambarincola holti* was emended to include variations in appearance of the peristomium, a description of the jaws, the shape of a normal spermatheca, and the correct description of the glandular atrium, granular rather than vacuolar. The glandular atrium contained a single type of secretion granule characterized as mucoprotein or glycoprotein, with an alkaline protein component. The vacuolar prostate gland secretions are uncharacterized as no stains would attach to the material. Spermatozoa were sometimes found in the spermathecal duct, but usually they were clumped together in the spermathecal bulb, or randomly distributed with each acrosome being in contact with a lining cell. The digitiform, glandular atrium of *Xironogiton victoriensis* is composed of nine regions based on the staining reaction of the granules. The granules were characterized as neutral mucoprotein, nonsulfated acid mucosubstance, glycoproteins, a combination of sulfated and nonsulfated acid mucosubstances, and totally non-staining granules. Spermatophores were observed only in *X. victoriensis* in depressions on the clitellum. The spermatozoa were separated from the cuticle by a layer of PAS staining material and covered by a lighter, PAS staining cap with a foamy outer layer that stains lightly with AB2.5. Further studies to characterize the secretions of the male reproductive organs may produce an additional character to add to the data matrices for phylogenetic analyses of these clitellates.

Introduction

Some of the major criteria used in the taxonomy of clitellate annelids are the location, anatomy, and to a lesser extent the histology, of the male and female reproductive systems. Comparing the location of these organs by simply using the segment number can lead to erroneous conclusions as the segments are numbered differently in oligochaetes, leeches and branchiobdellidans (Sawyer, 1986: 55; Holt & Opell, 1993: 252). The potential for confusion was removed by accepting a 'head' and numbering the body segments with arabic numerals for general use, then applying the oligochaete system, with the segmental ganglia being numbered with Roman numerals for comparative studies (Gelder & Brinkhurst, 1990). The major regions of the clitellate male reproductive system consist of sperm funnels, sperm tubes (vas efferens, vas deferens), glandular atri-

um, muscular atrium, bursa and penis. The glandular and muscular atria in branchiobdellidans were referred to as the spermiducal gland, following the term created by Beddard, and the ejaculatory duct, respectively (Holt, 1960). This usage has continued in branchiobdellidan studies by North American workers, but not by colleagues in Europe or Asia. Holt & Opell (1993: 253) state that 'spermiducal gland' is an incorrect construction and should be replaced by 'seminiducal gland.' This would appear to be an opportune time to return to the 'atrial' names for all branchiobdellidans.

Accessory structures to the glandular atrium of the male reproductive system in oligochaetes are generally referred to as the prostate. The morphology of the prostate varies greatly within the oligochaetes, and in many the cytology has not been described. The ultrastructural observations of prostate gland cells by Fleming (in Jamieson, 1981: 385) in *Tubifex tubifex*

Müller, 1774 show clearly that short ducts run from the cell bodies and then expand into cell-like storage structures that line the glandular atrium. No such cellular configurations have been reported in branchiobdellidans, but the term prostate is used for a localized swelling, 'prostate protuberance', or diverticulum, 'prostate gland', arising from the glandular atrium (after Holt, 1960). The cells forming these prostatic structures may contain similar (granular), or different (vacuolar) secretion bodies from those of the glandular atrium. The inclusion of the prostate in a character suite for a phylogenetic analysis of the clitellates including branchiobdellidans, therefore needs to be used with caution.

The secretions produced by the glandular atrium and prostate in branchiobdellidans have not been characterized histochemically. This preliminary study on two species of branchiobdellidans was designed to characterize the cell types and their secretions in the glandular atrium, and, where present, the prostate gland, using histochemical protocols. The information will enable the male secretions of the branchiobdellidans to be compared, and subsequently included in a character matrix for phylogenetic analysis.

Materials and methods

Specimens of live *Cambarincola holti* Hoffman, 1963 were removed from *Cambarus* (*Puncticambarus*) *robustus* Girard, 1852, *Orconectes* (*Gremicambarus*) *rhoadesi* Hobbs, 1949, *Orconectes* (*Procericambarus*) *placidus* Hagen, 1870, and *Orconectes* (*Procericambarus*) *rusticus* (Girard, 1852) collected at a total of five sites in Overton County, and *O.* (*P.*) *rusticus* also from one site in William County, Tennessee during September 1993. These are the first records of *C. holti* outside the type locality in Kentucky (Hoffman, 1963: 316). Live individuals of *Xironogiton victoriensis* Gelder & Hall, 1989 were removed from the chelipeds of *Pacifastacus* (*Pacifastacus*) *leniusculus leniusculus* (Dana, 1852) collected from the Stanislaus River, 40 km northeast of Modesto, Stanislaus County, California, by Mr Bruno Pernet in September 1992. Crayfish names, type citations and identifications were made following the nomenclature and keys in Hobbs (1989).

Twelve sexually mature specimens of *C. holti* and 16 of *X. victoriensis* were fixed with half their respective number in acetic acid-formalin-alcohol (AFA) (Brinkhurst & Gelder, 1991) and the other half in 10% formalin in 0.1 M phosphate buffer (pH 7.4).

These specimens were dehydrated in a graded ethanol series, cleared in xylene, infiltrated with Paraplast wax (m. pt. 56 °C, SIGMA) and sectioned serially for either a longitudinal or transverse aspect. Sections were stained by the Erhlich's hematoxylin (H) and eosin (E) method (denoted as H-E) for comparability with literature descriptions, and nuclear fast red (NFR) provided a suitable counterstain when needed. Proteins were visualized by mercuric bromophenol blue (MBPB) and eosin (Humason, 1979), alkaline Fast Green FCF (FG), and Light Green SF (LG) (James & Tass, 1984), with homoglycans and heteroglycans utilizing Alcian blue at pH 1.0 or 2.5 (AB1.0, AB2.5) and periodic acid-Schiff (PAS) methods (Sheehan & Hrapchak, 1980). The number of specimens examined for each species (*C.h.*, *X.v.*) are given after the respective technique: H-E (2,1), H-FG (2,1), MBPB-E (2,2), AB1.0-NFR (0,2), AB2.5-NFR (0,2), AB1.0-PAS (2,3), AB2.5-PAS (3,5), and PAS-LG (2,2). The interpretation and terms for the carbohydrate-containing compounds followed that in Sheehan & Hrapchak (1980).

Observations

A brief description of the male reproductive system and spermatheca of *C. holti* and *X. victoriensis* will be given only to place the histochemical data clearly in context. However, observations on *C. holti* that either contradicted or were omitted from the type description (Hoffman, 1963) will be given and then incorporated into an emended, brief description of the species.

Cambarincola holti

Habitat and anatomy
Specimens of *C. holti* were only found pressed into the folds and depressions of the ventral thoracic and abdominal surfaces of the host. The behavior of individuals was always sluggish, even when stimulated with a pin or fixative. Live specimens were easily identified both *in situ* on the host and *in vitro* because of the bright orange color of the stomach and intestine. Although the red blood and white reproductive organs contrasted with the orange pigment in live individuals, the orange color disappeared after fixation.

Relaxed, live specimens measured about 5.0 mm long but contracted to about 4.5 mm after fixation (Fig. 1). The length of the large head in relation to the body varied from a ratio of 1:2 to 1:3. The body is terete as used by Holt (1986) to mean that the cylindrical body tapers slightly at each end (pers. comm.). The four, dorsal peristomial tentacles each arise from a lobular base (Fig. 1). However, when alive and after fixation these appendages may be extended and appear as 'horns' (Fig. 2, left), or be contracted and hardly visible (Fig. 2, right). The jaws are similar in size with the dorsal (upper) being triangular and the ventral (lower) a truncated triangle with a dental formula of 5/4 (Fig. 3). Oral papillae are present. The male reproductive organs usually extend into the dorsal portion of the segment, but may be bent under the alimentary canal. The tubular, glandular atrium has prominent deferent lobes and its length is subequal to the diameter of the segment (Fig. 4). The gland is folded in its midregion. The prostate gland is subequal in length to the glandular atrium and lies against its upper surface. The prostate gland is lined with vacuolated, or 'differentiated', cells and lacks an ental bulb. The vacuolated cells extend beyond the confluence with the glandular atrium before merging into the muscular atrium (Fig. 4). The muscular atrium extends into the oval, muscular bursa to form a protrusible penis. The length of the spermatheca is subequal to the segment diameter (Fig. 5). The duct and elongate, oval bulb are about the same length, with an ental process usually present. In two sectioned specimens the process was absent in one, and inverted into the bulb's lumen in the other (Fig. 6). The reliability of the ental bulb as a taxonomic character must be viewed with caution.

An emended brief description for **Cambarincola holti** Hoffman, 1963

Body terete, about 4.5 mm long, low dorsal ridges, supernumerary muscles present; head to body length 1:2 to 1:3, head wider than segment 1, dorsal lip 4 small lobes each subtending a tentacle, 3 pairs of small lateral lobes; oral papillae present; jaws large, dorsal triangular and ventral truncated triangle, dental formula 5/4; male reproductive organs usually extend to dorsad of segment, glandular atrium length subequal to segment diameter, prominent deferent lobes; prostate gland length subequal to glandular atrium, differentiated, ental bulb absent; muscular atrium present; bursa ovoid, penial sheath forms ental half, protrusible penis; spermatheca length about 3/4 segment diameter, duct length about equal to elongate oval bulb, ental process usually present.

Histochemical characterization of reproductive secretions

The epithelium of the glandular atrium is composed of a single cell-type; ciliated, columnar secretory cells. The secretion granules, about 2.0 μm in diameter, are present throughout the cells in the gland (Fig 4., shaded). However, the granules in the distal half of the cells usually show the greatest degree of staining (Fig 4., stippled). The degree of granular staining varied slightly in different regions of the atrium, and probably reflects different stages in the synthesis cycle. The granules stain positively with PAS and LG (Fig. 6), MBPB and E (Fig. 7), and FG techniques. This suggests one or more mucoproteins and glycoproteins, with some alkaline (FG) components. In addition, the distally located granules in the mid and ectal regions of the atrium (Fig. 4, stippled) appear purple as a result of the dual staining reaction with both AB2.5 and PAS. This suggests that the granules in the latter stage of synthesis possess an acid non-sulfated mucosubstance moiety. Only one of the specimens examined was observed to have granular secretions (eosinophilic) in the lumen.

The prostate gland consists primarily of columnar, vacuolar secretory cells. Scattered between these cells are ciliated cells which appear to be located in the distal regions of the epithelium (Fig. 4). Such a situation was described by Jamieson (1981: 385). No staining reactions were obtained although secretory products could be observed in the vacuoles using differential interference contrast (DIC) illumination. Specimens were all collected within five days and so it is possible that in this phase of the reproductive physiological cycle the granules did not have any exposed staining sites, or the preparation procedures used may have blocked those sites.

To obtain a sample of spermatozoa from a branchiobdellidan, the body wall of segment 6 was incised and then the worm was fixed. An examination of the glandular atrium and prostate of three such treated specimen revealed no secretory granules in any of the cells. The trauma of the incision probably caused an ejaculation, providing a spontaneous secretion release as predicted by Jamieson (1992: 310).

The epithelium of the spermathecal duct is covered by the cuticle for only a short distance beyond the external pore (Fig. 5). The lining cells vary in height from cuboidal to columnar producing an irregular lumenar surface. The inner surface may become so irregular as to appear ciliated. The cells of the spermathecal

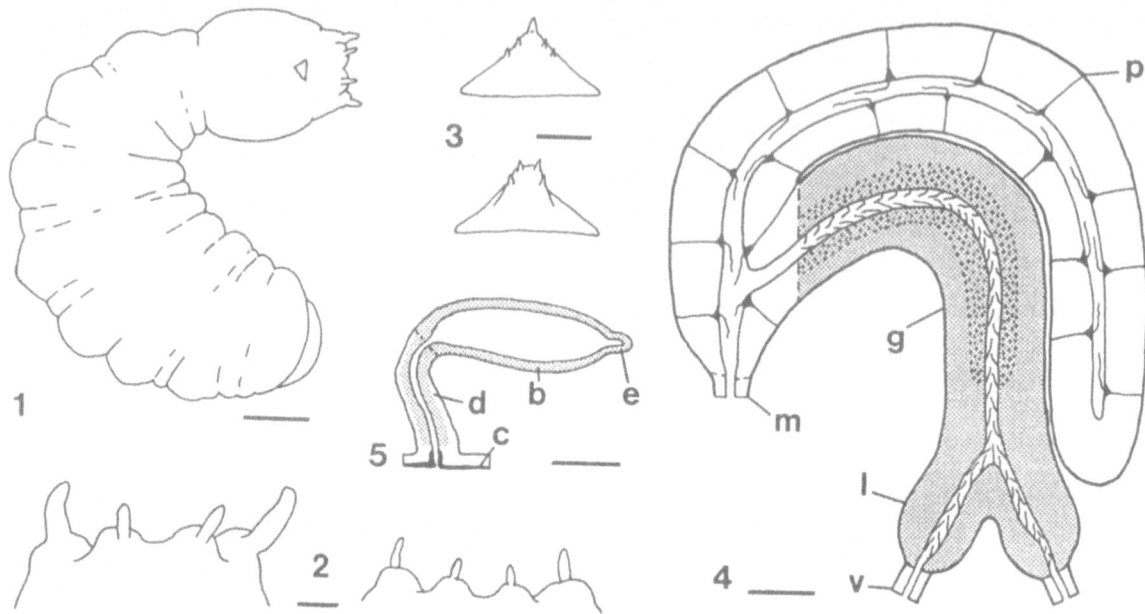

Figs 1–5. *Cambarincola holti*. (1) Dorsal view of live specimen; scale bar = 0.5 mm. (2) Dorsal view of peristomium with appendages extended (left) and partially withdrawn (right); scale bar = 0.1 mm. (3) Dorsal view of upper jaws (upper) and ventral jaw (lower); scale bar = 50 μm. (4) Diagram of a longitudinal section of the glandular atrium (shaded) with distal staining granules (black spheres), and prostate gland with ciliated cells; scale bar = 0.16 mm. (5) Diagram of a longitudinal section of the spermatheca; scale bar = 0.3 mm. Legend: b, spermathecal bulb; c, cuticle; d, spermathecal duct; e, ental process; g, glandular atrium; l, deferent lobes; m, muscular atrium; p, prostate gland; v, vas deferens.

bulb vary depending upon the distension caused by the lumenar contents. Usually the cell height is about half the diameter. No difference was observed in the cells lining the ental process.

Large numbers of spermatozoa sometimes are found in the lumen of the duct. The heads of the spermatozoa are usually parallel to each other with the tails extending entally. The cells adjacent to the spermatozoa contain PAS staining granules (Fig. 8). Live spermatozoa in the bulb may be clumped together, or randomly distributed with each acrosome being in contact with a lining cell.

Xironogiton victoriensis

Anatomy

Because of the dorsoventral flattening of segments of the posterior body, the ental tip of the tubular, glandular atrium is located between the mid-line and the lateral margin of the body. The gland traverses across the segment and then bends ventrad before merging into the muscular atrium. Deferent lobes and a prostate are absent (Fig. 10). The muscular atrium is about one quarter the length of the glandular atrium. The muscular tube extends into the subspherical, muscular bursa to form a protrusible penis. The length of the pyriform spermatheca (Fig. 11) is about half the height of the flattened segment, and is therefore considered small in relation to all other branchiobdellidan spermathecae.

Histochemical characterization of reproductive secretions

The epithelium of the glandular atrium is composed of ciliated, columnar secretory cells all containing granules about 2.0 μm in diameter. Based on the staining reaction of the granules to the AB2.5 and PAS technique, nine regions can be recognized and have been numbered starting from the ental end (Fig. 10).

Region 1: The granules stain very strongly with PAS, and lightly with MBPB and E, suggesting a neutral mucoprotein.

Region 2: The granules stain lightly with MBPB, and PAS or AB2.5, indicating a nonsulfated, acid mucosubstance (Fig. 12).

Region 3: Very similar to 2, except that no PAS staining has been observed (Fig. 13).

223

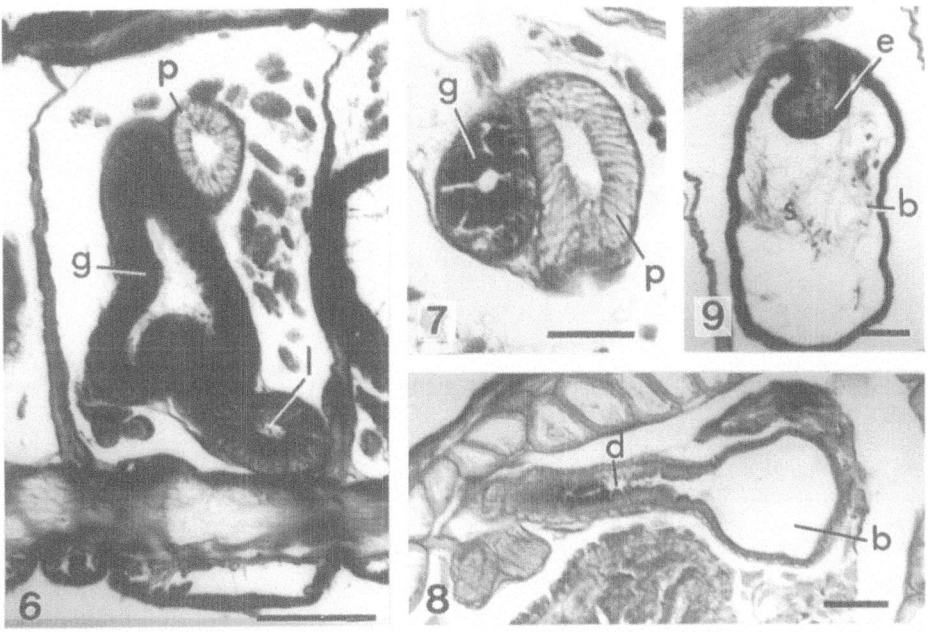

Figs 6–9. Cambarincola holti. (6) Longitudinal section through segment 6 to show the glandular atrium and prostate gland following the periodic acid-Schiff and light green SF staining method; scale bar = 50 μm. (7) Transverse section of the glandular atrium and prostate gland following the mercuric bromophenol blue and eosin staining method; scale bar = 50 μm. (8) Oblique longitudinal section through segment 5 to show the spermathecal duct and ectal bulb region following the Alcian blue (pH 1.0) and periodic acid-Schiff staining method; scale bar = 25 μm. (9) Longitudinal section through segment 5 to show the spermathecal bulb with the ental process inverted following the hematoxylin and eosin staining method; scale bar = 40 μm. Legend: b, spermathecal bulb; d, spermathecal duct; e, ental process; g, glandular atrium; l, deferent lobes; p, prostate gland; s, spermatozoa.

Region 4: Consists of the mid third of the gland with granules that stain very strongly with PAS (Figs 12 & 13), E, or MBPB, suggesting glycoproteins.

Region 5: Granules were observed to stain very lightly and only with MBPB.

Region 6: The granules in this small region stained lightly with MBPB, PAS or AB2.5 (Fig. 12), and very lightly with AB1.0. These granules contain sulfated and nonsulfated acid mucosubstances.

Region 7: Granules were characterized as those in region 5.

Region 8: The granules stained moderately with MBPB, very strongly with AB1.0, and also by AB2.5 (Fig. 12) or PAS, indicating their composition to be sulfated and nonsulfated acid mucosubstances.

Region 9: Was recognized only because of the occasional weak MBPB and PAS staining, or total non-staining reaction of the granules.

The low cuboidal cells of the spermathecal duct are covered with an invagination of the cuticle (Fig. 11) and possess no secretory granules. Granules in the distal layer of the spermathecal bulb cells stain with LG, PAS or AB2.5. When spermatozoa are present in the lumen, PAS or AB2.5 staining material forms a layer between the spermatozoa and the epithelium.

Spermatophores were observed on the dorsal, ventral and lateral surfaces of segments 5, 6, and 7. A spermatophore (Fig. 14) consisted of spermatozoa lying generally parallel to each other, but twisted to form a flattened spheroid which fits into a depressed portion of the body wall. The cuticle of the depression is covered by a thin layer of PAS material (Fig. 14, arrows) which adheres to a covering cap that stains lightly with PAS, LG and toluidine blue. This morphological description

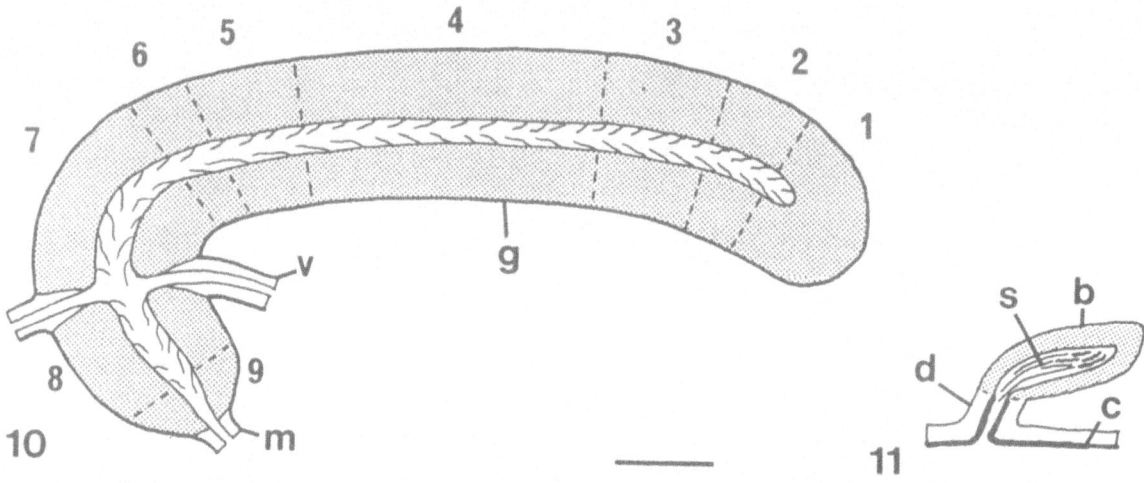

Figs 10–11. Xironogiton victoriensis. (10) Diagram of a longitudinal section of the glandular atrium (shaded) with the nine secretory regions numbered; scale bar = 0.2 mm. (11) Diagram of a longitudinal section of the spermatheca; scale bar = 0.2 mm. Legend: b, spermathecal bulb; c, cuticle; d, spermathecal duct; g, glandular atrium; m, muscular atrium; s, spermatozoa; v, vas deferens; 1–9, regions of secretory cells.

was confirmed by ultrastructural observations provided by Ferraguti (pers. comm.). A thin, incomplete layer of foam over the cap stains lightly with AB2.5.

Discussion

The male and female reproductive systems in most oligochaetes and branchiobdellidans are reasonably well-described because of their importance in taxonomy. However, few investigations have characterized the secretions produced by the reproductive organs in oligochaetes, and it is from these studies that the functions of the secretions have been inferred (Jamieson, 1992: 310).

Glandular atrium and prostate

The general method of insemination in branchiobdellidans is believed to consist of spermatozoa from one individual being deposited into the spermatheca of a recipient, a process that has been described only twice (Henle, 1835; Yamaguchi, 1934). From this, it is assumed that all branchiobdellidan spermatozoa require the same kinds of supporting secretions to produce successful inseminations, thus these secretions should be demonstrable in all 'typical' male systems. The glandular atrium is the secretory portion of the

system and may be supplemented by a prostate. Of 21 branchiobdellidan genera, a prostate is present only in seven, and all of these are in the Cambarincolidae. It is believed that the prostate cells simply increase the amount of secretions available to the glandular atrium, sometimes with regionalization of the secretory cell types. Of greater importance than the presence of a prostate to the process of insemination is the diminutive spermatheca in *Xironogiton* spp., and total absence of the organ in *Caridinophilus* sp. and *Ellisodrilus* spp. In these genera some variation in the secretory types present would be expected to reflect the modified, and as yet undescribed, method of spermatozoan transfer.

The cytological appearance of the prostate in a particular species of *Cambarincola* is believed to be constant, and provides an important taxonomic character. The present study found vacuolated cells in the prostate of *C. holti*, and not granular cells as reported in the type description (Hoffman, 1963: 301). A similar error was found and corrected in *Cambarincola macrocephala* Goodnight, 1943 following a study by Holt (1981). Where the taxonomic characters are few, such errors can have a significant impact on a phylogenetic analysis. It is generally assumed in *Cambarincola* spp. that the glandular atrium merges with the muscular atrium, and the prostate forms an ancillary diverticulum from the former. However, based on the cell types observed in *C. holti*, it is the prostate that connects with the mus-

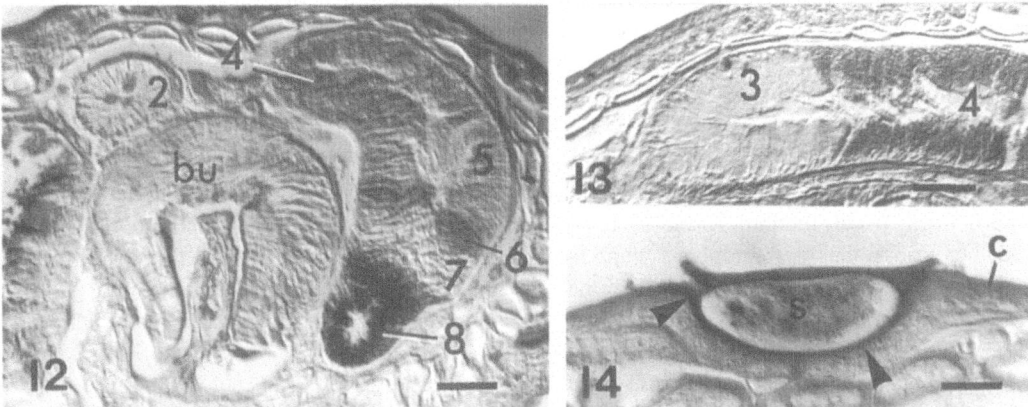

Figs 12–14. *Xironogiton victoriensis*: (12) Transverse section through segment 6 showing six regions of the glandular atrium and muscular bursa following the Alcian blue (pH 2.5) and periodic acid-Schiff staining method; scale bar = 50 μm. (13) Oblique longitudinal section through the glandular atrium showing regions 3 and 4 following the Alcian blue (pH 1.0) and periodic acid-Schiff staining method; scale bar = 50 μm. (14) Transverse section through a spermatophore on segment 5 showing "cap" with raised edges and attachment lining (arrows) following the periodic acid-Schiff and light green SF staining method; scale bar = 25 μm. Legend: bu, muscular bursa; c, cuticle; 1–9, regions of secretory cells.

cular atrium. This feature is not easy to recognize in whole mount preparations, and may have been over-looked in previous studies. A re-examination of the 47 species of *Cambarincola* will have to be made before this feature can be introduced into a character matrix.

As *C. holti* and *X. victoriensis* have essentially similar male genitalia, it was anticipated that at least two glandular secretions would be demonstrated in each species. This assumption was based on the presence of a glandular atrium and prostate gland in *C. holti*, and the eosinophilic and non-eosinophilic regions of the glandular atrium reported in *Xironogiton instabilis* by Holt (1949). Although this assumption was supported in *C. holti*, the nine secretory regions characterized in the glandular atrium of *X. victoriensis* was unexpected. The secretions from the glandular atrium in *C. holti* most closely resemble those of region 4 in *X. victoriensis*, and the "homogeneous globules" composed of a neutral mucopolysaccharide-protein complex in the oligochaete *T. tubifex*, by Fleming (in Jamieson, 1981: 386). Similarly, the granules in region 8 of *X. victoriensis*, in both size and histochemical characterization, are very similar to the "heterogeneous globules" consist-ing of carboxylated acid mucopolysaccharide with a distinct protein core (Fleming in Jamieson, 1981: 385) in *T. tubifex*. No further similarities have been found between the granules in regions 1–3, 5–7 and 9 of the glandular atrium in *X. victoriensis* or the vacuolar, non-staining secretions in the prostate gland of *C. holti* with any other clitellate male reproductive organs.

Spermatophores

Spermatophores, or 'sperm packets', in branchiobdell-idans were first reported by Holt (1949: 539) on speci-mens of *X. instabilis* collected in Virginia during June, 1947. The second report is in this study on specimens of *X. victoriensis* collected in September, 1992. Consider-ing the large numbers of specimens of *Xironogiton* spp. that both Holt and I have examined, only two reports of spermatophores would indicate that their production is restricted to a particular season.

Holt (1949: 550) proposed that the spermatophores were an adjunct to compensate for the small capac-ity of the spermatheca. Later, Holt (1974) also sug-gested the possibility of 'hypodermic' impregnation of the sperm masses. This explanation seems inap-plicable as the spermatophores are deposited on the surface of the cuticle, and not delivered directly into the coelom of the recipient as occurs in some poly-chaetes (Schroeder & Hermans, 1975). If the sper-matophores were to pass into the recipient's coelom by some process, then only those spermatophores deposit-ed on segment 7 would enter the female segment and result, presumably, in internal fertilization, a unique occurrence in the branchiobdellidans. Those sperma-tozoa from the spermatophores on the male segments

226

would be destroyed, presumably as non-self, following contact with the developing spermatids in the recipient's coelom.

The suggestion that spermatophores are an adjunct to compensate for the small capacity of the spermatheca is considered reasonable. The spermatophores being located on segments 5 to 7 would be covered by the cocoon secretions produced by the clitellum. The deposition of the ovum between the cuticle and secreted tube, could then cause the spermatophore 'cap' to be detached and release spermatozoa to fertilize the ovum in the usual external fashion. This process would ensure that a cocoon-containing zygote would be produced even if the diminutive spermatheca had run out of spermatozoa, or had not been inseminated. The same hypothesis could also explain how the spermatheca-lacking *Ellisodrilus* spp. and *Caridinophilus* sp. externally fertilize their ova. However, spermatophores have not been observed on specimens of either of these two genera. The production of spermatophores by *X. victoriensis* is certainly consistent with the large number of secretions being produced in the glandular atrium.

Very little is known of the reproductive biology of the branchiobdellidans, and this needs to be rectified, particularly the secretory aspects of the male glands and the mechanisms of the various types of penes. When more information has been gathered, then missing penes and spermathecae can be viewed in context. This study has shown that granular secretions in the glandular atrium are even more complex in *X. victoriensis* than indicated in a sister species using hematoxylin and eosin staining procedures (Holt, 1949). Although the granular and vacuolar secretions of the glandular atrium and prostate, respectively, are used in the taxonomy of *Cambarincola* spp., the 'prostate' cells connecting with the muscular atrium in *C. holti* are atypical. These observations, together with species descriptions of spermatozoa (Ferraguti & Gelder, 1991; Gelder *et al.*, 1994) are part of the search for new characters that will improve the resolution of future phylogenetic analyses on the branchiobdellidans.

The type description of many oligochaetes contain details of the male genitalia that were obtained from reconstructions using hematoxylin and eosin or Azan stained sections (Erséus, 1993). Therefore, a review of the literature, but more likely the deposited slides, could provide significant information on the secretory cells in the glandular atrium and prostate in these oligochaetes. From this base-line work, representative species could then be examined with suitable histo-chemical protocols for comparison with branchiobdellidans resulting in an expansion of the character matix for these clitellates.

Acknowledgements

Thanks to Dr Ralph O. Brinkhurst for assistance and facilities in collecting and examining the crayfish in Tennessee, to Keith Toomey for invaluable photographic help, and Dr Andrea M. Gorman and reviewer for constructive suggestions. Ms Jane Egler contributed laboratory assistance during her credit-fulfilling research project. Funding was provided by Mini-Grant and Faculty Development funds from the University of Maine at Presque Isle, and a private donation.

References

Brinkhurst, R. O. & S. R. Gelder, 1991. Chapter 12. Annelida: Oligochaeta and Branchiobdellida. In: J. H. Thorpe & F. Covitch (eds), Ecology and Classification of North American Freshwater Invertebrates. Academic Press, New York: 401–435.

Erséus, C., 1993. The marine Tubificidae (Oligochaeta) of Rottnest Island, Western Australia. In: F. E. Wells, D. I. Walker, H. Kirkman & R. Lethbridge (eds), Proceedings of the Fifth International Marine Biological Workshop: The Marine Flora and Fauna of Rottnest Island, Western Australia. Western Australian Museum, Perth: 331–390.

Ferraguti, M. & S. R. Gelder, 1991. The comparative ultrastructure of spermatozoa from five branchiobdellidans (Annelida: Clitellata). Can. J. Zool. 69: 1945–1956.

Gelder, S. R. & R. O. Brinkhurst, 1990. An assessment of the phylogeny of the Branchiobdellida (Annelida: Clitellata) using PAUP. Can. J. Zool. 68: 1318–1326.

Gelder, S. R. & L. A. Hall, 1990. Branchiobdellidans from southwestern British Columbia, Canada with a phylogenetic analysis of *Xironogiton* (Annelida: Clitellata) using PAUP. Can. J. Zool. 68: 2352–2359.

Gelder, S. R., M. Ferraguti & M. A. Subchev, 1994. A description of spermatozoan ultrastructure and some anatomical characters in *Branchiobdella kozarovi* Subchev, 1978 (Annelida: Clitellata), and a review of the spermatozoan morphology within the genus. Hydrobiologia 278: 17–26.

Goodnight, C. J., 1943. Report on a collection of branchiobdellids. J. Parasitol. 29: 100–102.

Henle, J. F. G., 1835. Über die Gattung *Branchiobdella*. Arch. Anat. Physiol. wiss. Medizin: 574–608.

Hobbs, H. H., Jr., 1989. An illustrated checklist of the American crayfishes (Decapoda: Astacidae, Cambaridae, and Parastacidae). Smithson. Contr. Zool. 480: 236 pp.

Hoffman, R. L., 1963. A revision of the North American annelid worms of the genus *Cambarincola* (Oligochaeta: Branchiobdellidae). Proc. US nat. Mus. 114: 271–371.

Holt, P. C., 1949. A comparative study of the reproductive systems of *Xironogiton instabilis* (Moore) and *Cambarincola philadelphica*

(Leidy) (Annelida, Oligochaeta, Branchiobdellidae). J. Morph. 84: 535–572.

Holt, P. C., 1960. The genus *Ceratodrilus* Hall (Branchiobdellidae, Oligochaeta) with the description of a new species. Virginia J. Sci. 11: 53–77.

Holt, P. C., 1974. The genus *Xironogiton* Ellis, 1919 (Clitellata: Branchiobdellida). Virginia J. Sci. 25: 5–19.

Holt, P. C., 1981. A resume of the members of the genus *Cambarincola* (Annelida: Branchiobdellida) from the Pacific drainage of the United States. Proc. biol. Soc. Wash. 94: 675–695.

Holt, P. C., 1986. Newly established families of the order Branchiobdellida (Annelida: Clitellata) with a synopsis of the genera. Proc. biol. Soc. Wash. 99: 676–702.

Holt, P. C. & B. D. Opell, 1993. A checklist of and illustrated key to the genera and species of the central and North American Cambarincolidae (Clitellata: Branchiobdellida). Proc. biol. Soc. Wash. 106: 251–295.

Humason, G. L., 1979. Animal Tissue Techniques. 4th ed. Freeman and Co., San Francisco, 661 pp.

James, J. & J. Tass, 1984. Histochemical Protein Staining Methods. Oxford Univ. Press, Oxford, U.K., 40 pp.

Jamieson, B. G. M., 1981. The Ultrastructure of the Oligochaeta. Academic Press, New York, 462 pp.

Jamieson, B. G. M., 1992. Chapter 3. Oligochaeta. In: F. W. Harrison & S. L. Gardiner (eds), Microscopical Anatomy of Invertebrates, Volume 7: Annelida. Wiley-Liss, New York: 301–316.

Sawyer, R. T., 1986. Leech Biology and Behaviour. Clarendon Press, Oxford, 1065 pp.

Schroeder, P. C. & C. O. Hermans, 1975. Annelida: Polychaeta. In: A. C. Giese & J. S. Pearse (eds), Reproduction of Marine Invertebrates. Academic Press, New York, Vol. 3: 1–213.

Sheehan, D. C. & B. B. Hrapchak, 1980. Theory and practice of Histotechnology. C. V. Morsby Co., St. Louis, 441 pp.

Yamaguchi, H., 1934. Studies on Japanese Branchiobdellidae with some revisions on the classification. J. Fac. Sci. Hokkaido Imp. Un. Ser. VI. (Zool.) 3: 177–219.

Hydrobiologia **334**: 229–239, 1996.
K. A. Coates, Trefor B. Reynoldson & Thomas B. Reynoldson (eds), Aquatic Oligochaete Biology VI.
© 1996 *Kluwer Academic Publishers.*

An ultrastructural investigation of *Hrabeiella* Pižl & Chalupský, 1984 (Annelida). I. Chaetae and body wall organization

Emilia Rota & Pietro Lupetti
Dipartimento di Biologia Evolutiva, Università di Siena, Via P.A. Mattioli 4, I-53100 Siena, Italy

Key words: *Hrabeiella*, cuticle, epidermal gland cells, muscles, peritoneum

Abstract

Specimens of *Hrabeiella* sp., collected for the first time in Italy (Tuscany), were investigated by transmission and scanning electron microscopy. Chaetae conform with previous descriptions in light microscopy, but appear considerably different in electron microscopy, as they show an unusual brush-like pattern. The first TEM analysis of the body wall layers was undertaken. The cuticle ultrastructure is similar to that found generally in very small-sized annelids, in that a coarse collagenous grid is absent, but it is peculiar for the paucity of ascending microvilli. Three different types of secretory cells are distinguishable in most of the 12 epidermal glandular blocks occurring around the body circumference. Such blocks regularly alternate with as many muscle fields, each containing four longitudinal fibres. A subepidermal layer of circular muscles is present. The somatopleure is cellular and complete.

Introduction

During a survey of the Italian terrestrial enchytraeids, specimens of *Hrabeiella* sp., a peculiar non-clitellate annelid, were extracted from coniferous, beech and chestnut wood soils taken at altitudes between 300–1320 m a.s.l. in Tuscany (sites 5, 8, 13 and 22 of Rota, 1994; sites 5, 8, 14, 25 and 31 of Rota, 1995). The new record for Italy is notable, since the genus was known to date only from South Bohemia (Pižl & Chalupský, 1984) and Germany (Jans & Römbke, 1989; Graefe, 1989, 1990, 1993a, 1993b) (Fig. 1).

The minute and slow-moving *Hrabeiella* worms are readily identified by the unique chaetal shape (Figs 2 & 3), and the conspicuous and regularly distributed epidermal gland cells, whose contents cause an intense white-spotting of the body surface. Such distinctive features and other major traits of the external and internal organization, were carefully investigated by light microscopy by Pižl & Chalupský (1984) in the original description of the only species so far known, *H. periglandulata*. Later, while reporting on the ecology and biology of southern German populations, Jans & Römbke (1989) provided scanning electron micrographs of the whole animal and its chaetae.

Italian specimens show no deviation from the primary literature in light microscopy, but differ in the chaetal ultrastructure from German specimens. Here we undertake the first ultrastructural studies of the body wall cuticle, epidermis and muscle layers.

Material and methods

Specimens described in this paper were obtained from brown sandy humus and coarse wood litter collected in a centuries-old *Abies alba* and beech forest at La Verna (Arezzo), 1120 m a.s.l., on 9 May 1992. Worms were extracted by a modified version of O'Connor's (1955) wet-funnel method (Healy & Rota, 1992) and examined alive by light microscopy. Photographs of chaetae were taken *in vivo* with Ilford Pan F Plus using a Leitz Aristoplan microscope.

For ultrastructural studies, specimens were fixed in Karnovsky's (1965) fluid for 1 h, washed twice in cacodylate buffer (pH 7.4) and post-fixed in a cacodylate buffered 2% osmium tetroxide solution (4 °C, for 90 min). For SEM, after rinsing twice in cacodylate buffer, they were dehydrated in ethanol, critical point dried, mounted on aluminium stubs and sputter-coated

230

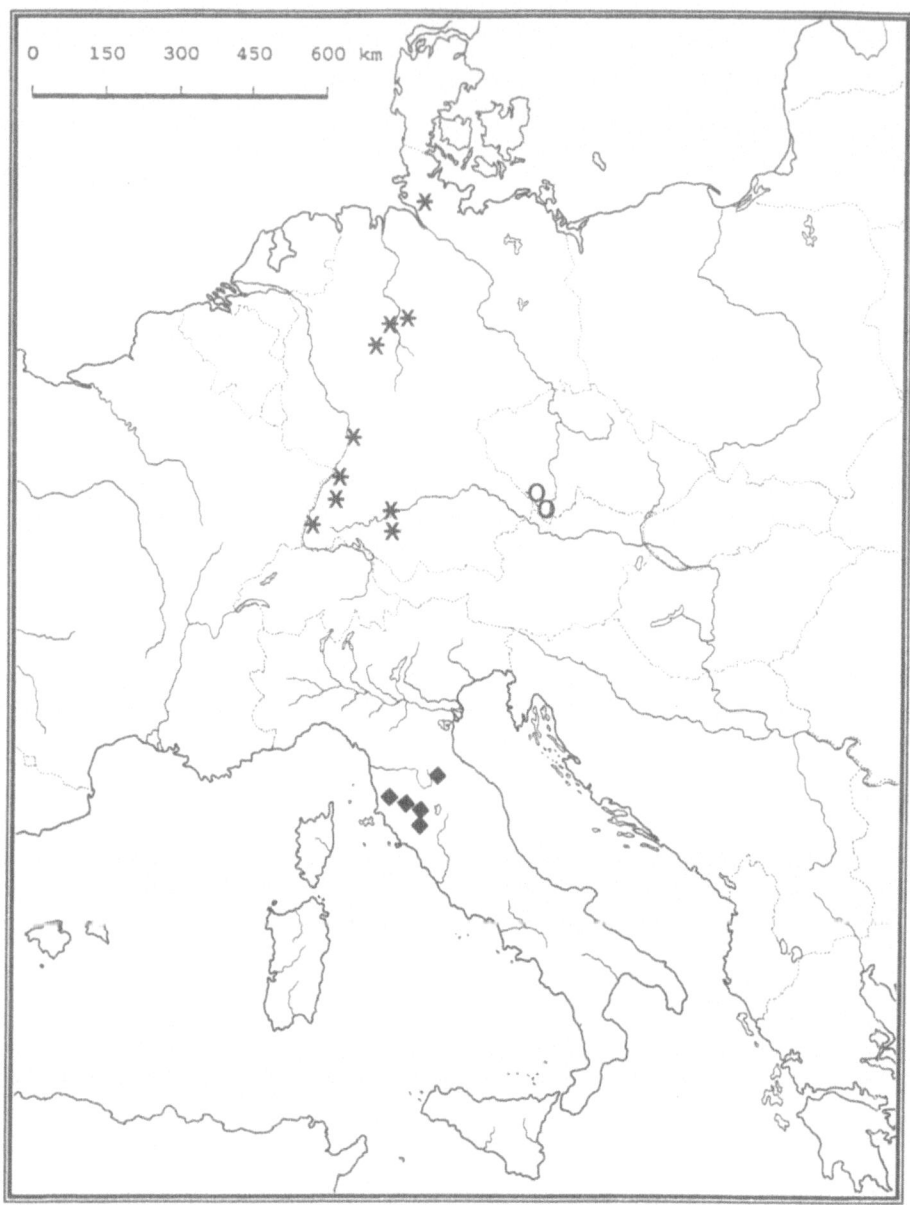

Fig. 1. Distribution map of *Hrabeiella*. Italian sites (diamonds) from present study. South Bohemian sites (circles) from Pižl & Chalupský (1984). German sites (stars) from Jans & Römbke (1989) and Graefe (1989, 1990, 1993a, 1993b).

with gold. They were examined and photographed with a Philips XL 20 scanning electron microscope. For TEM, after rinsing overnight in cacodylate buffer, they were dehydrated in ethanol and embedded in Epon-Araldite. All sections were cut with a diamond knife on a Reichert Ultracut E ultramicrotome. Semi-thin sections (0.35 μm thick) were stained with toluidine blue for light microscopy and photographed with Ilford Pan F Plus using a Leitz Aristoplan microscope. Ultra-thin sections were routinely stained with uranyl acetate and lead citrate; they were examined and photographed with a Philips CM 10 transmission electron microscope at 80 kV.

Observations

Chaetae

Chaetae of *Hrabeiella* are all of one kind and display a characteristic shovel-like appearance in light microscopy (Fig. 3). Measurements taken on Italian specimens fall within the range reported for S-Bohemian ones (Pižl & Chalupský, 1984) i.e. total length 30–34 μm, shaft diameter 1.4–1.6 μm, shaft length 15–17 μm plus 7–9 μm parallel to the flattened distal portion; the latter maximally 6–7 μm across.

Electron micrographs of Tuscan specimens reveal a quite unexpected ultrastructure: the shaft, which is devoid of nodulus, supports distally a flattened, quadrangular cushion, which gives off on one side numerous, densely packed 'bristles'. Each chaeta thus resembles a handled brush measuring a few μm (Figs 4–9). Serial cross-sections from the base to the top of the chaetae (Figs 10–16) show the characteristic canalicular pattern of annelid chaetae, each chaeta appearing composed of parallel tubules separated by walls of fibrillar material (cf. Richards, 1978). Along the shaft, all tubules are sealed together so as to form the solid cylindrical handle (Figs 9 & 13). More distally (Figs 14–16), two flattened, fibrous laminae appear, which merge at their borders: the posterior one, which is supported to a certain level by the terminal portion of the shaft (Figs 7, 14–16), does not contain tubules; the anterior one binds proximally the tubules which form the bristles of the brush. Along its lateral borders, the frame of the brush is thickest and rich in embedded tubules (Figs 16 & 17). This accounts for the fork-like appearence shown by chaetae under the light microscope (Fig. 4).

Near the chaetal follicle base (Fig. 11), tubules are seen in the chaetal medulla but not in the peripheral zone, which consists only of fibrillar material. More distally along the 'handle' (Figs 12 & 13), tubules appear distinctly formed throughout the section. In the medullar area they measure 190–290 nm in diameter, against 170–190 nm in the periphery. At increasingly higher levels (Figs 14–16, in an antero-ventral view), tubules are still distinguishable, even at the very top of the chaeta. The number of tubules is constant along the handle (about 85); it increases at the brush base by addition of tubules on the hairy side of the brush, and it comes to a maximum near the brush middle, where, beside the embedded tubules (located in the distal part of the shaft and on the hairy side and lateral borders of the brush), up to 120–150 'bristles' (free tubules) can be counted. In the distal part of the brush the tubule number decreases again. Free bristles measure 120–130 nm in thickness proximally.

Chaetae are distributed in four bundles per segment, in an equatorial ventral position (Fig. 2). Each bundle contains 2 or 3, exceptionally 4, chaetae. Internally the two bundles on the same side of the body merge together into a single chaetigerous bulb (Figs 10, 12).

Chaetae pertaining to the same bundle are produced in adjacent follicles and are moved by a common musculature (Fig. 10). Muscles are attached to the basal lateral cells *via* a 45 nm thick basal membrane. Tonofilaments of the lateral cell run to the basal lamina and anchor the chaeta to the muscle cells (Fig. 11). Synaptic knobs provide connections between nerves and muscles. Remotor muscles run from the anterior side of the follicle up to the epidermal surface. Promotor muscles are attached to the posterior side of the follicle and radiate posteriad to reach the epidermal surface. The contractions of such mutually antagonistic muscles cause the back and forth swinging movements of chaetae which are observed in living specimens. Transverse muscles are also seen near the follicle base, bridging the lateral and the ventral chaetal bundles (Fig. 10). In all chaetal muscle fibres, the maximum thickness of thick myofilaments ranges between 22–26 nm.

Body wall organization

The body wall is constructed on the usual plan of annelids, comprising a cuticle, the epidermis, a thin, amorphous basement membrane (basal lamina), circular and longitudinal muscle layers, and a complete, cellular, somatic layer (coelothelium) (Figs 17–25).

Cuticle. The cuticle (Fig. 17) is 1.6–2.1 μm thick and composed of: (1) an inner layer of low electron density, 1.2–1.4 μm thick, containing an irregular network of thin filaments; (2) two thin, moderately electron-dense bands of closely packed fibrils (epicuticle); (3) an outer discontinuous layer, 0.15–0.18 μm thick, consisting of electron dense, finger-like epicuticular projections, circular in surface view and coated with fine mucous strands. Occasionally, microvilli are seen to ascend from the underlying epithelium and penetrate the cuticle (Fig. 31).

Epidermis. The epidermis contains supporting cells, secretory cells and sensory cells (Figs 17–21). Supporting cells (Figs 17, 18, 20) prevail in number and surround and isolate all other cell types in the epidermis; their lateral surfaces are apically connected to the adjacent cells by zonulae adherentes and sep-

232

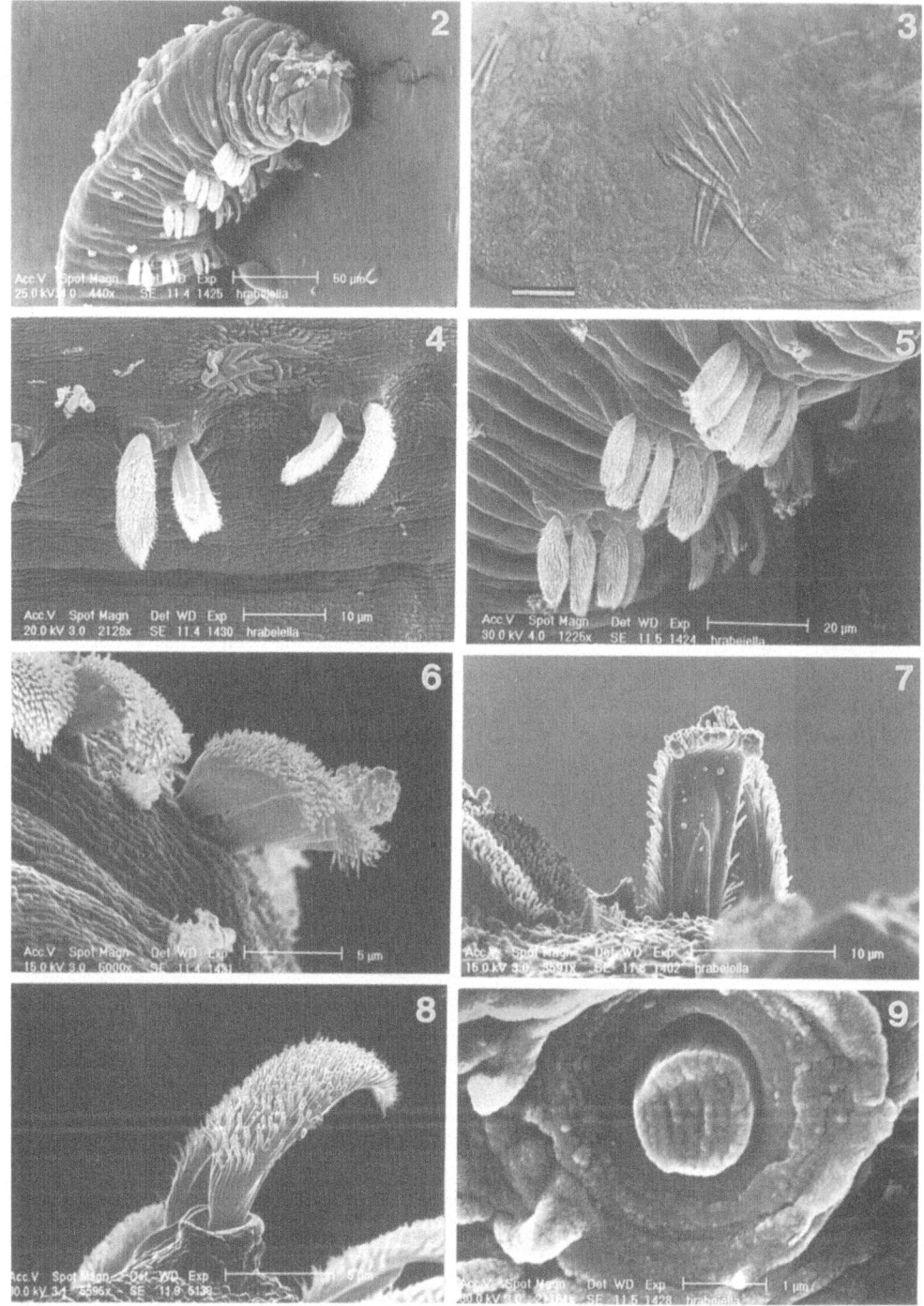

Figs 2–9. Chaetae of *Hrabeiella.* (2) Anterior region of the body showing ventral location of 'lateral' and 'ventral' chaetal bundles. SEM. (3) Chaetae in light microscopy in a live, almost collapsing specimen. Scale = 20 μm. (4) Chaetal row in a midventral view. SEM. (5) Detail of Fig. 2 showing the 'hairy' side of distal chaetal portion. SEM. (6)–(8) Highly magnified distal chaetal portions showing the shaft 'plugging in' the smooth side of the brush. SEM. (9) Chaeta broken near body wall, showing the fine internal structure of the shaft. SEM.

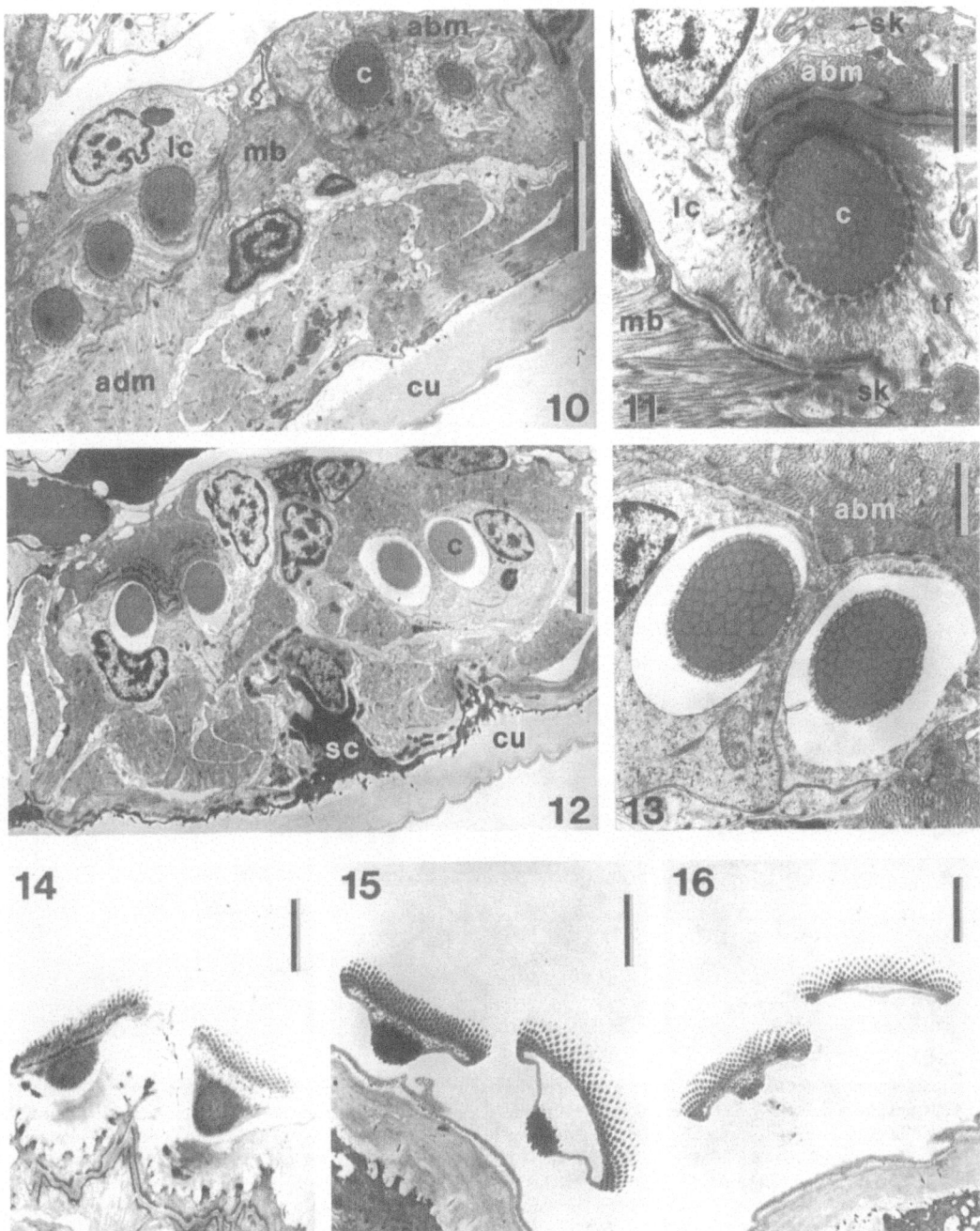

Figs 10–16. Chaetae of *Hrabeiella*. Cross sections at different chaetal levels (see text). TEM. (10) Chaetigerous bulb cut near base of follicles, showing the muscular connection between lateral and ventral chaetal bundles. Scale = 4 μm. (11) Detail of a follicle at same level as Fig. 10. Scale = 1 μm. (12) Chaetigerous bulb at a shallower level. Scale = 4 μm. (13) Detail of chaetae cut at same level as Fig. 12; note the gaps between the lateral cells and chaetal shafts. Scale = 1 μm. (14)–(16) Increasingly distal sections of the ectal brush. Scales = 2 μm. abm = remotor chaetal muscle; adm = promotor chaetal muscle; c = chaeta; cu = cuticle; lc = follicle lateral cell; mb = bridging muscle; sc = supporting cell; sk = synaptic knob; tf = tonofilaments.

234

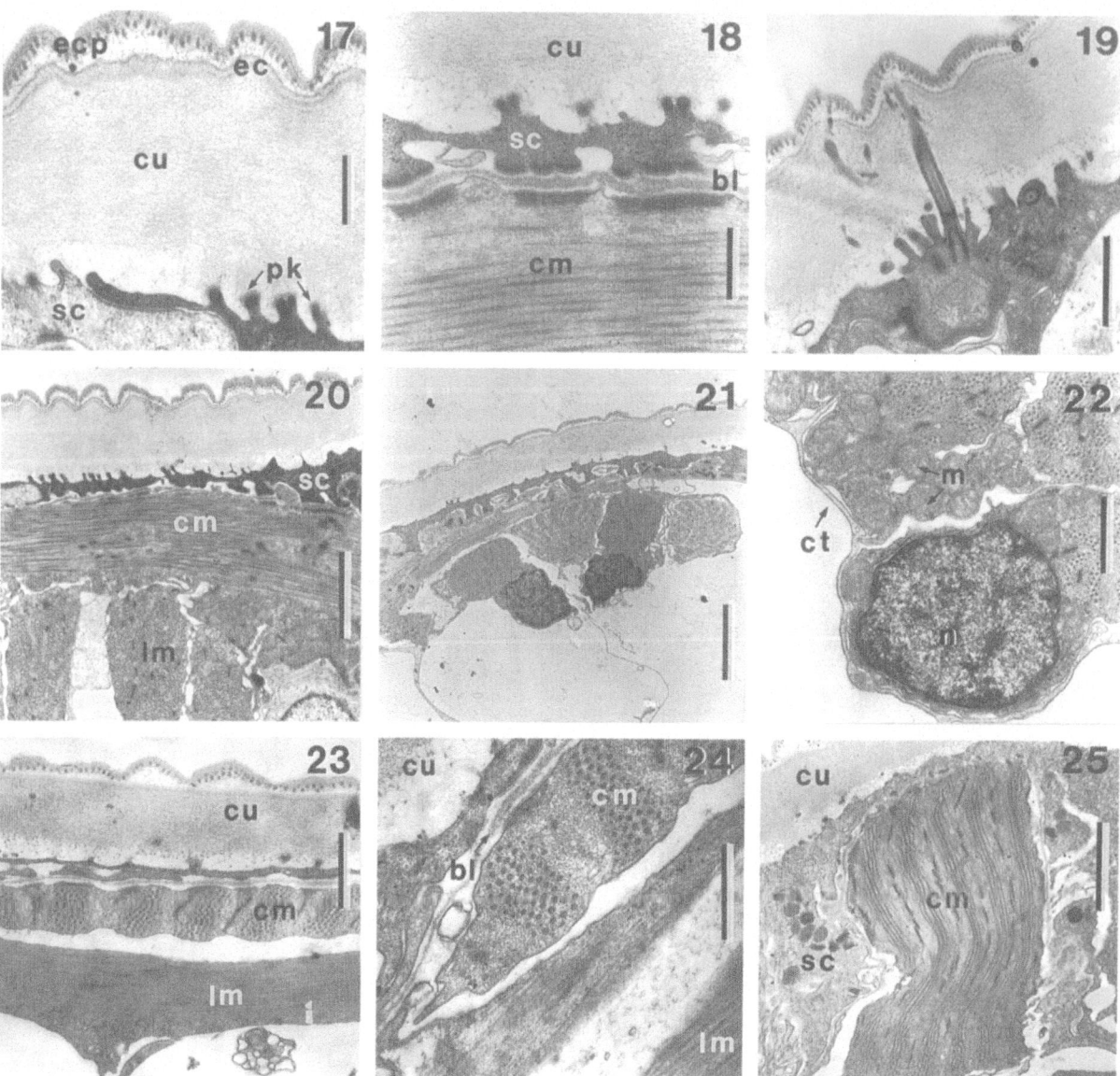

Figs 17–25. Body wall of *Hrabeiella*. TEM. (17) Cross section of the cuticle. Scale = 0.5 μm. (18) Relationships between cuticle, epidermal supporting cells and circular muscle layer. Scale = 0.5 μm. (19) Sensory cell penetrating the cuticle with two cilia. Scale = 1 μm. (20) Body wall organization in caudal segments. Scale = 2 μm. (21) Cross section of a midbody segment showing a group of four longitudinal muscle fibres. Note the different shade of adjacent fibers. Scale = 3 μm. (22) Coelomic location of nucleus and mitochondria in longitudinal muscle fibres. Scale = 1 μm. (23) Longitudinal frontal section through body wall showing relationships between cuticle, epidermis, and circular and longitudinal muscle layers. Scale = 1 μm. (24) Detail of a circular muscle fibre at its lateral edge. Scale = 0.5 μm. (25) Shallow, longitudinal section of body wall showing a circular muscle fiber cut oblique to its longitudinal axis. Scale = 1.5 μm. bl = basal lamina; cm = circular muscle fibre; ct = coelothelium; cu = cuticle; ec = epicuticle; ecp = epicuticular projection; lm = longitudinal muscle fibre; m = mitochondrion; n = nucleus; pk = papillary knob; sc = supporting cell.

tate desmosomes, while the attachment to the cuticle is ensured by tonofilament bundles terminating in the ascending microvilli or in shorter, 75–100 nm wide, hemidesmosome-like papillary knobs on their shoul-

ders. Basally, hemidesmosomes anchor the supporting cells to the underlying circular muscle layer (Figs 18 & 24). A thin basal lamina is interposed between epidermis and muscle fibres or faces directly the somatopleure where the musculature is interrupted (Figs. 18, 24, 32).

Mature secretory cells are completely filled with their products and bulge considerably into the coelomic cavity (maximum height of cells 9.5–11 μm), so as to represent the most conspicuous components of the body wall (Fig. 26). SEM micrographs (Figs 27–29) show the regular distribution of their pores over the body surface and the intense discharging activity that occurred at the time of fixation. In transverse sections (Fig. 26), the secretory cells appear to form 12 rounded blocks, regularly alternating around the body circumference to as many fields of muscle fibres. In most blocks, three different types of gland cells are distinguishable (Fig. 30).

The 'granular cells' (Fig. 33), which *in vivo* are opaque to transmitted light and responsible for the white-spotted appearance of *Hrabeiella* under incident light, after fixation appear to contain oval, membrane-bound granules characterized by a dense, tubular core, and small, cristalline, dense bodies embedded in globular (1.2–2.2 μm wide), transparent vesicles. The secretory apparatus and the nucleus, which is provided with an evident nucleolus, lie basally. The external orifice is surrounded by a double ring of pore microvilli arising from the secretory cell and from the adjacent supporting cells (Fig. 34).

The 'mucous cells' (Figs 30–31), that are hyaline and colourless in live specimens, are filled with large, membrane-bound vesicles (max. 1.9–2.6 μm wide), containing structureless, mucoid material deeply staining with toluidine and of high electron density. Ribosomes are abundant and mitochondria are often seen peripherally. The nucleus lies laterally or basally. The structure of the external orifice resembles that of the granular cell type. Pores of granular and mucous cells often open at a very short distance (Figs 27–29, 31).

The 'grey cells' (Fig. 32), which have a hyaline, brown-coloured appearance *in vivo* with transmitted light, are moderately electron dense; their granules, which are maximum 1.2–1.7 μm wide, appear irregularly shaped and formed by two unequal parts of slightly different electron densities surrounded by a single membrane. The nucleus is located laterally and contains a large nucleolus and conspicuous, mainly peripheral, clumps of chromatin. The structure of the pore corresponds with that of the 'granular' and 'mucous' cell types.

Sensory cells having two cilia surrounded by microvilli have been observed (Figs 19 & 33).

Body wall muscles

A basal lamina, 60 nm thick, separates the epidermis from the circular muscle layer of the body wall (Figs 18 & 24). Circular muscle fibres (Figs 20, 23–25) form discrete rings which are here and there displaced by epidermal gland cells, chaetigerous bulbs and other structures; the contractile material, which is organized in up to 8 adjacent sarcomeres across each fiber, faces the epidermis, while mitochondria and nuclei are located towards the coelomic cavity (platymyarian type, cf. Lanzavecchia *et al.*, 1986). In longitudinal sections (Figs 23–25), the contractile region has a spindle-shaped section, 580–670 nm high, 4.5–6 μm wide. The width of sarcomeres (from Z-rod to Z-rod) ranges between 600–800 nm. The diameter of the thick (myosin) filaments reaches maximally 40 nm. The thin-to-thick filament ratio is about 15/1.

Longitudinal muscle fibres are disposed in 12 groups of four around the body cavity, regularly alternating with the epidermal gland blocks (Fig. 26). They can be classified within the flattened circomyarian type (cf. Lanzavecchia *et al.*, 1988). The contractile region has a rectangular (2.1–2.4 by 3.4–3.6 μm) to trapezoidal cross-section (Fig. 21) and faces the circular fibers; the cytoplasmic region, which contains the nucleus and mitochondria, occupies the coelomic border (Fig. 22). Mitochondria are only occasionally seen in the contractile sarcoplasm. Sarcomeres are often 'V' or 'Y'-shaped and show a fairly irregular arrangement; their width ranges between 290–390 nm. The thick filaments reach maximally a diameter of 31–33 nm and the thin-to-thick filament ratio is about 12/1.

The coelomic surfaces of the longitudinal muscles and of bulging components of the epidermis are lined by coelothelium (Figs 22 & 32).

Discussion

SEM pictures of German specimens of *Hrabeiella* (Jans & Römbke, 1989) showed the ectal blades of chaetae as being curved (sail-like) and scaly, with flame-like 'fringes' coming off at two distinct levels of their convex side. Our specimens look quite different and even more peculiar in their chaetal ultrastructure, in that a hairiness appears to cover more or less equally and densely the whole of one side of the blades; such brush-like pattern was revealed both by SEM – no matter which fixative we used (e.g. Karnovsky's, Bouin's etc.) – and by TEM. Indeed, although the thickness

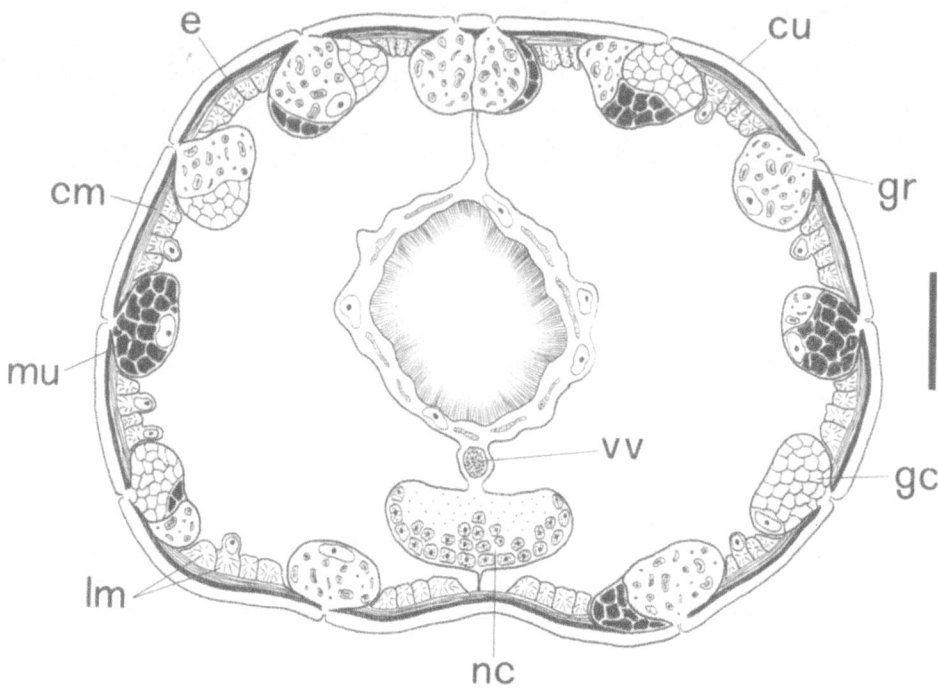

Fig. 26. Transverse section of a midbody segment of *Hrabeiella* as reconstructed from semithin sections. Semidiagrammatic with coelothelium omitted for clarity. Scale = 15 µm. cm = circular muscles; cu = cuticle; e = epidermis; gc = 'grey gland cell'; gr = 'granular gland cell'; lm = longitudinal muscles; mu = 'mucous gland cell'; nc = nerve cord; vv = ventral blood vessel.

of the single bristles is below the resolving power of light microscope, some roughness of the blade surface was detectable even in live worms in light microscopy (Fig. 3, bottom-right: central area of shovel-blade).

Chaetae bearing tufts of hair are found in some polychaetes (Specht, 1988) but have no equivalent among terrestrial annelids, in which, at most, they can be ornamented with fine spines, ridges or teeth especially in connection with the reproductive function. Also, whilst in terrestrial taxa (including the polychaete *Parergodrilus heideri* Reisinger, 1925, pers. obs.) the core of the formed chaeta appears relatively solid in its distal part, in *Hrabeiella* – as in some marine polychaetes (Kristensen & Nørrevang, 1982, quoted in Specht, 1988) – the single chaetal tubules are visible from the base to the distal tip.

The way such brush-like chaetae are produced is problematic, since the chaetal tip is formed first during chaetogenesis (cf. Bouligand, 1967; O'Clair & Cloney, 1974). Sequential modulations of the number, arrangement and orientation of the chaetoblast microvilli (cf. O'Clair & Cloney, 1974) may be involved in designing the core of the brush, while it can be suggested that

an uneven contribution by the lateral follicle cells to the peripheral coat of the chaeta (cf. Richards, 1978; Specht, 1988) is partly responsible for some of the tubules projecting distally as free bristles. Some preliminary observations on a growing chaeta (not shown herein) seem to confirm such hypothesis, in that a distinct cortical layer is lacking on the hairy side of the developing, still unemerged, portion of the brush.

The delicate ultrastructure of the cuticle is similar to that found generally in very small-sized annelids: polychaete larvae (Eckelbarger & Chia, 1978) and adults of interstitial forms (Rieger & Rieger, 1976; Westheide & Rieger, 1978), and aeolosomatids (Postwald, 1971), which also lack a coarse collagenous grid and show only a fine filamentous matrix. In *Hrabeiella*, however, the cuticle is at least twice as thick as in the above groups; it is also thicker than in some larger sized terrestrial oligochaetes (see, for instance, Goodman & Parrish, 1971; Burke, 1974). Furthermore, in contrast to larval and adult polychaetes (e.g. Rieger & Rieger, 1976; Eckelbarger & Chia, 1978; Westheide & Rieger, 1978), the cuticle of *Hrabeiella* is very poor in ascend-

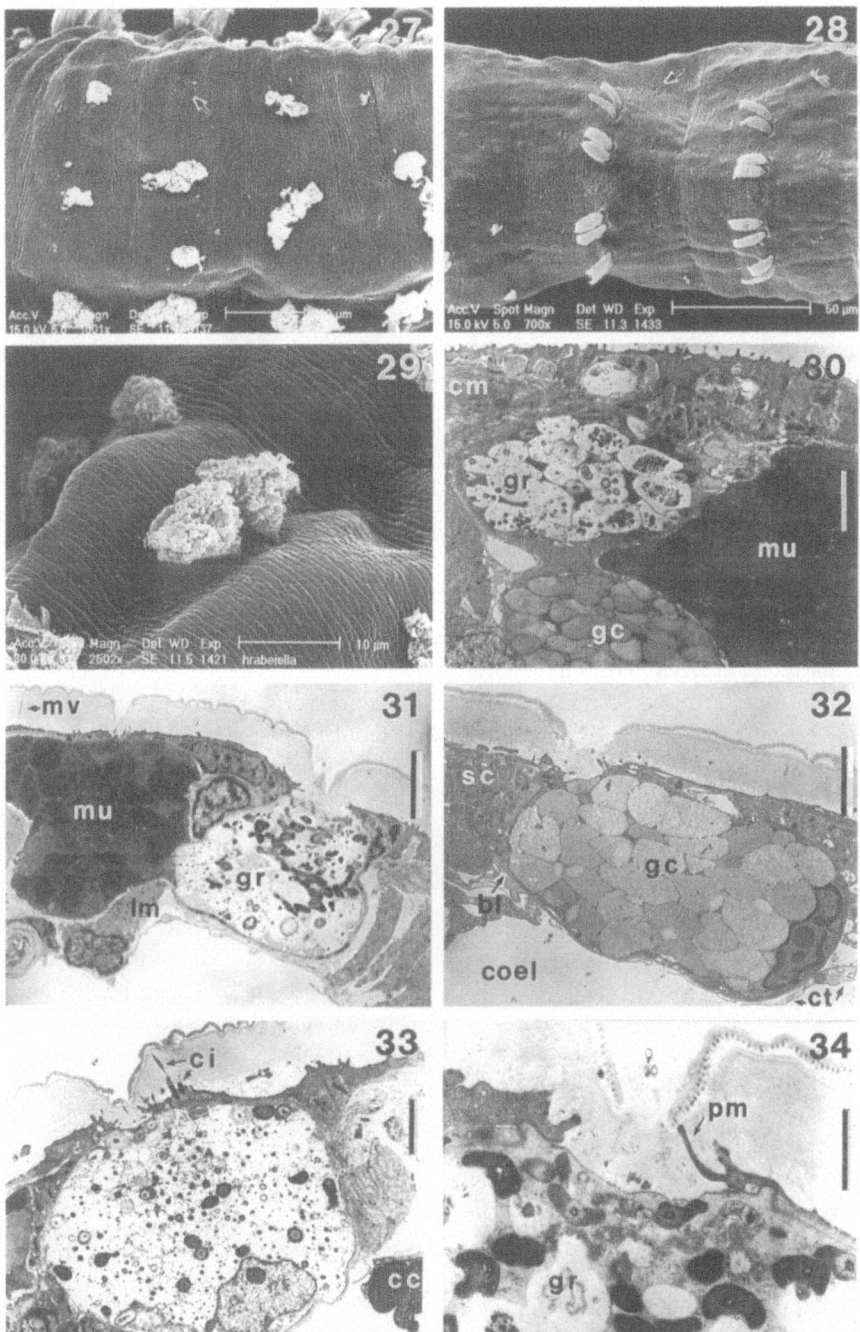

Figs 27–34. Epidermal gland cells of *Hrabeiella*. (27) & (28) Dorsal and ventral views of midbody segments. Arrows show coupled pores of granular and mucous gland cells. SEM. (29) Detail of the dorsal body surface. SEM. (30) Section of body wall across an epidermal glandular block. Scale = 2 μm. TEM. (31) Transverse section through mucous (left) and granular (right) gland cells in correspondence of their external openings. Scale = 1.5 μm. TEM. (32) Transverse section through a grey gland cell. Scale = 2 μm. TEM. (33) Transverse section through a granular gland cell. Note the sensory cilium near the gland pore. Scale = 2 μm. TEM. (34) Detail of the pore in a granular gland cell. Scale = 1 μm. TEM. bl = basal lamina; cc = coelomocyte; ci = cilium; cm = circular muscle fiber; coel = coelom; ct = coelothelium; gr = granular gland cell; gc = grey gland cell; lm = longitudinal muscle fibre; mu = mucous cell; mv = microvillus; pm = pore microvillus; sc = supporting cell.

ing microvilli, and the ones which are occasionally seen do not branch distally.

Platymyarian fibres similar to those of *Hrabeiella* are also found in the circular muscle layer of enchytraeids (Valvassori *et al.*, 1989).

Observations on live specimens make one conclude that *Hrabeiella* is unsuited for crawling over a compact surface while it is designed to move through soft substrata. Direct peristaltic waves are seen to travel along the body wall but do not seem to be sufficiently powerful to ensure by themselves progression through the substratum. This could be achieved, however, by using the chaetae as fastening, nonslip devices. On the other hand, if chaetae are used as the main propulsory devices, their hairiness could magnify the contact surface with the substratum during their power-stroke, as observed in other interstitial annelids (S. R. Gelder & O. Giere, pers. comm.).

Neither the partial/complete reduction of the coelomic space observed among the Psammodrilidae, Polygordiidae and Protodrilidae, nor the partial/complete reduction of the peritoneal lining reported for other, similarly small-sized, interstitial polychaetes (e.g. *Ctenodrilus serratus* Schmidt, 1857; *Stygocapitella subterranea* Knöllner, 1934; *Psammodrilus aedificator* Kristensen & Nørrevang, 1982) and the terrestrial *Parergodrilus heideri* (see Fransen, 1980, 1988), are observed in *Hrabeiella* which, instead, shows the typical annelidan organization, i.e. a complete, cellular peritoneum lining both somatic and splanchnic walls of a spacious body cavity.

Conclusion

The systematic position of *Hrabeiella* remains quite enigmatic. The absence of a clitellum precludes a position among the Euclitellata; this is also confirmed by ultrastructural data herein reported pertaining to the chaetae and cuticle, and unpublished findings concerning the sperm morphology (in prep.).

On the other hand, very few similarities are found with marine interstitial polychaetes – including the 'archiannelids', a polyphyletic group with simple body organization but complex reproductive characters (Westheide, 1985). From the studies accomplished so far, such similarities do not seem to go beyond certain aspects of the cuticle and/or chaetae, which, along with a small body size, could be structural adaptations to the interstitial life in marine or terrestrial environment.

Suggestions have been made (Pižl & Chalupský, 1984; Jans & Römbke, 1989) that the genus should be either allocated to a new annelid class or assigned to the Parergodrilidae, the polychaete family (Reisinger, 1960) which includes the terrestrial *Parergodrilus heideri* and the marine interstitial *Stygocapitella subterranea*. The latter solution appears now unsatisfactory because of the different coelomic organization of *Hrabeiella*. More research is in progress on both *Hrabeiella* and *Parergodrilus* and we hope it will contribute to unravelling the puzzle.

The histochemistry of the three types of epidermal gland cells deserves to be thoroughly investigated. Moreover, a reinvestigation of the anatomy and ultrastructure of the topotypes of *Hrabeiella periglandulata* is needed to clarify the extent of variation of chaetal morphology and to establish whether the Italian or the German populations might be recognized as a separate species rather than a variant of the Bohemian ones.

Acknowledgements

We are very grateful to Dr S. R. Gelder and Dr J. Römbke for helpful discussions and to Dr K. A. Coates for her valuable criticism on the manuscript. This work was supported by a 40% M.U.R.S.T. grant to Prof P. Omodeo and 40% and 60% M.U.R.S.T. grants to Prof R. Dallai. We wish to thank the staff of the Laboratory of Electron microscopy, Department of Evolutionary Biology, University of Siena, and particularly F. Anselmi and L. Gamberucci for technical assistance with sectioning and with processing and printing of photographs, respectively.

References

Bouligand, Y., 1967. Les soies et les cellules associées chez deux Annélides Polychètes. Z. Zellforsch. 79: 332–363.

Burke, J. M., 1974. An ultrastructural analysis of the cuticle, epidermis and esophageal epithelium of *Eisenia foetida* (Oligochaeta). J. Morph. 142: 301–320.

Eckelbarger, K. J. & F.-S. Chia, 1978. Morphogenesis of larval cuticle in the polychaete *Phragmatopupa lapidosa*: A correlated scanning and transmission electron microscopic study from egg envelope formation to larval metamorphosis. Cell Tissue Res. 186: 187–201.

Fransen, M. E., 1980. Ultrastructure of coelomic organization in Annelids. I. Archiannelids and other small polychaetes. Zoomorphologie 95: 235–249.

Fransen, M. E., 1988. Coelomic and vascular systems. In: W. Westheide & C. O. Hermans (eds), The Ultrastructure of Polychaeta. Microfauna marina 4: 199–213.

Goodman, D. & W. B. Parrish, 1971. Ultrastructure of the epidermis in the ice-worm, *Mesenchytraeus solifugus*. J. Morph. 135: 71–86.

Graefe, U., 1989. Der Einfluss von sauren Niederschlägen und Bestandeskalkungen auf die Enchytraeidenfauna in Waldböden. Verh. Ges. Ökol. 17: 597–603.

Graefe, U., 1990. Untersuchungen zum Einfluss von Kompensationskalkung und Bodenbearbeitung auf die Zersetzerfauna in einem bodensauren Buchenwald- und Fichtenforst-Ökosystem. In: J. Gehrmann (ed.), Umweltkontrolle am Waldökosystem. Forschung und Beratung, Reihe C, Münster 48: 232–241.

Graefe, U., 1993a. Die Gliederung von Zersetzergesellschaften für die standortsökologische Ansprache. Mitt. Dtsch. Bodenkundl. Ges. 69: 95–98.

Graefe, U., 1993b. Veränderungen der Zersetzergesellschaften im Immissionsbereich eines Zementwerkes. Mitt. Dtsch. Bodenkundl. Ges. 72: 531–534.

Healy, B. & E. Rota, 1992. Methods for collecting enchytraeids during expeditions. Soil Biol. Biochem. 24: 1279–1281.

Jans, W. & J. Römbke, 1989. Funde eines terrestrischen Polychaeten (Annelida) in Wäldern Baden-Württembergs. Carolinea 47: 158–162.

Karnovsky, M. J., 1965. A formaldehyde-glutaraldehyde fixative of high osmolarity for use in electron microscopy. J. Cell. Biol. 27: 137A–138A.

Knöllner, F., 1934. Die Tiere des Küstengrundwassers bei Schilksee (Kieler Bucht). 5. *Stygocapitella subterranea* nov. gen. nov. spec. Schr. naturwiss. Ver. Schleswig-Holstein 20: 468–472.

Kristensen, R. M. & A. Nørrevang, 1982. Description of *Psammodrilus aedificator* sp. n. (Polychaeta), with notes on the arctic interstitial fauna of Disko Island, W. Greenland. Zool. Scr. 11: 265–279.

Lanzavecchia, G., R. Valvassori & M. de Eguileor, 1986. Helical muscles from Bjorn Afzelius to the present state. In: M. Cresti & R. Dallai (eds), Biology of Reproduction and Cell Motility in Plants and Animals. University of Siena, Siena: 119–124.

Lanzavecchia, G., M. de Eguileor & R. Valvassori, 1988. Muscles. In: W. Westheide & C. O. Hermans (eds), The Ultrastructure of Polychaeta, Microfauna marina 4: 71–88.

O'Clair, R. M. & R. A. Cloney, 1974. Patterns of morphogenesis mediated by dynamic microvilli: chaetogenesis in *Nereis vexillosa*. Cell. Tiss. Res. 151: 141–157.

O'Connor, F. B., 1955. Extraction of enchytraeid worms from a coniferous forest soil. Nature 175: 815–816.

Pižl, V. & J. Chalupský, 1984. *Hrabeiella periglandulata* gen. et sp. n. (Annelida) – a curious worm from Czechoslovakia. Vest. cs. Spolec. zool. 48: 291–295.

Postwald, H. E., 1971. A fine ultrastructural analysis of the epidermis and cuticle of the oligochaete *Aeolosoma bengalense* Stephenson. J. Morph. 135: 185–212.

Reisinger, E., 1925. Ein landbewohnender Archiannelide. Zugleich ein Beitrag zur Systematik der Archianneliden. Z. Morph. Ökol. Tiere 13: 197–254.

Reisinger, E., 1960. Die Lösung des *Parergodrilus*-Problems. Z. Morph. Ökol. Tiere 48: 517–544.

Richards, K. S., 1978. Epidermis and cuticle. In P. J. Mill (ed), Physiology of Annelids. Academic Press, London: 33–61.

Rieger, R. M. & G. E. Rieger, 1976. Fine structure of the archiannelid cuticle and remarks on the evolution of the cuticle within the Spiralia. Acta zool. 57: 53–68.

Rota, E., 1994. Enchytraeidae (Annelida: Oligochaeta) of the Mediterranean region: a taxonomic and biogeographic study. Unpubl. Ph.D. Thesis. The National University of Ireland, 255 pp.

Rota, E., 1995. Italian Enchytraeidae (Oligochaeta). I. Boll. Zool. 62(2): 183–231.

Schmidt, O., 1857. Zur Kenntnis der Turbellaria Rhabdocoela und einiger anderer Würmer des Mittelmeeres. Sitzber. Akad. Wiss. Wien 23: 347–366.

Specht, A., 1988. Chaetae. In: W. Westheide & C. O. Hermans (eds), The Ultrastructure of Polychaeta, Microfauna marina 4: 45–59.

Valvassori, R., M. de Eguileor, G. Lanzavecchia & G. Scarì, 1989. Body wall organization in enchytraeids. Hydrobiologia 180: 83–89.

Westheide, W., 1985. The systematic position of the Dinophilidae and the archiannelid problem. In: S. Conway Morris, J. D. George, R. Gibson & H. M. Platt (eds), The Origins and Relationships of Lower Invertebrates. Syst. Assoc. Spec. Vol., Clarendon Press, Oxford 28: 310–326.

Westheide, W. & R. M. Rieger, 1978. Cuticle ultrastructure of hesionid polychaetes (Annelida). Zoomorphologie 91: 1–18.

Hydrobiologia **334**: 241–249, 1996.
K. A. Coates, Trefor B. Reynoldson & Thomas B. Reynoldson (eds), Aquatic Oligochaete Biology VI.
© 1996 *Kluwer Academic Publishers.*

Peripheral vascular apparatus in some aquatic oligochaetes with special references to haplotaxids

Roberto Valvassori[2], Giulio Lanzavecchia[1], Magda de Eguileor[1], Annalisa Grimaldi[1] &
Laura Colombo[2]
[1]*III Facoltà di Scienze, Università di Milano, via Ravasi 2, I-21100 Varese, Italy*
[2]*Dipartimento di Biologia, Università di Milano, Via Celoria 26, I-20133 Milano, Italy*

Key words: Haplotaxidae, vessels, ultrastructure, microdriles

Abstract

The organization of the peripheral vascular apparatus in two haplotaxids has been studied and compared with that of other microdriles. Considerable differences in the circulatory systems of *Pelodrilus leruthi* and *Haplotaxis gordioides*, especially in relationships to the body wall muscle fibers, separate and distinguish the two animals. Different organizations of the peripheral apparatus that can be observed in these microdriles are: in the first species, capillary vessels have no contact with the body wall; in the second species, capillaries extend between the longitudinal muscle fibers until they reach the body surface, thus approaching the situation in megadriles where circulation can become intraepithelial.

Generally, vessels hanging in the coelom are of a large diameter. When a capillary network related to the body wall muscle develops, vessels are small in diameter and their walls have variable numbers of contractile elements, ensuring the forced circulation of the blood.

Introduction

The vascular apparatus of microdrile oligochaetes is a closed system of longitudinal and lateral vessels (Cook, 1971; Hanson, 1949; Jamieson, 1981). The longitudinally oriented dorsal and ventral vessels are the main contractile trunks, connected by dorso-ventral connectives. Anterior commissural vessels in conjunction with the main pulsating dorsal vessel sometimes serve as hearts (Jamieson, 1981).

The capillary apparatus varies from taxon to taxon especially in relation to the body wall muscles (Stephenson, 1930; Ruppert & Carle, 1983; Jamieson, 1981). Microdriles, generally, lack perineural vessels, the lateral hearts and the intraepidermal circulation found in megadrile oligochaetes (earthworms) (Stephenson, 1930; Omodeo, 1942; Jamieson, 1981). Available data about ultrastructure of capillaries of microdriles are for *Tubifex tubifex* Müller, 1774 (Peters, 1977), *T. tubifex* and *Branchiura sowerbyi* Beddard, 1892 (Comolli & Ferraguti, 1975).

In this paper, we describe the structures and ultrastructure of the peripheral vessels of two haplotaxids, some tubificids, lumbriculids and enchytraeids and relate these to their locomotory behaviours and habitats.

Materials and methods

Specimens of four families were processed for electron microscopy. Haplotaxidae: *Haplotaxis gordioides* Hartmann, 1821, *Pelodrilus leruthi* Hrabe, 1953; Tubificidae: *Monopylephorus* sp., *T. tubifex*; Enchytraeidae: *Enchytraeus* sp., *Enchytraeus albidus* Henle, 1837, *Enchytraeus minutus* Nielsen and Christensen, 1961; Lumbriculidae: *Rhynchelmis limosella* Hoffmeister, 1843, *Bythonomus lemani* Grube, 1879, *Lumbriculus variegatus* Müller, 1774. The animals were cut into small pieces, fixed in 2% glutaraldehyde in 0.1 M cacodylate buffer (pH 7.2) for two hours, and washed overnight in 0.1 M cacodylate buffer (pH 7.2).

The samples were postfixed for two hours with 1% osmic acid in the same buffer (pH 7.2), washed in distilled water, and then stained for two hours in the dark in 2% aqueous uranyl acetate. After standardized dehydration in an ethanol series, specimens were embedded in an Epon 812-araldite mixture. Sections were obtained with an LKB Ultrotome V. Semi-thin sections (1 μm thick) were stained with crystal violet and basic fuchsin according to Moore *et al.* (1960) and then observed with a Jenaval light microscope. Thin sections were stained with uranyl acetate and lead citrate and were observed with a JEOL 100B electron microscope.

Results

Haplotaxidae

The peripheral vascular apparati in the two species appear to be organized differently and are therefore described separately.

Haplotaxis gordioides: Numerous (up to nine per section) large vessels of round profile occupy most of the body cavity (Fig. 1), internal to the longitudinal muscle layer; they are laterally placed (Figs 1 & 2). The vessels hang into the coelomic cavity (Figs 4 & 5) and in some places are compressed by the chloragogen cells (Figs 3 & 8). Vessels are enveloped by a coelothelium (Figs 4, 5 & 7) which connects to the body wall muscles and forms a thin mesentery (Figs 4 & 5). A continuous coelothelium also covers the longitudinal muscle layer (Figs 2 & 4). The flat coelomic cells are characterized by roundish bodies with a granular structure (Fig. 7) (see de Eguileor *et al.*, 1990). The vessel walls are lined by a vascular lamina (Friedman & Weiss, 1979; Coggeshall, 1965) and a layer of flattened cells (myoepithelial cells) which are very thin (0.2 μm) (Figs 6, 7 & 8) except in the nuclear area (Fig. 6). These cells contain a few (15–20) myofilaments (Figs 7 & 8). No capillaries can be seen among the muscle fibers.

Pelodrilus leruthi. The walls of all capillaries are composed by thin myoepithelial cells (Figs 12 & 15) with little contractile material (Fig. 15). Inside the vessels a vascular lamina 50 nm thick (Fig. 14) is visible.

Anterior cross-sections: Numerous capillaries can be found in addition to the vessels hanging into the coelomic cavity (Fig. 10). These capillaries penetrate into the longitudinal muscle layer (Fig. 9), dividing it into fiber portions of varying size (de Eguileor *et al.*,

1990). All capillaries filling the coelomic cavity are enveloped in a coelothelium (Figs 10 & 13).

Posterior cross-sections: The caudal region of the worm, involved in gas exchange, is characterized by a number of vessels pushing through the longitudinal and the circular muscle layer (Figs 11 & 16) and even beyond the circular layer, up to the epidermis (Fig. 17).

Lumbriculidae

The organization of the peripheral vascular system is similar in the three lumbriculid species examined.

Anterior cross-sections: In the first 9–10 metameres, there is a well-developed vascular network (Fig. 18). Some longitudinal vessels have thick walls in the area protruding into the coelom and thin walls in the area of muscle fibers (Figs 18 & 19). These vessels, which have a specialized contractile system (Fig. 21), can be compared to pulsating hearts (Stephenson, 1930). There are also numerous longitudinal capillaries inserted in an orderly fashion among the ribbon-shaped fibers (Fig. 18), and consequently, these capillaries are in contact with fibers of the longitudinal and circular muscle layers (Figs 18 & 23). The myoepithelial cells vary in thickness. They contain few myofilaments, organized in tiny functional units (Fig. 24), in which the myofilaments are positioned at 90° to one another, as described for *T. tubifex* by Comolli & Ferraguti (1975).

Posterior cross-sections: In contrast, the body wall muscle of the posterior body is very thick, with ribbon-shaped fibers that form a 'fence' over the entire body area behind the reproductive segments (Fig. 20). In this region, vessels are present predominantly in the coelom (Fig. 20) with a few branches between the ribbon-shaped fibers. In the most posterior metameres, corresponding to the growth area of the worm, there are large vessels investing large portions of the body wall (Fig. 22).

Enchytraeidae

The three enchytraeid species examined have the same peripheral vascular organization. The vessels are small and few. They are distributed in the coelomic cavity and among the cytoplasmic pouches of the longitudinal muscle fibers (Fig. 25). The walls of the vessels consist of robust myoepithelial cells with well-organized contractile material (Figs 26–28). Sarcomeres are clearly distinguishable.

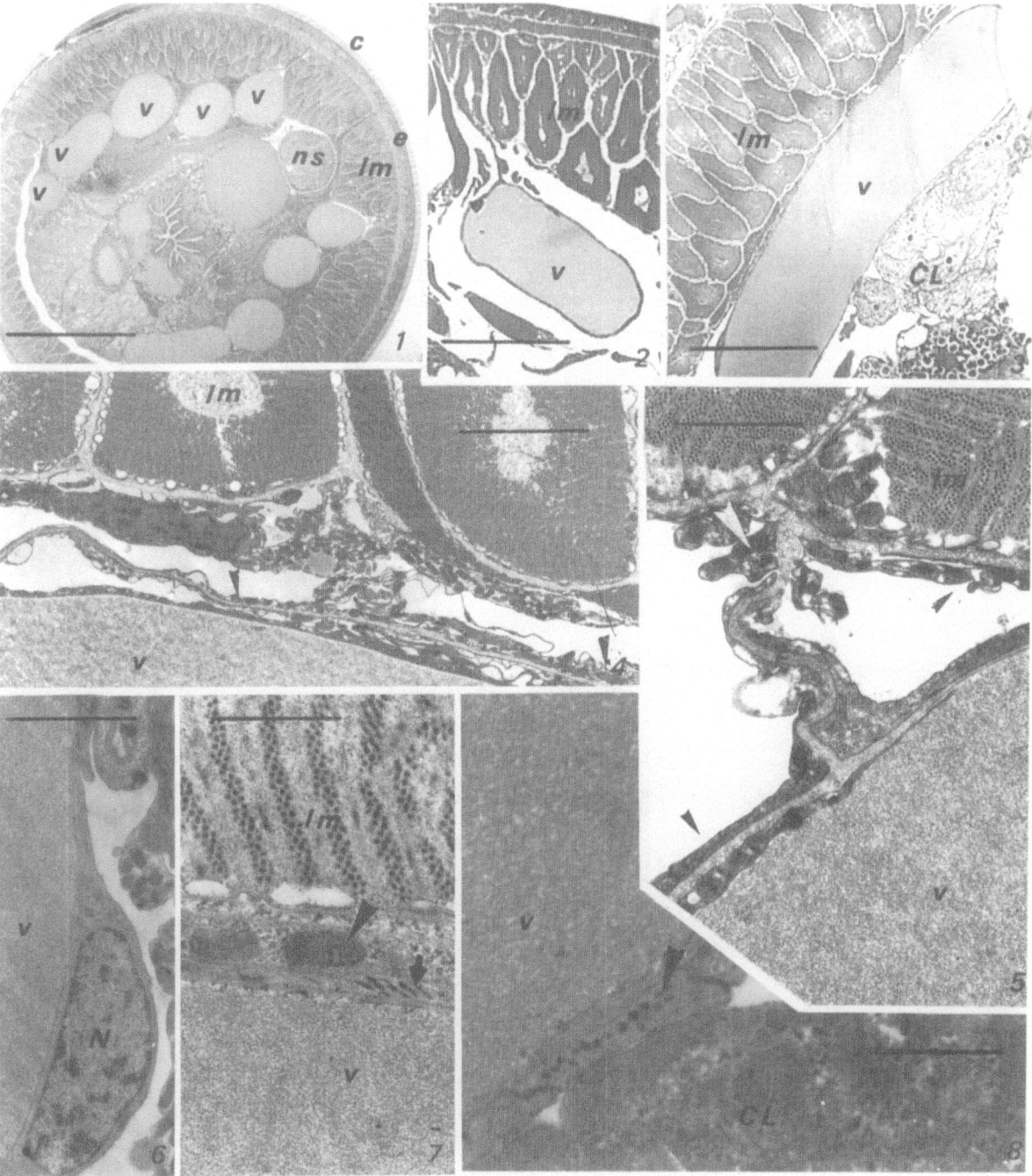

Figures 1–8. Haplotaxidae. *Haplotaxis gordioides*. Anterior cross sections.
Fig. 1. Light micrograph of body wall in cross section. Large vessels (v) are located under the longitudinal muscle layer. c: cuticle; e: epidermis; lm: longitudinal muscles; ns: nervous system. Scale bar: 100 μm. *Fig. 2*. Light micrograph of body wall in cross section. Vessel (v) hangs in the coelomic cavity. lm: longitudinal muscle. Scale bar: 25 μm. *Fig. 3*. Light micrograph of body wall in cross section. The vessel (v) lies between longitudinal muscle fibers (lm) and chloragogen cells (CL). Scale bar: 50 μm. *Figs 4 & 5*. The vessel (v), underneath the longitudinal muscle layer (lm), hangs in the coelomic cavity. The coelothelium (arrowheads) envelops the vessels; the fibrillar sheath (F) is lined by peritoneal cells filled with granular bodies (white arrowheads). Scale bars: 6 μm for Fig. 4; 2.3 μm for Fig. 5. *Fig. 6*. The vessel (v) wall is extremely thin except in the nuclear (N) area. Scale bar: 3 μm. *Fig. 7*. A vessel (v) is separated from a longitudinal muscle fiber (lm) by a coelothelium with characteristic granular bodies (arrowhead). The myoepithelial cells contain a few myofilaments (arrow). Scale bar: 1.2 μm. *Fig. 8*. A vessel (v) in close contact with the chloragogen tissue (CL). The vessel wall is composed of myoepithelial cells containing a few myofilaments (arrowhead). Scale bar: 0.77 μm.

244

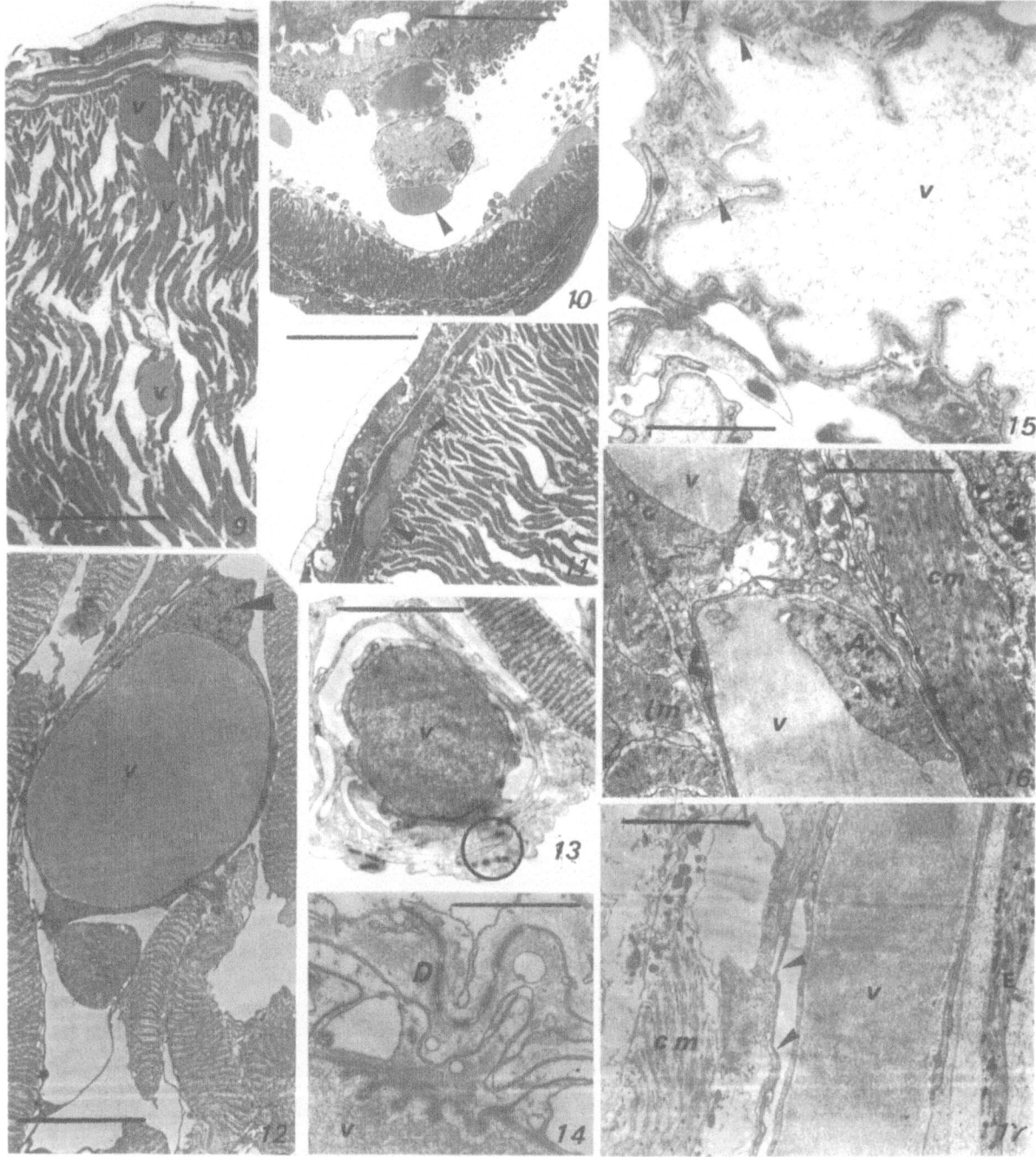

Figures 9–17. Haplotaxidae. *Pelodrilus leruthi.*

Figs 9–11. Light micrographs of body wall in cross section. Peripheral vessels (v) localized among longitudinal fibers in anterior sections (Fig. 9) and between circular and longitudinal muscle layer in posterior sections (Fig. 11) (arrowheads). There is also a subneural vessel in posterior section (Fig. 10) (arrowhead). Scale bars: 33 μm for Fig. 9; 133 μm for Fig. 10; 50 μm for Fig. 11. *Fig. 12.* Posterior section. A vessel (v) among the longitudinal fibers seen in an anterior cross section has a very thin wall except in the nuclear area (arrowhead). Scale bar: 6 μm. *Fig. 13.* Anterior section. A vessel of the anterior body hanging in the coelom is surrounded by a continuous sheath. This sheath (encircled) is formed by folds of the coelothelium stiffened by desmosomes. Scale bar: 6.6 μm. *Fig. 14.* Anterior section. Detail of the sheath enveloping the vessel (v) hanging in the coelomic cavity. D: desmosomes. Scale bar: 2.3 μm. *Fig. 15.* Posterior sections. The wall of the vessel (v) contains myoepithelial cells with a few myofilaments (arrowheads). Scale bar: 1.8 μm. *Fig. 16.* Posterior section. Some vessels (v) are placed between circular (cm) and longitudinal (lm) muscle fibers. An amoebocyte is visible inside the vessel (A). Coelothelial cells (arrowheads). Scale bar: 1.2 μm. *Fig. 17.* Posterior section. A capillary (v) invades the area close to the epidermis (E). This type of vessel is partially covered by a coelothelium (arrowhead). cm: circular muscle. Scale bar: 3.6 μm.

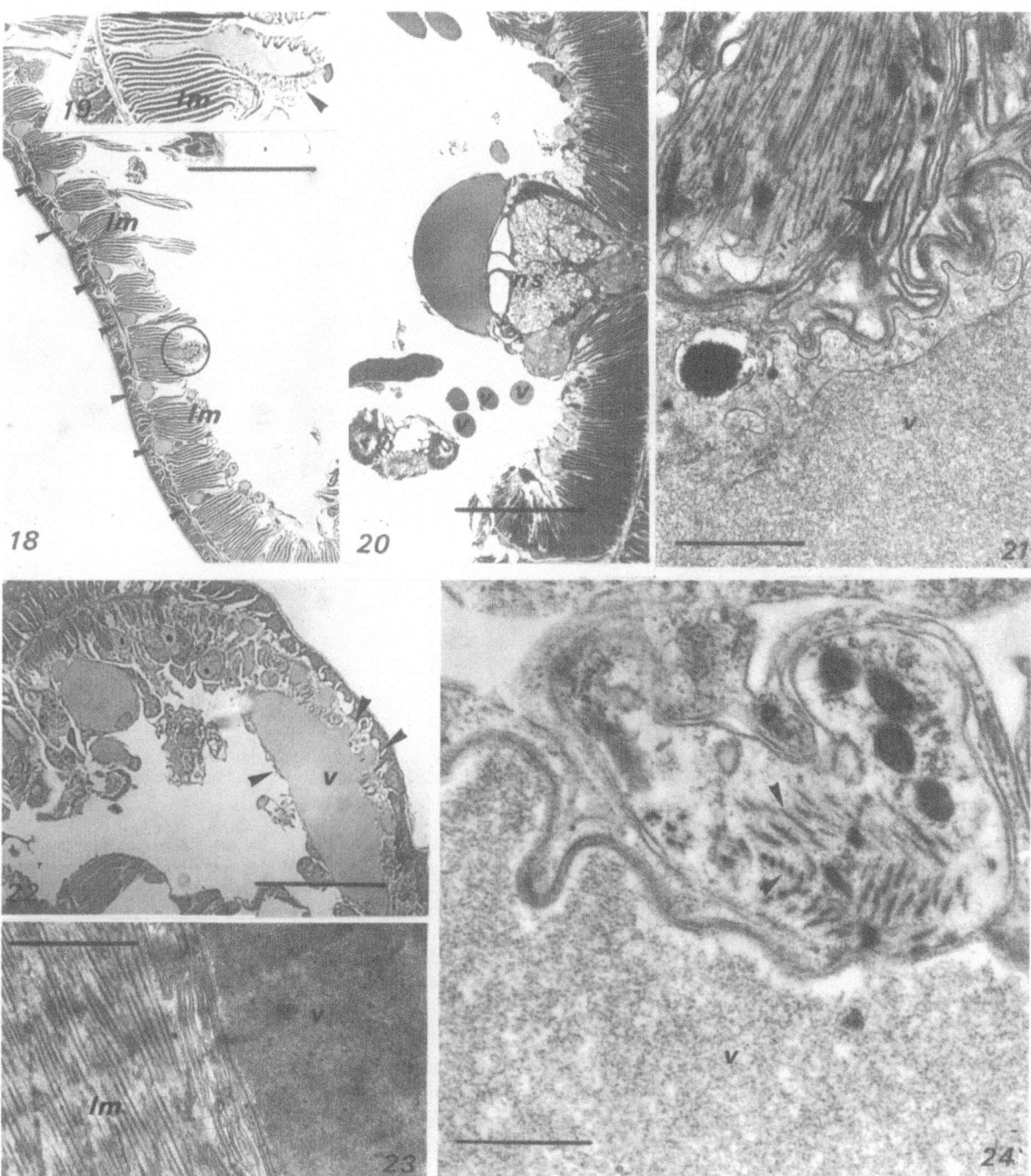

Figures 18–24. Lumbriculidae.

Figs 18–21. Rhynchelmis limosella. Fig. 18. Anterior cross section. Capillaries (arrowheads) are recognizable among the ribbon-like longitudinal muscles (lm). Some of these blind vessels have muscular walls where they protrude into the coelom (encircled) and very thin walls close to muscle fibers. Scale bar: 110 μm. *Fig. 19.* Enlargement of Fig. 18. lm: longitudinal muscles; muscular vessel wall (arrowhead). Scale bar: 55 μm. *Fig. 20.* Cross section of posterior area of the body wall. Vessels (v) are most often in the coelom. ns: nervous system. Scale bar: 90 μm. *Fig. 21.* Anterior cross section. Detail of the thick wall of a blind vessel (v). The myoepithelial cells are characterized by organized contractile material (arrowhead). Scale bar: 1.7 μm. *Fig. 22. Bythonomus lemani.* Light micrograph of the pre-pygidial growth zone. The vessels (v) are very wide (arrowheads). Scale bar: 100 μm. *Fig. 23. Lumbriculus variegatus.* Anterior section. A capillary (v) seen in anterior cross section is in close contact with the muscle fiber. lm: longitudinal muscle. Scale bar: 0.7 μm. *Fig. 24. Lumbriculus variegatus.* Anterior section. The myoepithelial cells have sarcomeres in which the filaments are positioned 90° to one another (arrowheads). v: vessel. Scale bar: 0.9 μm.

246

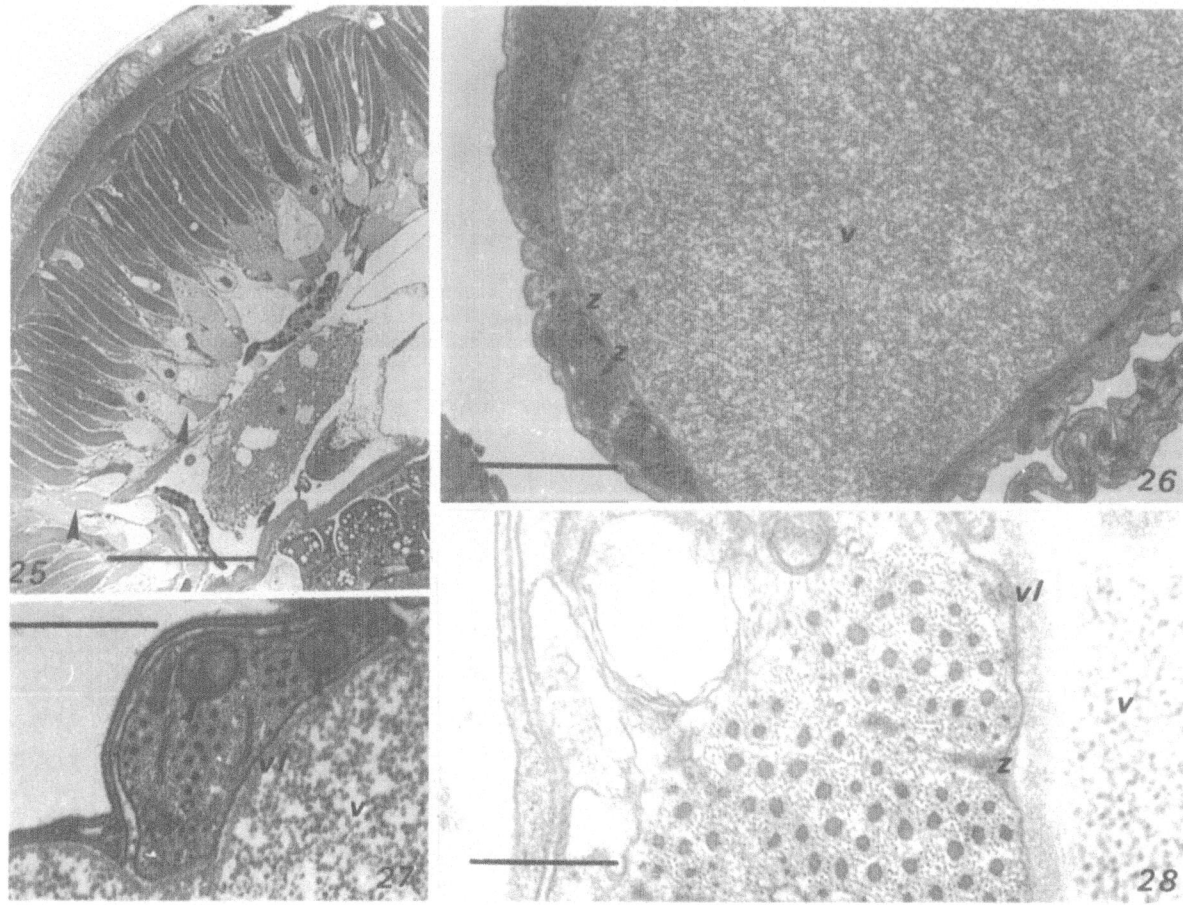

Figures 25–28. Enchytraeidae.
Fig. 25. Enchytraeus albidus. Light micrograph of cross sectioned anterior body wall. The vessels are few and small (arrowheads). Scale bar. 20 μm. *Figs 26–28. Enchytraeus* sp. Mid-body cross sections. The wall of the vessels (v) is formed of thick myoepithelial cells characterized by myofilaments organized in small sarcomeres. In these contractile functional units, Z elements (Z) and mitochondria (arrowheads) are visible. vl: vascular lamina. Scale bars: 1.8 μm for Fig. 26; 0.6 μm for Fig. 27; 0.3 μm for Fig. 28.

Tubificidae

The organizations of the vascular systems of the two species of tubificid examined are almost the same.

Anterior cross-sections: In the anterior segments the larger vessels are free in the coelomic cavity and are surrounded by chloragocytes (Fig. 30).

Posterior cross-sections: The smaller vessels of the posterior body regions lie between cytoplasmic portions of the longitudinal muscle fibers (Fig. 29). The myoepithelial cells have few thick and thin filaments (Figs 32 & 33). The vessels are surrounded only by the cytoplasmic region of the muscle fibers and are not covered by coelothelial cells (Fig. 31).

Discussion

The peripheral vascular systems of microdriles have been shown to be heterogeneous with regard to the distribution of capillaries in the worms and to the thickness of myoepithelial cells lining the capillary. The vessels of *H. gordioides* are large (up to 90 μm) and protrude into the coelomic cavity (Figs 1 & 2). They hang from the mesenteries (Figs 2, 4 & 5) and/or are kept in that position by the chloragogen tissue (Figs 3 & 8). All these vessels have very thin walls. The myoepithelial cells are quite flat, endowed with a few (15–20) myofilaments (Figs 7 & 8). As a consequence, it is probable that the vessels of *H. gordioides* have a modest contractile activity of their own. Movements of these vessels

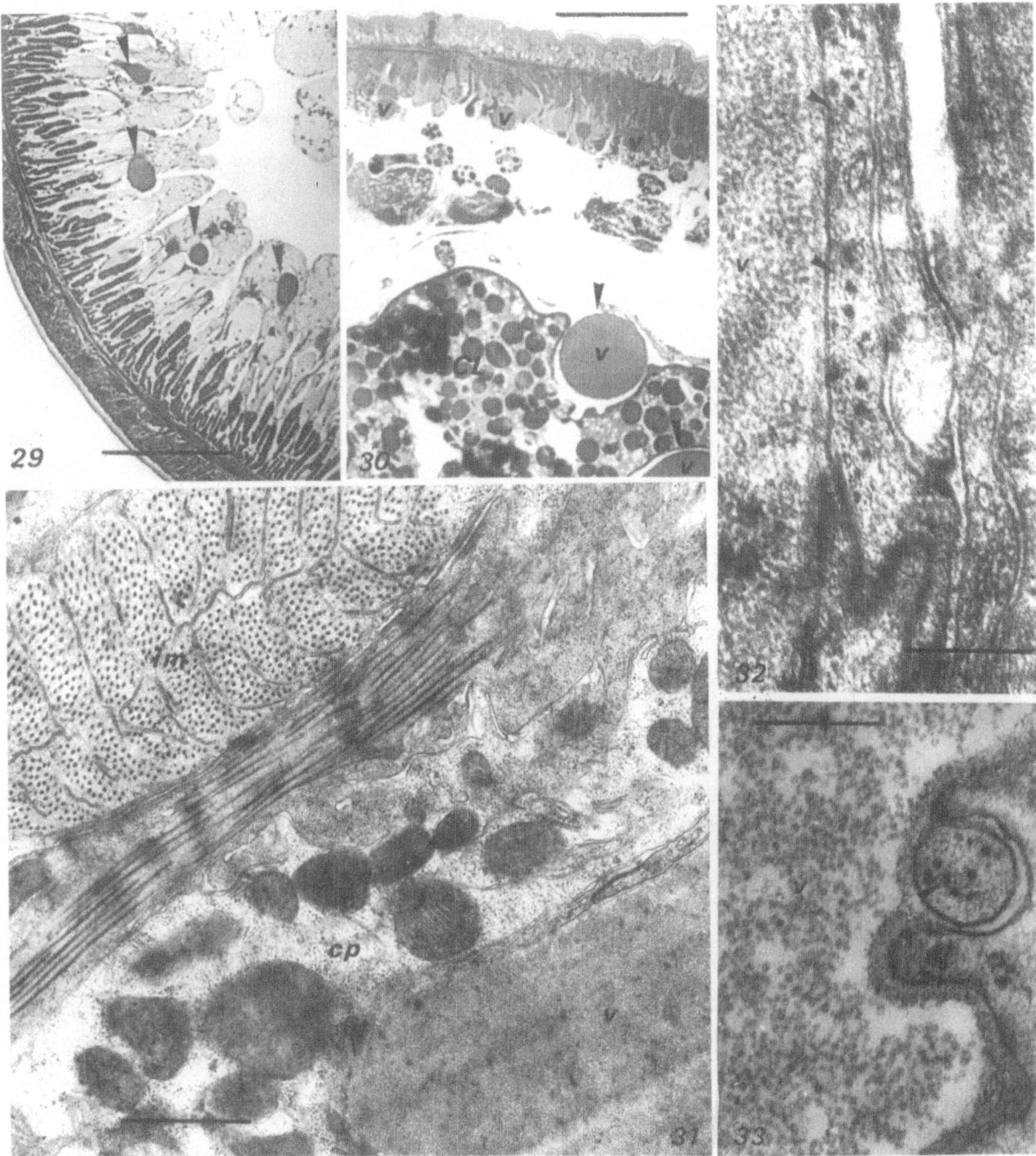

Figures 29–33. Tubificidae.

Figs 29 & 30. Tubifex tubifex (posterior section), and Fig. 30 *Monopylephorus* sp. (anterior section). Light micrographs of posterior (Fig. 29) and anterior (Fig. 30) cross-sectioned body wall. The vessels (v and arrowheads) are located between the cytoplasmic portions of muscle fibers or are surrounded by chloragocytes (CL). Scale bars: 28 μm for Fig. 29; 50 μm for Fig. 30. *Fig. 31. Monopylephorus* sp. Posterior cross section. The vessel (v) is in close contact with the cytoplasmic portion (cp) of the longitudinal muscle fiber (lm). Scale bar: 1.2 μm. *Figs 32 & 33. Tubifex tubifex.* Posterior sections. The flattened myoendothelial cells show a few myofilaments (arrowhead). v: vessel; vl: vascular lamina. Scale bars: 0.35 μm for Fig. 32; 0.3 μm for Fig. 33.

248

may be supported by the movements of the entire body wall (P. Omodeo, pers. comm.).

The peripheral vessels never penetrate the body wall. They just rest against the innermost muscle fibers of the longitudinal layer, from which they are separated by the coelothelium (Figs 4 & 7). The surface of exchange between vessels and adjacent tissues is limited. It is important to bear in mind that *H. gordioides* can live either in oxygenated waters or in asphyctic mud. Epidermal exchanges must be sufficient to supply oxygen to the blood of a thin worm with a given surface area/volume ratio. In *H. gordioides*, the vascular apparatus seems to function as a reservoir which gradually releases oxygen to the tissues by diffusion.

The vascular system in *P. leruthi* is different. The vessels are smaller in diameter (15 μm) and can be subdivided into two classes according to position and probable function. (1) Vessels with thin walls and incomplete coelothelium lying between the muscle fibers (Figs 12 & 16). At times these vessels reach the fibers of the circular muscle layer (Figs 11 & 16) or even the bases of the epidermal cells (Fig. 17). This arrangement could facilitate oxygen transport to the active tissues. (2) Vessels lying within the coelomic cavity, enveloped in a thick sheath of coelothelial folds joined by junctions (Figs 13 & 14). The junctions confer higher mechanical resistance. It should be noted that in *P. leruthi*, the superficial capillaries facilitate gas exchange between vessels and muscle fibers, on one side, and the external environment on the other. In order to exploit the small amount of oxygen dissolved in cave water, these animals, which sink their anterior ends into the mud, sway their free posterior ends, which are highly vascularized. This is like the well-known behaviour of tubificids. The subdivision of the muscle fibers into blocks by capillaries may further ensure the supply of oxygen and nutrition to their flat circomyarian fibers, which are capable of rapid, coordinated contractions. *Haplotaxis gordioides* has pseudocircomyarian fibers which undergo slower and less coordinated contraction and the peripheral vascularization of the body is less.

The division of muscle fibers into sectors by the capillaries is irregular in *P. leruthi* but highly regular in the anterior area of lumbriculids. In lumbriculids, small vessels and small groups of longitudinal fibers alternate, up to the circular muscle layer. This organization could adequately supply energy to highly specialized fibers. In addition, the blood flow rate may be increased by vessels with contractile portions on their coelomic sides (Figs 18, 19 & 21).

In the enchytraeids examined, peripheral vascularization is not as extensive as in lumbriculids. However, although these peripheral vessels have a smaller diameter, the other blood vessels have highly contractile walls, which might enhance their circulatory efficiency (Figs 26–28).

In tubificids, the peripheral vascular network is extensive, especially in the caudal area, thus ensuring efficient gas exchange in these small animals. Moreover, these capillaries are in close contact with the cytoplasmic part of the muscle fibers (Fig. 31).

Two types of vascular system were found in the microdriles examined: in the first, the capillaries hang in the coelom and, generally, their walls are lined by myoepithelial cells which are characterized by myofilaments more or less organized into sarcomeres; in the second, the capillaries extend among the longitudinal fibers, reaching to the epidermis. The latter arrangement is similar to that of megadriles, in which circulation can become intraepithelial (Stephenson, 1930; Omodeo, 1942).

In addition to the differences already emphasized in the organization of vascular systems in the two species of haplotaxids, the presence in *P. leruthi* of a subneural vessel is worth noting (Fig. 10). There is no subneural vessel in microdriles or in the most primitive megadriles (*Alma*) (Rota & Omodeo, 1992) but it is generally present in megadriles. Consequently, the taxonomic separation and distinction between *P. leruthi* and *H. gordioides* seems even more plausible (Brinkhurst, 1982, 1988, 1992; de Eguileor *et al.*, 1990; Lanzavecchia *et al.* 1994; Kasprzak, 1984; Omodeo, 1987; Timm, 1981).

Acknowledgments

This work was supported by CNR and MURST grants. The authors are very grateful to P. Omodeo for his very helpful discussion and suggestions.

References

Brinkhurst, R. O., 1982. Evolution in the Annelida. Can. J. Zool. 60: 1043–1059.

Brinkhurst, R., 1988. A taxonomic analysis of the Haplotaxidae (Annelida, Oligochaeta). Can. J. Zool. 66: 2243–2252.

Brinkhurst, R. O., 1992. Evolutionary relationships within the Clitellata. Soil Biol. Biochem. 24: 1201–1205.

Coggeshall, R. E., 1965. A fine structural analysis of the ventral nerve chord and associated sheath of *Lumbricus terrestris* L. J. Comp. Neurol. 125: 393–438.

Comolli, A. & M. Ferraguti, 1975. An unusual obliquely striated muscle pattern: the myoendothelial cell of the blood vessels of Tubificidae (Annelida Oligochaeta). Monit. Zool. Ital. 9: 25–36.

Cook, D. G., 1971. Anatomy: Microdriles. In R. O. Brinkhurst & B. G. M. Jamieson (eds), Aquatic Oligochaeta of the World. Oliver and Boyd, Edinburgh: 8–41.

de Eguileor, M., R. Valvassori, G. Lanzavecchia & S. Giorgi, 1990. Body wall muscles in haplotaxids *Haplotaxis gordioides* and *Pelodrilus leruthi* (Annelida, Oligochaeta). Zoomorphology 110: 27–36.

Freidman, W. & L. Weiss, 1979. The fine structure of blood follicles in the earthworm genera *Amynthas* and *Lumbricus* (Annelida, Oligochaeta). J. Morphol. 161: 123–144.

Hanson, J., 1949. The histology of the blood system in Oligochaeta and Polychaeta. Biol. Rev. Cam. phil. Soc. 24: 127–173.

Jamieson, B. G. M., 1981. The Ultrastructure of the Oligochaeta. Academic Press, London, 462 pp.

Kasprzak, K. 1984. The previous and contemporary conceptions on phylogeny and systematic classifications of Oligochaeta (Annelida). Annales Zoologici 389: 205–223.

Lanzavecchia, G., R. Valvassori & M. de Eguileor, 1994. Body wall muscles in oligochaetes. Hydrobiologia 278: 179–188.

Moore, R. D., V. Mumaw & M. D. Shoenberg, 1960. Optical microscopy of ultrathin tissue sections. J. Ultrastruct. Res. 4: 113–116.

Omodeo, P., 1942. Contributo alla conoscenza della circolazione in *Allolobophora complanata* Sav. Archivio zool. ital. 30: 1–37.

Omodeo, P., 1987. Some new species of Haplotaxidae (Oligochaeta) from Guinea and remarks on the history of the family. Hydrobiologia 155: 1–13.

Peters, W., 1977. Possible site of ultrafiltration in *Tubifex tubifex* Müller (Annelida, Oligochaeta). Cell Tissue Res. 179: 367–375.

Rota, E. & P. Omodeo, 1992. Phylogeny of Lumbricina: reexamination of an authoritative hypothesis. Soil Biol. Biochem. 24: 1263–1277.

Ruppert, E. & K. J. Carle, 1983. Morphology of metazoan circulatory systems. Zoomorphology 103: 193–208.

Stephenson, J., 1930. The Oligochaeta. Clarendon Press, Oxford, 978 pp.

Timm, T., 1981. On the origin and evolution of aquatic Oligochaeta. Eesti NSV Teaduste Akadeemia Toimetised 30. Köide Bioloogia 3: 173–181.

Hydrobiologia **334**: 251–261, 1996.
K. A. Coates, Trefor B. Reynoldson & Thomas B. Reynoldson (eds), Aquatic Oligochaete Biology VI.
© 1996 *Kluwer Academic Publishers.*

Osmoregulation in two aquatic oligochaetes from habitats with different salinity and comparison to other annelids

Olaf Generlich & Olav Giere
Zoological Institute and Zoological Museum, University of Hamburg, Martin-Luther-King-Platz 3, D-20146 Hamburg, Germany

Key words: osmoregulation, Oligochaeta, marine, terrestrial

Abstract

The osmoregulatory capacity of two oligochaete species, *Enchytraeus albidus* Henle, 1837, and *Heterochaeta costata* (Claparède, 1863), was investigated by direct measurements of the osmolality of the coelomic fluid. Terrestrial and marine (28‰ S) populations of *Enchytraeus albidus* and a brackish water population (14‰ S) of *H. costata* were used in the study. The range of salinity acclimation investigated was 0–40‰. The response to osmotic stress was measured (a) after a long-term maintenance (>14 days) in various salinities (*E. albidus* only), and (b) after a hyperosmotic shock as a short-term time-course sequence. The rate of water loss following a hyperosmotic shock was measured for *E. albidus*. *Long-term acclimation. E. albidus* maintained a hyperosmotic coelomic fluid over all salinities tested. In low salinities the osmolality of the coelomic fluid of the marine population was significantly higher than that of the terrestrial population. Possible genetic discrepancies or long-term acclimation may account for this difference. The coelomic fluid of *H. costata* was hyperosmotic at 15‰ S and isoosmotic at 30‰ S. *Short-term acclimation* (hyperosmotic shock). Both species investigated, kept at 15‰ S and then exposed to a salinity of 30‰, showed fast responses: within the first two hours the internal concentrations were adjusted to the new external condition with only small subsequent changes. Regulation of the body-water content after an exposure to a hyperosmotic shock was much slower: individuals of terrestrial *E. albidus*, acclimated for two weeks to either 0‰ or 15‰ S, had the same water content; hence, they showed a 100% regulation. However, after exposure to 30‰ S, a 100% regulation was still not attained 4 days after the hyperosmotic shock. *Enchytraeus albidus* is capable of actively reducing water loss following the hyperosmotic shock: the observed loss of water was only 40% of that expected for a passive osmotic flow. The observed reactions are compared with those in other annelids. It seems that an active transport of ions combined with a changeable permeability of the body wall play a major role in the regulation of body fluids.

Introduction

Oligochaetes are common inhabitants of intertidal habitats (Giere & Pfannkuche, 1982) and, thus, often exposed to salinity fluctuations. Many studies have stressed the relevance of abiotic factors to the distributional pattern of oligochaetes (Jansson, 1967; Giere, 1970, 1971; Lasserre, 1970, 1971, 1975). A good correspondence between salinity preferences or tolerances and the distribution of worms in the field was reported by Jansson (1962), Tynen (1969), Birtwell & Arthur (1980) and Giere (1980).

In contrast little information is available on the direct impact of a changing salinity regime on the osmotic concentration of the body fluid of aquatic oligochaetes (Oglesby, 1978). Pertinent experiments refer mainly to polychaetes. Among the eulittoral oligochaetes, *Enchytraeus albidus* Henle, 1837, is best investigated. It has a broad salinity tolerance (Schulz, 1955; Drawert, 1968; Kähler, 1970), and a capacity for regulation in low salinities (Schöne, 1971). Richter & Gersch (1967) found changing potentials in 'P'- and 'Q'-cells of the brain resulting from osmotic stress and there is probably a neurosecretory influence on the water permeability of the body wall (Drawert, 1968).

The aim of the present study was to investigate the reaction of oligochaetes to changing salinities by measurements of the osmotic concentration and volume of the coelomic fluid. We used as experimental animals a widely distributed terrestrial and coastal species, *E. albidus* (Enchytraeidae) and a truly marine species, *Heterochaeta costata* (Claparède, 1863) (Tubificidae) (Holmquist, 1985). Two aspects of osmoregulatory behaviour subsequent to osmotic stress were investigated: (a) long-term reaction of animals acclimated to various salinities; and (b) reaction after a hyperosmotic shock as a short-term time-course sequence. Osmolality and corresponding regulation of the body-water content were calculated.

Materials and methods

Collection and maintenance of animals. Specimens of *E. albidus* belonged to two different populations, one from under decaying seaweed on the shore of the island of Sylt (North Sea; 28–29‰ S), the other from a compost heap in Hamburg. *Heterochaeta costata* was collected at the coast of the western Baltic Sea in Wismar Bight (13–14‰ S). Worms were kept in plastic jars with a thin layer of azoic sediment in aerated sea water at 16 °C and fed with dry fish food. Prior to use they were allowed to acclimate to the different salinities for at least two weeks, a period of time that was established in previous experiments (unpub. obs.).

Determination of osmolality of body fluids. All determinations were made with a Clifton Cryostat Nanoliter Osmometer, which works best with volumes from 2 to 5.5 nanoliters (Frick & Sauer, 1973; Hoffacker, 1990). Prior to puncture, the worms were anaesthetised in a 1‰-solution of MS 222 Tricaine (Serva) in sea water of the appropriate salinity and then put under paraffin oil to prevent desiccation and contamination of the coelomic fluid. A thin glass capillary (tip diameter 15–20 μm) was inserted into the coelomic cavity with the aid of a micromanipulator and a sample of coelomic fluid was taken for immediate measurement. For long-term acclimation, the populations of *E. albidus* were kept at 0, 10, 20, 30, 40‰ S for at least two weeks prior to experimental use. Worms maintained at 15‰ S were exposed to a hyperosmotic shock of 30‰ S for a short-term time-course sequence. Measurements were done 0.1, 0.25, 0.5, 1, 2, 4, 8, 16, and 96 hours after

exposure. These measurements where carried out both with *E. albidus* (compost heap) and *H. costata*.

Determination of body water content. Single worms were removed from the medium, blotted on filter paper and weighed to the nearest 0.001 mg in a pre-weighed aluminium tray (wet weight; WW). Animals were then dried to constant weight (dry weight; DW) at 80 °C (24 h), cooled in a desiccator and weighed again. The calculation of water content should allow for variations in dry weight due to salinity-related differences in cellular solute content. Total body water content was therefore calculated as follows (see Diem, 1962; Ferraris & Schmidt-Nielsen, 1982):

Specific gravity: $SG = 1 + 0.035 Osm$

The water content (F_{H_2O}) was calculated after:

$$F_{H_2O} = \frac{WW - DW}{WW}.$$

The dry weight (DW_{corr}) was corrected after:

$$DW_{corr} = 1 - F_{H_2O}(1 + 0.035 Osm)$$

and was then used to correct the water content (F_{corr}):

$$F_{corr} = \frac{F_{H_2O}}{DW_{corr}}$$

given in this study as a corrected percentage relation W:

$$W = \frac{F_{corr}}{(1 + F_{corr})} 100 [\%].$$

The water content was determined (a) for specimens of a terrestrial population of *E. albidus* (compost heap), (b) for specimens of the same population acclimated to 15‰ S for two weeks, and (c) for specimens acclimated to 15‰ S and then exposed to 30‰ S. Subsequent to this hyperosmotic shock the water content was determined in a time series 1, 2, 4, 16 and 96 h after exposure.

The decrease in water content ($W_2 W_1^{-1}$) in relation to the simultaneously increasing osmolality of the coelomic fluid ($O_1 O_2^{-1}$) was used to assess the occurrence of volume regulation during a given exposure period. W_1 and W_2 are the body water contents at the beginning and end of a given time period. O_1 and O_2 is the osmolality of the coelomic fluid for the same time period. If the change in water content was entirely osmotic, the merely passively induced water flow would follow an essentially inverse relationship between the change in water content ($W_2 W_1^{-1}$) and

253

Enchytraeus albidus

Salinity [‰]

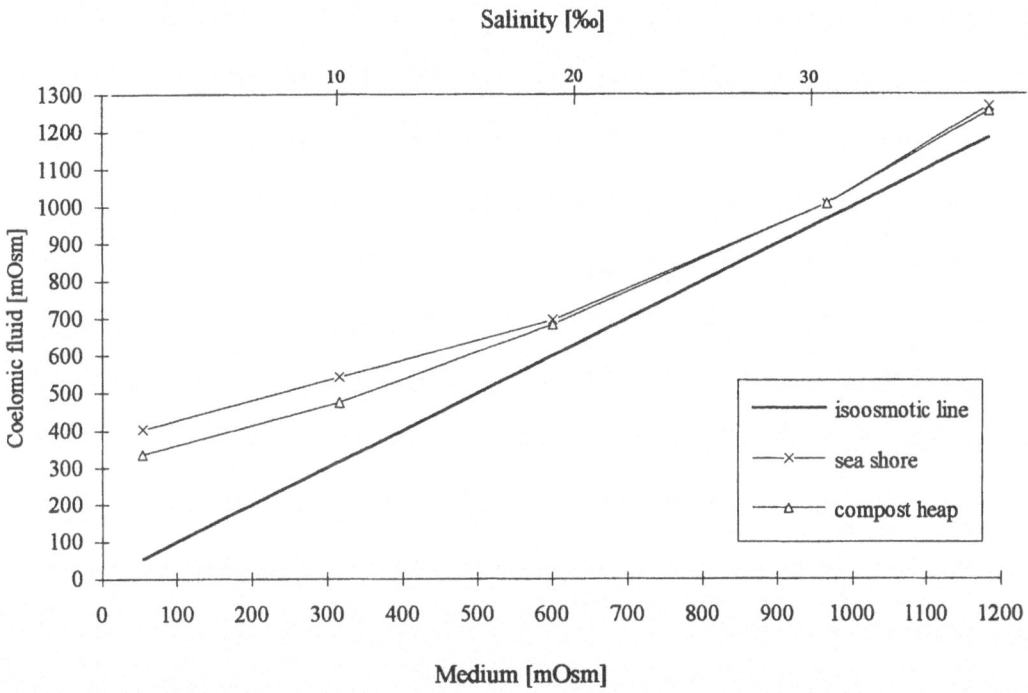

Fig. 1. Osmolality of the coelomic fluid of *E. albidus* (sea shore and compost heap). Worms were acclimated for 2 weeks to the various salinities. For standard deviations see Table 1.

that in coelomic osmolality ($O_1O_2^{-1}$). Volume regulation may occur in a first phase as a limitation of water loss while cells are shrinking, and, in a second step, as a net volume gain. Both phases are accompanied by an increase in solute content. If during a given time period $W_2W_1^{-1}$ is greater than $O_1O_2^{-1}$ (i.e., $W_2W_1^{-1}$ is closer to unity than is $O_1O_2^{-1}$), water content would have changed less than expected which would correspond to a limitation of water loss. However, if subsequent to shrinkage, $W_2W_1^{-1}$ is greater than both unity and $O_1O_2^{-1}$, a regulation towards a net volume gain after initial shrinkage would be indicated.

Statistical treatment. To compare the osmolality of the coelomic fluid with the constant osmolality of the medium, 99% confidence limits for the mean were calculated using

$$\overline{x} \pm t \cdot s \cdot \sqrt{n}^{-1}.$$

Thus, there is a significant difference ($p<0.01$), if the osmolality of the medium does not fall in this

interval (Sachs, 1978; Lozán, 1992). The means of the different populations were compared by T-test and controlled by Mann-Whitney U-test. Confidence limits for significant differences were calculated after Lozán (1992). The data of the short-term reaction were compared using one-way analysis of variance followed by multiple comparisons of means by the Tukey HSD method (Sokal & Rohlf, 1981). Results in which a statistically significant difference ($p < 0.05$) occurred are noted as 'significantly different' or the significance levels are designated.

Results

Long-term osmotic reactions – Enchytraeus albidus

All specimens of *E. albidus*, regardless of their terrestrial or marine origin, that were acclimated for at least two weeks to salinities between 0 and 40‰ showed in all salinities tested a significantly ($p < 0.01$) hyper-

Table 1. Osmolality of coelomic fluid of *Enchytraeus albidus* populations over a range of salinities. Mean ($\bar{x}$) in milliosmol kg^{-1} [mOsm], standard deviation (*S*), 99% confidence interval of the mean (K 99%) and number of measurements (*N*).

Medium		Coelomic fluid (Sea shore)				Coelomic fluid (Compost heap)			
‰	mOsm	$\bar{x}$	*S*	K (99%)	*N*	$\bar{x}$	*S*	K (99%)	*N*
0.9	27	404	55	380–427	41	336	53	308–364	30
10.2	316	542	57	518–567	40	475	50	451–499	32
19.3	601	695	14	689–701	44	683	36	668–698	43
30.9	966	1006	54	991–1022	84	1007	48	987–1027	43
38.0	1183	1268	34	1256–1279	64	1255	20	1248–1262	50

osmotic coelomic fluid compared to the medium (Table 1).

In salinities below 20‰ (601 mOsm) a clear hyper-regulation was obvious (Fig. 1) and there existed a significant difference ($p < 0.01$) between the two populations: in worms from the North Sea shore the osmolality of the coelomic fluid remained higher than in those from the compost heap (Fig. 1). For acclimation at higher than 20‰ S no differences were found between the populations. The changes in osmolality of the coelomic fluid paralleled the isoosmotic line with a linear regression of $f(x) = 0.97x + 95.2$.

Short-term osmotic reactions – Enchytraeus albidus

In order to find out how fast the response to a changed salinity proceeds, *E. albidus* from the compost heap, acclimated to 15‰ S (458 mOsm), was subjected to a hyperosmotic shock in a doubled salinity of 30‰ (940 mOsm). After only five minutes of exposure, a significant ($p < 0.01$) increase in osmolality of the coelomic fluid was detectable. The osmolality continued increasing significantly until a slight, but significant ($p < 0.01$) hyperosmoticity to the medium was reached after two hours (Fig. 2). Thereafter, with the exception of a dip at sixteen hours, no change was recorded.

Short-term osmotic reactions – Heterochaeta costata

The osmolality of the coelomic fluid of specimens acclimated to 15‰ S (458 mOsm) was significantly ($p < 0.01$) hyperosmotic to the medium. Animals exposed to 30‰ S (940 mOsm) showed a significant ($p < 0.01$) increase in osmolality of their coelomic fluid during the first two hours (Fig. 3). After two hours the coelomic fluid became isosmotic to the medium but increased again significantly ($p < 0.05$) after eight

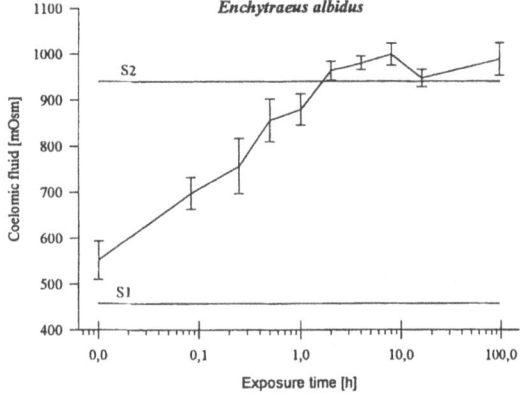

Fig. 2. Osmolality of coelomic fluid of *E. albidus* (compost heap) acclimated to 458 mOsm = S1 and then exposed to a hyperosmotic shock (940 mOsm = S2). $N = 20$. The bars indicate ± 1 standard deviation. Time scale logarithmic.

hours, to become hyperosmotic. This condition did not stay constant; at the next time interval (16 h) a significant ($p < 0.01$) decrease in osmolality was recorded. Later (96 h), no further changes were detectable and the coelomic fluid remained isosmotic to the medium.

Water content and water loss – Enchytraeus albidus

The relative water content of the population from the compost heap (0‰ S) was 82.3% (± 1.8 SD) of the body weight. In specimens acclimated to a salinity of 15‰, it stayed at 82.4% (± 2.07 SD). After exposure to a hyperosmotic shock of 30‰ S, a significant decrease in water content was recorded after one hour ($p < 0.01$) and an even greater decrease was noted after four hours ($p < 0.05$). In the following time intervals (16 h and 96 h) no significant changes were measured (Fig. 4).

Table 2. Concentration of body fluids of annelids in different salinities (steady-state reactions). CFM^{-1} = Ratio coelomic fluid/medium.

Species	Concentration of medium [% sea water]	CF M^{-1}	Author
A. POLYCHAETA NEREIDAE			
Hediste diversicolor	1.4%	11–15	Hohendorf (1963)
	14%	2.16	
	29%	1.27	
	50%	1.199	Schlieper (1929)
	50%	1.121	Fletcher (1974c)
	70%	1.086	Fletcher (1974c)
	97%	1.022	Fletcher (1974a)
	100%	1.049	DeLeersnyder (1971)
	102%	1.000	Hohendorf (1963)
	106–109%	1.026-1.098	Karandeeva (1965)
B. OLIGOCHAETA ENCHYTRAEIDAE			
Enchytraeus albidus	2.6%	14.96 or 12.44	this study
	31%	1.715 or 1.503	
	44%	1.205	
	58%	1.146	
	94%	1.042	
	115%	1.066	
	<75%	hyperosmotic	Schöne (1971)
	112–120%	isosmotic	
Marionina achaeta	<75%	hyperosmotic	Lasserre (1975)
TUBIFICIDAE			
Heterochaeta costata	44%	1.087	this study
	91%	isosmotic (0.991–1.009)	
Clitellio arenarius	47%	1.083	Ferraris & Schmidt-Nielsen (1982); Ferraris (1984)
	62%	1.062	
	92%	isosmotic	
NAIDIDAE			
Nais elinguis	<20%	hyperosmotic	Little (1984)
	20%	isosmotic	
	20–57%	hyposmotic	
MEGASCOLECIDAE			
Pontodrilus bermudensis	14%	2.62	Subba Rao (1978b); Subba
	43%	0.99	Rao & Ganapati (1986)
	86%	1.013	
LUMBRICIDAE *Lumbricus terrestris*	0.3%	62.8	Prush & Otter (1977)
	1%	10.6	Ramsay (1949a)
	1%	17.1	Dietz & Alvarado (1970)
	13%	1.27	
	15%	1.44	Ramsay (1949a)
	26%	1.055	Dietz & Alvarado (1970)
	45%	1.1	Ramsay (1949a)
C. HIRUDINEA			
Hirudo medicinalis	33%	1.075	Nieczaj & Zerbst-Boroffka (1993)

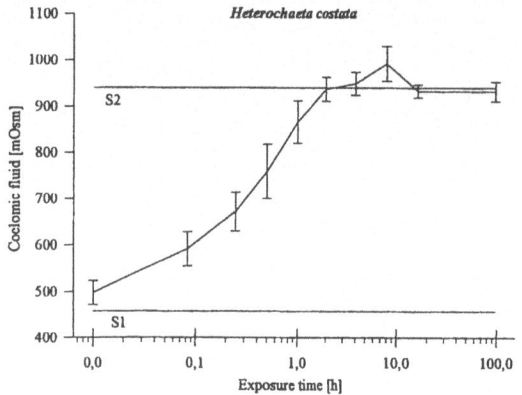

Fig. 3. Osmolality of coelomic fluid of *H. costata* acclimated to 458 mOsm=S1 and then exposed to a hyperosmotic shock (940 mOsm=S2). $N=20$. The bars indicate ± 1 standard deviation. Time scale logarithmic.

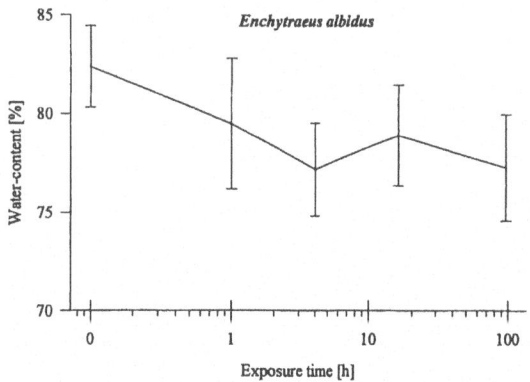

Fig. 4. Relative water-content (*W*) of *E. albidus* (compost heap) acclimated to 458 mOsm and then exposed to a hyperosmotic shock (940 mOsm). $N=20$. The bars indicate ± 1 standard deviation. Time scale logarithmic.

For worms exposed to 30‰ S, after one hour a clear limitation of water loss could be recognized: the decrease in water content was smaller than the increase in coelomic fluid osmolality ($W_2 W_1^{-1} = 0.96$; $O_1 O_2^{-1} = 0.63$). In the next time period (1–4 h) this regulatory trend continued: again water loss was low ($W_2 W_1^{-1} > O_1 O_2^{-1}$) while osmolality increased. In the following time intervals the relation between water content and osmolality remained constant with the only exception at sixteen hours where a slight, but not statistically significant, increase in water content was accompanied by a decrease in osmolality (Figs 2 & 4).

The decrease in water content was not merely a passive process, instead it is evident that *E. albidus* was capable of reducing water loss following a hyperosmotic shock. If the worms would react passively without any regulation like a simple osmometer, the water content, after doubling the salinity, would be halved. This would reduce the original water content after

$$100 \frac{0.5\text{WW}\%}{0.5\text{WW}\% + \text{DW}\%}$$

to a water content of $100 \cdot 41/(41+18) = 69.5\%$. After 96 hours of exposure *E. albidus* had a water content of 77.3%, which is equivalent to a reduction of water loss to 40% of that expected for a passive reaction. Animals losing no water whatsoever would have a 100% regulation (Oglesby, 1975). The above values for the degree of regulation may not be very accurate, but they are helpful for comparing different species (see Oglesby, 1981).

Discussion

Osmoregulation

Enchytraeus albidus is often described as a cosmopolitan holeuryhaline species (Schulz, 1955; Nielsen & Christensen, 1959; Giere, 1970). It is well known for its ability to survive a wide salinity range. Kähler (1970) experimented with salinities between 0 and 60‰, Schöne (1971) found these worms surviving at salinities up to 70‰ S. Earlier studies (Krizenecky, 1916; Schulz, 1955; Drawert, 1968) used even higher salinities but only for very short exposure times. In salinities of 200‰, the worms lost their ability to move. The salinity range used in this work (0–40‰) was tolerated by the worms without any problems. The other species investigated here, *H. costata*, is described as an euryhaline marine species (Brinkhurst, 1964; Timm, 1980) with a tolerance range of 0–34‰ S (Birtwell & Arthur, 1980).

Low salinity range. While there exist some studies on the tolerance range of euryhaline aquatic oligochaetes (see above), most annelid investigations on osmotic reaction after exposure to various salinities were performed with polychaetes, while very few concerned oligochaetes (Table 2). The marine tubificid *Clitellio arenarius* (Müller, 1776) maintains a hyperosmotic coelomic fluid in 15 and 21‰ S (Ferraris & Schmidt-Nielsen, 1982; Ferraris, 1984). The coastal marine enchytraeid *Marionina achaeta* (Hagen, 1951) also is hyperosmotic in salinities below 25‰. A close linkage between salinity and respiration rate and the inhibitory effect of Ouabain on osmoregulation demonstrates the active physiological processes involved (Lasserre 1969, 1970, 1971, 1975). The marine megadrile *Pontodrilus bermudensis* Beddard, 1891, is hyperosmotic in salinities below 15‰ and regulates sodium and chloride (Subba-Rao, 1978a, 1978b; Subba-Rao & Ganapati, 1986). *Nais elinguis* Müller, 1774 shows hyperregulation below 7‰ S (Little, 1984). The medical leech, *Hirudo medicinalis* Linnaeus, 1758, was found to occasionally invade brackish water habitats. Beyond 7‰ S it behaves like a hyperosmotic osmoconformer probably as a result of accumulation of various short-chain carboxylic acids (Nieczaj & Zerbst-Boroffka, 1993).

Our measurements for *E. albidus* (both of terrestrial and marine origin) and for *H. costata* are in agreement with this, apparently, general hyperosmoticity, that is maintained at least in lower salinities. Many studies point out that the body fluid of other marine invertebrates is also normally slightly hyperosmotic to the surrounding medium (Dice, 1969; Pierce, 1970; Oglesby, 1973, 1981). Remmert (1969) considered all poikilosmotic organisms to be slightly hyperosmotic to their average habitat salinity.

Physiological mechanisms of osmoregulation are well investigated in the earthworm, *Lumbricus terrestris* Linnaeus, 1758, which also has a hyperosmotic behaviour in salinities up to 15‰ (Ramsay, 1949a; Kamemoto *et al.*, 1962; Dietz & Alvarado, 1970; see Table 2). *Lumbricus terrestris* is able to produce a urine hypotonic to the body fluids (Wolf, 1940; Bahl, 1945; Ramsay, 1949b). The production of this hypotonic urine is due to reabsorption of sodium (active) and chloride (passive) in the proximal tubules of the nephridia, which are impermeable to water, so that a concentration gradient can be built up (Boroffka, 1965). Other studies have shown that the water-permeability of the body wall is small (Dietz & Alvarado, 1970; Carley, 1975) or may be reduced under stress (Carley, 1978). Furthermore, an active uptake of sodium and chloride across the body wall against an electrochemical gradient could be demonstrated (Dietz & Alvarado, 1970; Dietz, 1974; Prush & Otter, 1977).

Investigations on another well-studied annelid, the polychaete *Hediste diversicolor* (Müller, 1776) (see Table 2) suggest that three basic pathways are involved in osmoregulation at low salinities (Oglesby, 1969a, 1969b, 1970, 1972; Smith, 1970a, 1970b, 1970c;

Fletcher, 1974a, 1974b, 1974c): (1) production of a urine hypotonic to the body fluids, (2) differing permeability of the body wall to water and to salts, and (3) active absorption of salts from the medium.

High salinity range. Here, *E. albidus* maintained a fairly constant, slightly higher, concentration in its coelomic fluid than measured in the ambient medium ('hyperconformity', see Table 2). Oglesby (1978) claimed the same capacity occurs widely in annelids but the two species tested in the present study clearly contrast in their osmotic behaviour in higher salinities. When acclimated to 30‰ S, *E. albidus* showed hyperconformity while *H. costata* was isosmotic. Other aquatic oligochaetes also show this divergence: specimens of *Clitellio arenarius*, acclimated to 30‰ S, were isosmotic (Ferraris & Schmidt-Nielsen, 1982) while *Nais elinguis* showed a totally different reaction: it was hyposmotic in salinities from 10 to 20‰ (Little, 1984).

Hence, in higher salinities there is no uniform osmotic reaction of euryhaline annelids. Their regulatory processes are poorly understood, but it seems that it is not the electrolyte content, which is the biggest portion of osmotically active substances, that is regulated in high salinities (Hohendorf, 1963; Oglesby, 1970, 1978). Pierce (1970) suggested that osmotically active, but indiffusible substances (e.g. proteins) were responsible for a Gibbs-Donnan equilibrium, causing the hyperosmoticity of the body fluid. However, direct measurements of the colloid osmotic pressure did not support this hypothesis (Mangum & Johansen, 1975). Siebers (1976), Siebers & Bulnheim (1977) and Siebers & Ehlers (1978) found an active, salt dependent, uptake of amino acids across the body wall of *E. albidus* (from soil culture). Concentration gradients of up to $1:2 \cdot 10^4$ between medium and coelomic cavity were built up. Amino acid uptake was not observed in freshwater, but was observable between 10 and 20‰ S, and had a maximum in salinities of 30–40‰. Amino acids were taken up under the same salinity range in which *E. albidus* showed hyperconformity in the present study. Direct measurements of free amino acids in the coelomic fluid parallel to its osmolality could confirm the role of active uptake of amino acids in osmoregulation at high salinities.

Water loss and regulation

Specimens of *E. albidus*, acclimated to 0 or 1 5‰ S, maintained their water content, which means they had a

100% regulation up to mesohaline salinities. If exposed to a salinity of 30‰, the specimens were capable of reducing the water loss down to 40% of that typical for a passive osmotic reaction. This is equivalent to a 60% regulation of water content in the first four days after a hyperosmotic shock. Our data indicate that the water regulation in *E. albidus* was not completed within these four days and a net volume gain could be expected later in the time course. Other euryhaline species, e.g. the polychaetes *H. diversicolor* or *Nereis limnicola* Johnson, 1903, were capable of a 75–90% regulation of water content (Oglesby, 1975). More stenohaline species, e.g. *Arenicola marina* Linnaeus, 1758 or *Abarenicola pacifica* Healy & Wells, 1959, were capable of only 30–40% regulation (Oglesby, 1975; Reitze & Schöttler, 1989).

Our data confirm the general time course of acclimation to a new haline environment. According to Kinne (1964, 1971) and Oglesby (1978) it can be separated into an immediate reaction which lasts from seconds to hours, a 'plateau phase' (4–8 h) and a 'regulation phase' (days to weeks) until the new equilibrium is reached. The regulatory reaction following a hyperosmotic shock is slower than after a hyposmotic shock (Oglesby, 1978). In *Marionina* it lasted 5–15 days following a hyperosmotic shock, but only 8–14 hours after a hyposmotic shock (Lasserre, 1975).

Physiological processes leading to volume regulation are much slower compared with the regulation of osmolality: Both in *E. albidus* and *H. costata* the osmoregulation was completed after about 2 h. *Clitellio arenarius* showed a decrease of the osmolality and an increase of the water content 5 min after a hyposmotic shock (Ferraris & Schmidt-Nielsen, 1982; Ferraris, 1984). While osmoregulation was completed after 6 h with the greatest changes in the first hour, volume regulation started later and was under neurosecretory control. In this case water was lost by reduction of sodium and chloride in the coelomic fluid. A similar temporal divergence between osmoregulation and water regulation was found for the polychaetes *H. diversicolor* and *A. marina* (Fletcher, 1974c; Reitze & Schöttler, 1989) and for the medical leech, *H. medicinalis* (Nieczaj & Zerbst-Boroffka, 1993). Hohendorf (1963) found the time course of acclimation to be temperature dependent: at 1 °C, *H. diversicolor* needed at least 7 days for acclimation after a hyposmotic shock, but at 20 °C only 1 day.

Conclusions

Differences between disjunct populations

The present study demonstrates a conspicuous difference between the two *E. albidus* populations in their reaction to low salinity (0 and 10‰) with the terrestrial population maintaining a much lower osmolality. Hohendorf (1963) was of the opinion that the minimal osmotic concentration of the coelomic fluid is species-specific. With osmotic values of 330 and 400 mOsm respectively, *E. albidus* regulates on a relatively high level compared to other species (see below). Schöne (1971) studied a Baltic sea population of *E. albidus* and found 306–403 mOsm to be the range of concentration. This wide range does not allow for conclusions as to whether or not the Baltic population possessed a specific adaptation to its local conditions. Further values of osmotic concentrations are given in Table 2.

The different reaction of geographically disjunct populations of the same species, also seen in *Mytilus edulis* (Bivalvia) from the Baltic Sea and the North Sea, were attributed to long-term acclimation (Theede & Lassig, 1967). In the genus *Tisbe reticulata* Bocquet, 1951 (Copepoda, Harpacticoida), genetic differences were found to parallel the different salinity tolerance of geographically separate populations (Battaglia, 1959). Studies on the oligochaete *Lumbricillus rivalis* Levinsen, 1883 revealed several morphologically identical sibling species in parallel to various salinity ranges (Christensen & Jelnes, 1976). At present, such differences among genetically heterogeneous populations of *E. albidus* (Westheide & Brockmeyer, 1992) have not been investigated.

The ecological significance

Oglesby (1969a) assumed that exposure to regular tidal variations of salinity would never allow for a time span sufficient for worms to reach osmotic equilibrium with the medium. The present investigation shows that acclimation to osmotic changes can be sufficiently fast to follow the rapid salinity fluctuations in these habitats. Further studies should focus on the adaptive benefit of the marked difference between isosmotic and hyperosmotic regulation found here and known from many animals living under varying salinity conditions. For this problem it seems a prerequisite to analyse the incompletely understood mechanisms of osmo- and volume regulation.

The phylogenetic implications

The above notion directly relates to the evolutionary advantage for oligochaetes as a limnogenic group to settle in marine environments. Regarding the similar osmotic values in the coelomic fluids of the species studied here, direct conclusions on a more limnic nature of *E. albidus* compared to a more marine characterisation of *H. costata* cannot be drawn and seem too simple to be helpful for phylogenetic speculations on oligochaete descent. For these aspects, further osmotic studies on truly marine groups like the gutless species among the tubificids or the genus *Grania* among enchytraeids would be more promising.

Acknowledgments

We are much indebted to the Biologische Anstalt Helgoland, Hamburg, for lending us the osmometer. We thank Dr D. Siebers, Biologische Anstalt Helgoland, Hamburg, for his interest in this work and advice in osmometric methods.

References

Bahl, K. N., 1945. Studies on the structure, development, and physiology of the nephridia of Oligochaeta. VI. The physiology of excretion and the significance of the enteronephric type of nephridial system in Indian earthworms. Quart. J. Micr. Sci. 85: 343–389.

Battaglia, B., 1959. Il polymorfismo adattivo e i fattori della selezione nel copepode *Tisbe reticulata* Bocquet. Arch. Oceanogr. Limnol., Venezia 11: 305–355.

Birtwell, I. K. & D. R. Arthur, 1980. The ecology of tubificids in the Thames estuary with particular reference to *Tubifex costatus* (Claparède). In R. O. Brinkhurst & D. G. Cook (eds), Aquatic Oligochaete Biology. Plenum Press, New York: 331–381.

Boroffka, I., 1965. Elektrolyttransport im Nephridium von *Lumbricus terrestris*. Z. vergl. Physiol. 51: 25–48.

Brinkhurst, R. O., 1964. Observations on the biology of the marine oligochaete *Tubifex costatus*. J. mar. biol. Ass. U.K. 44: 11–16.

Carley, W. W., 1975. Effects of brain removal on integumental water permeability and ion content of the earthworm *Lumbricus terrestris* L. Gen. Comp. Endocrinol. 27: 509–516.

Carley, W. W., 1978. Water economy of the earthworm *Lumbricus terrestris* L. Coping with the terrestrial environment. J. exp. Zool. 205: 71–78.

Christensen, B. & J. Jelnes, 1976. Sibling species in the oligochaete worm *Lumbricillus rivalis* (Enchytraeidae) revealed by enzyme polymorphisms and breeding experiments. Hereditas 83: 237–244.

DeLeersnyder, M., 1971. Sur la régulation ionique du milieu intérieur de *Nereis diversicolor* O. F. Müller. Cah. Biol. Mar. 12: 49–55.

Dice, J. F., 1969. Osmoregulation and salinity tolerance in the polychaete annelid *Cirriformia spirabrancha* (Moore, 1904). Comp. Biochem. Physiol. 28: 1331–1343.

Diem, K., 1962. Documenta Geigy Scientific tables, 6th edn. Geigy Pharmaceuticals, Div. Geigy Chemical Corp., Ardsley, New York.

Dietz, T. H., 1974. Active chloride transport across the skin of the earthworm, *Lumbricus terrestris* L. Comp. Biochem. Physiol. 49A: 251–258.

Dietz, T. H. & R. H. Alvarado, 1970. Osmotic and ionic regulation in *Lumbricus terrestris* L. Biol. Bull. 138: 247–261.

Drawert, W., 1968. Histophysiologische Untersuchungen zur Beziehung zwischen Osmoregulation und Sekretionstätigkeit im Nervensystem von *Enchytraeus albidus* Henle. Zool. Jb. Abt. allg. Zool. Physiol. 74: 292–318.

Ferraris, J. D., 1984. Volume regulation in intertidal *Procephalothrix spiralis* (Nemertina) and *Clitellio arenarius* (Oligochaeta). II. Effects of decerebration under fluctuating salinity conditions. J. comp. Physiol. 154B: 125–137.

Ferraris, J. D. & B. Schmidt-Nielsen, 1982. Volume regulation in an intertidal oligochaete, *Clitellio arenarius* (Müller). I. Short term effects and the influence of the supra- and subesophageal ganglia. J. exp. Zool. 222: 113–128.

Fletcher, C. R., 1974a. Volume regulation in *Nereis diversicolor*. 1. The steady state. Comp. Biochem. Physiol. 47A: 1199–1214.

Fletcher, C. R., 1974b. Volume regulation in *Nereis diversicolor*. II. The effect of calcium. Comp. Biochem. Physiol. 47A: 1215–1220.

Fletcher, C. R., 1974c. Volume regulation in *Nereis diversicolor*. III. Adaptation to a reduced salinity. Comp. Biochem. Physiol. 47A: 1221–1234.

Frick, J. H. & J. R. Sauer, 1973. Examination of a Biological Cryostat/Nanoliter osmometer for use in determining the freezing point of insect hemolymph. Ann. Ent. Soc. Am. 66: 781–783.

Giere, O., 1970. Untersuchungen zur Mikrozonierung und Ökologie mariner Oligochaeten im Sylter Watt. Veröff. Inst. Meeresforsch. Bremerh. 12: 491–529.

Giere, O., 1971. Beziehungen zwischen abiotischem Faktorensystem, Zonierung und Abundanz mariner Oligochaeten in einem Küstengebiet der Nordsee. Thalassia jugosl. 7: 67–77.

Giere, O., 1980. Tolerance and preference reactions of marine Oligochaeta in relation to their distribution. In R. O. Brinkhurst & D. G. Cook (eds), Aquatic Oligochaete Biology. Plenum Press, New York: 385–409.

Giere, O. & O. Pfannkuche, 1982. Biology and ecology of marine Oligochaeta. A review. Oceanogr. Mar. Biol. Ann. Rev. 20: 173–308.

Hoffacker, M., 1990. Entwicklung und Anwendung neuer Methoden zur Darstellung schadstoffbedingter Aberrationen an Fischembryonen in der südlichen Nordsee. Dissertation, Rheinische Friedrich-Wilhelms Universität, Bonn.

Hohendorf, K., 1963. Der Einfluß der Temperatur auf die Salzgehaltstoleranz und Osmoregulation von *Nereis diversicolor* O. F. Müller. Kieler Meeresforsch. 19: 196–218.

Holmquist, C., 1985. A revision of the genera *Tubifex* LAMARCK, *Ilyodrilus* EISEN, and *Potamothrix* VEJDOVSKY & MRÇZEK (Oligochaeta, Tubificidae), with extensions to some connected genera. Zool. Jb. Syst. 112: 311–366.

Jansson, B.-O., 1962. Salinity resistance and salinity preference of two oligochaetes *Aktedrilus monospermathecus* Knöllner and *Marionina preclitellochaeta* n. sp. from the interstitial fauna of marine sandy beaches. Oikos 13: 293–305.

Jansson, B.-O., 1967. The importance of tolerance and preference experiments for the interpretation of mesopsammon field distributions. Helgoländer wiss. Meeresunters. 15: 41–58.

Kähler, H., 1970. Über den Einfluß der Adaptationstemperatur und des Salzgehaltes auf die Hitze- und Gefrierresistenz von *Enchytraeus albidus*. Mar. Biol. 5: 315–324.

Kamemoto, F. I., A. F. Spalding & S. M. Keister, 1962. Ionic balance in blood and coelomic fluid of earthworms. Biol. Bull. 122: 228–231.

Karandeeva, O. G., 1965. Correlation of rates of individual processes participating in the initial osmoregulating reaction in invertebrates. Byull. Mosk. O.-Va. Ispyt. Prir. Otd. Biol., 1965: 144–145.

Kinne, O., 1964. The effects of temperature and salinity on marine and brackish water animals. II. Salinity and temperature combinations. Oceanogr. Mar. Biol. Ann. Rev. 2: 281–339.

Kinne, O., 1971. Salinity. Animals – invertebrates. In O. Kinne (ed), Marine Ecology, Vol. 1, Environmental Factors, Part 2. Wiley Interscience, New York: 821–925.

Krizenecky, J., 1916. Beitrag zum Studium der Bedeutung osmotischer Verhältnisse des Mediums für Organismen. Versuche an Enchytraeiden. Pflügers Arch. ges. Physiol. 163: 325.

Lasserre, P., 1969. Relations énergétiques entre le métabolisme respiratoire et la régulation ionique chez une annélide oligochète euryhaline. C.R. Acad. Sci. Paris 268: 1541–1544.

Lasserre, P., 1970. Action des variations de salinité sur le métabolisme respiratoire d'oligochètes euryhalines du genre *Marionina* Michaelsen. J. exp. mar. Biol. Ecol. 4: 150–155.

Lasserre, P., 1971. Donnés écophysiologiques sur la répartition des oligochètes marins meiobenthiques. Incidence des paramètres salinité, température, sur le métabolisme respiratoire de deux espèces euryhaline du genre *Marionina* Michaelsen 1889 (Enchytraeidae, Oligochaeta). Vie Milieu (Suppl.) 22: 523–540.

Lasserre, P., 1975. Métabolisme et osmorégulation chez une annélide oligochète de la méiofaune: *Marionina achaeta* Lasserre. Cah. Biol. Mar. 16: 765–798.

Little, C., 1984. Ecophysiology of *Nais elinguis* (Oligochaeta) in a brackish-water lagoon. Estuar. coast. Shelf. Sci. 18: 231–244.

Lozán, J. L., 1992. Angewandte Statistik für Naturwissenschaftler. Parey Verlag, Berlin, 237 pp.

Mangum, C. P. & K. Johansen, 1975. The colloid osmotic pressures of invertebrate body fluids. J. exp. Biol. 63: 661–671.

Nieczaj, R. & I. Zerbst-Boroffka, 1993. Hyperosmotic acclimation in the leech, *Hirudo medicinalis* L.: Energy metabolism, osmotic, ionic and volume regulation. Comp. Biochem. Physiol. 106A: 595–602.

Nielsen, O. & B. Christensen, 1959. Enchytraeidae. Critical revision and taxonomy of European species. Natura jutl. 8–9: 1–160.

Oglesby, L. C., 1969a. Salinity stress and desiccation in intertidal worms. Am. Zool. 9: 319–331.

Oglesby, L. C., 1969b. Inorganic components and metabolism; ionic and osmotic regulation. Annelida, Sipuncula, and Echiura. In M. Florkin & B. T. Scheer (eds), Chemical Zoology. Academic Press, New York, Vol. 4: 211–310.

Oglesby, L. C., 1970. Studies on the salt and water balance in *Nereis diversicolor*. I. Steady-state parameters. Comp. Biochem. Physiol. 36: 449–466.

Oglesby, L. C., 1972. Studies on the salt and water balance in *Nereis diversicolor*. II. Components of total sodium efflux. Comp. Biochem. Physiol. 41A: 765–790.

Oglesby, L. C., 1973. Salt and water balance in lugworms (Polychaeta. Arenicolidae) with special reference to *Abarenicola pacifica* in Coos Bay, Oregon. Biol. Bull. 145: 180–199.

Oglesby, L. C., 1975. An analysis of water-content regulation in selected worms. In F. J. Vernberg (ed), Physiological Ecology of Estuarine Organisms. University of South Carolina Press: 181–204.

Oglesby, L. C., 1978. Salt and water balance. In P. J. Mill (ed), Physiology of Annelids. Academic Press, New York: 555–658.

Oglesby, L. C., 1981. Volume regulation in aquatic invertebrates. J. exp. Zool. 215: 289–301.

Pierce, S. K., 1970. The water balance of *Modiolus* (Mollusca: Bivalvia: Mytilidae): osmotic concentrations in changing salinities. Comp. Biochem. Physiol. 36: 521–533.

Prush, R. D. & T. Otter, 1977. Annelid transepithelial ion transport. Comp. Biochem. Physiol. 57A: 87–92.

Ramsay, J. A., 1949a. The osmotic relations of the earthworm. J. exp. Biol. 26: 46–56.

Ramsay, J. A., 1949b. The site of formation of hypotonic urine in the nephridium of *Lumbricus*. J. exp. Biol. 26: 65–75.

Reitze, M. & U. Schöttler, 1989. The time dependence of adaptation to reduced salinity in the lugworm *Arenicola marina* L. (Annelida: Polychaeta). Comp. Biochem. Physiol. 93A: 549–559.

Remmert, H., 1969. Über Poikilosmotie und Isoosmotie. Z. vergl. Physiol. 65: 424–427.

Richter, K. & M. Gersch, 1967. Das Ruhepotential neurosekretorischer Zellen bei *Enchytraeus albidus* Henle (Oligochaeta-Plesiopora). Zool. Jb. Abt. allg. Zool. Pysiol. 73: 386–397.

Sachs, L., 1978. Angewandte Statistik. 5. Auflage. Springer Verlag, Berlin, 552 pp.

Schlieper, C., 1929. Über die Einwirkung niederer Salzkonzentration auf marine Organismen. Z. vergl. Physiol. 9: 478–514.

Schöne, C., 1971. Über den Einfluß von Nahrung und Substratsalinität auf Verhalten, Fortpflanzung und Wasserhaushalt von *Enchytraeus albidus* Henle. Oecologia (Berlin) 6: 254–266.

Schulz, W., 1955. Zur Biologie von *Enchytraeus albidus* (Henle). I. Ökologie. Z. wiss. Zool. 158: 31–78.

Siebers, D., 1976. Absorption of neutral and basic amino acids across the body surface of two annelid species. Helgoländer wiss. Meeresunters. 28: 456–466.

Siebers, D. & H. P. Bulnheim, 1977. Salinity dependence, uptake kinetics, and specificity of amino acid uptake across the body surface of the oligochaete annelid *Enchytraeus albidus*. Helgoländer wiss. Meeresunters. 29: 473–492.

Siebers, D. & U. Ehlers, 1978. Transintegumentary absorption of acidic acids in the oligochaete annelid *Enchytraeus albidus*. Comp. Biochem. Physiol. 61A: 55–60.

Smith, R. I., 1970a. Chloride regulation at low salinities by *Nereis diversicolor* (Annelida, Polychaeta). 1. Uptake and exchanges of chloride. J. exp. Biol. 53: 75–92.

Smith, R. I., 1970b. Chloride regulation at low salinities by *Nereis diversicolor* (Annelida, Polychaeta). II. Water fluxes and apparent permeability to water. J. exp. Biol. 53: 93–100.

Smith, R. I., 1970c. Hypo-osmotic urine in *Nereis diversicolor*. J. exp. Biol. 53: 101–108.

Sokal, R. S. & F. J. Rohlf, 1981. Biometry. Freeman & Co., New York, 859 pp.

Subba Rao, B. V. S. S. R., 1978a. Volume regulation in a euryhaline oligochaete, *Pontodrilus bermudensis* Beddard. Proc. Indian Acad. Sci. 87B: 339–348.

Subba Rao, B. V. S. S. R., 1978b. Osmotic regulation in a brackish water oligochaete, *Pontodrilus bermudensis* Beddard. Indian J. mar. Sci. 7: 132–134.

Subba-Rao, B. V. S. S. R. & P. N. Ganapati, 1986. Regulation of chlorides, sodium and potassium in a brackish-water oligochaete, *Pontodrilus bermudensis* Beddard. In M. F. Thompson, R. Sarojini & R. Nagabhushanam (eds), Biology of Benthic Marine Organisms. Techniques and Methods as Applied to the Indian Ocean 12: 19–34.

Theede, H. & J. Lassig, 1967. Comparative studies on cellular resistance of bivalves from marine and brackish waters. Helgoländer wiss. Meeresunters. 16: 119–129.

Timm, T., 1980. Distribution of aquatic oligochaetes. In R. O. Brinkhurst & D. G. Cook (eds), Aquatic Oligochaete Biology. Plenum Press, New York: 55–77.

Tynen, M. J., 1969. Littoral distribution of *Lumbricillus reynoldsoni* Backlund and other Enchytraeidae (Oligochaeta) in relation to salinity and other factors. Oikos 20: 41–53.

Westheide, W. & V. Brockmeyer, 1992. Suggestions for an index of enchytraeid species (Oligochaeta) based on general protein patterns. Z. zool. Syst. Evolut.-forsch. 30: 89–99.

Wolf, A. V., 1940. Paths of water exchange in the earthworm. Physiol. Zool. 13: 294–308.

Hydrobiologia **334**: 263–267, 1996.
K. A. Coates, Trefor B. Reynoldson & Thomas B. Reynoldson (eds), Aquatic Oligochaete Biology VI.
© 1996 Kluwer Academic Publishers.

Cinematographic documentation of enchytraeid morphology and reproductive biology

W. Westheide & M. C. Müller
Spezielle Zoologie, Fachbereich Biologie/Chemie, Universität Osnabrück, D-49069 Osnabrück, Germany

Key words: Oligochaeta, *Enchytraeus*, food uptake, morphology, copulation, cocoon-shedding, development, hatching

Abstract

This paper summarizes the processing of small enchytraeid species for cinematographic documentation and covers the main details of a film on their locomotion, food uptake, anatomy and reproductive biology. This includes copulation, shedding of the cocoon and hatching of the young, characteristic reproductive patterns for all clitellate worms.

Introduction

Only few invertebrate taxa can be recognized as clearly and simply as a monophyletic group by a single autapomorphic character as is the Clitellata (Purschke *et al.*, 1993). The glandular epidermal region producing cocoons in which the embryos develop (the clitellum) occurs in all the hermaphroditic oligochaetous and hirudinean annelids. One can assume that this character is connected with the general Bauplan of these organisms: the relatively smooth body surface and the general absence of parapodia and other appendages in Clitellata can be interpreted as specific adaptations to the cocoon shedding process. Although it would be highly desirable for teaching purposes to demonstrate the function of the clitellum and the reproductive biology in detail, this is only very rarely successful. Thus, few zoologists have had the opportunity to observe this process – crucial to a proper understanding of this taxon – and even many clitellate specialists have never seen copulation or cocoon formation. With the exception of earthworms, there are relatively few descriptions and graphic illustrations of this process in the literature (Brumpt, 1900; Grove & Cowley, 1926; Oishi, 1930; Hirao, 1965a, 1965b). This is even more true of photo series (Albert, 1975) or film sequences. Since other life history details of clitellate species, e.g., the process of food uptake, are frequently not well documented or are

even unknown, we searched for species which would feed, copulate and shed cocoons under the harsh conditions of strong lighting, low humidity and total visual access.

Filming of the reproductive biology has been successful for some aquatic leeches such as *Erpobdella octoculata* (Westheide & IWF, 1980), *Piscicola geometra* and *Theromyzon tesselatum* (Westheide & IWF, 1981a). Feeding was documented on film for *Piscicola geometra* (Westheide & IWF, 1978, 1981b), *Hirudo medicinalis*, *Haemopis sanguisuga*, *Glossiphonia* sp. and *Helobdella stagnalis* (Westheide & IWF, 1981b). Another film of leech behaviour was produced by Malecha (1976). All types of film documentation have proved much more difficult for terrestrial oligochaetes and, to our knowledge, filming has probably only been done for earthworms (Graff & IWF, 1993a, 1993b), but without showing oviposition properly. Here, the discovery that some species of the genus *Enchytraeus* (Enchytraeidae) are easy to culture on an artificial medium helped. Problem-free observation of the entire set of life cycle parameters of individuals kept in agar proved to be the ideal prerequisite for a cinematographic documentation of the most interesting episodes of their life histories. This applied above all to those species which have, at the relatively high temperature of 21 °C, a generation time of only 2 weeks, e.g., *Enchytraeus crypticus* Westheide & Graefe, 1992 (Westheide & Graefe, 1992). Film sequences were made to show a series of details of the life history of two enchytraeid annelids which have never been filmed before.

Material and methods

Individuals and cocoons used for filming were taken from large soil cultures of *Enchytraeus albidus* Henle, 1837, and soil and agar cultures of *E. crypticus*. In all cultures the worms were fed oat meal. The length of the *E. albidus* individuals was about 20 mm, that of *E. crypticus* about 7 mm. *Enchytraeus albidus* was utilized for the copulation sequences because this species often copulates on the surface of the soil. Oviposition and shedding of the cocoon by *E. crypticus* were observed with a compound microscope. Those animals were sorted out of the cultures whose shape and colour of the clitellum led us to expect sudden formation of a cocoon. Nevertheless, many hours of microscope observations were often necessary.

A special method was developed for microscope observation which prevented the animals from drying out and also slowed down their movements as well. The individual worm was placed on a slide in a drop of water, then covered with a thin layer (1–2 mm) of agar and a small coverslip. This procedure was also used for filming developmental processes, which took nine days from the first cleavage to hatching of the young; in this case the coverslip was additionally sealed with wax to avoid evaporation and prevent bacterial or fungal infection. Cleavage was filmed with a time-lapse of 30 pictures h^{-1}. Filming was performed with Zeiss Axioplan and Axioskop microscopes. For several sequences, additional incident cold light sources were used. The camera was a specifically adapted Arritechno 35 working with 32 mm Kodak-Eastman ECH 5296 colour film.

The film with the title 'Organisation und Fortpflanzung von Enchytraeen (Oligochaeta)' is available from the IWF, Göttingen (it can be borrowed, or bought for ca. 80 US$). It is available with a German or English soundtrack. An accompanying publication with a German and English version of the spoken commentary is also available (Westheide & Müller, 1995).

Content of the film

The first sequences show crowded enchytraeids in soil cultures. Their cocoons are covered with soil particles and are therefore barely distinguishable from the substrate. On agar it becomes evident that a worm is able to move over a moist, smooth surface with only a few peristaltic contractions. With higher magnification under a compound microscope, back and forth movement of the short simple chaetae and the way they facilitate locomotion is demonstrated. The almost complete transparency of the organisms allow recognition of the ciliary activity of the preseptal metanephridial funnels which continually flush coelomic fluid into the excretory canal.

Feeding habits on agar appear the same as in soil. During food uptake the dorsal pharyngeal pad of the foregut is everted from the mouth and retracted again in rapid, rhythmical cycles; acting like a labellum, to which food particles adhere before being transported into the alimentary canal.

Copulation takes place by mutual transfer of sperm between two hermaphroditic individuals. The anterior sections of the copulating partners are aligned facing opposite direction in such a way that the penial organs of each individual are anchored in the spermathecal openings of the partner (Fig. 1A). The spermathecae are then reciprocally filled with seminal fluid – the latter process, however, is not visible in the film. When separating, the male copulatory organs of one partner may remain longer in the spermathecae of the other, as shown in one of the sequences. After copulation the penial bulbs are momentarily visible. In a series of scenes the hermaphroditic genital system is presented: spermathecae, seminal funnels with attached mature spermatozoa, sperm ducts, penial bulbs, yolky oocytes, and the clitellum.

Immediately before cocoon formation, the cocoon wall, which is secreted from the surface of the clitellar epidermis, becomes clearly distinguishable. Then ripe yolky eggs – three in the first sequence – are pressed out of the female genital pore into the fluid between the body wall and the cocoon wall (Fig. 1B, C). After oviposition the animal sheds the cocoon by simply withdrawing its anterior end from it as from a serviette ring (Fig. 1D). On a level with the exterior spermathecal pore this operation is interrupted for about a minute (Fig. 1E), while the partner's sperm flow from the spermathecae into the cocoon fluid and fertilize the eggs. Shedding of the cocoon concludes with retraction of the anterior end. The cocoon closes up at both poles

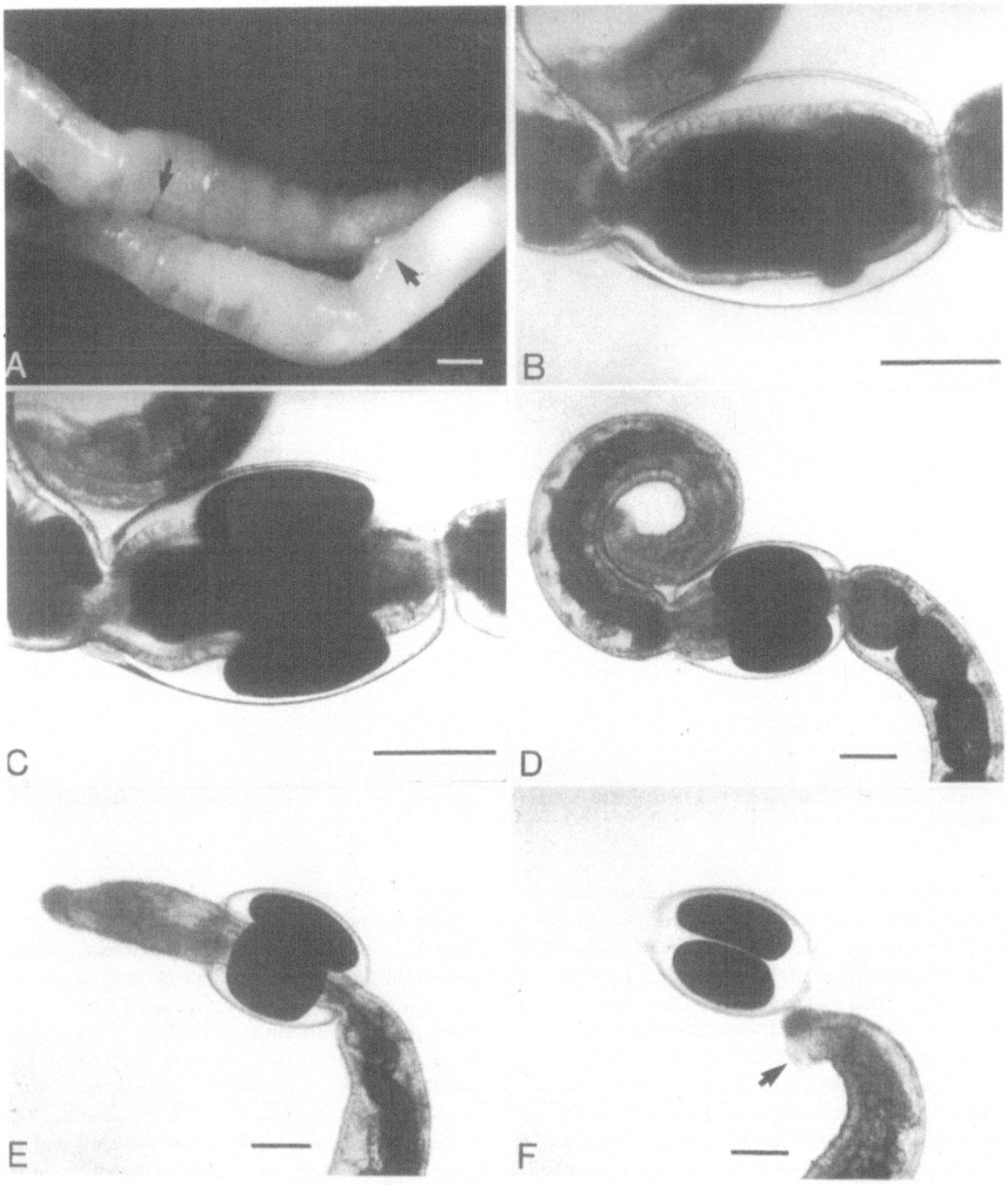

Fig. 1. Reproductive biology. A. *Enchytraeus albidus;* copulation. Arrows indicate where the penial organs anchor in the spermathecal openings. B–F. *Enchytraeus crypticus.* B, C. Oviposition. Eggs are pressed into the fluid of the cocoon between clitellum and cocoon wall. D–F. Shedding of the cocoon. Withdrawing of anterior end from the cocoon (D), process interrupted on a level with the spermathecal pore (E). Pharyngeal bulb extruded (arrow); worm camouflages the cocoon with particles (F). Scale bars: A. 0.5 mm; B–F. 0.2 mm (Micrographs taken from the film).

266

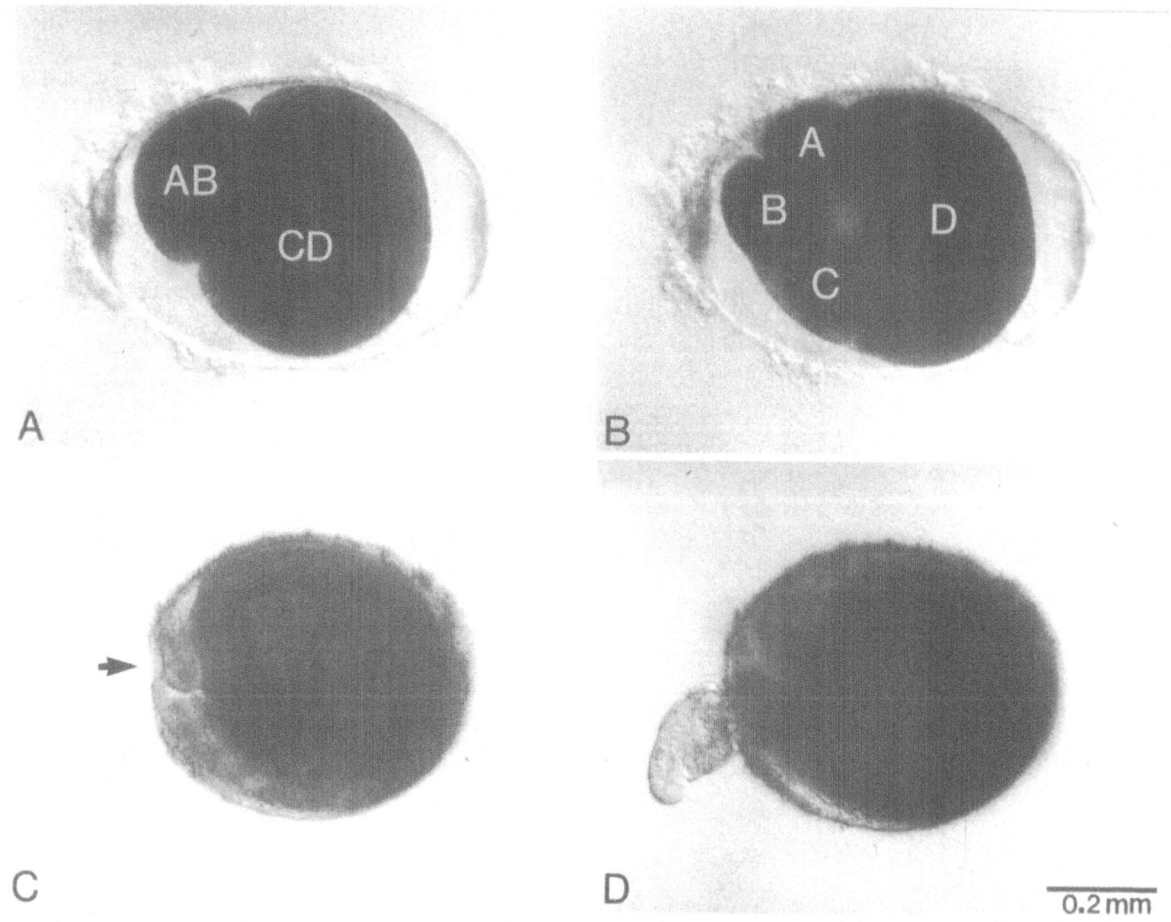

Fig. 2. Enchytraeus crypticus; first cleavages and hatching. A. First division; the blastomere *AB* is much smaller than *CD*. B. Second division. C. Juvenile attacks the cocoon wall with its dorsal pharyngeal pad (arrow). D. One juvenile leaves the cocoon. Same scale bar for A–D. (Micrographs taken from the film).

and remains loosely on the substrate. The animal then executes several movements as if to camouflage the cocoon by covering it with soil particles (Fig. 1F), as it would probably do in natural substrates (see also Christensen, 1958). This characteristic procedure of oviposition in the cocoon, which is basically identical in all oligochaetes, was recapitulated in other sequences.

Well fed individuals of *E. crypticus*, the species shown in these sequences, continually produce cocoons from the onset of sexual maturity until shortly before death, their number being dependent on the ambient temperature. The egg count also varies, in this species up to 32, subject to food supply and population density. Development starts immediately after formation of cocoon–shown in time-lapse within a cocoon having only one egg. Spiral cleavage produces unequal blastomeres even at the first division

(Fig. 2A, B). Development is direct and proceeds to a young enchytraeid, which has nearly all the outer attributes of an adult. Immediately before the young hatch from the cocoon, they attack the wall of the cocoon with the dorsal pharyngeal pad (Fig. 2C). This results in short consecutive denting, as if the wall of the cocoon was being sucked inward by the pharyngeal pad. This occurs principally at the location where the cocoon received its pointed seal after shedding. The opening may be finally achieved mechanically and the young crawl through the cocoon wall to the outside (Fig. 2D), which is shown in time-lapse.

At a temperature of 21 °C the time from deposition of the cocoon to hatching of the young for *E. crypticus* takes about nine days. The hatching rate is high – the proportion of eggs that develop in cultures always exceeds 90%. After a further eight days the worm can reach sexual maturity.

Acknowledgments

Cooperation with the Institute for Scientific Films (IWF) in Göttingen (a central, non-profit service institution which provides the basic audio-visual requirements for science and teaching in Germany) was crucial for the adequate realization of this film. Only with the excellent personnel and technical facilities of this institute was it possible to surmount the difficulties in filming these relatively small organisms. We are grateful to Dr G. Lotz and several of the staff members of IWF, Göttingen, especially J. Kaeding and S. Hornig.

References

Albert, R., 1975. Zum Lebenszyklus von *Enchytraeus coronatus* Nielsen & Christensen, 1959 (Oligochaeta). Mitt. Hamburg. Zool. Mus. Inst. 72: 79–90.

Brumpt, E., 1900. Reproduction des Hirudinées. Mem. Soc. Zool. Fr. 13: 286–430.

Christensen, B., 1958. Studies on Enchytraeidae 6. Technique for culturing Enchytraeidae, with notes on cocoon types. Oikos 7: 302–307.

Graff, O. & IWF, 1993a. *Lumbricus terrestris* (Lumbricidae) – Fortbewegung und Ernährungsweise. Film E 2714 des IWF, Göttingen.

Graff, O. & IWF, 1993b. *Lumbricus terrestris* (Lumbricidae) – Paarung. Film E 2715 des IWF, Göttingen.

Grove, A. J. & L. F. Cowley, 1926. On the reproductive processes of the brandling worm *Eisenia foetida*. Quart. J. Micr. Sci. 70: 559–581.

Hirao, Y., 1965a. Cocoon formation in *Tubifex*, with its relation to the activity of the clitellar epithelium. Jour. Fac. Sci. Hokkaido Univ. Ser. 6: 625–633.

Hirao, Y., 1965b. A method of the observation of oviposition in the fresh water oligochaete, *Tubifex hattai*. (in Japanese with English abstract) Zool. Mag. (Tokyo) 74: 283–285.

Malecha, J., 1976. Film: 'Biologie des Sangsues'. Realization: M. Guillon, distribution: Service du Film de Recherche Scientifique (16 mm, 25 min) (English version available).

Oishi, M., 1930. On the reproductive process of the earthworm *Pheretima communissima*. Sci. Rep. Tohoku imp. Univ. 4: 509–524.

Purschke, G., D. Rhode, R. O. Brinkhurst, & W. Westheide, 1993. A morphological reinvestigation and phylogenetic relationship of *Acanthobdella peledina* (Annelida, Clitellata). Zoomorphology 113: 91–101.

Westheide, W. & U. Graefe, 1992. Two new terrestrial *Enchytraeus* species (Oligochaeta, Annelida). J. Nat. Hist. 26: 479–488.

Westheide, W. & IWF, 1978. *Piscicola geometra* (Hirudinea) – Befall von Wirtstieren. Publikation von W. Westheide, Publ. Wiss. Film., Sekt. Biol., Ser. 11, Nr. 36/E 2484, 7 pp.

Westheide, W. & IWF, 1980. *Erpobdella octoculata* (Hirudinea) – Spermatophorenübertragung, Kokonablage, Schlüpfen der Jungtiere. Film E 2562 des IWF, Göttingen. Publikation von W. Westheide, Publ. Wiss. Film., Sekt. Biol., Ser. 13, Nr. 27/E 2562, 12 pp.

Westheide, W. & IWF, 1981a. Nahrungsaufnahme bei Egeln (Hirudinea). Film C 1416 des IWF, Göttingen. Publikation von W. Westheide, Publ. Wiss. Film., Sekt. Biol., Ser. 14, Nr 20/C 1416, 17 pp.

Westheide, W. & IWF, 1981b. Fortpflanzung bei Egeln (Hirudinea). Film C 1394 des IWF, Göttingen. Publikation von W. Westheide, Publ. Wiss. Film., Sekt. Biol., Ser. 14, Nr. 6/C 1394, 14 pp.

Westheide, W. & M. C. Müller, 1995. Organisation und Fortpflanzung von Enchytraeen (Oligochaeta). Publ. Wiss. Film, Biol. 22: 153–170.

Hydrobiologia **334**: 269–276, 1996.
K. A. Coates, Trefor B. Reynoldson & Thomas B. Reynoldson (eds), Aquatic Oligochaete Biology VI.
© 1996 *Kluwer Academic Publishers.*

The first two cleavages in *Tubifex* involve distinct mechanisms to generate asymmetry in mitotic apparatus

Takashi Shimizu
Division of Biological Sciences, Graduate School of Science, Hokkaido University, Sapporo 060, Japan

Key words: Tubifex, cleavage, asymmetric mitotic spindles, centrosomes

Abstract

We have investigated factors which determine inequality of the first two cleavages in *Tubifex hattai*. A mitotic spindle for the first cleavage, which is located at the center of the egg, possesses an aster at one pole, but not at the other pole. Inequality of the first cleavage is determined by the asymmetric organization of the spindle poles, rather than by the spindle position in the egg. A centrosome which appears as a dot stained with an anti-γ-tubulin antibody is found at one pole (at the center of the aster) of the spindle, but not at the other pole. This centrosome appears to be maternal in origin. In contrast to the first cleavage, the poles of the second cleavage spindle are not different from each other either in their ability to form asters or in γ-tubulin distribution. As a result of an interaction of one of the spindle poles with the cell cortex, however, an asymmetric spindle is formed in the cell CD, giving rise to unequal division in this cell. Thus, factors generating asymmetry in spindle organization are intrinsic to the mitotic spindle in the first cleavage, but not in the second cleavage.

Introduction

Since Penners (1922) first reported the cell lineages of *Tubifex rivulorum* (*T. tubifex*), various aspects of embryogenesis in this animal have been studied (see Lehmann, 1956; Shimizu, 1982a, for review). One of the most important findings derived from these studies is that if a *Tubifex* egg, which normally divides unequally, undergoes *equal* division, it produces an embryo with duplicated heads and/or duplicated tails (Penners, 1924; Inase, 1960). This result has been interpreted to mean that cytoplasmic determinants, which are normally segregated to one of two daughter cells, might be allotted to both daughter cells. Apparently in *Tubifex*, inequality of the first two cleavages is one of the factors which determine the body pattern of worms.

In this paper, we compare the mechanisms for the first and second cleavages in *Tubifex hattai*, reviewing recent findings from our laboratory. In addition, we discuss the possibility of parthenogenetic development in this animal in relation to inequality of the first cleavage.

Early development of *Tubifex*

A brief review of early development in *Tubifex* is presented here as background for the observations described below (for details, see Shimizu, 1982a). *Tubifex* eggs are oviposited at metaphase of the first meiosis, and are believed to be fertilized during cocoon deposition by spermatozoa which are released from spermatozeugmata stored in the spermathecae (Hirao, 1965; Braidotti & Ferraguti, 1982). Following sperm penetration (or activation), the vitelline membrane is separated from the egg surface. Activated *Tubifex* eggs extrude polar bodies twice, and then divide meridionally into two blastomeres, a smaller cell AB and a larger cell CD. During this period of time, eggs undergo three episodes of microfilament-dependent surface contractile activity. The first two are observed during polar body formation, and give rise to deformation movement; the last activity brings about ooplasmic segregation of pole plasms which takes place following the second polar body formation (Shimizu, 1982b, 1984, 1986).

The second cleavage is also meridional, and yields cells A, B, C and D (four-cell stage): The AB-cell divides into cells A and B, and the CD-cell into a smaller C and a larger D. The AB-cell divides 40 min later than the CD-cell. The AB-cell at the two-cell stage and the B-cell at the four-cell stage are located at the future anterior end of the embryo. Thereafter the four quadrants divide in a spiral cleavage pattern, producing micromeres and yolky macromeres. The pole plasms, which have been inherited by the CD-cell, are segregated into the D-cell and subsequently into its descendant cells 2d and 4d, both of which exclusively possess the ability to generate germ bands of the embryo. Pole plasm inheritance into blastomeres is mediated by actin microfilaments up to eight-cell stage and by microtubules thereafter (Shimizu, 1988, 1989). The pole plasms appear to be involved not only in the developmental fate specification of blastomeres but also in controlling cleavage cycles (Shimizu, 1993, 1994, 1995).

Inequality of the first cleavage in *Tubifex* is determined by asymmetric organization of the mitotic spindle

The first cleavage begins with the formation of meridionally running furrows at the two points of the egg's equator which are about 90° apart from each other (Fig. 1b). These furrows deepen toward the center of the egg (i.e., the egg axis), and eventually partition the egg into two parts of unequal size (Figs 1c–f).

Eggs which had been treated with 5 μg ml^{-1} cytochalasin D failed to divide, suggesting that as in other animal eggs (Mabuchi, 1986), actin microfilaments are involved in the furrow formation in the *Tubifex* egg. To gain an insight into the mechanism by which the pattern of furrow formation mentioned above is generated, organization of cortical F-actin was analyzed in cortices which had been isolated and stained with 825 ng ml^{-1} rhodamine-phalloidin according to Shimizu (1986). Cortices isolated upon furrow formation at the equator show two meridional bundles of F-actin crossing the equator. These bundles are structurally continuous to the actin networks at the animal and vegetal poles (see Shimizu, 1984, 1986). A careful examination suggests that these actin bundles are located at the bottom of the cleavage furrow. Before the onset of furrow formation, such actin bundles are not seen at the egg's equator, though a coarse network of F-actin is present throughout the equator. These results suggest that cortical F-actin reorganizes to form actin bundles which link the polar actin networks. It is not difficult to imagine that contraction of these bundles together with the polar actin networks could bring about an ingress of the equatorial surface toward the egg's axis since cortical actin is physically connected to the oolemma.

In immunocytochemical whole-mounts labeled with anti α- and β-tubulin antibodies, a mitotic spindle is found at the center of the egg; it possesses an aster at one of its poles but not at the other. An array of astral microtubules covers a sector of 300°, and reaches the cell surface during anaphase and telophase. The resulting CD-cell appears to correspond to the domain of this array. The anastral spindle pole also extends some microtubules toward the cell periphery, but it is not until early telophase that a substantial number of microtubules become recognizable. In view of the proposed role for asters in cleavage plane determination (Rappaport, 1986), it is conceivable that the position of cleavage furrows in the first division of *Tubifex* might be determined by the aster localized at one pole of the spindle. First cleavage spindles depicted in papers by Penners (1922), Woker (1944), Lehmann (1946) and Huber (1946), all resemble that described in this paper (see Fig. 5a).

The spindles labeled with the anti γ-tubulin antibody exhibit a granular structure (i.e., centrosome) at the astral but not at the anastral pole. Even at the onset of spindle assembly, eggs individually exhibit only one centrosome. The asymmetric organization of poles of the first cleavage spindle would result from the intrinsic difference in microtubule-organizing activity between the spindle poles.

In summary we suggest that inequality of the first cleavage in *Tubifex* is generated by the asymmetric organization of the spindle and, in fact, by the presence of a single centrosome at one of the spindle poles only (Ishii & Shimizu, 1995; Shimizu, 1996).

Origin of centrosomes. It seems unusual in animal development that only one centrosome, rather than two, is involved in the assembly of cleavage spindles. How is such a situation brought about in *Tubifex*? There are two possible ways to generate it: (1) a centrosome duplicates at the transition from meiosis to mitosis, however, one of the two exclusively loses its integrity and the capacity to organize spindles; (2) a centrosome which has been in eggs during meiosis participates in cleavage spindle assembly without its duplication. Recent observation that a centrosome in the spindle of the second meiosis does not duplicate

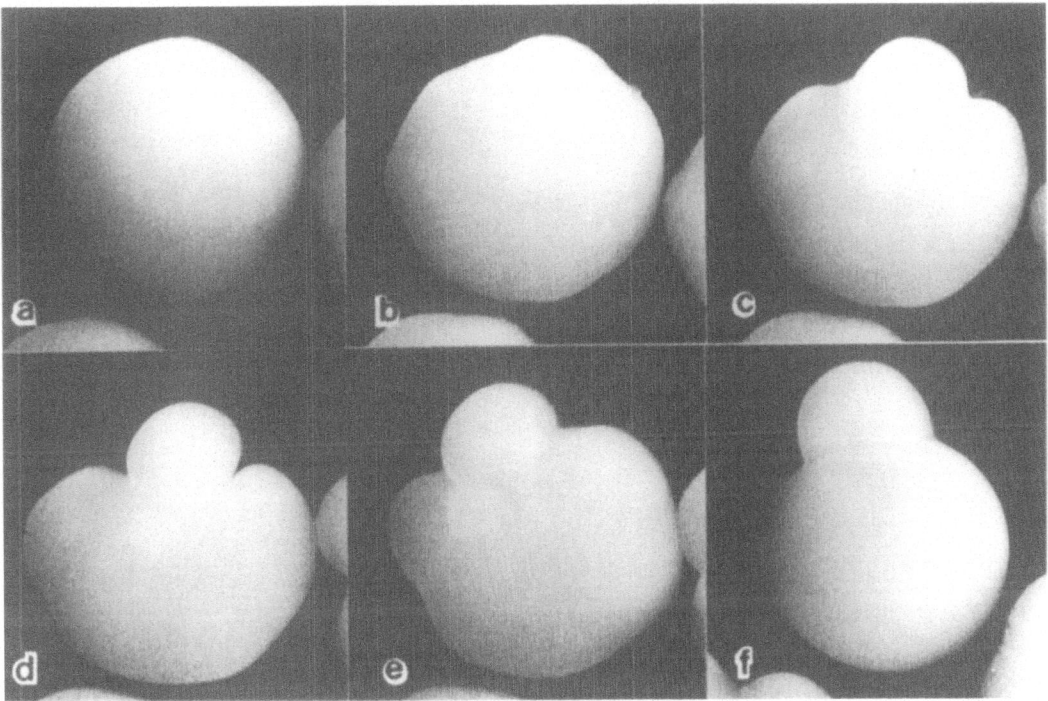

Fig. 1. Photomicrographs of a living *Tubifex hattai* egg undergoing the first cleavage. Animal pole view. This egg is about 450 μm in diameter. Two furrows form at the egg's equator (b); during the following 10 min they deepen toward the egg axis (c, d). After another 10 min the egg is partitioned into two parts of unequal size (e). As a result of this unequal division, the two-cell embryo of *T. hattai* is comprised of smaller AB-cell and larger CD-cell (f).

during S-phase of the first mitotic cycle appears to favor the second possibility. Furthermore, this observation suggests that the maternal centrosome survives at the meiosis/mitosis transition and is responsible for the first cleavage. This is supported by the following experimental results. When first polar body formation was suppressed by centrifugation or cytochalasin-treatment, both of the two centrosomes in the first meiotic spindle persisted in eggs throughout meiosis, and took part in organizing the first cleavage spindle, which possessed equal poles. These results suggest that maternal centrosomes persist even after the second meiosis.

Examination of the chromosome numbers also favors the possibility of the maternal origin of centrosomes. As described below, it seems unlikely that in eggs used in this study, paternal genome participates in embryogenesis. At the turn of the century, Gathy (1900) reported that the number of bivalent chromosomes in the metaphase I spindle of *Tubifex rivulorum* is about 110 and that the number of chromosomes in the metaphase spindle of the first cleavage is not more

than 100. In the present study, we counted the chromosome numbers in squashed preparations of eggs stained with DNA-specific fluorescent dye Hoechst 33258. Although we could not determine the precise number of chromosomes because of their insufficient spreading, we nevertheless confirmed Gathy's observation. That is, the numbers of bivalents in metaphase I spindles, dyads in metaphase II spindles, and chromosomes (chromatid pairs) in first cleavage spindles, were all $\sim$ 120 (Figs 2, 3). We also found that the first polar body received $\sim$ 120 dyads, and the second polar body $\sim$ 120 chromatids. A cytological examination of eggs undergoing meiosis showed that segregation of chromosomes into polar bodies occurs normally at both the first and second meiosis. These results suggest that maternal chromatids of $\sim$ 120 which are inherited by the egg proper at the second meiosis might duplicate and exclusively take part in mitosis for the first cleavage. Apparently the embryo develops parthenogenetically which makes a male contribution to either the set of chromosomes or the centrosome most unlikely.

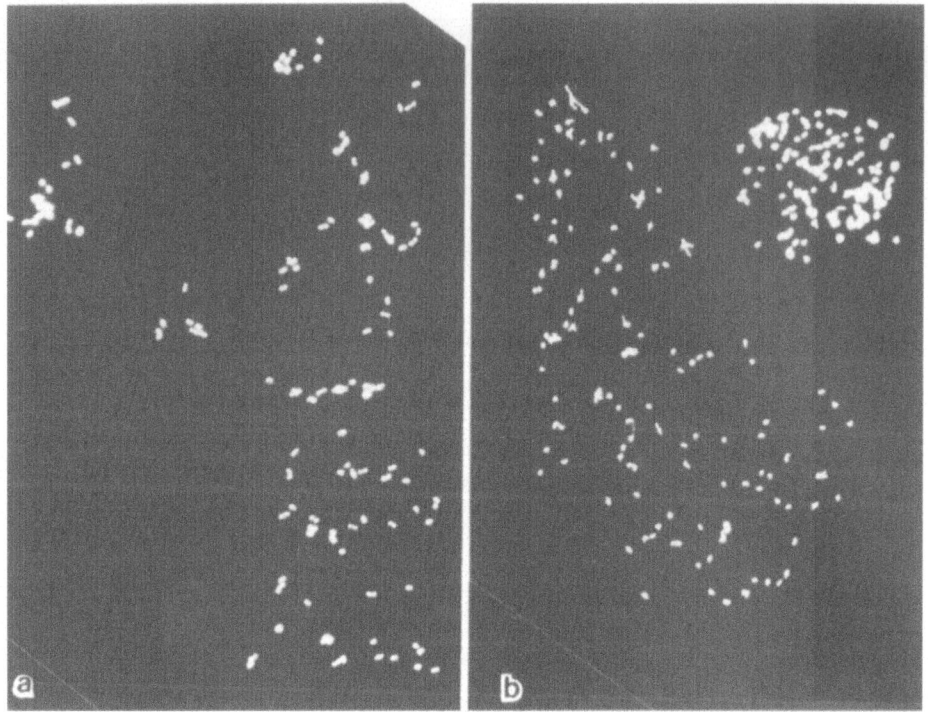

Fig. 2. Fluorescence micrographs of nuclear regions of an ovisac (metaphase I) egg (a) and a metaphase II egg (b) which were squashed and stained with Hoechst 33258. Bivalent chromosomes in the metaphase plate of the first meiotic spindle (a) and dyads in the second meiotic spindle (b) both count around 120. A clump of chromosomes seen upper right in (b) belongs to the first polar body. × 600.

Cell cortex plays a role in determining cleavage pattern in the CD-cell

As in the first cleavage, CD-cells of two-cell embryos divide unequally. Upon the onset of the second mitosis, the CD-cell protrudes toward the side of its sister cell AB (Fig. 4b, c). It is constricted by cleavage furrows which appear around the proximal region of the protrusion (Fig. 4d), and is eventually partitioned into two blastomeres, a smaller C and a larger D (Fig. 4e, f). Besides the manner of division of CD-cells, the second cleavage is different from the first cleavage in three more respects. First, both poles of the spindle in the CD-cell are symmetrically organized up to metaphase, but acquire asymmetric properties during anaphase. Second, centrosomes are present at both poles of the spindle; they differ neither in size nor by γ-tubulin staining. And the third aspect is that the right spindle pole, but not the left (when viewed from the animal pole and CD oriented below AB), is always associated with the cortex facing the AB-cell. Astral microtubules are absent between the cortex and the (right) spindle

pole; it looks as if this spindle pole possesses a 'half aster' (see Fig. 5b).

These results suggest that the unequal cleavage in CD-cells results from the asymmetric organization of spindle poles, which is brought about via interaction of one of the spindle poles with the cell cortex. To test this possibility, we examined the morphology of ectopically formed spindles and the division pattern of CD-cells with such spindles. When two-cell embryos (before the beginning of the second cleavage) were centrifuged at $180 \times g$ with the AB-cells centrifugally, the nuclei in CD-cells were moved to the side opposite to the AB-cells, and as the second mitosis started, spindles were assembled ectopically there. Asters at both poles of the ectopic spindles were not only equal in size, but also grew symmetrically. If centrifuged embryos were allowed to develop further, their CD-cells were found to divide equally. Interestingly similar symmetric organization of spindle poles were observed in CD-cells which had been treated with 50 μg ml^{-1} cytochalasin D for 10 min, although these cells failed to divide. These results suggest that the interaction of the spindle pole with the cell cortex, which may be

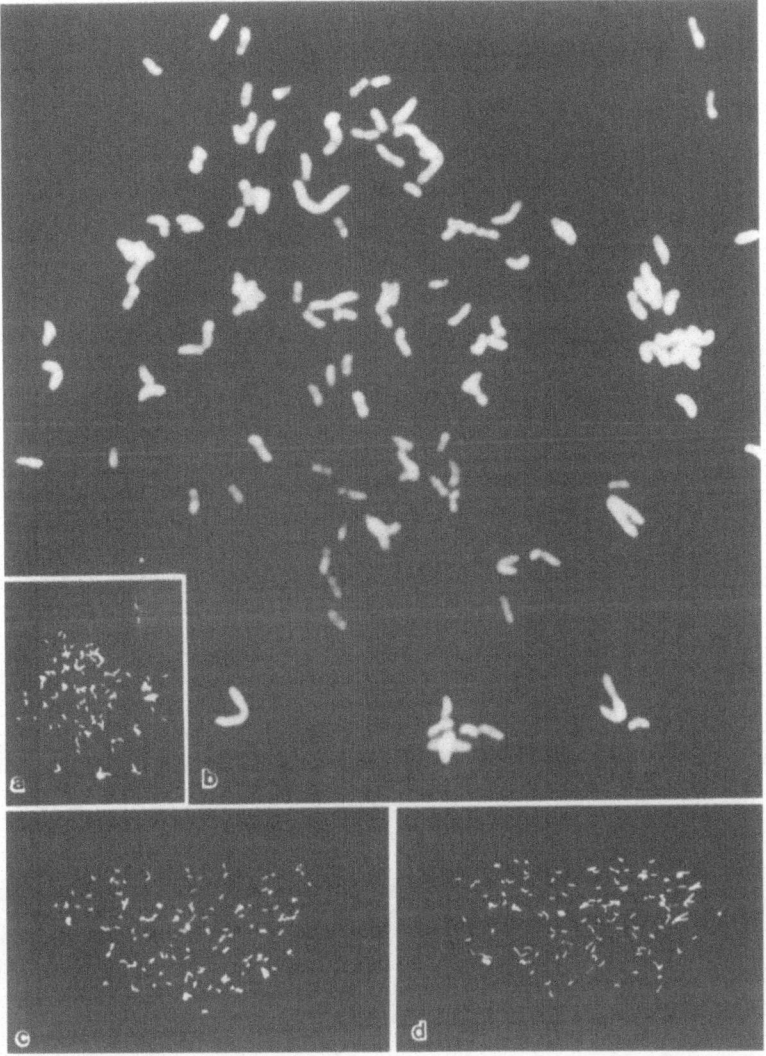

Fig. 3. Fluorescence micrographs of metaphase chromosomes (stained with Hoechst dye) in a single egg at the first cleavage (a, b) and single blastomeres 1D (c) and 2d (d). (b) is an enlargement of (a). Chromosomes count around 120 at either of these stages. a, c, d: × 600. b: × 3000.

mediated by F-actin, plays an important role in generating an asymmetric organization of the spindle poles. Another centrifugation experiment also indicates that the spindle pole is physically connected to the cortex.

The molecular basis of this spindle pole/cortex interaction remains to be elucidated. Based on the morphology of the spindle pole which is to be associated with the cortex, it may be envisaged that the cortex of the CD-cell facing the AB-cell contains at least two kinds of microtubule-associated proteins. First, astral microtubules are shortened between the spindle pole and the cortex, suggesting the presence of molecules which 'cap' or depolymerize the plus-ends of astral

microtubules. If some molecules homologous to those present in kinetochores are present in the cortex, the plus-ends of astral microtubules would be depolymerized as in anaphase chromosome movement. Second, there should be molecules which link the microtubules to the cortex. In addition, if molecules such as cytoplasmic dyneins (McIntosh & Porter, 1989) are present in the cortex, translocation of the spindle pole toward the cortex would be facilitated. Identification of such molecules might contribute to developmental studies, since the interaction between the spindle pole and the cortex is not unique to *Tubifex* eggs, but widespread

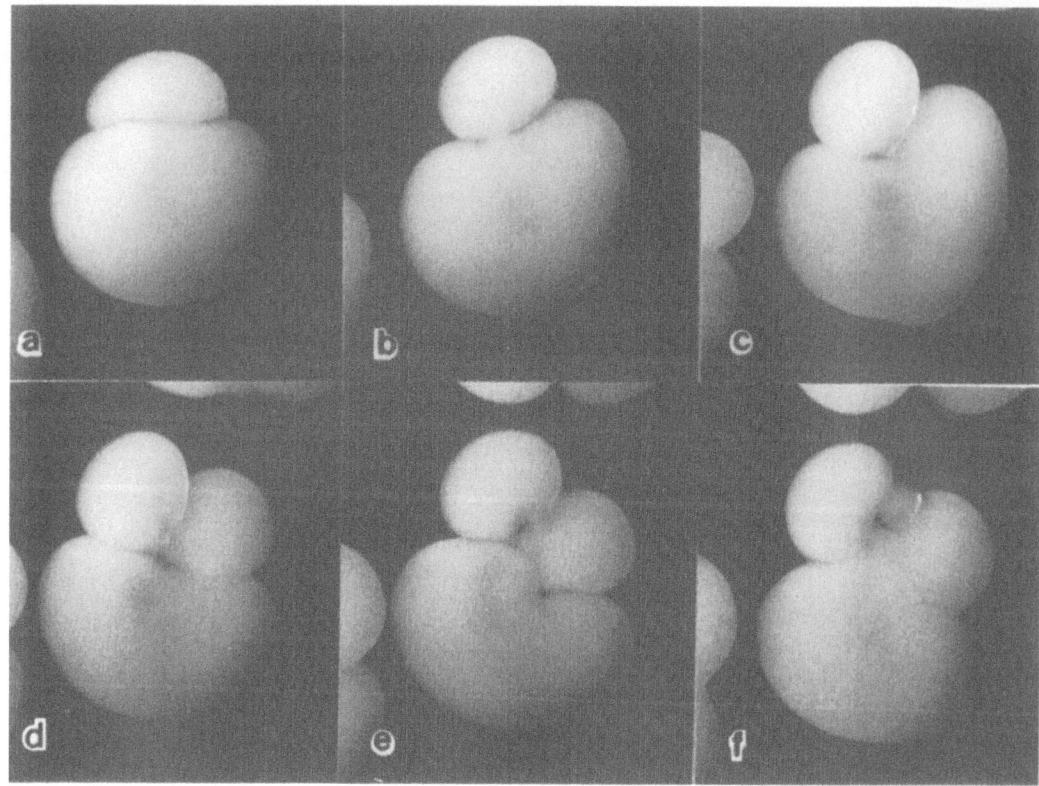

Fig. 4. Photomicrographs of a living *T. hattai* embryo undergoing the second cleavage. Animal pole view. The larger blastomere CD divides first (a–e); then approximately 45 min later the smaller blastomere AB divides (f).

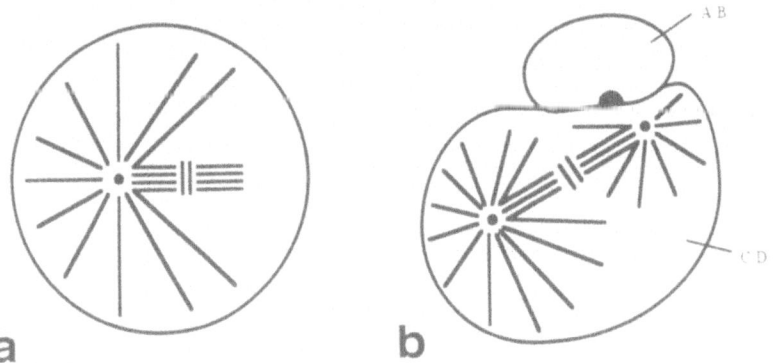

Fig. 5. Diagrammatic summary of the mitotic spindle organization at the first (a) and second (b) cleavages in *T. hattai*. Here early anaphase spindles are depicted; a pair of short lines in the middle of the spindle represent separating chromatids. (a) A centrosome (filled circle) is present at the left spindle pole, but not at the right. An array of astral microtubules grow radially from the left pole, but not from the right. (b) Centrosomes are located at both poles of the second cleavage spindle. The right spindle pole exhibits an asymmetric aster. The mitotic spindle in the AB-cell is anastral; details of its organization are not illustrated here (see Shimizu, 1993).

in animal development (Dan & Ito, 1984; Schroeder, 1987; Hyman, 1989).

Role for AB-cells. When CD-cells were cultured in isolation, they divided equally. Again, mitotic spin-dles in these isolated cells were found to be orga-nized symmetrically. If isolated CD-cells were cul-tured in contact with AB-cells which had been isolated from synchronously developing embryos, they divid-ed unequally to the same extent as the intact CD-cells.

strongly suggest that AB-cells play an important role in determining the cleavage pattern of CD-cells. More specifically, AB-cells may maintain the cortical structures of the CD-cell which are responsible for the partial disassembly of the nearby spindle pole or guarantee the interaction between the spindle pole and the cortex in CD-cells. These possibilities have yet to be tested. Furthermore, there are still unanswered questions as to the interaction between these two sister cells, that is: whether or not blastomeres other than AB-cells could exert similar effects on CD-cells; whether or not AB-cells possess a specialized surface region which affects CD-cells; and whether or not any part of the CD-cell's surface could receive information from other cells. All of these questions could be tested experimentally.

Discussion

The first two cleavages in *Tubifex*, which are both unequal, are brought about by asymmetrically organized mitotic apparatus (Fig. 5). The factors which generate this asymmetry in spindle organization are intrinsic to the mitotic spindle in the first cleavage, but not in the second cleavage. As in *Tubifex*, the first two cleavages in other oligochaetes and leeches so far studied are known to be unequal (see Freeman & Lundelius, 1992, for review). In a leech *Helobdella,* it has been reported that mechanical constraints imposed by cell AB play an important role in determining inequality of cleavage in cell CD (Symes & Weisblat, 1992). This may suggest that *Tubifex* and *Helobdella* involve similar mechanisms for unequal cleavages in larger cells of two-cell embryos. At present, however, nothing is known about organization of mitotic spindles in *Helobdella* embryos. Furthermore, the mechanisms for unequal cleavages have not been analyzed yet in other species of oligochaetes and leeches. Therefore, it remains to be elucidated whether the cleavage mechanisms described in this paper are unique to *Tubifex*, or widespread in oligochaetes and leeches.

We have suggested the possibility that the centrosome which is involved in *Tubifex* embryogenesis is maternal in origin. Furthermore, a single centrosome exclusively participates in the first cleavage. In other animals, it has been known that the paternal centrosome duplicates to take part in the assembly of the first cleavage spindle (Sluder, 1992). In this respect the behavior of the centrosome in the *Tubifex* egg appears to be unusual. It has also yet to be determined whether

or not such features of the centrosome as are seen in *Tubifex* are 'usual' in oligochaetes.

Acknowledgments

I am pleased to express my gratitude to my collaborators R. Ishii and H. Takahashi for the substantial contribution they have made to the work described here. This work was supported in part by Grant-in-Aid for Scientific Research from the Ministry of Education, Science and Culture of Japan.

References

Braidotti, P. & M. Ferraguti, 1982. Two sperm types in the spermatozeugmata of *Tubifex tubifex* (Annelida, Oligochaeta). J. Morph. 171: 123–136.

Dan, K. & S. Ito, 1984. Studies on unequal cleavages in molluscs: I. Nuclear behavior and anchorage of a spindle pole to cortex as revealed by isolation technique. Dev. Growth & Differ. 26: 249–262.

Freeman, G. & J. W. Lundelius, 1992. Evolutionary implications of the mode of D quadrant specification in coelomates with spiral cleavage. J. evol. Biol. 5: 205–247.

Gathy, E., 1900. Contribution à l'étude du développement de l'oeuf et de la fécondation chez les annelides. Cellule 17: 7–62.

Hirao, Y., 1965. Cocoon formation in *Tubifex* with its relation to the activity of the clitellar epithelium. J. Fac. Sci. Hokkaido Univ. Ser VI, Zool. 15: 625–632.

Huber, W., 1946. Der normale Formwechsel des Mitoseapparates und der Zellrinde beim Ei von *Tubifex*. Rev. suisse Zool. 53: 468–474.

Hyman, A. A., 1989. Centrosome movement in the early divisions of *Caenorhabditis elegans*, a cortical site determining centrosome position. J. Cell Biol. 109: 1185–1193.

Inase, M., 1960. On the double embryo of the aquatic worm *Tubifex hattai*. Sci. Rep. Tohoku Univ. Ser. IV, Biol. 26: 59–64.

Ishii, R. & T. Shimizu, 1995. Unequal first cleavage in the *Tubifex* egg: involvement of a monastral mitotic apparatus. Dev. Growth & Differ. 37: 687–701.

Lehmann, F. E., 1946. Mitoseablauf und Bewegungsvorgänge der Zellrinde bei zentrifugierten Keimen von *Tubifex*. Rev. suisse Zool. 53: 475–480.

Lehmann, F. E., 1956. Plasmatische Eiorganisation und Entwicklungsleistung beim Keim vom Tubifex (Spiralia). Naturwissenschaften 43: 289–296.

Mabuchi, I., 1986. Biochemical aspects of cytokinesis. Int. Rev. Cytol. 101: 175–213.

McIntosh, J. R. & M. E. Porter, 1989. Enzymes for microtubule-dependent motility. J. biol. Chem. 264: 6001–6004.

Penners, A., 1922. Die Furchung von *Tubifex rivulorum* Lam. Zool. Jb. Abt. Anat. Ontog. 43: 323–367.

Penners. A., 1924. Experimentalle Untersuchungen zum Determinationsproblem an Keim vom *Tubifex rivulorum* Lam. I. Die Duplicitas cruciata und Organbildende Keimbezirke. Arch. Mikrosk. Abt. Entwick. Mechan. 101: 51–100.

Rappaport, R., 1986. Establishment of the mechanism of cytokinesis in animal cells. Int. Rev. Cytol. 105: 245–281.

276

Schroeder, T. E., 1987. Fourth cleavage of sea urchin blastomeres: microtubule patterns and myosin localization in equal and unequal cell divisions. Dev. Biol. 124: 9–22.

Shimizu, T., 1982a. Development in the freshwater oligochaete *Tubifex*. In F. W. Harrison & R. R. Cowden (eds), Developmental Biology of Freshwater Invertebrates. A. R. Liss, New York: 283–316.

Shimizu, T., 1982b. Ooplasmic segregation in the *Tubifex* egg: mode of pole plasm accumulation and possible involvement of microfilaments. Roux's Arch. dev. Biol. 191: 246–256.

Shimizu, T., 1984. Dynamics of the actin microfilament system in the *Tubifex* egg during ooplasmic segregation. Dev. Biol. 106: 414–426.

Shimizu, T., 1986. Bipolar segregation of mitochondria, actin networks, and surface in the *Tubifex* egg: role of cortical polarity. Dev. Biol. 116: 241–251.

Shimizu, T., 1988. Localization of actin networks during early development of *Tubifex* embryos. Dev. Biol. 125: 321–331.

Shimizu, T., 1989. Asymmetric segregation and polarized redistribution of pole plasm during early cleavages in the *Tubifex* embryo: role of actin networks and mitotic apparatus. Dev. Growth & Differ. 31: 283–297.

Shimizu, T., 1993. Cleavage asynchrony in the *Tubifex* embryo: involvement of cytoplasmic and nucleus-associated factors. Dev. Biol. 157: 191–204.

Shimizu, T., 1994. The prevention of smaller blastomeres of early *Tubifex* embryos from entering mitosis by unreplicated DNA. Dev. Biol. 161: 274–284.

Shimizu, T., 1995. Lineage-specific alteration in cell cycle structure in early *Tubifex* embryos. Dev. Growth & Differ. 37: 263–272.

Shimizu, T., 1996. Behaviour of centrosomes in early *Tubifex* embryos: asymmetric segregation and mitotic cycle-dependent duplication. Roux's Arch. dev. Biol. 205: 290–299.

Sluder, G., 1992. Control of centrosome inheritance in echinoderm development. In V. I. Kalnins (ed), The Centrosome. Academic Press, San Diego: 235–259.

Symes, K. & D. A. Weisblat, 1992. An investigation of the specification of unequal cleavages in leech embryos. Dev. Biol. 150: 203–218.

Woker, H., 1944. Die Wirkung des Colchicins auf Furchungsmitosen und Entwicklungsleistungen des *Tubifex*-Eies. Rev. suisse Zool. 51: 109–170.

Hydrobiologia **334**: 277–285, 1996.
K. A. Coates, Trefor B. Reynoldson & Thomas B. Reynoldson (eds), Aquatic Oligochaete Biology VI.
© 1996 *Kluwer Academic Publishers.*

Leeches (Oligochaeta?: Euhirudinea), their phylogeny and the evolution of life-history strategies

Mark E. Siddall & Eugene M. Burreson
School of Marine Science, Virginia Institute of Marine Science, College of William and Mary, P.O. Box 1346, Gloucester Point, Virginia 23062, USA

Key words: leech, oligochaete, parasite, evolution, phylogeny

Abstract

Powerful tests of adaptational hypotheses can be made in the context of well-supported cladograms by investigating the most parsimonious transformation of intrinsic or extrinsic factors to explain their distribution across taxa in a cladogram. Such tests are used here to discover patterns of life-history evolution in leeches; in particular in relation to exploitation of terrestrial and aquatic habitats, parental care and resource utilization. Moreover, the relationships among leeches, acanthobdellids and branchiobdellids is reaffirmed as is their collective placement within the oligochaetes.

> '*Poor innocent* Branchiobdella
> *Urgently asks us to tell her*
> *Why* she *doesn't reach*
> *The rank of a leech*
> *Instead of equal with that lumbriculid mud-dweller.*'
>   Reinmar Grimm (1994 – 6th Int'l Symposium on Aquatic Oligochaetes).

Introduction

The phylogenetic and taxonomic relationships among the clitellates historically has been a matter of some debate. One issue contributing to this debate involves the relationships of leeches to acanthobdellids, branchiobdellids and oligochaetes, as well as the relationships of leeches to each other. With respect to the former, an hypothesis of phylogenetic affinity between leeches (Euhirudinea) and *Acanthobdella peledina* Grube, 1851 is traceable to Livanow (1906) who considered *A. peledina* to be an ancient hirudinean. His reasoning was that, like leeches, *A. peledina* has a fixed number of body somites, a caudal sucker for attachment to its fish host, an ectocommensal life-style, partial reduction of the coelomic spaces, loss of somatic (though not cephalic) setae, fused male gonopores, and oblique musculature, among other characteristics. Michaelsen (1919) countered these arguments, classi-

fying *A. peledina* as an oligochaete instead, citing the presence of the cephalic setae and the oligochaete-like seminal funnel, and arguing that the observed similarities between leeches and *A. peledina* must therefore be convergently acquired in relation to the adoption of ectocommensalism. Livanow (1931) later reaffirmed his hypothesis which has since been accepted by most workers (e.g., Mann, 1962; Elliott & Mann, 1979; Klemm, 1982; Sawyer, 1972, 1986). Conversely, Davies (1991) has since resurrected Michaelsen's (1919) position of an independent origin. As was cogently indicated by Purschke *et al.* (1993) this issue is inextricably linked with the question of the phylogenetic position of the branchiobdellids. These clitellates also have an ectocommensal life-history, constant (though different) number of body somites, a caudal attachment organ, loss of setae and fused gonopores. Whereas some have considered the branchiobdellids to be leeches (Odier, 1823; Sawyer, 1986) others

278

have argued for separate origins for the two (Holt, 1953, 1989; Ferraguti & Gelder, 1991; Brinkhurst & Gelder, 1989) citing arguments for convergence similar to those forwarded by Michaelsen (1919) and by Davies (1991).

Phylogeny

Contributing factors to this debate, perhaps, have been confusion surrounding the relative *ingroup* phylogenetic relationships of the leeches themselves (i.e., Euhirudinea) and identification of allied sister-group relationships. Leeches exhibit a remarkable scope of diversity including ground-dwelling, ectocommensal, blood-feeding and predatory life-history strategies; a variety of themes in parental care and cocoon deposition behaviors; as well as habitats that range from terrestrial to freshwater to marine environments. Consideration of leeches outside of a rigorous phylogenetic context has tended towards argumentation from an evolutionary progressivism that has its root in eighteenth century Goetheian *Naturphilosophie* long since abandoned by phylogeneticists (Gould, 1977; Hull, 1988). Thus, ectocommensalism, and then blood feeding, are seen as being more specialized and later events in leech evolution. Similarly, the membranous cocoons and less extensive annulation in somites of glossiphoniid leeches have been viewed as the simpler antecedents of the hardened proteinaceous cocoons and more numerous annuli in other leeches. In this light, the arguments presented for widespread convergence among leeches, acanthobdellids and branchiobdellids would appear reasonable.

Brinkhurst (1994) proposed a phylogenetic hypothesis for the clitellates rooted with an ancestral archetype (Fig. 1a). Although we were unable to reconstruct the same tree with his data, whether characters were ordered (Fig. 1b) or unordered (Fig. 1c) the results are all in agreement in two respects: all three analyses agree on a sister-group relationship for the leeches and the branchiobdellids and they all support the notion that this clade is *within* the Oligochaeta as opposed to standing apart from that larger clade. Similarly, Purschke *et al.* (1993) have hypothesized an extensive list of synapomorphies for a sister-group relationship of acanthobdellids and euhirudinids, as well as other synapomorphies in support of a sister-group relationship for the branchiobdellids to that clade. We have recently investigated the phylogenetic position of taxa (Table 1) representative of the major families and subfamilies of leeches from a cladistic viewpoint (Siddall

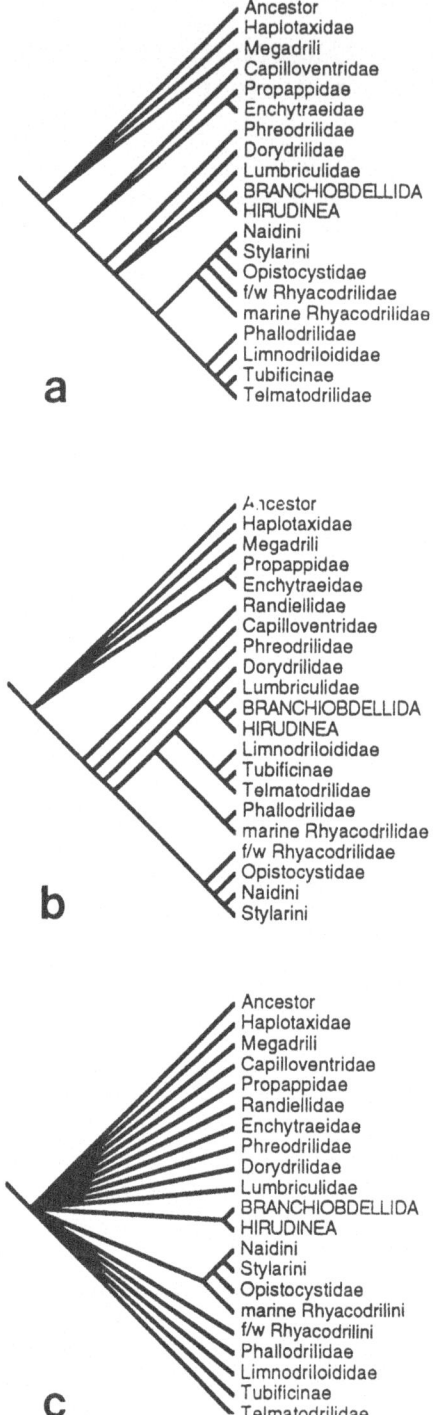

Fig. 1. Phylogenetic relationships among clitellates (a) redrawn from Brinkhurst (1994), (b) consensus of 5 trees resulting from reanalysis with ordered characters and (c) consensus of 310 trees resulting from reanalysis with unordered characters.

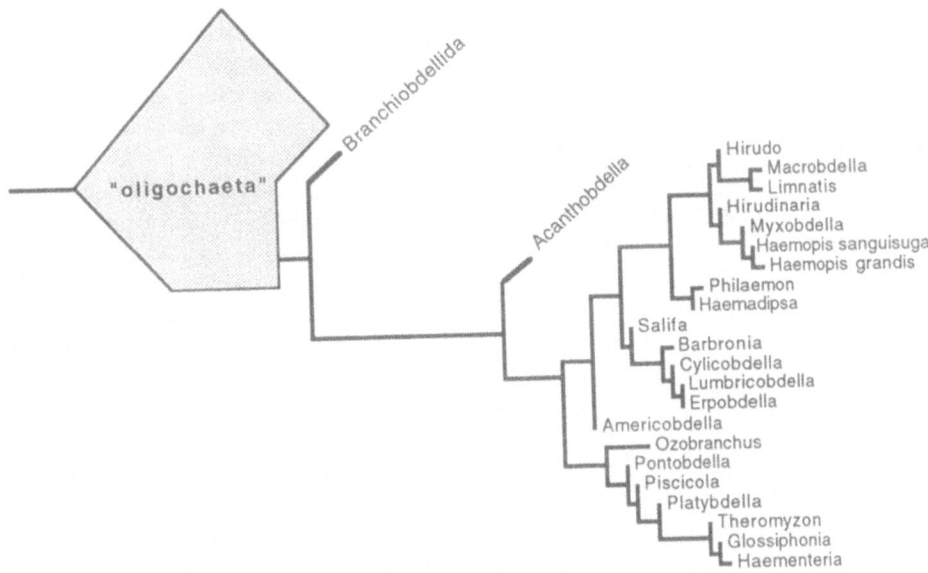

Fig. 2. Combined phylogenetic hypothesis of the position of the Branchiobdellida, *Acanthobdella*, and the Euhirudinea relative to the oligochaetes. Branch lengths are proportional to amount of morphological change. The taxon 'oligochaeta' is a paraphyletic assemblage and not a natural group unless it includes the other listed taxa as well. Based on Purschke *et al.* (1993) and Siddall & Burreson (1995).

& Burreson, 1995). The combination of these phylogenetic developments, that is the combined relationships of the foregoing, is presented in Fig. 2.

There are a number of interesting phylogenetic components to this hypothesis. First, although Purschke *et al.* (1993) presented 11 synapomorphies for the Clitellata as a whole, none can be unambiguously proposed for monophyly of the taxon 'Oligochaeta' unless that taxon is expanded to redundancy with Clitellata. Additionally, with branch lengths drawn in proportion to the amount of hypothesized morphological change (Fig. 2), one quickly sees the source of prior reluctance to group the branchiobdellids with the leeches. The branchiobdellids are much more similar overall to the 'oligochaetes' than to *Acanthobdella* or the euhirudinids. Purschke *et al.* (1993) explained all five morphological synapomorphies of branchiobdellids and euhirudinids *post-hoc* in terms of convergence in spite of the phylogenetic evidence to the contrary. Thus, arguments proposing the inclusion of branchiobdellids within the oligochaetes and exclusion of the leeches are purely phenetic and have no phylogenetic support. The relationships in Fig. 2 are further strengthed with the consideration of life-history strategies in this paper, wherein supposed apomorphic characteristics are seen to be, in fact, plesiomorphic

and fully consistent with the relationships presented in Fig. 1.

The order of appearance of various characteristics or life-history strategies such as those described above can be more rigorously assessed in the context of a well-supported cladogram as opposed to resorting to speculative arguments about convergence or baseless adaptationist scenarios (Coddington, 1988; Brooks & McLennan, 1991). Some argue that the characteristic or phenomenon in question should be mapped onto a cladogram a posteriori to avoid circularity (Coddington, 1988; Brooks & McLennan, 1991), while others suggest that it should be included in the reconstruction as part of the overall phylogenetic hypothesis (Kluge & Wolf, 1993). Regardless, these optimization procedures seek to locate historical transformation of intrinsic or extrinsic traits by discovering their most parsimonious distribution across organisms to determine to what degree the observed similarities are due to common ancestry. Here, life-history strategies are traced on the phylogenetic hypothesis of the Euhirudinea (Siddall & Burreson, 1995) together with the sister-group relationships proposed for the acanthobdellids and branchiobdellids (Purshke *et al.*, 1993; Brinkhurst, 1994). In each of the following macroevolutionary hypotheses, both optimization (Coddington, 1988; Brooks & McLennan, 1991) and character con-

Table 1. Taxonomic heirarchy for taxa considered in the phylogeny of the Euhirudinea (Siddall and Burreson, 1995). Paraphyletic taxa are delimited by quotation marks.

Arhynchobdellida	Rhynchobdellida
Hirudiniformes	Ozobranchidae
"Hirudinidae"	*Ozobranchus branchiatus*
Hirudo medicinalis	
Macrobdella decora	"Piscicolidae"
Limnatis nilotica	*Piscicola geometra*
Hirudinaria javanica	*Pontobdella muricata*
Myxobdella africana	*Platybdella anarrhichae*
Haemopidae	
Haemopis sanguisuga	Glossiphoniidae
Haemopis grandis	*Glossiphonia complanata*
	Theromyzon pallens
Haemadipsidae	*Haementeria ghilianii*
Haemadipsa zeylanica	
Philaemon pungens	
Americobdellidae	
Americobdella valvidiana	
Erpobdelliformes	
"Salifidae"	
Salifa perspicax	
Barbronia rouxi	
Erpobdellidae	
Erpobdella octoculata	
Lumbricobdella shaefferi	
Cylicobdellidae	
Cylicobdella joseenensis	

gruence (Kluge & Wolf, 1993) methodologies yielded the same results.

Results and discussion

Land and sea

As in other oligochaetes, there are both aquatic and terrestrial leech taxa. Some, such as the haemadipsids *Haemadipsa* spp. and *Philaemon* spp., are strictly terrestrial, occurring on the leaves and stems of terrestrial vegetation distributed from the Indian subcontinent and Southeast Asia through Wallacea to Australia. Others, including all of the Rynchobdellida are strictly

aquatic while the hirudinids exhibit varying degrees of amphibious habits involving foraging and cocoon deposition on land while spending most of their time in water. The distribution of terrestrialism across taxa (Fig. 3a) suggests that the plesiotypic condition was a fully aquatic life-history. The early split of the euhirudinids into the Rhynchobdellida and the Arhynchobdellida apparently involved a move to land in the latter whereas the strictly aquatic distribution in the rhynchobdellids requires no explanation save that of historical constraints present in an ancestor common with the branchiobdellids and acanthobdellids. Conversely, if the phylogeny is correct, the aquatic habits of the erpobdellids appear as a secondary return to water. Similarly, it appears as though the hirudinids have more or less returned to water while retaining some aspects of terrestrialism, particularly those involving cocoon deposition and early ontogeny.

Whereas aquatic habits appear to have been fixed in the history of the rhynchobdellids but not the arhynchobdellids, the reverse appears to be true with respect to the colonization of marine environments (Fig. 3b). That is, for those arhynchobdellids with aquatic or semiaquatic lifestyles, their distributions are restricted to freshwater conditions, the plesiotypic condition retained from the original ancestral oligochaete from which the branchiobdellids and acanthobdellids are derived. Among the rhynchobdellid leeches, however, the situation appears less straightforward. Whereas the pontobdellids are strictly marine, ozobranchids, piscicolids and platybdellids are found in salinities that span the range of freshwater, estuarine and wholly marine. As such the original condition for the ancestral rhynchobdellid is equivocal. Part of this incongruence may be due to the level of taxonomic analysis. If, for example, the platybdellids prove to be a paraphyletic assemblage with plesiotypically marine taxa and with freshwater taxa proving to be monophyletic with the glossiphoniids, a portion of the apparent ambiguity described above would be resolved. In any case the most parsimonious solution appears to entail early colonization of a marine environment by some ancestral piscicolid followed by subsequent reinvasion of freshwater habitats leading to the origins of the glossiphoniids.

Sawyer (1986) suggested that piscicolids dispersed into marine environments as early as the late Mesozoic with the breakup of Pangaea. We would argue that this dispersal probably predated the Mesozoic because colonization of marine environments predates the origins of the strictly freshwater glossiphoniids in our phy-

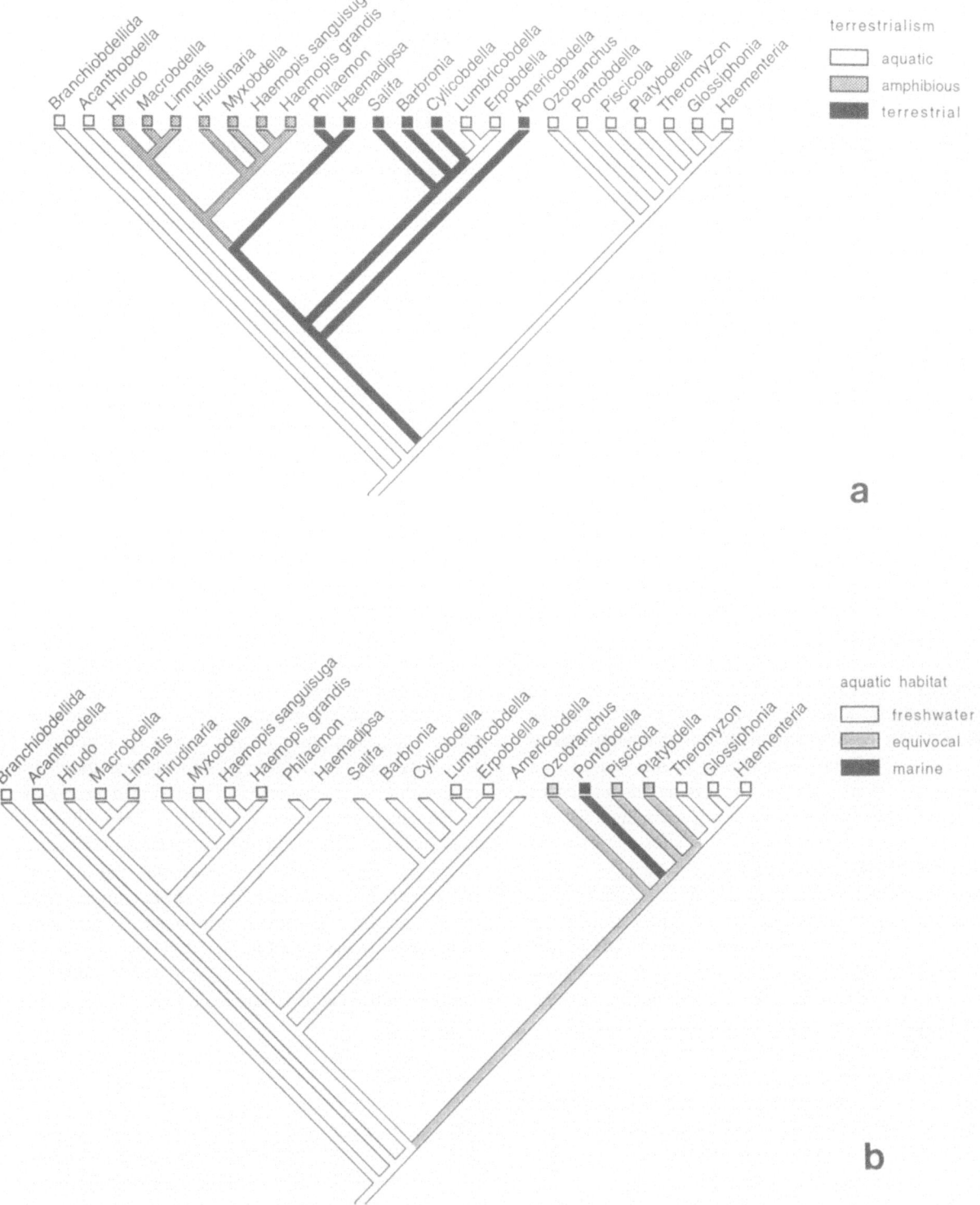

Fig. 3. Evolution of ancestral habitat preferences. (a) terrestrialism; (b) aquatic habitats. Missing boxes indicate inapplicable or unknown.

logeny. The latter have a global distribution suggestive of considerable diversification prior to the early Mesozoic.

Cocoonology

Both cocoon secretion from the clitellum by leeches and the fate of cocoons, in terms of parental care, vary across leech taxa. Unlike all other oligochaetes, glossiphoniids do not produce a cocoon that is resistant to environmental influences. Rather, the cocoon consists of a very simple and fragile membraneous bag into which the fertilized eggs are deposited (Fig. 4a). Schmidt (1944) suggested that the hardened cocoons of oligochaetes and leeches were acquired convergently. Perhaps even more surprisingly, he suggested that among leeches, the hirudinids and the piscicolids evolved hardened cocoons independently (Schmidt, 1944). In the context of our hypothesis these suggestions are untenable; the hardened covering on cocoons is most parsimoniously explained in terms of a simple plesiomorphic condition reflective of their origins within the oligochaetes. Similarly in error is the progressivist contention by Mann (1962) that glossiphoniid cocoons, by virtue of their simplicity, are necessarily the original state.

Leeches also encompass some diversity in terms of degrees of parental care of cocoons. Although some, such as the hirudinids, exhibit a more oligochaete-like behavior of simply secreting a cocoon among vegetation or soil and then abandoning it, this is seen as a secondary reversal among the arhynchobdellid leeches (Fig. 4b). Instead, the behavior of secreting a protective cocoon and cementing it to a substrate of some sort appears to be the original state in leech evolution, exhibited by branchiobdellids and acanthobdellids, and retained in piscicolids and erpobdellids alike. The substrate to which cocoons are attached ranges from the animate (e.g., crustaceans for branchiobdellids and for some piscicolids) to the inanimate (e.g., rocks and other submerged objects for the erpobdellids and most piscicolids). The argument that this attachment of cocoons represents a convergent adaptation to ectocommensalism is unfounded in light of the same behavior among strictly predaceous erpobdellids which exhibit no commensalistic habits at all. The unique behavior among glossiphoniids to remain with the cocoons, covering and fanning them with their ventral body surface (Fig. 4b) is closely correlated with the loss of the protective hardened cocoon (Fig. 4a).

Blood and parasites

Leeches are notorious for their blood feeding. In most freshwater environments, however, both in terms of species richness and absolute abundance, those that are not sanguivorous are better represented than those that are (Sawyer, 1986; Davies, 1991). In the context of leech phylogeny, Sawyer's (1986) assertion that the common ancestor of euhirudinids was not a blood feeder and that sanguivory has been independently acquired in the hirudinids and in the rhynchobdellids appears to have been prescient (Fig. 5a). In fact, there appears to be an origin of sanguivory in the ancestral rhynchobdellid whereas the ancestral arhynchobdellid is hypothesized to be neither sanguivorous nor even ectocommensal like the branchiobdellids and acanthobdellids (Fig. 5a). Inasmuch as there are two hypothesized origins of sanguivory, there are three hypothesized origins of predatory behavior, one in the ancestral arhynchobdellid, one in the common ancestor of the haemopids and one for the distribution of this characteristic in the glossiphoniids. Thus, Mann's (1962) suggestion that the haemopids abandoned sanguivory appears correct, though his suggestion that erpobdellids did as well, does not.

There are two independent lines of evidence adding to this hypothesis of separate origins of blood feeding among leeches. First, the sanguivorous hirudinids feed by making tripartite incisions in the skin of their hosts from which they draw on the upwelling blood (telmophagy sensu Laviopierre, 1965) whereas the rhynchobdellids feed by means of inserting a proboscis (selenophagy sensu Laviopierre, 1965). Secondly, only the rynchobdellids are known to transmit blood parasites (Fig. 5b) such as trypanosomatids and apicomplexans. In fact, for some of these, the haemogregarine blood parasites of fishes and chelonians, rhynchobdellid leeches are the definitive ancestral hosts and vertebrates are intermediate hosts (Barta, 1989; Siddall & Desser, 1990; Siddall & Burreson, 1994).

Conclusions

Notably, in each of the macroevolutionary comparisons made above, the position of the acanthobdellids as sister-taxon to the Euhirudinea and of the branchiobdellids as sister-taxon to those is fully consistent with the observed transformations. With respect to the issues of terrestrialism (Fig. 3a), aquatic habitat (Fig. 3b), cocoons (Fig. 4a), parental care (Fig. 4b), and

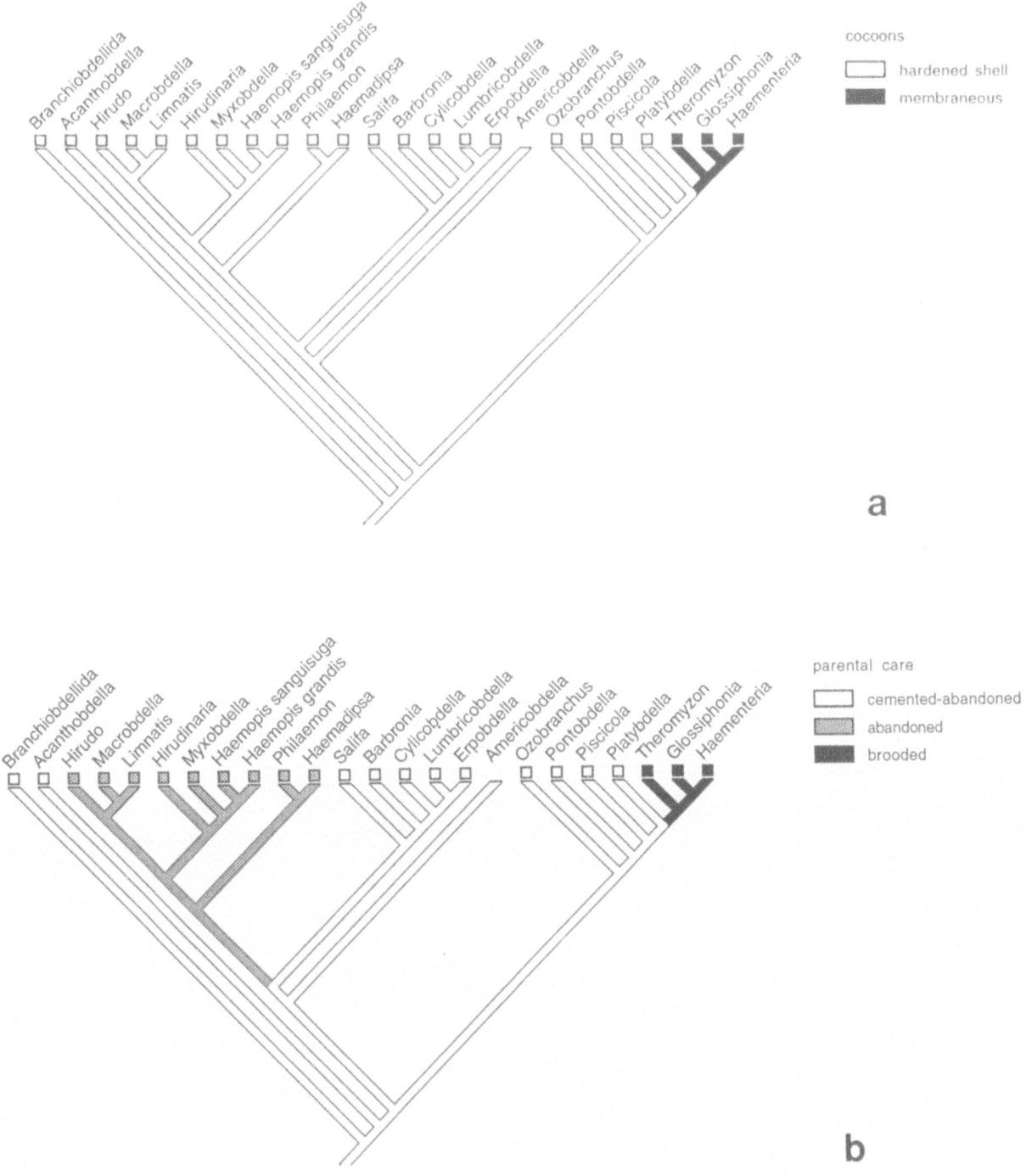

Fig. 4. Evolution of cocoon deposition characteristics. (a) cocoons; (b) parental care. Missing boxes indicate inapplicable or unknown.

blood parasites (Fig. 5b), both the branchiobdellids and *A. peledina* exhibit a plesiotypic condition representative of traditional 'oligochaetes', and yet these plesio-typic states are carried through into the euhirudinids too. With respect to trophisms (Fig. 5a), ectocommensalism is likewise carried into the leeches though

284

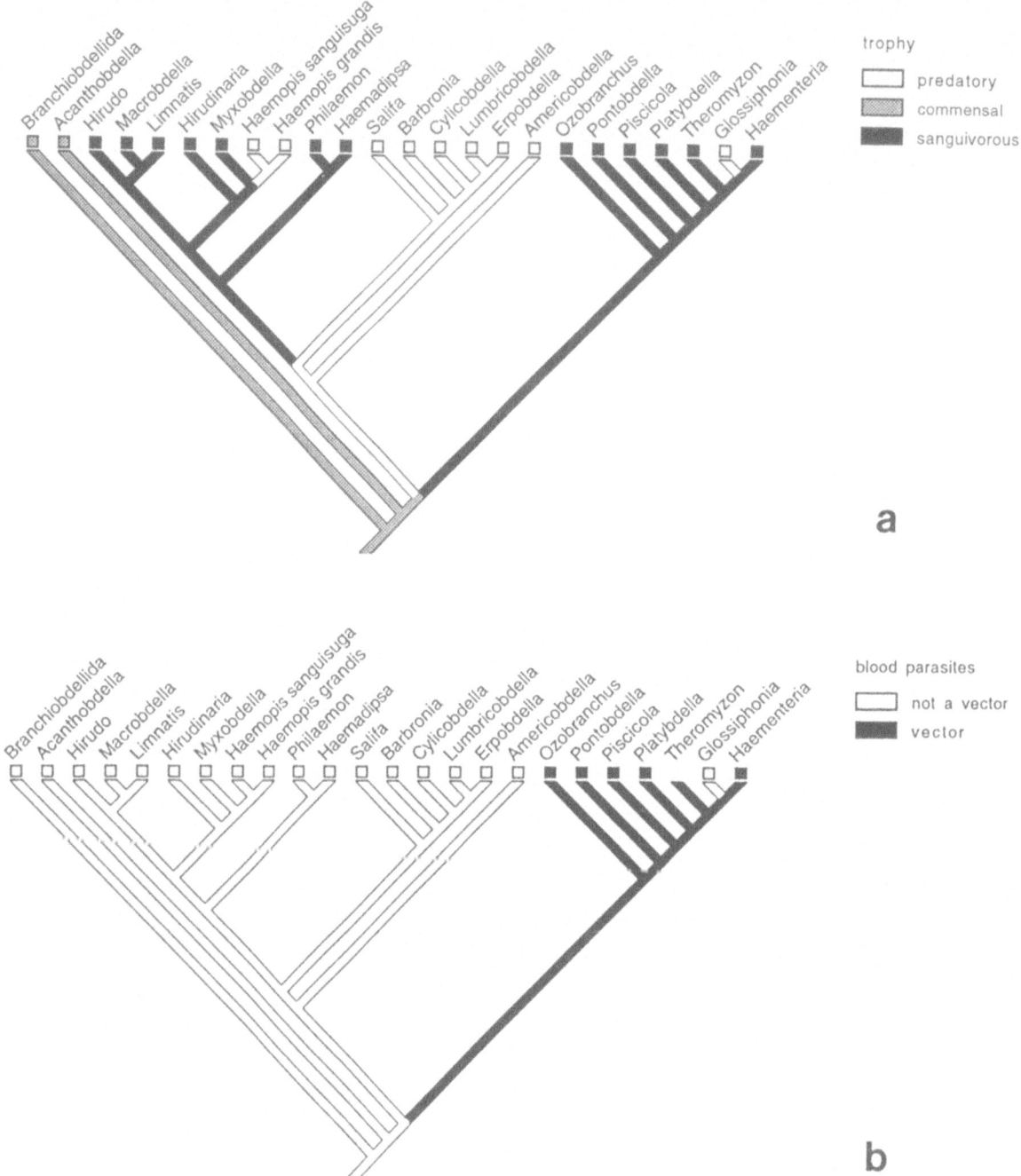

Fig. 5. Evolution of modes of nutrition. (a) trophy; (b) transmission of blood parasites. Missing boxes indicate inapplicable or unknown.

modified to blood parasitism in the rhynchobdellids. As such, and in contrast to the traditional view, there appears to be vanishingly little support for the notion of widespread convergence in life history strategies or in morphological and behavioral phenomena associated with ectocommensalism.

Acknowledgments

Lively discussions at the 6th International Symposium on Aquatic Oligochaetes surrounding leech phylogeny in the context of the clitellates stimulated our desire to prepare this contribution; in particular those discussions with Ralph Brinkhurst, Kathy Coates, Christer Erséus, and Wilfried Westheide. We thank Sandy Blake, Lisa Calvo and Brenda Flores for their reading of earlier drafts. This work was supported by a postdoctoral fellowship from the Natural Sciences and Engineering Research Council of Canada and by a Florence Christie fellowship from the New Brunswick Museum to MES. This is VIMS Contribution #2013.

References

Barta, J. R., 1989. Phylogenetic analysis of the class Sporozoea (Phylum Apicomplexa Levine, 1970): Evidence for the independent evolution of heteroxenous life cycles. J. Parasitol. 75: 195–206.

Brinkhurst, R. O., 1994. Evolutionary relationships within the Clitellata: an update. Megadrilogica 5: 109–112.

Brinkhurst, R. O. & S. R. Gelder, 1989. Did the lumbriculids provide the ancestors of the branchiobdellidans, acanthobdellidans and leeches? Hydrobiologia 180: 7–15.

Brooks, D. R. & D. A. McLennan, 1991. Phylogeny, Ecology and Behaviour: A Research Program in Comparative Biology. University of Chicago Press, Chicago, 434 pp.

Coddington, J. A., 1988. Cladistic tests of adaptational hypotheses. Cladistics 4: 3–22.

Davies, R. W., 1991. Annelida: Leeches, Polychaetes, and Acanthobdellids. In Ecology and Classification of North American Freshwater Invertebrates. Academic Press, New York: 437–479.

Elliott, J. M. & K. H. Mann, 1979. A key to the British freshwater leeches. Freshwater Biological Association Scientific Publication No. 40, 72 pp.

Ferraguti, M. & S. R. Gelder, 1991. The comparative ultrastructure of spermatozoa from five branchiobdellidans (Annelida: Clitellata). Can. J. Zool. 69: 1945–1956.

Gould, S. J., 1977. Ontogeny and Phylogeny. Belknap Press, Cambridge, Mass., 501 pp.

Holt, P. C., 1953. Characters of systematic importance in the family Branchiobdellidae (Oligochaeta). Virginia J. Sci. 4: 57–61.

Holt, P. C., 1989. Comments on the classification of the Clitellata. Hydrobiologia 180: 1–5.

Hull, D. L., 1988. Science as a Process. University of Chicago Press, Chicago, 586 pp.

Klemm, D. J., 1982. Leeches (Annelida: Hirudinea) of North America. U.S. Environmental Protection Agency Bulletin EPA-600/3-82-025, Cincinnati, Ohio, 177 pp.

Kluge, A. G. & A. J. Wolf, 1993. Cladistics: What's in a word? Cladistics 9: 183–199.

Laviopierre, M. M. J., 1965. Feeding mechanisms of blood-sucking arthropods. Nature 208: 302–303.

Livanow, N., 1906. *Acanthobdella peledina* Grube, 1851. Zool. Jb. Anat. 22: 637–866.

Livanow, N., 1931. Die Organisation der Hirudineen und die Beziehungen dieser Gruppe zu den Oligochaeten. Erg. Fortschr. Zool. 7: 378–484.

Mann, K. H., 1962. Leeches (Hirudinea) Their Structure, Physiology, Ecology and Embryology. Pergamon Press, N.Y., 201 pp.

Michaelson, W., 1919. Über die Beziehungen der Hirudineen zu den Oligochaeten. Mitt. hamb. Zool. Mus. Inst 36: 131–153.

Odier, A. M., 1823. Mémoire sur les branchiobdelle, nouveau genre d'annelides de la famille des hirudinées. Mém. Soc. Hist. Nat. Paris 1: 69–78.

Purschke, G., W. Westheide, D. Rhode & R. O. Brinkhurst, 1993. Morphological reinvestigation and phylogenetic relationship of *Acanthobdella peledina* (Annelida: Clitellata). Zoomorphology 113: 91–101.

Sawyer, R. T., 1972. North American Freshwater Leeches, Exclusive of the Piscicolidae, with a Key to All Species. Illinois Biological Monographs #46. University of Illinois Press, Urbana, Illinois, 107 pp.

Sawyer, R. T., 1986. Leech Biology and Behaviour. Oxford University Press, Oxford, 1065 pp.

Schmidt, G. A., 1944. Adaptive significance of peculiarities of the cleavage process in leeches. J. Gen. Biol. Moscow 5: 284–303.

Siddall, M. E. & E. M. Burreson, 1994. The development of a hemogregarine of *Lycodes raridens* from Alaska in its definitive leech host. J. Parasitol. 80: 569–575.

Siddall, M. E. & E. M. Burreson, 1995. Phylogeny of the Euhirudinea: Independent evolution of blood feeding by leeches? Can. J. Zool. 73: 1048–1064.

Siddall, M. E. & S. S. Desser, 1990. Gametogenesis and sporogonic development of *Haemogregarina balli* (Apicomplexa: Adeleina: Haemogregarinidae) in the leech *Placobdella ornata*. J. Protozool. 37: 511–520.

Hydrobiologia **334**: 287–295, 1996.
K.A. Coates, Trefor B. Reynoldson & Thomas B. Reynoldson (eds), Aquatic Oligochaete Biology VI.

Index

The manufacturer's authorised representative in the EU is Springer
Nature Customer Service Centre GmbH, Europaplatz 3, 69115 Heidelberg,
Germany. If you have any concerns regarding our products, please
contact ProductSafety@springernature.com

Printed and bound by CPI Group (UK) Ltd, Croydon, CR0 4YY

29/04/2026

02099554-0001